Communication Skills

SECOND EDITION

SANJAY KUMAR

Consultant
English and Soft Skills Development Training
Formerly Faculty, BITS Pilani and
Reader and Chairperson, Department of English, CDLU, Sirsa

PUSHP LATA

Associate Professor of English
Department of Humanities and Social Sciences
BITS, Pilani

OXFORD

UNIVERSITY PRESS

OXFORD
UNIVERSITY PRESS

Oxford University Press is a department of the University of Oxford.
It furthers the University's objective of excellence in research, scholarship,
and education by publishing worldwide. Oxford is a registered trade mark of
Oxford University Press in the UK and in certain other countries.

Published in India by
Oxford University Press
22 Workspace, 2nd Floor, 1/22 Asaf Ali Road, New Delhi 110 002

First Edition published in 2011
Second Edition published in 2015
13th impression 2022

ISBN-13: 978-0-19-945706-9
ISBN-10: 0-19-945706-9

Typeset in Times New Roman
by Cameo Corporate Services Limited, Chennai
Printed in India by Rakmo Press, New Delhi 110 020

For product information and current price, please visit www.india.oup.com

Third-party website addresses mentioned in this book are provided
by Oxford University Press in good faith and for information only.
Oxford University Press disclaims any responsibility for the material contained therein.

Dedicated
to all those who aspire to sculpt in time
through the words they write and speak

Features of

Learning Objectives

Learning objectives set the theme for all the chapters. Some chapters also begin with an interesting introduction that narrates a short story, which helps set the mood for the chapter.

Illustrations

Illustrations, interspersed in the chapters, make the book a more lively and interesting read. Besides adding a dash of humour, they also highlight some common mistakes made while speaking English.

PRACTICE TEST 2.4

Can you now choose the correct form of the word (singular/plural) so that they are correctly used in the sentences given below?

1. The girl was a stunner; she had blonde **hairs/hair**.
2. He loves listening to quality **music/musics**.
3. The speaker was appreciated with loud **applause/applauses** from the audience.
4. **Cutlery/cutleries has/have** become quite stylish of late.
5. **Businesses/business** cannot grow in times of recession.
6. The company is planning to purchase more **equipment/equipments**.
7. **Times have come/Time has come** when we need to be serious about climate change.
8. Housewives always have lots of **household work/household works** to do.
9. There is no point in visiting Rajasthan during summer; since it is hot during that period, **sightseeings/sightseeing** cannot be enjoyed.
10. In the laughter show the audience had lots of **funs/fun**.

Practice Tests

Chapters in Parts 1, 2, and 6 of the book are packed with practice tests that help us master the nuances of the English language. Answers can be checked with the key given at the end of the chapters.

Wisewell Quips

Wisewell and Madcap will entertain and educate us at the same time with their humorous and interesting conversations at the end of the chapters.

the Book

Recapitulation

The recapitulation at the end of each chapter revisits all the important points discussed in the chapter making for a fine guide for revision before exams.

Exercises

Different kinds of review questions, such as concept review questions and critical thinking questions, enumerated at the end of the chapter, aim at testing readers on their understanding of the topic.

Samples

Samples of reports, proposals, résumés, memos, letters, etc. will equip readers in preparing these documents with finesse and confidence.

Preface to the Second Edition

The book in your hand is the revised edition of *Communication Skills*, which has so far been worthy of your attention and has helped you hone your communication and professional skills in English for the last couple of years. We intend to engage you further in this enriching journey of conquering a language that has become essential for growth, success, and respect in the professional world.

In this edition, most of the chapters have been extensively revised and updated with relevant exercises and recent examples. This has been done keeping in view the feedback from the teachers, experts, students, and other users of the first edition of the book. Some of these additions have also been incorporated in consonance with the changes recently observed in the syllabi of different universities, most of which are gearing towards the rapidly increasing and complex global communication needs of the professional world. Therefore, the text also includes some of the latest topics such as *Role of Creative and Critical Thinking for Effective Communication* and *Inter-cultural Communication*. Further, keeping in mind the academic and professional needs of students, the section on *Barriers to Communication* has been enriched by adding *Physical, Mechanical, and Psychological Barriers*.

Since professional success depends on a person's ability to speak, the sections on *Developing Extempore and Story Telling Skills* and *Elocution* have also been added. Further, since students require writing different types of résumés for varied job positions and descriptions, a section on different *Types of Résumés* has been incorporated. An elaborate discussion on the different types of interviews is also included so that the readers may feel more comfortable while appearing at one. These days, blog writing has become very popular as a way of expressing one's creativity and views. So much so that the recruiters have started looking at the blogs of the candidates with a view to assess their personality and communication skills before shortlisting them for interviews. To help readers gain an insight into the world of blogging, a few practical tips for effective *blog writing* have been provided.

It has been observed that making students speak and write on the books and movies of their liking is an effective way of developing their language skills. A chapter exclusively on writing effective book and movie reviews has been included in this edition to this end. By bringing into sharp focus the various aspects of books and movies, the chapter on *Book and Movie Reviews* is aimed at enabling the learners' review critically and professionally the books they read and the movies they watch. The other new chapter in the edition, *Art of Negotiation,* will help students understand how negotiation skills play an important role in shaping their professional demeanour and improving their communicative ability.

New to This Edition

- Two new chapters on *The Art of Negotiation* and *Book and Movie Reviews*
- New sections such as blog writing; role of creative and critical thinking for effective communication; role of emotions in communication; inter-cultural communication; physical, psychological, and mechanical barriers; types of interviews; and types of résumés

- Concise content with simple and interactive style of writing
- New practice tests and chapter-end exercises
- New multiple-choice questions for practice, flashcard glossary, and additional exercises in the Students' Resource website

Organization of Content

The content of the book has been divided into six parts.

Part I: Essentials of Communication

Chapter 1 introduces the basic aspects of communication such as the process, forms, role of creative and critical thinking, emotions, and networks. It also covers the various barriers to communication and the ways to overcome them.

Part 2: Developing English Language Skills

Chapter 2 covers the essentials of grammar such as parts of speech, articles, modals, and sentences, whereas Chapter 3 enables the readers to understand the usage of subject–verb concord, verbs, tenses, active/passive voice, clauses, and punctuation marks.

Chapter 4 discusses the common errors and suggests how to avoid making them in your speech and writing. Chapter 5 deals with unscrambling the jumbled sentences and related matters. The commonly used erroneous words, expressions, phrases, and other linguistic structures in Indian English as well as their Standard English expressions are covered in Chapter 6. The basics of phonetics, such as pronunciation, individual consonant and vowel sounds, word stress, intonation, etc. are dealt in Chapter 7. Chapter 8 focuses on helping students improve their vocabulary and thus includes discussions and exercises on word formation, synonyms, antonyms, homonyms, phrasal verbs, one word substitution, etc.

Part 3: Listening Skills

Chapter 9 focuses on helping the reader develop effective listening skills by learning the advantages, process, types, and techniques of listening. It also discusses the various barriers to effective listening and suggests how to deal with them.

Part 4: Speaking Skills

The non-verbal aspect of communication such as body language, paralinguistic features, proxemics, and haptics are included in Chapter 10. Chapter 11 deals with the dynamics of professional presentations related to combating stage fright, preparing PowerPoint slides, and delivering JAM sessions. The features and types of group discussion (GD), group etiquette and mannerisms, and opening and summarizing a GD are covered in Chapter 12.

Chapter 13 elaborates on job interviews, which is an important part of the recruitment process. Public speaking, art of persuasion, story-telling techniques, and making different types of speeches are included in Chapter 14. Chapter 15 focuses on effective conversations, dialogues, and debates. Chapter 16 covers the art of negotiation, including the types of negotiation styles, different stages of negotiation process, and elements of successful negotiation.

Part 5: Reading Skills

Chapter 17 highlights the need for developing efficient reading skills. It also covers the basic steps to effective reading, types and methods of reading, and how to overcome common obstacles that a reader might face. Chapter 18 emphasizes on the reading comprehension part, how different reading skills can be employed as per the reading comprehension tasks, how to identify the central idea in a passage, its nature, structure, and tone.

Part 6: Writing Skills

Chapter 19 focuses on the art of condensation by discussing précis, summary, abstract, synopsis, and paraphrasing. Chapter 20 discusses the elements of writing an effective paragraph, while essay writing has been elaborated in Chapter 21. Chapter 22 covers the elements, layout, and types of business letters, as well as résumé. Chapter 23 analyses business report writing, explaining in detail the features, types, structure, and style of reports. Writing effective technical proposals is discussed in Chapter 24. Chapter 25 explains the nuances of email writing, such as common errors, guiding principles, as well as etiquettes to be followed while communicating through it. The essentials of blog writing are also discussed in this chapter. Other business writings, such as itinerary, memo, circulars, instructions, notice, agenda, and minutes, are explained in Chapter 26. Chapter 27 instructs how to interpret a book and a movie and helps students compose effective book and movie reviews.

Online Resources

The following resources are available to support the faculty and students using this text.

For Faculty

- PowerPoint Slides

For Students

- Multiple Choice Questions
- Flashcard Glossary
- Additional Practice Exercises on English Grammar
- Additional Reading Material
- Videos on group discussions, interviews, and professional presentations
- Audio clips related to effective conversations and phonetics
- Text supplements on formal documents

The videos on professional presentations, interviews, and group discussion can be viewed with VLC Media Player or Windows Media Player.

The file 'Text Supplements' is a PDF (Portable Document Format) document. You will need either Adobe Acrobat or its Reader to view it.

Acknowledgements

Writing and revising this book has been an arduous as well as an exciting task. It demanded from us a great deal of research, effort, hard work, and commitment. All this required a lot of motivation and professional efficacy which always came from our publishers, Oxford University Press India. We would like to thank all those working in OUP India who helped us during the publication and revision of this book. We also take this opportunity to thank the reviewers, fellow academicians, students, and other users of our book who helped us with their useful comments, observations, and suggestions from time to time.

We also acknowledge with gratitude the motivation and support we received from the management of BITS, Pilani—Prof. L.K. Maheshwari, former Vice Chancellor, Prof. B.N. Jain, Vice Chancellor, and Prof. G. Raghurama, Director, BITS, Pilani. We are also grateful to Dr Devi Singh, Vice Chancellor, Dr S.K. Majumdar, Director, IM, and Dr Anupam K. Singh, Director, IET, JK Lakshmipat University, Jaipur for their encouragement and support.

Writing a work of such magnitude would not have been possible without invaluable inputs, inspiration, pushing, and nudging from many of our fellow creative intellectuals. In this category, we are indeed grateful to Prof. Krishna Mohan for his guidance and encouragement that we received time and again. We are also grateful to Prof. Meenkashi Raman for her constant help and motivation through the project. We express our gratitude to Dr Binod Mishra at IIT, Roorkee, for his support, motivation, and cooperation. We are also indebted to our faculty friends Dr Sangeeta Sharma, Dr Geeta B., Dr Devika, Dr Sanjeev Kumar Chaudhary, Dr Sushila Rathore, Dr Virender Singh Nirban, Dr Joy Anuradha, Dr K. Aruna, Ms Ruchika Sharma, and Ms Poonam Vyas at BITS, Pilani, and Dr Satya Paul, Dr Sanjeev Kumar, Dr Umed Singh, Dr Lalit Sharma, Dr Veerender Mishra, Dr Meena Kumari, Dr Randeep Rana, Dr Randeep Hooda, Dr G.S. Chauhan, Mr Admya Veer Dagar, Mr Raj Kumar Saini, Mr Sonu Lohat, Mr Manoj Sharma, and Ms Manusmriti, associated with different academic institutes in different parts of the country, for their interest in our work and the support that they extended from time to time.

The occasion also makes us fondly recall the loving memory of Late Prof. M.K. Bhatnagar, whose intellectual wisdom, profound knowledge, and exemplary humility will continue to inspire us throughout our life. We also take this opportunity to thank our teachers, Prof. D.V. Dagar, Prof. Rajul Bhargava, Prof. Sudhir Kumar Saxena, Late Prof. Meera Banerji, Prof. Rajni Badlani, Prof. V.D. Singh, Prof. C.K. Sharma, Prof. R.P. Bhatnagar, Dr Satish Arya, Prof. Shyam Avasthi, and Prof. S.S. Sangwan, whose guidance, affection, and warmth have always guided us through trying situations. We would also like to express our indebtedness to Dr R.P. Patnaik, Dr G.P. Srivastava, Dr Upinder Dhar, and Dr B.V. Babu who all helped us learn and grow in life.

Further, we are delighted to compliment and express a sense of gratitude to our students Rishabh Gupta, Rashi, Ishan Sood, and Arushi Prasad who have admirably provided the voice-over in the audio segment of online resources. The fact that they accomplished the task in a short span of time makes their contribution all the more special. We would also like to thank our other students Rahul, Ganesh Soni, Mayur Karthik, H. Karthick, Rajat Jain, Pradeep, and Hina Jain for their help and support.

In addition, we thank all our seniors, well-wishers, family members, and friends whose silent but invaluable support we have always appreciated.

Finally, we most affectionately express a deep sense of appreciation for our son Siddharth and daughter Snigdha who bravely spared us from their care so that we could finish the project on time!

We would be glad to receive any comments on our book to improve future editions. Readers can send their suggestions to drarorasanjay@gmail.com.

Sanjay Kumar
Pushp Lata

Preface to the First Edition

Today, it is possible to communicate with people staying far away through various means such as mobile phones, SMS, online chatting, social networking, and videoconferencing. Therefore, we can say that we live in an age of communication, not just because of decisive breakthroughs in communication technology but also because with each passing day, the impact of the words we say and write seems to grow manifold. In an age of fragile hopes and tenuous human relations, it is how we present ourselves verbally and non-verbally that gives us a sense of adequacy and certainty. At the professional front also, it is communication and its related skills that decide a person's career curve. It is so because in the professional world, what professionals do most of the time is communicate.

With technology bringing us closer and English becoming a global language, proficiency in English is considered essential for a person's personal and professional growth. Today, it is effective communication in English that fetches students their dream jobs; helps a professional surge ahead of others; keeps afloat a multinational organization; elevates a common mortal to the dizzy heights of achievements; and in a way, defines and redefines our existence in this competitive world.

This book, *Communication Skills*, aims at helping readers acquire the skills required for communicating effectively in professional situations and learn the nuances of the English language.

About The Book

Written in consonance with the latest syllabi prescribed in universities across the country, the book attempts to cover the entire gamut of communication in English—its shades, shapes, colours, and nuances. Though primarily meant to be a textbook for the undergraduate students of engineering and arts, science, and commerce, the book will also serve as a reference guide for engineers, managers, scientists, teachers, trainees, administrative officers, and other professionals who need to use English as a tool for communicating in a professional environment.

Keeping in mind the requirements of students, the book revisits English grammar in detail. The chapters on grammar are heavily interspersed with examples that clarify the concepts adequately and also provide practice tests and exercises that reinforce what has been learnt. Each chapter clearly states the learning objectives at the beginning of the chapter and also recaps the important points at the end of the chapter.

Pedagogical Features

Listed below are some pedagogical features that make this book both interesting and highly educative:

- *Comprehensive text written in an interactive style* The comprehensive coverage, annotated examples, and the practice material given both embedded in the text and at the end of sections, chapters, and parts are likely to help students deal efficiently with any communication task they are required to confront in their professional careers. In order to keep the learner entertained, a warm, interactive, and personal style of writing is followed throughout the book.

- *Practice-oriented approach* Each chapter contains a large number of practice tests and chapter ending exercises, both with explanatory answers, and other practice material so that the learners not only gain conceptual clarity but also develop confidence by frequently exposing themselves to the variegated real-life situations in the arena of communication.
- *Focus on skill development* The chapters on writing skills include numerous samples and practice tests and exercises for business letters, reports, proposals, paragraph, essay, and email writing as well as tips for editing and proofreading. This helps students in improving their written communication skills.
- *Engaging illustrations and cartoon strips* A lot of interesting illustrations are introduced in the book to break the monotony and provide the readers some light but thought-provoking moments. There is also a cartoon strip at the end of the chapters titled 'Wisewell Quips' wherein fictitious characters Wisewell and Madcap enlighten readers with some proverbs, idiomatic expressions, phrasal verbs, and words often confused in the English language.
- *A fascinating audio–video accompaniment* Another highlight of the book is the companion audio–video CD that presents videos on group discussions (GDs), interviews, and professional presentations and audios on sample conversations. Students are expected to not only enjoy watching the CD but also to relate to the situations and gain enormously from the situations presented in the video and audio sections.

Content and Coverage

For the convenience of the reader, the book has been divided into six logical parts.

The first part covering Chapter 1 on fundamentals of communication lays the foundation for the need, importance, and features of effective communication.

The second part, comprising Chapters 2 to 8, focuses on developing English language skills. Chapter 2 covers the essentials of grammar, such as the parts of speech, pronouns, adjectives, verbs, adverbs, prepositions, articles, modals, etc. Chapter 3 on applied grammar and usage deals with subject–verb concord, tenses, moods, direct and indirect speech, clauses, tag questions, and punctuation marks. Chapter 4 on common errors and misappropriations helps readers identify common errors made while speaking and writing English and suggests how to avoid such errors. Chapter 5 on jumbled sentences discusses unscrambling jumbled sentences and other related matters. Chapter 6 discusses with examples the common divergences from standard English observed in the English language used by us in our day-to-day expressions. This chapter also provides the corresponding standard English usage for many such expressions popularly referred to as *Indianisms*. Chapter 7 on phonetics stresses on correct pronounciation, intonation, and word stress. Chapter 8 stresses on the importance of having a good vocabulary. It covers technical words, one word substitution, homonyms, homophones, phrasal words, and idiomatic expressions among others.

The third part on listening skills contains Chapter 9 on developing effective listening skills. It talks about how listening is an art and further covers the barriers to effective listening, advantages of good listening, techniques of effective listening, note taking, etc.

The fourth part on speaking skills covers Chapters 10 to 15. Chapter 10 discusses non-verbal communication. It talks about body language and paralinguistic features. Chapter 11 deals with the dynamics of professional presentations. It covers topics such as combating stage fright, JAM sessions, and preparing PowerPoint presentations. Chapter 12 is on group discussions (GDs). The different types of GDs, opening and summarizing a GD, dynamics of group behaviour, and some tips for GDs are all covered in this chapter. Chapter 13 discusses in detail the process, stages, and types of job interviews.

This chapter also discusses the desirable qualities required in an employee and provides tips for achieving success in interviews. It also lists some commonly asked questions in interviews. Chapter 14 on public speaking discusses how to prepare and deliver interesting speeches. Chapter 15 on conversations, dialogues, and debates provides tips on improving conversations, sample short conversations, sample telephonic conversations, and situational dialogues.

Since reading is perceived as the mother of all speaking, the fifth part of the book comprising Chapters 16 and 17 aims at imparting effective reading skills to the learner. It intends to achieve this objective by first introducing the steps required in inculcating the art of effective reading in students and then by giving them extensive practice in reading comprehension exercise. Chapter 16 on the art of effective reading covers the types and methods of reading. It also provides tips for effective reading. Chapter 17 focuses on comprehension passages. It discusses how by employing different reading skills, understanding discourse features, and inferring lexical and contextual meanings, the different types of reading comprehension passages can be attempted in an exam.

The last part of the book on writing skills will teach the learner to compose effectively the various forms of professional communication such as business reports, business letters, technical proposal, research papers, itinerary writing, and emails. Chapter 18 on the art of condensation discusses précis writing, synopsis, paraphrasing, abstracts, etc. Chapter 19 on note making covers the importance, the various stages, and useful tips for note making. Chapter 20 on paragraph writing covers the structure, construction, and features of paragraphs. It also discusses argumentative and analytical paragraphs. Chapter 21 on essay writing dwells on the dimensions of essay writing discussing in detail narrative, descriptive, reflective, expository, and imaginative essays. Chapters 22 and 23 cover business letters, résumé writing, and business reports. These chapters provide a lot of sample reports and business letters. Chapters 24 and 25 cover technical proposals and technical papers. Chapter 26 explains the nuances of email writing and Chapter 27 discusses different types of business writings such as circulars, notice, minutes, agenda, and memos. Chapter 28 provides useful tips for editing and introduces to students the standard proofreading symbols so that they are able to edit, revise, and proofread with precision all that they write as part of their professional engagements.

Two appendices have been given at the end of the book. Appendix A provides a list of common business terms with their usage. Appendix B provides a list of words from science and technology.

Another feature of the book is the additional material given in the CD accompanying the book. The CD gives an illustrative instruction on how to perform in various professional communication situations such as GDs, job interviews, and professional presentations. The CD also has an audio section which includes some sample conversations and role-plays and thus helps students converse better in day-to-day situations.

Sanjay Kumar
Pushp Lata

Brief Contents

PART 6   WRITING SKILLS

Detailed Contents

Fundamentals of Communication

Learning Objectives   After reading this chapter, you will be able to

- understand the importance of communication in the professional world
- familiarize yourself with the features of successful professional communication
- identify the various purposes for which communication is used in professional situations
- learn about the role of creative and critical thinking, as well as emotions in communication
- get acquainted with the different flows of communication in an organization
- understand the barriers in communication and learn the ways to overcome these barriers

1.1 INTRODUCTION

Mr A.R. Bajaj is the Chairman of Mirch Masala Foods Pvt. Ltd, New Delhi. Since the opening of its first outlet in Delhi three years back, five more outlets of Mirch Masala have come up in other cities. Customers walking in at a Mirch Masala outlet are received with a smile, entertained courteously, and explained the highlights of different food items listed in an aesthetically-designed menu. Those who visit Mirch Masala appreciate the courteous behaviour of the attending staff. 'Everyone around seemed so much willing to serve you,' was how Mr Parsanna, who recently visited Mirch Masala, quipped. On the other hand, Hot & Sweet Pvt. Ltd, which started almost four years back, is struggling hopelessly. Customers often complain about the indifferent attitude of waiters. 'They have named it Hot & Sweet but nobody is sweet there and nothing is served hot; waiters prefer to keep you waiting before paying you any attention and when they do, they speak to you as though they were the bosses around! You feel so unwelcomed there,' said Mrs Suhasini Roy, who recently dined at Hot & Sweet.

What do you think is the fate of these two fancy food outlets? Would Mirch Masala and Hot & Sweet move on their way to growth, success, and popularity? We seriously doubt it. It indeed is quite natural that Mirch Masala is growing but Hot & Sweet seems on its way out.

Yes, we believe that with the passage of time, Mirch Masala is going to grow owing to its good communication network within and outside the organization. Hot & Sweet, on the other hand, is most likely to ebb away into anonymity due to lack of openness, motivation, and proper communication flow.

The case in discussion highlights the deterministic power of communication in the professional world. In an age where one has to either communicate or collapse, communication holds the key to the growth of an organization. The system that fails to recognize the need for an effective communication fails to sustain itself for a long time.

Let us learn further why it is so.

1.2 COMMUNICATION—AN OVERVIEW

A blow with a word strikes deeper than a blow with a sword.

–Robert Burton

Living in an atomic world, we are well past an age when the blow of a sword would pose any considerable threat to us. Today, verbal blows appear mightier than the physical ones. Not just that, words—both written and spoken—have acquired a significantly crucial and an almost decisive force in contemporary times.

Essentially, ours is a society that moves on the wheels of communication. Particularly in the professional world, it is communication and its related skills that decide a person's career curve. The better one's communication skills, the higher are the chances for him/her to touch the zenith of success. The poorer one's communication skills, the greater is the possibility of not achieving one's goals.

Actually nothing happens in the professional world without communication, though it is only a means, and not the end. Still, it is communication that propels the management process and serves as a lubricant for its smooth operation. It helps professionals in their five major managerial tasks, namely planning, organizing, executing, staffing, and controlling. Since every organization is a social system that involves interaction among people working at different levels, proper communication among them becomes necessary for achieving the goals of an organization. In a way, it is communication that defines the existence of an organization in contemporary times. When communication crumbles, the organized action comes to an end. Communication, thus, is vital to the survival, sustenance, and growth of an organization.

It is so because in a professional world, what professionals do most of the time is communicate. The necessity and importance of communication skills can be gauged from the fact that professionals spend nearly three-fourths of their working time in communicating their ideas, views, and plans to others. Communication in the professional world occupies such a pivotal position that there hardly exists an activity in the business and industry that does not require communication to play any role. Consequently, the organization that disregards its importance cannot compete and survive in a demanding professional world that more than anything else demands its incumbents to be articulate, expressive, and communicative.

Understandably therefore, while selecting a new recruit, one of the first things that companies look for in an individual is the person's ability to communicate effectively with others. Our communication skills thus have the potential to make or ruin our fortune.

It is precisely to address this professional need that we need to master the various aspects of communication skills. However, before we proceed further, let us make an effort to acquaint ourselves with the other nuances of the term *communication*, starting with its definition, process, and features.

1.3 DEFINITION OF COMMUNICATION

Communication, the buzz word in today's world, originates from the Latin word *communico* or *communicare*, which means 'to share'. Various researchers and analysts define the term 'communication' in their own way. Despite their different versions, it can be briefly summed up that 'communication essentially means the transfer of ideas, feelings, plans, messages, or information from one person to another'. Obviously however, *communication is considered effective only when it gets the desired action or response*.

Let us explore some of the essentials of communication with the help of the discussion that follows.

1.4 PROCESS OF COMMUNICATION

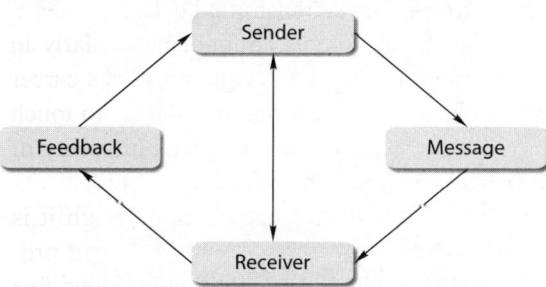

Fig. 1.1 The Communication Process

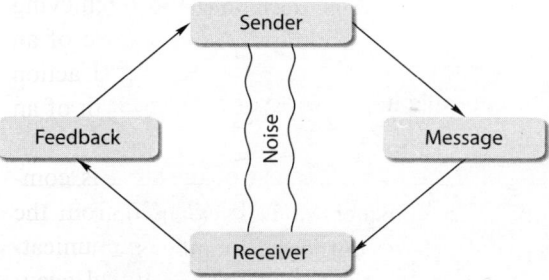

Fig. 1.2 Noise Hinders Communication Process

Communication is a process whereby information is encoded, channelled, and sent by a sender to a receiver via a medium. The receiver then decodes the message and gives the sender a feedback. All forms of communication require a sender, a channel, a message, a receiver, and the feedback that effectively winds up the process. Communication requires both the sender and the receiver to have an area of communicative commonality. The process can be well understood with the help of Fig. 1.1.

However, sometimes there occurs a hindrance in the communication process, which is called *noise*. Noise can be defined as an unplanned interference in the communication environment, the one that causes hindrance to the transmission of the message (Fig. 1.2). It may mainly occur due to two reasons: disturbance in the channel/medium and/or some kind of error in the message sent.

Before we go further, it is important for us to understand how general purpose communication differs from professional communication. Table 1.1 highlights the basic differences between these two types.

Table 1.1 Differences between General and Professional Communication

	General Communication	**Professional Communication**
Content	Contains general message	Contains a formal and professional message
Nature	Informal in style and approach	Mostly formal and objective
Structure	No set pattern of communication	Follows a set pattern such as sequence of elements in a report
Method	Mostly oral	Both oral and written

(Contd)

Table 1.1 *(Contd)*

	General Communication	**Professional Communication**
Audience	Not always for a specific audience	Always for a specific audience, e.g., customers, banks, etc.
Language	Does not normally involve the use of technical vocabulary, graphics, etc.	Frequently involves jargon, graphics, etc. for achieving the professional purposes

1.5 FEATURES OF SUCCESSFUL PROFESSIONAL COMMUNICATION

Since communication matters a lot in the professional world, it is quite important for us to get acquainted with the most important features of successful professional communication.

1. Communication is a two-way process by which information is transmitted between individuals and/or organizations so that an understanding may develop among them.
2. Communication is a continuous process of meaningful interactions among persons in an organization that results in meanings being perceived and understood in a desired way.
3. The role of the receiver and the sender keeps changing in the entire communication activity.
4. Communication broadly includes both verbal and non-verbal forms. Therefore, it also includes lip reading, finger-spelling, sign language, and body language used in face-to-face communication.
5. It is a process which transmits and disseminates important ideas, thoughts, feelings, plans, etc.
6. Communication skills are generally understood to be an art or technique of persuasion through the use of oral, written, and non-verbal features.

1.6 IMPORTANCE OF COMMUNICATION

The following factors make communication indispensable in the world of business.

1.6.1 Growth

Due to the emergence of multinational companies, large business houses usually operate both within and outside the country. The head office of a large corporate maintains a thorough and up-to-date knowledge of the various activities at each of its branch offices. It keeps them well-acquainted with the activities of all the centres, which in turn establishes a link among its various branches and leads to the growth and smooth running of the entire business.

1.6.2 Complexity

This is an age of specialization and therefore, even in a single organization different activities, such as planning, production, sales, stores, advertising, financing, accounts, welfare, etc., are handled by different departments. If these departments do not communicate with one another as well as with the management, there will be no coordination among them. For instance, when production is fully geared up, stores may report shortage of raw materials. Similarly, the finance department has to apprise the other departments regarding its constraints, which might have been the result of recent fluctuations in the market. In fact, all the departments and units of an organization have to go hand in hand to achieve its goals and for that, they need to keep communicating with one another.

1.6.3 Competitiveness

Items of common consumption, such as tea, cigarettes, soaps, blades, clothes, etc., are available in a dozen brands today. Marketing research suggests that firms which communicate better sell better. The better the communication skills of a salesperson, the larger the number of customers he/she can attract. Salesmanship is primarily an art of communication. Besides, companies keep competing with one another through advertisements and other propagandist strategies for securing a higher position in the market. All this involves communication at every step.

1.6.4 Harmony

Trade unions believe in bargaining with the management and insist on the protection of the rights and dignity of the workers. The management and such unions share a delicate relation and without proper communication between the two, no harmony can be expected to exist in an organization.

1.6.5 Understanding and Cooperation

If there exists good communication between the management and employees, it will bring about an atmosphere of mutual trust and confidence. Only when the employees know exactly what is expected of them, can the management utilize their potentialities and make up for their limitations. Through effective communication, employees get job satisfaction and develop a sense of belongingness with the enterprise which ultimately helps the organization grow well.

1.7 PURPOSE OF PROFESSIONAL COMMUNICATION

Professional communication aims at achieving the following objectives:

Advising   However competent a professional may be, he/she cannot have specialized knowledge of all the branches such as licensing, taxation, publicity, engineering, etc. To succeed in his/her job, he/she will have to seek frequent advice. Also, the junior employees need to be advised by the supervisory staff on how to go about doing their jobs. A proper and timely interaction with experts in the related areas helps the management take wise steps and grow.

Counselling   Even an efficient employee may become slow and indifferent if he/she is facing personal problems at home. Such employees are encouraged to consult the counselling department. Through effective communication, employees and workers share their concerns, ventilate their problems and thus are restored to their mental and physical health.

Giving orders   Order is an authoritative communication. It is a directive to somebody, always a subordinate, to do something, to modify or alter the course of something he/she is already doing, or not to do something. Whatever be the nature and size of an organization, orders are absolutely essential. Ordering without bullying, however, is an art that requires effective communication skills.

Providing instructions   Instruction is a particular type of order in which the subordinate is not only ordered for a particular job, but also given guidance on how to go about doing it. All instructions are orders, but all orders need not be instructions. Regardless of the fact that the management intends to give instructions or issue orders, effectiveness in communication is mandatory if the right impact is desired to be created.

Marketing   Just as marketing is crucial to all business, effective communication holds the key to marketing itself. Since the entire function of marketing rests on communication, it is hard to imagine any of its operations getting through without effective communication. Sometimes companies also hire consultants for obtaining right suggestions related to its various operational aspects. If not properly communicated, suggestions can be turned down.

Persuading   Persuasion may be defined as an effort to influence the attitudes, feelings, or beliefs of others or to induce action based on that. Buyers have to be persuaded to buy products. In factories or offices, the lazy, the incompetent, and the disgruntled workers have to be persuaded to do their work. It is effective communication alone that can inspire indolent and uninterested people and keep them persuaded towards achieving the common objectives of an organization.

Giving warnings   If employees do not abide by the norms of the organization or violate rules, it may become necessary to warn them. Warning is a forceful means of communication for it carries with it a sense of urgency. Sensitive in nature, warnings need to be communicated well so that impact is properly created and not exaggerated.

Raising morale   Morale stands for the mental health of all individuals and hence is important for the growth of an organization. Morale—to be maintained only through effective communication among professionals and hence within the organization—actually is a powerful and intangible factor representing the sum of several qualities, such as courage, determination, clarity, and confidence. It acts as a kind of lubricant among people, binds them with a sense of togetherness, and impels them to work in cooperation with one another in the best interest of their organization.

Staffing   Communication is needed in the recruitment process to rope in potential employees of merit to work for the enterprise. The recruits are told about the company's organizational structure, its policies, and practices. This way, proper communication helps the new entrants associate themselves with the organization and utilize their potential effectively. This also promotes proper delegation of work among employees. Thus, in such situations too, communication helps in building a good image of an organization.

Projecting image   Communication is of vital importance in projecting the image of an enterprise in the social environment that is affected by the information which elite groups and wider public have acquired about its goals, activities, and accomplishments. One can hardly wonder at the meticulousness with which all the brochures, advertisements, notices, announcements, and circulars—that are made public—are written and designed by the organizations. Understandably, all important documents are ruthlessly revised and edited linguistically until they help the organization achieve the objective of carving and retaining a positive image.

Preparing advertisements   No matter how good a product is, it cannot succeed without effective advertising. Advertising is done through newspapers, magazines, television, billboards, Internet, pamphlets, cards, etc. Without effective communication, persuasive and catchy advertisements and publicity material can neither be conceived nor created to achieve the desired objective.

Making decisions   Communication also has an important function in solving both simple and complex problems, and making accurate decisions to positively influence organizational performance. If not properly communicated, even a good decision may sound like a bad slip.

Getting feedback The receiver's reaction to the message is also a form of communication back to the sender. Through this mechanism, companies know how much they sell, what public opinion has been formed about their product, and what the customers feel about their products. Good companies also seek feedback with regard to the satisfaction of the employees serving them. It helps them remove the unnecessary cobwebs arising out of the ills of hierarchy and achieve the purpose within an organization.

Thus, communication is like a two-way street that entails the relation between the sender and the receiver both in day-to-day and business communication. In this process, a cycle of communicating messages is formed between the sender and the receiver.

Having discussed the purpose and importance of communication in the business world, let us discuss its different forms.

1.8 ROLE OF CRITICAL AND CREATIVE THINKING IN EFFECTIVE COMMUNICATION

Critical thinking and creative thinking are considered high order skills which are essential for professionals. *Critical thinking* is the active, persistent, and careful consideration of beliefs or knowledge keeping in view the available evidence whereas *creative thinking* is the generation of new ideas. Both are fundamental to human intellectual progress and instrumental in the development of the society. Depending on context and purpose, critical and creative thinking skills can be interdependent or separately applied.

Critical thinking, in fact, is a self-reflective process that involves elements of conceptualization, reasoning, analysis, interpretation, and evaluation of the available information upon which judgement is based. This involves a wide variety of skills that must be used in order to form that opinion/decision. A few of these include:

- making careful observations
- being inquisitive and asking the relevant questions
- challenging the beliefs, examining assumptions, and probing opinions which may even be against already established facts
- recognizing the problems and issues that may appear in future
- assessing the validity of statements and understanding the logic and strength of arguments given
- making workable decisions and finding valid solutions

The ideal critical thinker is habitually inquisitive, well-informed, dependent on reason, open-minded, flexible, objective in evaluation, honest in resolving biases, prudent in making judgements and willing to reconsider the judgements made earlier. As a professional, you will always come across new problems and aberrations to the existing practices, your ability to think critically will help you convert the problem into an opportunity. In order to be a critical thinker, you need to be

- Inquisitive
- Systematic
- Analytical
- Open-minded

- Judicious
- Truth seeking
- Confident in reasoning

Creative thinking, on the other hand, is the generation of new ideas within or across domains of knowledge. It requires preparation, incubation, insight, evaluation, elaboration, and communication. In order to develop this, you must try to put aside the common assumptions, look beyond the conditioning that creates stereotypes, prejudices, and parochial thinking. An unconditioned response to a challenge, an inquisitive approach, an insightful penetration, and a passionate commitment to the task helps us in:

- bringing the existing ideas together into new configurations;
- developing new properties or possibilities for something that already exists; and
- discovering or imagining something entirely new.

Given below are a few basic principles for inculcating creative thinking:

- Be open to new thoughts, ideas, and facts
- Keep your reading and listening faculty actively engaged in observation
- Regard the difficulty or a problem as an opportunity
- Enjoy the process of trying, learning, and evolving
- Avoid jumping to conclusions; follow deferred judgements
- Believe in cross-fertilization of ideas
- Be your worst critic

Thus, if you sharpen your creative and critical thinking, these will equip you with the skills which later in your professional life will provide you an edge in the competitive world.

1.9 ROLE OF EMOTIONS IN COMMUNICATION

Emotions are vital to human life. An integral part in all human interactions, emotions decide the very essence of it, characterizing the tone, colour, purpose, and intent of a message. Since emotions play a major role in the entire communication process, we should know how important or intrusive they can be. Emotions are felt intrapersonally, expressed interpersonally, and have a lasting impact on the entire communicative tapestry we weave with the help of words. Whether we wish to display them or not, emotions radiate our thoughts, express our feelings, reflect our perspectives, and reveal our prejudices more often than not. Functioning as a stimulus, they also trigger a reaction or response from others. Since emotions are central to all interpersonal relationships, it's important to know what causes and influences emotions so we can better understand our own emotions and respond appropriately to others when they display their emotions.

Understanding ours and others' emotions becomes an important skill in the overall process of communication. Such emotional intelligence helps us develop as an effective communicator. Remember, underplaying our emotions may expose as a bland communicator, while overplaying them may suggest lack of equanimity in you. This may, in turn, characterize the response of your co-communicator and hence may affect the quality of your interactions, and even personal or professional relations with them.

Emotionally aware people experience greater success in their careers and a greater sense of well-being in their personal lives. Studies have shown that success doesn't lead to emotional health and happiness, but rather the other way around. The emotionally healthy people experience positive moods, feel more confident, more optimistic, more energetic, and more sociable. These factors lead to greater success in many different aspects of life.

Though emotions are created by us, managing them is not a simple task. In fact, managing emotions largely depends on managing stress. Since modern day living keeps triggering unwarranted stressful responses, negotiating emotions become an uphill task. Therefore, healthy thinking, thought-provoking reading, taking regular exercises, socializing, and meditating can work as effective stress busters. Eventually one can achieve a stage when getting to understand the creation of stressful emotions within us becomes possible, which in turn can help us negotiate our emotions in a better way.

When one knows how to maintain a relaxed, focussed state of emotional awareness even in trying circumstances, one can maintain emotional equanimity and engage oneself in the task of communication in an emotionally controlled and poised manner.

1.10 ROLE OF INTER-CULTURAL COMMUNICATION

The way we communicate is determined strongly by the culture we are groomed in. There are several aspects of communication which differ from culture to culture. Such cultural differences may determine how loud or low we talk; the directness with which we speak; the amount of emotions we express in various situations; the use or avoidance of silence; the prevalence or absence of a particular non-verbal or verbal peculiarity and a series of defining signal which we may emit through our manners, facial expressions, posture, eye contact, tone, and pitch of our speech. Interestingly, all this may be misconstrued in an altogether different manner, depending upon the respondent's own bringing up and cultural variety. It is because of this variegated cultural confluence at work place that creates significant challenges to effective communication beyond the obvious barriers.

These days, companies are doing business more and more in a global context. The people that matter in any business including the suppliers, the clients, and the employees may belong to different cultures and may even be located in foreign countries. The need for effective and clear intercultural communication is becoming vital in securing success in today's global workplace. Greater understanding of intercultural differences, etiquette, protocol, and communication will certainly lead to a much higher probability of achieving business goals.

Another interesting aspect of multiculturalism is the inter-culture of people from both the hemispheres of the earth which throw up unique communication challenges owing to their diverse cultural nuances. Some of these cultures, such as the English-speaking and the Northern European cultures may be regarded as belonging to individualistic cultures, with each of whom enjoying simultaneous memberships in numerous overlapping, informal, loose groups that they join, and leave when convenient. Churches, companies, business associations, social clubs, sports clubs, civic associations, political groups, etc., today actually become manifestation of a culturally kaleidoscopic world. In such a flux, obligations for associations and bonding to other groups are weak, and loyalty is neither required nor highly valued. Common rules of polite behaviour apply equally to group members and non-group members. Relationships with strangers are easily formed and dissolved, and friendship groups are casually replaced and re-formed. Individuals assume as primary their rights to self-expression, self-realization, and self-protection.

On the other hand, highly group-oriented cultures, such as most East and South Asian, South American, Middle Eastern, Eastern European, and sub-Saharan countries, can be seen as a collection of strong groups, starting with close family ties and extending to other blood relatives, school groups, work and military units, community groups. In-group interaction is heavily circumscribed. Individuals are bound to their groups by heavy obligations and strict rules of intra-group relationships; loyalty is required and highly prized. Friendships exist primarily within groups, are formed with serious intent, and imply increasing reciprocal obligation.

Given such cultural diversity, it becomes important for a professional to be aware of all such culturally triggered behavioural differences and communicate accordingly.

1.11 DIFFERENT FORMS OF COMMUNICATION

Communication is generally classified into the following types:

- Verbal communication
  - Oral communication
  - Written communication
- Non-verbal communication
- Intrapersonal communication
- Interpersonal communication
- Extrapersonal communication
- Mass communication
- Media communication

Figure 1.3 provides a skeletal view of professional communication.

Let us briefly get acquainted with some of these varieties of communication.

1.11.1 Verbal Communication

Since a professional has to spend a large amount of his/her working time in speaking and listening to others besides reading and writing, most of the time he/she has to use language as a vehicle of communication. This type of communication is termed as verbal communication.

Verbal communication thus stands both for the spoken and the written word used in the communication process. It can further be divided into oral and written communication.

Oral communication   A face-to-face interaction between the sender and the receiver is called oral communication. In this type of communication, there could be two or more than two persons who use spoken language as a medium of communication. For instance, whenever we

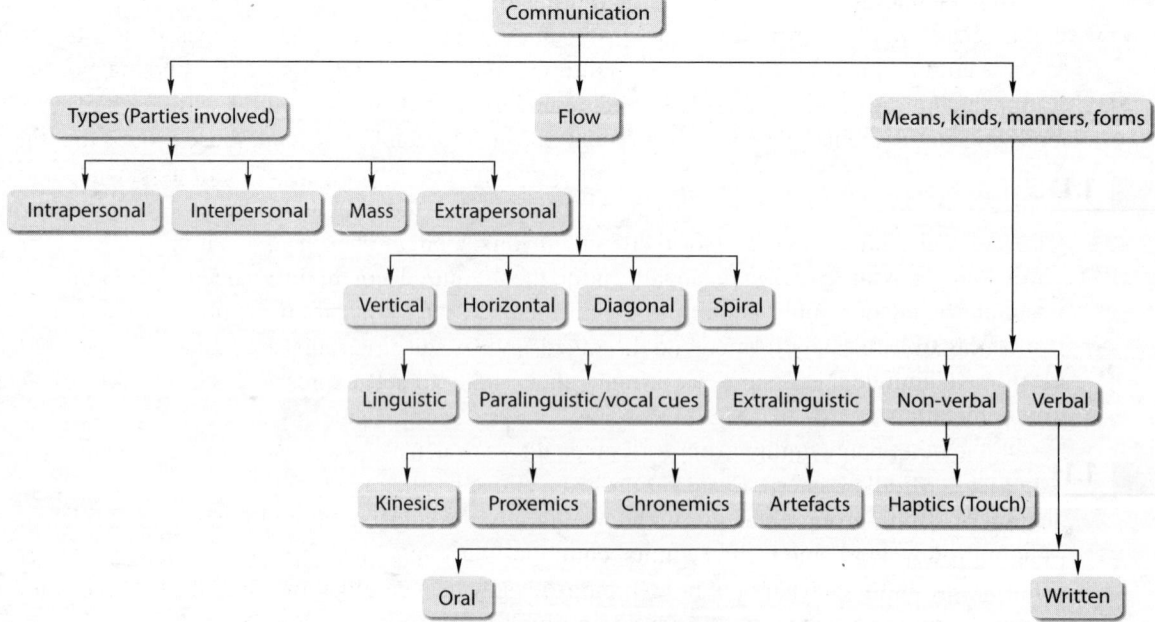

Fig. 1.3   Communication at a Glance

make presentations, deliver speeches, participate in group discussions, appear for interviews, or simply interact with somebody, we are involved in oral communication.

Written communication   In this type of communication, the sender uses the written mode to transmit his/her messages. Reports, proposals, books, handbooks, letters, emails, etc. come in this category. Written communication is routinely used for documentation purposes in business and government organizations.

1.11.2 Non-verbal Communication

When a message is communicated without using a word, the process requires non-verbal cues to be transmitted and received. Non-verbal communication forms an important part in the world of professional communication. It can be further categorized into two parts—body language and paralinguistic features. Body language involves aspects such as personal appearance, walk, gestures, facial expressions, hand movements, posture, and eye contact. The paralinguistic features include a person's voice, volume, pitch, rate, pauses, articulation, voice modulation, etc.

1.11.3 Intrapersonal Communication

This implies individual reflection, contemplation, and meditation. So, whenever communication takes place within one's own self, it is termed as intrapersonal communication. One example of this form of communication is transcendental meditation. It is also believed that this type of communication encompasses communicating with the divine and with spirits in the form of prayers and rites and rituals.

1.11.4 Interpersonal Communication

This is a direct, written, or oral communication that occurs between two or more persons. The oral form of this type of communication, such as a dialogue or a conversation between two or more people, is personal and direct, and permits maximum interaction through words and gestures.

1.11.5 Extrapersonal Communication

Communication does not take place only among human beings. If we observe carefully, we find that sometimes we do communicate with non-human entities, such as animals, birds, etc. For instance, whenever we command our pet dog or cat to sit, stand, or go, they immediately follow our orders. Whenever we caress them or pat on their back for doing something good, they are elated and they start wagging their tails. This type of communication is known as extrapersonal communication.

1.11.6 Mass Communication

This is generally identified with tools of modern mass media, which include books, the press, cinema, television, radio, the Internet, etc. It is a means of conveying messages to an entire populace. This also includes the speeches delivered by a prophet or a political leader.

1.11.7 Media Communication

It includes communication that takes place only with the help of electronic media, such as computer, cell phones, LCD, video, television, etc. Of these, the Internet has become a major means for all sorts of official or personal communication.

1.12 COMMUNICATION NETWORK IN AN ORGANIZATION

Communication serves as an instrument to measure the success or growth of an organization. The success of an organization is recognized by the quality and quantity of information flowing through its personnel. In today's business enterprises, information must flow faster than ever before. Even a little delay might cause a great loss. In large organizations, to keep oneself informed about the smooth operation of the various departments and for performing excellently in the market, it is essential for a professional, who plays a key role in the organization, to know and understand the different forms of communication that constitute the network in an organization.

1.12.1 Different Types of Communication Flow

Communication in a professional organization flows at different layers and levels which regulate, guide, and propel its flow. The different types of communication flow in an organization are as follows:

- Horizontal
- Vertical
  - Upward
  - Downward
- Crosswise
- Spiral

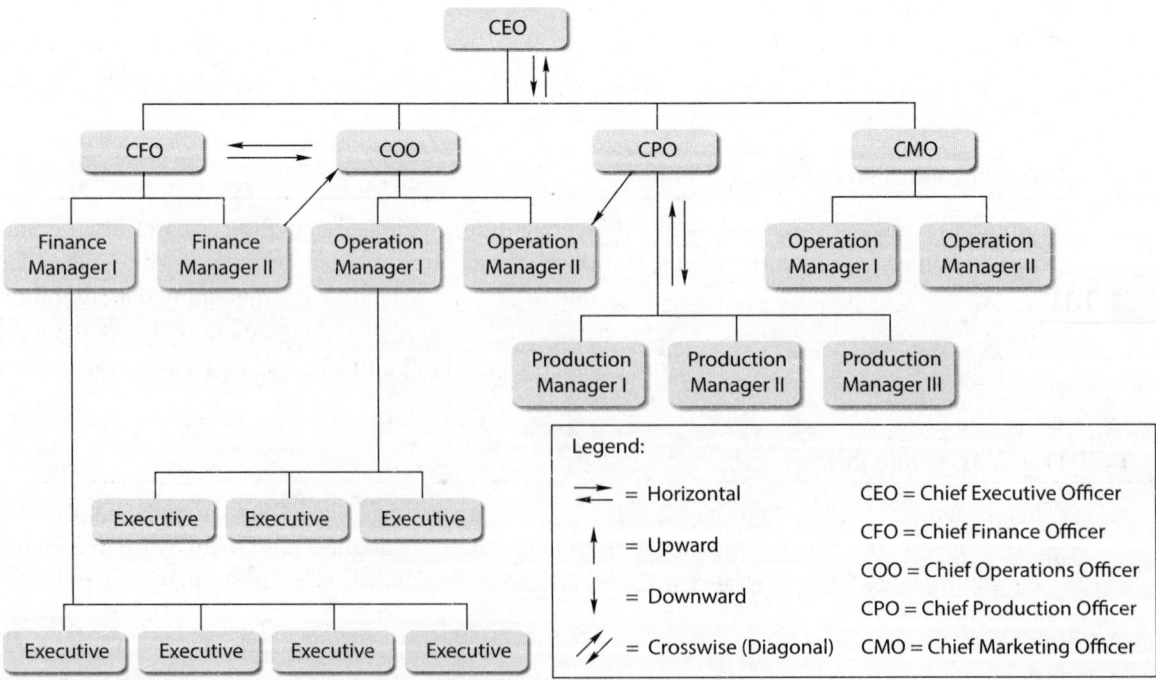

FiG. 1.4 Information Flow in an Organization

The information flow in an organization can be seen with the help of Fig. 1.4, which shows the communication flow that streams in various directions within an organization. Communication may flow horizontally, i.e., among people of the same rank in an organization. This is for better coordination among various departments and for effective decision-making purposes. In order to achieve the production target for a particular month, the discussion held between the production manager and the supply manager will certainly be a perfect instance of *horizontal communication*. However, when the production manager imparts certain instructions to the workers and supervisors for the same purpose, it will be *downward communication*. Here the information moves from the higher authority to its subordinates. And in the same context, if the supervisor reports to the production manager regarding the present state of production, it will be a case of *upward communication*.

If the management circulates a copy of new bonus and incentive scheme among all the employees, it will be called *spiral* communication in the organization. Sometimes however, communication flows between persons who belong to different levels of hierarchy and who have no direct reporting relationships. This is used generally to quicken the information flow, improve understanding and coordinate efforts for the achievement of organizational objectives. Such a movement of an information flow is termed as *diagonal communication*.

In organizations, informal communication also permeates the personal and professional lives of employees. This informal flow of communication is called *grapevine*. For instance, rumours about the company's expansion, promotion of an employee, relations between two colleagues are some of the examples of grapevine. It flows in all directions. Grapevine may create both negative and positive impact on the environment within an organization.

1.13 BARRIERS TO COMMUNICATION

If we look around carefully, we will see that there are people who do not listen to others at meetings; quite a few of them write incomprehensible memos; some do not value other's opinions; some others are unable to seek cooperation from their colleagues and subordinates. Moreover, even a thoroughly prepared communicator commits mistakes, even though the errors are rarely as frequent or as serious as the ones made by a careless communicator. When the flow of communication gets impeded, it is termed as communication barrier.

Let us study the different types of barriers that affect the effectiveness of communication in an organization.

1.13.1 Verbal Barriers

More often than not, most people consider themselves to be good and effective communicators simply because they feel they can speak fluently. While speaking fluently is an important aspect of communicating, yet it is not the only requirement. One should be able to listen effectively, speak fluently and clearly, write well, and read in the language(s) others are familiar with. Thus, there are some verbal barriers such as verbal attack, speaking loudly unnecessarily, and using complex words and phrases. Some of the following reasons may create verbal hindrances in the communication environment.

Lack of proper planning Too often, we find that people start talking or writing without thinking or planning. They do not clearly state the purpose of the message. This often results in miscommunication or partial breakdown in the communication process.

Cruel Jokes that Badly Encoded Messages Play on People

Selection of a wrong variety of language For various human interactions, we choose different varieties of expressions. Imagine a lawyer flaunting his courtroom gambits and exploiting his argumentative skills to convince his wife to give him a cup of tea! The chances are that he won't get that! The reason is obvious: it is a wrong variety of language chosen and hence it fails to elicit the desired result. Because of a wrong variety of language, even a strategy, otherwise effective and useful, may fail miserably.

Badly encoded or wrongly decoded messages The sender may have clarity about the message that is to be conveyed, but it may still not reach the receiver because the message might have been improperly expressed. Choice of wrong words, absence of punctuation marks or wrongly-timed pauses, poor organization of ideas, use of unnecessary jargon, etc. bring vagueness in the message. At times it is the listeners who may contribute to the messy world of misinterpretation by decoding the message in a startlingly different way.

Let us consider the following situation.

The Manager of Sieve Internationals refers to the report submitted by his Marketing Trainee as *bombastic*. The young amateurish employee with a limited vocabulary assumes from the form of the word that it must mean something like fantastic and bursts into a grateful *Thank you very much, Sir!* So a badly or wrongly decoded message can lead to a hilarious situation for the onlookers but an embarrassing one for the people involved.

Semantic gap Semantic gap or distortion might be deliberate or accidental, for example, an advertisement saying, 'We sell for less,' raises the question, 'Less than what?' Is the product sold to less number of people or offered at a less price? Thus, a message such as this may have different connotations and may leave the recipient of the message wondering about the real intention of the message. Thus, whenever there is a gap between the message sent and the message received, it might have arisen due to the language employed and the way it has been interpreted.

Differences in perception of a message Different people perceive a single message in different ways. Consider a situation wherein there occurs a dispute between a worker and his/her superior and a third person is asked to inquire into the matter. During investigation, it is very likely that both the worker and the superior will not recall the situation in exactly the same way. This is mainly because their perception levels are not the same. To overcome this problem, one needs to communicate from various perspectives, try to verify the matter from different points of view, and then come to a conclusion.

Similarly, in a business situation, one should be very careful in the choice of words. It is because like many things in the world, the words we utter are also open to a variety of interpretations. Words such as *good*, *bad*, *proper*, *inappropriate*, *character*, *nature* and a large number of other words are essentially subjective in nature and can be interpreted in various ways and hence need to be used carefully in order to avoid ambiguity in the message.

Variation in language   Sometimes certain words and idiomatic expressions are culture specific. If we do not use them appropriately in the respective cultural context, it might lead to miscommunication or non-communication. For instance, what is called *sidewalk* in the US, in Britain it is called *pavement*, whereas in India, it is called *platform*. Similarly, it is *apartment* in the US, *flat* in Britain, and *house* in India. In the same way, we find in Britain *to table a proposal* means *to act on it*, whereas in America it means *to postpone*.

Therefore, if we use such expressions indiscriminately, it would lead to confusion and consequently may lead to the breakdown of communication.

1.13.2 Non-verbal Barriers

Apart from the basic aspects of communication stated earlier, one needs to keep in mind the non-verbal aspects too in order to be considered adept in communication skills. Moreover, when verbal and non-verbal messages clash, receivers tend to trust the non-verbal messages. Sometimes even flashing eyes, rolling eyes, quick movements or very slow movement, or avoiding eye contact may also cause *non-verbal barriers* to effective communication. *Raising eyebrows* constantly suggests that the speaker is not convinced about the information that he/she has shared. Bulging eyes leave the audience alienated as the speaker unnecessarily sounds arrogant. Even when a presenter keeps his/her hands or thumbs constantly in the pockets of his/her trousers, he/she will certainly appear snobbish, scared, or deceptive to his/her audience. Similarly, if some awkward gestures are constantly used by the speaker, these will create blocks in the smooth flow of communication. Chapter 10 on non-verbal communication deals with the topic in greater detail.

Mobile Games: New-age Barrier to Effective Listening

1.13.3 Listening Barriers

Poor listening results in incomplete, incorrect, and inconsistent responses. Sometimes people do not listen to others properly and patiently because rather than listening to others' views, they may just be waiting anxiously for the speaker to keep quiet so that they may articulate their own views. Experience suggests that those who listen to others with dwindling attention fail to speak properly. It is so because listening is the mother of all speaking.

Listening requires concentration, patience, and focus; the turbulence characterizing a quintessential twenty-first century mind, however, renders the whole task extremely challenging. And although we all pretend to listen to others while sitting in a meeting or attending some oral presentation, we usually are occupied with the idea of speaking at the earliest opportunity. It is so because speaking is a human urge, while listening is a compulsion. We all aspire to be speakers but not necessarily all of us crave to be a listener. That is why, there are many speakers but just a few listeners. Moreover, listeners interrupting the speakers or avoiding an eye contact with them also causes barriers to effective listening.

Besides the above, the following reasons may also cause listening barriers in the communication environment:

- Making the speaker feel as though he/she is wasting the listener's time
- Being distracted by something that is not part of the ongoing communication
- Getting ahead of the speaker and completing his/her thoughts
- Topping the speaker's story with one's own set of examples
- Forgetting what is being discussed
- Asking too many questions for the sake of probing

These barriers have been elaborately discussed in Chapter 9 on developing effective listening skills.

1.13.4 Physical and Mechanical Barriers

Closed office doors, physical distance between the communicators, disturbance in transmission channels, etc. create physical barriers. Research shows that one of the most important factors in building cohesive teams is proximity. Nearness to others helps people communicate effectively as it helps them get to know one another better. Distance therefore, is a barrier that often affects communication. At times, poor printing, badly indented text, lack of space or margins, very small font sizes, and crowded paragraphs can lead to barriers. Similarly, erratically functioning machines, faulty devices and systems are some of the mechanical barriers which affect communication between two parties. Failure of microphone during a speech may be seen as a mechanical barrier, for instance.

1.13.5 Psychological Barriers

Compared to physical barriers, psychological barriers are far more damaging and cause frequent disruptions in the process of communication. Psychological barriers are closely linked to emotional barriers; emotional barriers play an important role in the communication of a message. Some such psychological barriers may arise due to the presence of fear, distrust, absent mindedness, emotional apathy, intensity of feelings, lack of interest, obsessive temperament, multiple distractions, and so on between the communicating parties. For instance, on receiving a message, we may not be able to interpret it properly as we may be caught up in some emotional turmoil of our own.

1.13.6 Miscellaneous Barriers

Besides the barriers discussed so far, there are other barriers too for which a detailed discussion is provided below

Premature evaluation of message It is an undesirable human tendency to jump to hasty conclusions, approve or disapprove what is being said or written, and generalize the ideas without being convinced. This tendency often leads to failure in communication because the listeners and readers evaluate the message without fully understanding its real essence.

Information overload In various organizations, the employees in the key positions get unrestricted flow of information. Struggling with an information overload thus, they often tend to ignore the important information unconsciously while processing the information. For instance, the person concerned might miss the word 'not' in a message, which reverses the intended meaning. It is also observed that people respond to information overload by simply escaping from the task of communication, for example, wherever authorities demand detailed documentation for procuring government contracts; people sometimes shirk the task and simply furnish unauthentic or incomplete information.

Distrust, threat, and fear   Inconsistent and unpredictable behaviour of the superior leads to arousing distrust and fear among the subordinates. If a subordinate has been punished for presenting unfavourable but true facts in the past, he/she will not express his/her ideas however innovative and crucial they may be due to real or imagined fear. Eventually, the communication process in a scenario like this will break down.

Less time for orientation and for adjustment to change   Changes affect people in different ways and it may take some time to adjust to the implication of changes occurring regularly both in our personal and professional worlds. Some communication situations indicate a need for further training, career adjustment, or status identification. If the employees do not get sufficient time to adjust, alter, or prepare themselves to face the changes around them, it hampers communication and can severely affect their performance in an organization.

Emotional reaction   On his retirement day, a senior executive of a leading firm received many sincere tributes during a special dinner hosted in his honour. When finally asked to speak, he got up from his seat, spoke a few words but could not continue. He was choked with emotion. Depending on the intensity of our emotions thus, our response or reaction may create a barrier in communication. Emotions such as fear, suspicion, anger, and joy may act as hurdles in making ourselves clear to our audience.

Rigid attitude   Human communication is all about sharing and conveying emotions, ideas, and attitudes. A stubborn attitude on the part of the listener or the speaker may lead to a failure of communication. It is a well-known fact that we cannot learn anything unless we purge ourselves of our pre-conceived notions.

1.14 SOME REMEDIES

Following are some of the ways to overcome the different types of barriers we confront both in our personal and professional lives:

1. Send the data only to the people who require that.
2. Emphasize the major ideas.
3. Delete unwanted details.
4. Maintain transparency in policy matters.
5. Ensure clarity in message and look for a genuine feedback.
6. Understand others' emotions.
7. Understand other cultures and language variations and use the appropriate variety in the given context.
8. Make sure that information overload does not affect the communication environment adversely.
9. Maintain openness and acknowledge that people have different perceptions and views regarding things.
10. Encourage innovative ideas and views so that people do not unnecessarily live in fears.
11. Listen attentively to others.
12. Speak with clarity and conviction.

To sum up, barriers which are caused due to fear, ecstasy, joy, threat, etc. can easily be overcome by increasing self-awareness, careful listening, and a desire to share and build empathy towards others. Moreover, by knowing more about the receiver's background and the level of knowledge or language proficiency, we can achieve the desired result in communication.

RECAPITULATION

✓ Communication refers to an exchange of information, ideas, feelings, and emotions.

✓ The purpose of communication is to get one's message across to others. This is a process that involves both the sender of the message and the receiver. In fact, a message is successful only when both the sender and the receiver perceive it in the same way.

✓ Communication in business usually involves interpersonal communication, communication between management and staff as well as other business contacts.

✓ There are different forms of communication that exist. These are known as verbal, non-verbal, interpersonal, intrapersonal, mass communication, media communication, and extrapersonal communication.

✓ Every organization has a proper network of communication, which is established due to different types of communication flow in the organization. These are horizontal, vertical, downward and upward, diagonal, and spiral communication.

✓ An informal flow of communication, that is, grapevine, also exists in organizations.

✓ Sometimes the message received is not the same as the message sent. This is called breakdown of communication. It is caused by barriers in communication due to various verbal, non-verbal, listening, or some other factors; overcoming these barriers is highly important for making the entire communication process successful.

✓ Proper planning, rightly encoded messages, understanding other cultures, conducive communication environment, that is, an environment devoid of fear and distrust, may help in overcoming these barriers.

WISEWELL QUIPS

EXERCISES

Objective Questions

I. State whether the following statements are true (T) or false (F).

1. Effective communication leads to better work production.

2. When verbal and non-verbal messages clash, receivers tend to believe the non-verbal messages.

3. External communication often consists of email, memos, and voice messages; internal communication consists of letters.

4. Good listening skills are inherent and cannot be inculcated.

5. To improve communication and to compete more effectively, many of today's companies encourage teamwork and better interpersonal communication.

6. Business communication is both highly formal and unstructured.

7. Before the sender completes his/her message, the listener thinks, 'I know what he/she is going to talk

about.' Such type of listener is sharp, intelligent, and good at interpersonal communication skills.

8. Grapevine is a formal communication flow in an organization which has both positive and negative impact on the environment.

9. 'I had personal problems, so I could not prepare the budget efficiently. I am sorry for this. We cannot submit the details to the client today.' Such utterances reflect the lack of commitment and sincerity on the part of a professional.

10. Communication helps management only to make accurate decisions to influence organizational performance positively.

11. The observance of the receiver's reaction to the message is a kind of tool to maintain smooth communication flow between or among individuals.

12. The discussion held between the production manager and the Head, HRD, is a perfect example of horizontal communication.

13. If the management circulates a memorandum regarding change in working hours among all the employees, it will be a case of spiral communication.

14. It is imperative to listen carefully to others in order to avoid confusion regarding instructions, advice, proposals, reminders, etc.

15. Information overload strengthens the communication network in an organization.

16. Every workday, every employee frequently sends and receives messages and, as the size and complexity of the organization increases, so do the number of messages and the possibilities for communication-related problems.

17. Rigidity of thought helps the officer in maintaining a good rapport with his/her subordinates.

18. Badly encoded message leaves its receiver confused and not well informed.

Concept Review Questions

1. What does the term 'communication' imply? Why is effective communication vital in today's world?

2. How important is 'effective communication' in today's business word? Discuss a few aspects of business where communication is very important.

3. Counselling, instructing, giving orders, persuasion are some of the purposes of communication. Discuss.

4. Discuss the different levels of communications in detail.

5. Explain the importance of communication in business in about 300 words.

6. What steps would you follow if you have a communication problem?

7. What are barriers to communication? Do you remember any case of poor communication? Specify what went wrong in the case that resulted in poor communication.

8. How does a receiver influence the sender's communication skills? Substantiate your answer with appropriate examples.

9. Discuss any four barriers to communication and substantiate your answer with one example for each.

10. 'My father sees me reading newspapers everyday. Today he asked me about 'the Big Bang—the biggest experiment', but I could not give any information. I read only sports news.' Identify the communication barrier involved in this statement and discuss how it affects communication negatively.

11. 'A free flow of information ensures the success of an organization.' Elaborate this statement in the light of the flow of communication in any organization.

12. 'Growth and success of an organization broadly lies in continuous, multi-directional, and multi-level flow of communication.' Elaborate the statement citing suitable examples from your own experience.

13. 'Whether an organization is small or large, it is communication that binds the organization together.' Discuss in detail the formal flow of communication in an organization in the light of the above statement.

14. How does a receiver influence the sender's communication skills? Substantiate your answer with appropriate examples.

ANSWER KEY

Objective Questions

1. T	**2.** T	**3.** F	**4.** F	**5.** T	**6.** F
7. F	**8.** F	**9.** T	**10.** F	**11.** T	**12.** T
13. T	**14.** T	**15.** F	**16.** T	**17.** F	**18.** T

Essentials of Grammar

Learning Objectives   After reading this chapter, you will be able to

- get to know nouns, pronouns, and their types
- familiarize yourself with the different types and degrees of adjectives
- learn in detail about the functions of verbs and adverbs and their placement in a sentence
- learn about prepositions and their usage
- recognize different connectives and their usage
- learn about articles and their usage
- understand the usage of different modals in a sentence
- learn in detail about sentences and their various types

2.1 INTRODUCTION

Listen to the speech of a young, aspiring, and passionate engineering graduate:

> My respective Sirs and my dear most friends, myself is Hitesh Tyagi, and I am right now standing in front of you to give a farewell speech. Well friends, our stay in the college was momentous for all those days, months, and years. I still remember a day when I had entered into the gate of this college. I had been feeling extreme nervous. Of course everyone as a fresher feel that. But things changes very quickly.
>
> Today I am passing away. But it is not just only me, all of us are passing away. I have tears in my eyes. I had cried when I had come here. I have cried again today, when I am going away from here. It was fantastical. All that we had. All the funs. Thanks for all of you—all the teachers of mine, those who are junior than me. I loves you all. And will remember you for ever and ever.
>
> Thanks a lots....

How many marks would you give this guy for his passion and intensity of emotions? Well, hundred out of hundred. But for grammar? Probably, zero out of a hundred. Of course, the speech is an exaggeration of the errors and you may well say that graduating students do not make such hopeless errors. We know that it is an exaggeration, but only of a vital truth. It is so because though students are more communicative than they used to be, they are hardly better as learning a language does not mean just speaking it fluently. It is also required that whatever we speak or write is grammatically correct. It is for this purpose that we intend to take you along with us on this grammar voyage. We plan to help you learn the different concepts of grammar and also make you understand how to use them correctly. In order to help you learn faster, we have tried to keep this section as interactive, interesting, and rewarding as possible.

Let us first learn the essentials of grammar.

2.2 PARTS OF SPEECH

In English, there are certain elements such as noun, pronoun, adjective, verb, adverb, preposition, conjunction, etc. which are considered essential parts of speech. Let us discuss them one by one. We earnestly hope that you are acquainted with these essential elements of grammar. Therefore, the discussion, rather than defining and theorizing these aspects, focuses on how to use them correctly and effectively in your written and spoken English.

2.2.1 Nouns

Look at the following sentences:

1. **Jack** is a stupid **boy**.
2. **India** is a great **country**.
3. The **jury** found the **prisoner** guilty.
4. **Beauty** needs no **ornaments**.
5. I have one **sister**.
6. **Milk** is good for **health**.

Can you figure out the part of speech that the highlighted words 'Jack', 'boy', 'India', 'country', 'jury', 'prisoner', 'beauty', 'ornaments', 'sister', 'milk', and 'health' refer to? They are all nouns.

Common and proper nouns As you know, noun is a word that refers to the name of a person, place, or thing. In this context, the word 'thing' means anything we can think of. Therefore, all the highlighted words are nouns. They, however, are not the same types of nouns. The word 'Jack' refers to a particular person, whereas 'boy' can stand for any other person as well. So, 'Jack' is an example of a proper noun and 'boy' is an example of a common noun. Similarly, 'India' is a proper noun, whereas 'country' is a common noun.

PRACTICE TEST 2.1

Can you tell which of the following words are nouns and whether they are proper nouns or common nouns?

1. Jaipur is a fascinating city.
2. Kathak is a famous dance.
3. Mohmmad Rafi was a great singer.
4. Nokia is a mobile.
5. Delhi is the capital of India.
6. Pilani is a small town.
7. Oranges can be had from Reliance Fresh.
8. *The Tribune* is a good newspaper.
9. *The God of Small Things* is written by Arundhati Roy.
10. Steve Waugh was an inspirational captain.

To see how accurate you were, you can refer to the answers given in the Answer Key at the end of the chapter. In fact, you can cross-check your answers for all the subsequent Practice Tests with the Answer Key.

Collective nouns   Sometimes, a number of persons or things are taken together and spoken of as one. Look at the words written in bold letters in the following sentences:

1. **The army** has besieged the town.
2. **The police** went for a cane charge.
3. **The jury** gave its verdict.
4. Suddenly, **the mob** started chasing us.
5. **The committee** comprises three members.

The words highlighted above refer to groups of soldiers, policemen/policewomen, judges, people, and members respectively. Such nouns are known as collective nouns. Now, go ahead and see how common and collective nouns can easily be distinguished from each other:

1. The boys are sitting in the class.
2. The sheep has lost touch with the herd.
3. There are three girls in the family.
4. Eleven players constitute a cricket team.
5. The minister was ridiculed in the parliament.

Obviously, the nouns 'class', 'herd', 'family', 'team', and 'parliament' are the collective nouns in the above set of sentences and 'boys', 'sheep', 'girls', 'players', and 'minister' are the common nouns.

Abstract nouns   To go further, let us consider the following extract from a speech:

> Ladies and Gentlemen, you know why is it that Gandhiji was a man of extraordinary **stature** and **standing**? It is simply because he had many complex and rare **virtues** rolled into one single human being. Well, Gandhi had **wisdom**; he had **honesty**; he possessed a **vision** which is rare and demonstrated an unimpeachable **integrity**. His **insight** was exemplary and his **depth** immeasurable. But what impresses me as an ardent admirer of this great man was the **magnitude** of his **concern** for the entire **mankind**.

Mark the words written in bold letters. All these express the notions that one cannot touch, smell, hold, hear, or see. Obviously, you cannot catch hold of things like honesty, magnitude, mankind, etc. Hence, all these words express abstract notions. The nouns that denote abstract, hidden, and intangible notions are known as abstract nouns.

Do you know how abstract nouns are formed? Look at the following set of sentences:

1. **Laughter** is a good medicine; so, you must **laugh**.
2. Though common mortals like you and me **die**, people like Mother Teresa become immortal; **death** cannot kill them.
3. **Friendship** is a great blessing. We must be proud of our **friends**.
4. It is moving to see the **poverty** of the **poor**, but sickening to see the **richness** of the **rich**.
5. Though **ignorance** is bliss, being **ignorant** is a curse.

Can you make out what has happened? We have formed abstract nouns 'laughter' and 'death' from the verbs 'laugh' and 'die'. We made more abstract nouns, such as 'poverty', 'richness', and 'ignorance'.

The abstract nouns *laughter* and *death* are made of verbs *laugh* and *die*. *Poverty*, *richness*, and *ignorance* are made of adjectives *poor*, *rich*, and *ignorant*, while *friendship* is made of another noun *friend*. Hence, we can say that abstract nouns are formed from adjectives, verbs, and other nouns.

PRACTICE TEST 2.2

Form abstract nouns from the following words:

1. Choose	5. Starve	9. Woman	13. Hate
2. Judge	6. Captain	10. Quick	14. Think
3. Broad	7. Good	11. Dark	15. Bond
4. Sane	8. Proud	12. Hero	16. Vacant

Countable and uncountable nouns   Consider the following sentences:

1. I have one **sister**.

2. **Milk** is good for **health**.

See, we can write 'one sister', but can we write 'a milk' or 'one health'? That brings us to understand the fact that there are certain nouns which can be counted and some others that cannot be counted. The nouns that we cannot count, such as *milk, oil, water, bravery, beauty, dedication,* etc., are known as uncountable nouns and those which can be counted, such as *cup, orange, book, engineer, donkey,* etc., are called countable nouns.

PRACTICE TEST 2.3

Find out whether the nouns given below are countable or uncountable:

1. Girl	5. Style	9. Paper	13. Chair	17. Integrity
2. Wisdom	6. Composure	10. Tub	14. Magazine	18. Movie
3. Idea	7. Kite	11. Pass	15. Seminar	19. Cricket
4. Imagination	8. Intuition	12. Title	16. Crime	20. Speech

Common grammatical errors in noun usage   Having travelled with us thus far, we feel by now you must have picked up the fundamentals of nouns. The most important thing regarding nouns is to figure out the errors that we normally commit while using them in different grammatical structures. Look at the following exercise and see how we sometimes go wrong while using nouns:

1. India has won both **the one-day and the test serieses**.
2. The gift cost me **twenty thousands rupees**.
3. I bought **three dozens bananas**.
4. Indian Air Force is planning to buy **twenty new aircrafts**.
5. If we ignore **the advices** of our parents, we cannot grow in life.
6. Where should I keep my **luggages**?
7. The **evidences** prove that he is guilty.
8. **Employments** are not easy to fetch these days.
9. **Furnitures** have become quite costly in recent times.
10. The government **machineries** are employed in the rescue operation.

Remember that some nouns are normally used only in the singular form and hence are followed by a singular verb. Look at the corrected version of this exercise.

1. India has won both **the one-day and the test series**.
2. The gift cost me **twenty thousand rupees**.
3. I bought **three dozen bananas**.

Sir, two advices I will give: please change all the old furnitures and machineries because old things block our growths.

Excuse me???

Don't Use Plurals Unnecessarily

4. Indian Air Force is planning to buy **twenty new aircraft**.
5. If we ignore **the advice** of our parents, we cannot grow in life.
6. Where should I keep my **luggage**?
7. The **evidence proves** that he is guilty.
8. **Employment is** not easy to fetch these days.
9. **Furniture has** become quite costly in recent times.
10. The government **machinery is** employed in the rescue operation.

Some nouns can mean different in singular and plural forms   Some nouns mean one thing when used in the singular form and another when used in the plural form. Look at the following expressions:

1. **People** (persons) in Europe are very broad-minded.
2. There are many different **peoples** (nations) in Europe.
3. **Rooms** (dwelling place) are available in the guest house.
4. There is no **room** (scope) for further discussion with them.
5. The beautiful statue is made of **stone** (material).
6. He had **stones** (chemical depositions) in his stomach.
7. I have broken my **glasses** (reading spectacles).
8. He filled his **glass** (tumbler) with more wine.
9. There was no help in **sight** (available, visible).
10. We are going to Paris for the weekend to see the **sights** (scene, view).

PRACTICE TEST 2.4

Can you now choose the correct form of the words (singular/plural) so that they are correctly used in the sentences given below?

1. The girl was a stunner; she had blonde **hairs/hair**.
2. He loves listening to quality **music/musics**.
3. The speaker was appreciated with loud **applause/applauses** from the audience.
4. **Cutlery/cutleries has/have** become quite stylish of late.
5. **Businesses/business** cannot grow in times of recession.
6. The company is planning to purchase more **equipment/equipments**.
7. **Times have come/Time has come** when we need to be serious about climate change.
8. Housewives always have lots of **household work/household works** to do.
9. There is no point in visiting Rajasthan during summer; since it is hot during that period, **sightseeings/sightseeing** cannot be enjoyed.
10. In the laughter show the audience had lots of **funs/fun**.

Following are words which give different meanings in the singular and plural form. Observe the correct form used in the following set of sentences.

Choices	Correct Usage
• Rushdie is a man of **letter/letters**.	• Rushdie is a man of **letters**.
• Thousands of people gathered to pay their last **respect/respects** to the departed leader.	• Thousands of people gathered to pay their last **respects** to the departed leader.
• The armed **force/forces** can be seen on a move along the border.	• The armed **forces** can be seen on a move along the border.
• We have received the **good/goods** sent by you.	• We have received the **goods** sent by you.
• There are so many pollutants in the **air/airs** of the city.	• There are so many pollutants in the **air** of the city.
• Oh, please don't run him down; I have immense **respect/respects** for the man.	• Oh, please don't run him down; I have immense **respect** for the man.
• Those who have illusions about their capabilities often remain in **air/airs**.	• Those who have illusions about their capabilities often remain in **airs**.

At times certain nouns end with 's' in spellings but they are treated as singular nouns. Look at the following expressions:

1. Billiards (not billiard) is the game of the rich.
2. Mathematics (not mathematic) is an interesting subject.
3. News is (not new) being telecast right now.
4. Rabies is (not raby) a dangerous disease.
5. Language is a means (not mean) of communication.

Some other nouns, on the other hand, end with 's' in spellings and are used only as plurals. Look at the following expressions:

1. In the annals (not annal) of history, there is no one like Ashoka, the Great.
2. Obsequies (not obsequy) will be performed on Monday.
3. Where are my scissors (not scissor)?
4. His assets are (not asset) meagre.
5. Thanks are (not thank) due to each and every member of the party.

At times, errors are caused as we are not sure how to show the possessive case of a noun. Look at the following sentences. Do you find them correctly expressed?

1. That is Tagore's the poet's house.
2. Where is Lalu's and Balu's bakery?
3. The book's cover is torn.
4. The chair's leg is broken.
5. Last night, the interview of Amitabh Bachchan was telecast.

As a rule, we do not use **apostrophe s ('s)** to denote the possessive case of nouns that suggest non-living objects. So, ideally these expressions should be rewritten as under:

1. That is Tagore, the poet's house.
2. Where is Lalu and Balu's bakery?
3. The cover of the book is torn.
4. The leg of the chair is broken.
5. Last night, Amitabh Bachchan's interview was telecast.

Can you now make out how to show the possessive forms of nouns? Remember to use 's only with the second noun if both the nouns refer to the same person. Similarly, if two nouns possess a common thing, only the second noun should be shown in the possessive case. With nouns denoting non-living objects, 's is not normally used and the possessive form is shown by using the **preposition** 'of'. But to show the possessive case of the nouns suggesting living beings, 's is preferred to of.

The complexity at times gets aggravated when we come across sentences similar to the ones cited below:

1. I love Keats' poetry. (or Keats's?)
2. Do it, for conscience's sake! (or conscience' sake?)
3. The boys' mobiles are lying there. (or boys's mobiles?)
4. Girls' tantrums are hard to understand. (or girls's tantrums?)

As a rule, we do not add another 's' when we have to show the possessive of a plural noun ending in 's'. Therefore, it is wrong to say girls's tantrums or boys's mobiles because both these are common nouns shown in plural. So, what we need to do is to put an **apostrophe** (') without adding another 's' after nouns to show the possessive of plural nouns.

As far as showing the possessive case of proper nouns is concerned, we can add another 's' even if the name itself ends in 's'. For example, it is perfectly all right to say *I love Keats' poetry*. But the same can be written as *I love Keats's poetry* as well. However, adding another 's' to such nouns depends on the way it is pronounced. If adding another 's' makes it difficult for the speaker to pronounce the nouns properly, do not repeat it. For example, if you write *conscience's sake*, it would be fairly clumsy for the speaker to articulate it.

See how to use the following expressions correctly:

Incorrect Usage	Correct Usage
• We are at **Gods'** mercy.	• We are at **God's** mercy. (God is a singular noun.)
• **Cow's** horns are often painted.	• **Cows'** horns are often painted.
• **Sarah Water's** style is quite literary.	• **Sarah Waters'** style is quite literary.
• Keep quiet, for **goodness's** sake!	• Keep quiet, for **goodness'** sake!

Avoid errors while denoting plural noun forms   Sometimes, we make errors while denoting the plural of certain nouns. See the following expressions and choose the correct plural form of the nouns as highlighted below:

Choices	Correct Usage
• **Volcanoes/Volcanos** can keep simmering on for hundreds of years before they burst.	• **Volcanoes** can keep simmering on for hundreds of years before they burst.
• The first three **cantoes/cantos** of the book are wonderfully written.	• The first three **cantos** of the book are wonderfully written.
• The **thiefs/thieves** made off with the entire jewellery.	• The **thieves** made off with the entire jewellery.

(Contd)

(*Contd*)

Choices	Correct Usage
• **Photoes/Photos** clicked in the broad daylight are generally not very clear.	• **Photos** clicked in the broad daylight are generally not very clear.
• At the end of the programme, **mementoes/ mementos** were distributed.	• At the end of the programme, **mementoes/ mementos** were distributed. (Both **mementos and mementoes** are used.)
• In Shakespearean world, even **handkerchieves/handkerchiefs** can spell a tragedy.	• In Shakespearean world, even **handkerchiefs** can spell a tragedy.
• The **gooses/geese** were cackling around joyously.	• The **geese** were cackling around joyously.
• The **chiefs/chieves** of different states were to attend the funeral.	• The **chiefs** of different states were to attend the funeral.
• Hurriedly, he swallowed couple of **loafs/ loaves** and dashed out.	• Hurriedly, he swallowed couple of **loaves** and dashed out.
• Governments that impose unnecessary **taxs/ taxes** are not very popular.	• Governments that impose unnecessary **taxes** are not very popular.

Remember, therefore, that some of the nouns would take -s or -es and some others would take -ves in their plural form.

Choosing plural forms appropriately?   Sometimes, we go wrong while making the plural of compound nouns. See the sentences written in the following exercise and choose the correct form for each of the compound nouns shown in the plural.

By the way, we hope that you are aware of the term 'compound noun'. Actually, in order to give some specific information about something or someone, we use one defining type of noun ahead of another, for example, a book rack (a + noun + noun). However, forming the correct compound noun forms can be confusing at times.

Let us see how to use such nouns correctly:

Incorrect Usage	Correct Usage
• The **good trains** derailed on its way to Delhi.	• The **goods train** derailed on its way to Delhi.
• The **cloth shop** is just round the corner.	• The **clothes shop** is just round the corner.
• The **runner-ups** trophy goes to St Xavier's School, Jaipur.	• Both the **runners-up** trophies go to St Xavier's School, Jaipur.
• Like **true bird of preys**, vultures have penetrating eyes.	• Like **true birds of prey**, vultures have penetrating eyes.
• Her **daughter-in-laws** have made her life miserable.	• Her **daughters-in-law** have made her life miserable.

(*Contd*)

(Contd)

Incorrect Usage	Correct Usage
• We need to have a **saving account** in some bank to start transactions with traders.	• We need to have a **savings account** in some bank to start transactions with traders.
• Recently, he attended a **three-days workshop**.	• Recently, he attended **a three-day workshop**.
• You are required to give **a twenty-minutes speech**.	• You are required to give a **twenty-minute speech**.
• Several miners were trapped inside the **coals mine**.	• Several miners were trapped inside the **coal mines**.
• On the way, I saw **a five-years-old child**.	• On the way, I saw a **five-year-old child**.

With different compound nouns, plurals are shown in different ways. Sometimes the first noun takes the plural form (for example, the instructors-in-charge and not the instructor-in-charges) and at times the second noun takes plural form, for example, *tea leaves* and not *teas leaf*. In fact some of the compound nouns denote different meanings when we use -s with the first noun and when we do so with the second noun. For example, *the glasses case* means the case for spectacles/reading glasses whereas *the glass cases* refers to the cases made of glass.

The compound nouns in which two nouns are joined by *of* or *in*, we use -s with the first noun and not with the second, for example, *commanders-in-chief* and not *commander-in-chiefs*.

Further, we normally use -s/-es to denote the plural forms of the nouns. But some of the nouns have unusual plural forms.

PRACTICE TEST 2.5

Choose the correct forms of the following nouns:

1. India and America have signed quite a few memorandums/memoranda.
2. What is the criterion/criteria for selection in this organization?
3. Our country is facing several types of crisis/crises.
4. The phenomenon/phenomena of Indian engineers and doctors going abroad has to change.
5. The parenthesis/parentheses shown within the text are to be removed.

2.2.2 Pronouns

You know that pronoun is a word that replaces a noun. Without using pronouns, we actually cannot write in a manner that would be viewed as polished and proper. For example, see how a passage devoid of pronouns appears to have been written almost ridiculously:

> Adela is a good girl but Adela is not hard working. Adela keeps sitting in front of the television. In fact, Adela is quite a bit of a couch potato. Adela actually loves watching television to such an extent that it does not matter what the idiot box has to show to Adela. Whatever the idiot box has to offer to Adela, Adela is there to accept it and watch the whole lot of that.

The above passage is quite horrible to read. Though it uses pronouns to replace other nouns including the impersonal pronoun 'it' for the noun 'television/idiot box', it does not use personal pronouns (she, her) for Adela. As a result, the whole passage appears to be repetitive, inelegant, and verbose. See, how the same passage begins to express the idea properly in the presence of pronouns replacing the noun Adela wherever required:

> Adela is a good girl but she is not hard working. She keeps sitting in front of the television. In fact, she is quite a bit of a couch potato. She actually loves watching television to such an extent that it does not matter what the idiot box has to show to her. Whatever the idiot box has to offer to her, Adela/she is there to accept and watch the whole lot of that.

In fact, most of us who read and write English are normally aware of the importance of pronouns and use them quite frequently. The real issue with pronouns is their correct usage and it is this aspect of pronouns that we are willing to share with you.

It is observed that though many of us use pronouns wherever they are required to be chosen, lots of grammatical errors creep in when we are not sure of how to use them. Look at the examples given below. Try and figure out the correct usage of pronouns in these expressions:

1. It is I/me who protested the move in the meeting.
2. We are not so stupid as they/them are.
3. Ladies and Gentlemen, myself is/I am the Senior Sales Executive of Horrendous Automobiles Pvt Ltd.
4. Let I/me speak for a while.
5. Mildred and I/me are childhood friends.
6. If you also do the same as he has done, what is the difference between you and he/him?
7. No problem, let he/him come and see our work.
8. At the concert, both he and me/I gave a performance.
9. It was I/me who first informed you about him/his whereabouts.

The general rule for making choices between *I* and *me*, *he* and *him*, *she* and *her*, *they* and *them*, etc. should be to see the function of the pronoun. If the pronoun has to be the subject of the sentence, it should be in the subjective case, i.e., it should be written as *I, he, she, they*, etc. On the other hand, if the pronoun has be the object of the sentence, it should be in the objective case, i.e., it should be written as *me, him, her, them*, etc.

To understand the point in the easiest way, look at the following sentences:

1. I slapped him.
2. He slapped me.

In the first sentence, the pronoun is the propeller of the verb—the doer of the action; therefore, *I* should be in the subjective case. Likewise, the other pronoun *him* is the recipient of the action and hence is correctly in the objective case. Since the situation reverses in the second sentence, so does the structure. Just see how idiotic and ungrammatical the entire structure sounds the moment we injudiciously use the pronoun cases:

1. Me slapped he.
2. Him slapped me.

Therefore, it is always important to know whether we have to use the pronoun in the **subjective case** (*I, we, you, they, he, she, it, etc.*); in the **objective case** (*me, us, you, them, him, her, it, etc.*) or in the **possessive case** (*my, our, your, their, his, her, its, etc.*). Going by the discussion we have had so far, you can conveniently choose the correct forms of pronouns in the sentences listed above something like this:

1. It is **I** who protested the move in the meeting.
2. We are not so stupid as **they** are.
3. Ladies and Gentlemen, **I am** the Senior Sales Executive of Horrendous Automobiles Pvt Ltd.
4. Let **me** speak for a while.
5. Mildred and **I** are childhood friends.
6. If you also do the same as he has done, what is the difference between you and **him**?
7. No problem, let **him** come and see our work.
8. At the concert, both he and **I** gave a performance.
9. It was **I** who first informed you about **his** whereabouts.

As you can observe, the sentences where the pronoun needed to direct the verb and govern the action in a sentence, we have chosen pronouns in the subjective case. Look at the first, second, third, fifth, eighth, and ninth sentences. However, objective case is required in the fourth, sixth, and seventh sentences because after the verb 'let' and preposition 'between', the pronoun that immediately follows is normally in the objective case.

Before we venture further into discussing the different nuances with regard to the usage of pronouns, let us see the different types of pronouns. Well, they all are listed below:

1. Personal pronouns (he, she, they, I, we, you, etc.) – Impersonal pronouns (It)
2. Demonstrative pronoun (this, those, these, etc.)
3. Distributive pronouns (each, either, neither, etc.)
4. Indefinite pronouns (some, many, everyone, someone, etc.)
5. Relative pronouns (who, which, whose, that)
6. Reflexive and emphatic pronouns (myself, yourself, themselves, herself, himself, etc.)

Look at the sentences given below to find out more about how to use pronouns correctly.

Correct Usage

Incorrect Usage	Correct Usage
• **Its** my duty to serve my parents.	• **It's** my duty to serve my parents.
• The dog wagged **it's** tail vigorously.	• The dog wagged **its** tail vigorously.
• These problems are **our** and let we solve it.	• These problems are **ours** and let us solve them.
• My dog is better than **Ramesh**.	• My dog is better than **that of Ramesh**.
• Everyone has come, **hasn't he**?	• Everyone has come, **haven't they**?
• The streets of London are wider than **Delhi**.	• The streets of London are wider than **those of Delhi**.
• This is the Taj Mahal **whom** everyone admires.	• This is the Taj Mahal **that** everyone admires.

The first two sentences are wrong as the distinction between *its* and *it's* is not maintained. See, *its* is the possessive case of the impersonal pronoun *it*. Therefore, there is no place for *its* in the first sentence. *It's*, on the other hand, is the contracted version of either *it has* or *it is*. See the following sentences:

1. **It's** raining quite hard this time. (**It is** raining quite hard this time.)
2. **It's** been ages since I met him. (**It has** been ages since I met him.)

So, we should use *it's* only when the impersonal pronoun *it* and the verb *is* or *has* are contracted. *Its*, on the other hand, is the possessive case of the impersonal pronoun *it* and should only be used to replace some noun. See the following example.

1. **It's** a dog. It can wag **its** tail.

The third sentence has three errors. We can say *it is our problem*. But can we say *this problem is my*? Naturally, a sentence like this has to take *mine*. So, when a possessive pronoun has to come at the end of the sentence, it should be written like *mine, ours, yours, theirs, hers*, etc. Moreover, the verb *let* is always followed by a pronoun in the objective case and *let we* is grammatically incorrect. Further, since the reference is to the plural noun *problems*, it should be followed by the plural pronoun *them* and not the singular pronoun *it*.

The fourth and sixth sentences have similar errors. When we compare two nouns, the structure should be parallel as only comparable things can be compared. For instance, it would be atrocious on our part to compare *my dog* to *Ramesh*. So, we need demonstrative pronouns *that, those*, etc. to carry out the comparison correctly.

The fifth sentence is incorrect as the indefinite pronouns *someone, everyone, everybody, somebody*, etc. take a singular verb but are referred to as *they* in tag questions. The last sentence wrongly uses *whom* for the non-living object Taj Mahal. For non-living things, the relative pronouns can be *which* or *that* but not *who* or *whom*.

Demonstrative pronouns and demonstrative adjectives   Sometimes, the confusion arises between demonstrative pronouns and demonstrative adjectives. Can you figure out which of these are demonstrative pronouns and which are the demonstrative adjectives? Look at the sentences given below and decide:

1. **This** dictionary is mine.
2. I don't like reading **such** books.
3. **That** boy is very brave.
4. **This** girl is quite bright.
5. **This** is the room that I like the most.
6. Uneasy lies the head **that** wears a crown.
7. **Such** is the situation that we can't do anything.

In the first four sentences the words *this, that, such* are demonstrative adjectives. Look carefully at them; they all are followed by some nouns, such as *dictionary, books, boy, girl*. In the remaining three sentences, the words *this, that, such* are followed by verbs—*is, wears, is*. Since a noun can be replaced by a pronoun and not by an adjective, it is a pronoun that can govern a verb. An adjective, on the other hand, is only a modifier and cannot govern a verb. It can only qualify a noun or a pronoun.

Distributive pronouns and distributive adjectives   Similar confusion arises between distributive pronouns and distributive adjectives. Let us see how we can sort out one from the other:

1. **Neither** girl was speaking the truth.   (Adjective)
2. **Neither** of the girls was speaking the truth.   (Pronoun)
3. **Either** scientist will be awarded the prestigious award.   (Adjective)
4. **Either** of the scientists will be awarded the prestigious award.   (Pronoun)
5. **Each** girl will sing a song.   (Adjective)

When we use adjectives, they are to be naturally followed by nouns.

So, the words *neither, either*, and *each* in the first, third, and fifth sentences are respectively distributive adjectives. In the remaining sentences, these words are followed by *of* and a plural noun. So, they actually replace the noun in these sentences. They, therefore, are distributive pronouns.

Relative pronouns and interrogative adjectives   At times, the distinction between relative pronouns and interrogative adjectives is also blurred. Are you in a position to figure out the difference between these two classes of words? Let us see if you can guess correctly:

1. **Which** colour do you like?   (Adjective)
2. **Whose** house is this?   (Adjective)
3. **What** nonsense!   (Adjective)
4. The house **which** is next to ours is a haunted one.   (Relative pronoun)
5. The boy **whose** nose is broken is an athlete.   (Relative pronoun)

PRACTICE TEST 2.6

Fill in the blanks with appropriate pronouns:

1. _____ am the one who cares for _____.
2. When _____ came to the room, _____ was locked.
3. He said to his wife, '_____ will buy a silk saree for _____ on _____ birthday.'
4. The boy has broken _____ bat and is asking _____ to get a new one for _____.
5. Here is _____ book, take _____ away.
6. I thought over _____ plan and I feel I do not agree to _____.
7. _____ never intervened between us. _____ was you who began to quarrel with _____.
8. He loves _____ wife and cannot live without _____.
9. _____ has lent _____ scooter to _____ for a week.
10. _____ knew that _____ deserved punishment. So, _____ did not object to _____.

PRACTICE TEST 2.7

Fill in the blanks with appropriate pronouns:

1. We often deceive _____.
2. There are silver doors in this palace; all of _____ are locked.
3. This watch is for _____.
4. David fell on the road and broke _____ arm.
5. My friend has invited _____ to dinner.
6. The jury were divided in _____ opinion.
7. Today is 11th November. _____ is celebrated as National Education Day.
8. Birds build _____ nests in trees.
9. The crew will reach _____ destination in a week.
10. _____ is easy to find faults with others.

2.2.3 Adjectives

Look at the following sentences:

1. Amita is a **clever** girl.
2. Rohan gave me **five** books.
3. There is **little** time for preparation.
4. **Each** boy must wait for his turn.
5. **Neither** statement is true.
6. Give me **some** advice.
7. There are **enough** books in **this** bookstore.
8. Don't give me **such** ideas.
9. Jenny was a **brilliant** scientist.
10. **Which** colour do you like?
11. I saw him doing this with my **own** eyes.
12. The **upper** floor of the house is meant for guests.
13. The villagers gave us **sweetish-brownish** tea.
14. The **latest** reports on global warming confirm our **worst** fears.

Look carefully at the words written in bold letters. All these words precede certain nouns. They add to the meaning of words, namely *girl, books, time, boy, statement, advice, bookstore, ideas, scientist, colour, eyes, floor, tea, reports*, and *fears*, all of which are nouns.

A word used to add to the meaning of a noun or pronoun is an adjective. It would be interesting to observe the different types of adjectives which are used in the above set of sentences.

To begin with, the word *clever* that qualifies the word *girl* tells the quality of the person. Such adjectives are known as **adjectives of quality**. Hence, words such as *good, bad, wonderful, stupid, beautiful, ugly, proper, stylish, ideal, capable, great, ordinary, extraordinary*, and a large number of such similar words are to be seen as adjectives of quality for they usually denote some nouns or pronouns.

In the second sentence, the word *five* tells how many books are being referred to. This is an **adjective of number**. The adjectives of number are broadly divided into two categories—the **definite numeral adjectives** such as *one, two, three*, etc. (cardinal numbers) and *first, second, third*, etc. (ordinal numbers) and the **indefinite numeral adjectives** such as *all, few, many, some, certain, enough*, etc.

When words such as *each, either, neither*, etc. are used with some nouns, they are called **distributive adjectives**, whereas when words such as *this, that, those*, etc. precede some nouns, they are called **demonstrative adjectives**. Similarly, when words like *which, what*, etc. precede a noun or a pronoun, they are known as **interrogative adjectives**. At times, expressions such as *very, own*, etc. too function as adjectives. For example, look at the following sentences:

1. I need this *very* book.
2. The father killed his son with his *own* hands.

Such adjectives are known as **emphasizing adjectives**.

Types of Adjectives	Examples
Adjectives of quality	*good, bad, wonderful, stupid, beautiful, ugly*
Definite numeral adjectives – Cardinal – Ordinal	 *one, two, three* *first, second, third,*
Indefinite numeral adjectives	*all, few, many, some, certain, enough*
Demonstrative adjectives	*this, that, those*
Distributive adjectives	*each, neither*
Interrogative adjectives	*which, what*
Emphasizing adjectives	*very, own*

Now, let us see if you can figure out the words that act as adjectives in the following sentences. Also tell the type of adjectives each of these words belongs to:

1. I entered a dark room.
2. On the way, we saw a huge snake.
3. What nonsense!
4. This boy needs to be taught a lesson.
5. May I have some colours?
6. Einstein is one of the greatest scientists of all times.
7. *Hamlet* is an immortal Shakespearean tragedy.
8. The stupid boy looked at me with an idiotic smile on his face.
9. I don't like reading such books.
10. They were defeated at their own game.
11. Neither movie is well made.
12. Thirty passengers were killed in the accident.
13. Red rose has a unique quality.
14. The little girl gave a mesmerizing performance on stage.
15. He often narrates interesting anecdotes.

We trust the exercise should not have been too difficult for you. In any case, this is how you can identify the adjectives in these sentences. The type of adjectives is mentioned in parentheses.

1. I entered a **dark** room. (Quality)
2. On the way, we saw a **huge** snake. (Quality)
3. **What** nonsense! (Interrogative)
4. **This** boy needs to be taught a lesson. (Demonstrative)
5. May I have **some** colours? (Adjective of indefinite numerals)
6. Einstein is one of the **greatest** scientists of all times. (Quality)
7. *Hamlet* is an **immortal Shakespearean** tragedy. (Quality)
8. The **stupid** boy looked at me with an **idiotic** smile on his face. (Quality)
9. I don't like reading **such** books. (Demonstrative)
10. They were defeated at their **own** game. (Emphasizing)
11. **Neither** movie is well made. (Distributive)
12. **Thirty** passengers were killed in the accident. (Quantity)
13. **Red** rose has a **unique** quality. (Quality)
14. The **little** girl gave a **mesmerizing** performance on stage. (Quality)
15. He often narrates **interesting** anecdotes. (Quality)

The real issue with adjectives is not exactly how to identify them or to identify their class. It is their usage that requires proper attention and care. For example, look at the following sentences:

1. Shakespeare is the greatest of all other dramatists.
2. She is senior than me in this office.
3. Don't worry! He is more better today.
4. India's economy is more stronger than Pakistan.
5. He is my oldest brother.
6. He is taller to me.
7. Mumbai is busier than all the Indian cities.
8. Suresh is cleverer than wiser.
9. As a dramatist, Pinter is superior than Beckett.
10. Old Mr Sada Jeevan Ram is the eldest person in our society.

Remember, we do not use comparative structures such as *other* when we use the superlative forms of adjectives as is wrongly done in the very first sentence here. Words like *senior, junior, inferior, superior,* etc. are followed by *to* and not *than*. Further, *more* suggests comparison and so does *better*. So, they cannot be chosen to qualify each other.

Difference between elder and older; eldest and oldest Also remember that 'elder' and 'eldest' are used to show *comparisons in the family*, whereas 'older' and 'oldest' are used to *draw comparison with other persons* and can also be applied to things.

Again, a part cannot be compared to the whole as is shown in the fourth sentence. This is how exactly these sentences should have been written in the first place:

1. Shakespeare is the greatest of all dramatists.
2. She is senior to me in this office.
3. Don't worry! He is better today.
4. India's economy is stronger than that of Pakistan.
5. He is my eldest brother.
6. He is taller than me.
7. Mumbai is busier than the other Indian cities.
8. Suresh is more clever than wise.
9. As a dramatist, Pinter is superior to Beckett.
10. Old Mr Sada Jeevan Ram is the oldest person in our society.

Look at the following sentences to observe how other errors creep in while we use adjectives and also observe how to avoid them by choosing the correct usage:

Incorrect Usage	Correct Usage
• Wow! This is the **most perfect** answer.	• Wow! This is the **perfect** answer.
• Of Ponting, Sachin, and Lara, he admires **the latter**.	• Of Ponting, Sachin, and Lara, he admires **the last**.
• They are witnessing the **most extreme** situation in Pakistan.	• They are witnessing the **extreme** situation in Pakistan.
• Teaching becomes all the more interactive when there are **few** students in the class.	• Teaching becomes all the more interactive when there are **a few** students in the class.
• The patient may kick the bucket any time; there is **a little** hope of his survival.	• The patient may kick the bucket any time; there is **little** hope of his survival.
• The higher you go, **cooler** you feel.	• The higher you go, **the cooler** you feel.
• Of coffee and tea, I prefer **the last**.	• Of coffee and tea, I prefer **the latter**.
• He is likely to occupy the room vacated by his **oldest** sister.	• He is likely to occupy the room vacated by his **eldest** sister.
• **A few** actresses who performed on stage in ancient times felt awkward while doing so.	• **The few** actresses who performed on stage in ancient times felt awkward while doing so.

 Certain adjectives such as unique, chief, square, round, complete, ideal, perfect, etc. are not used in comparative and superlative forms.

So, the first and the third sentences should have 'perfect' and 'extreme' instead of 'most perfect' and 'most extreme'. Further, 'last' is used when there are more than two things and 'latter' is used when there are two. So, the second and seventh sentences are also wrongly written.

Difference between few, a few, and the few; little, a little, and the little *Few, a few*, and *the few* give different meanings. *Few* actually suggests hardly anything or anybody. *A few* stands for some, whereas *the few* is used to refer to the nouns in the context. Similarly, *little* is almost nothing; *a little* means some; and *the little* means whatever little amount of something. Therefore, the fourth, fifth, and ninth sentences need to be rewritten as suggested in the preceding table.

Avoid using adverbs in place of adjectives   At times, errors are caused because in place of adjectives, adverbs are used and vice versa. Look at the following sentences and identify the problems in them by observing the wrong usage vis-à-vis the correct ones:

Incorrect Usage	Correct Usage
• He seems **very happily** in his married life.	• He seems **very happy** in his married life.
• He appears to be **quite genuinely**.	• He appears to be quite **genuine**.
• The bride looked **fairly daintily** in her wedding dress.	• The bride looked **fairly** dainty in her wedding dress.
• Their new book is likely to appear **short**.	• Their new book is likely to appear **shortly**.
• Her new dress looks very **prettily**.	• Her new dress looks very **pretty**.
• Before resuming his speech, the speaker cleared his throat **loud**.	• Before resuming his speech, the speaker cleared his throat **loudly**.
• The weather has finally turned **coldly**.	• The weather has finally turned **cold**.
• On hearing the shriek, he turned **quick**.	• On hearing the shriek, he turned **quickly**.
• The cake smells **sweetly**.	• The cake smells **sweet**.
• The audience applauded the speaker quite **warm**.	• The audience applauded the speaker quite **warmly**.

 NOTE   Remember, an adjective and not an adverb usually follows the verb, such as *seem*, *appear*, *look*, *feel*, etc. However, when a word has to add to the meaning of the action, an adverbial structure is required to be chosen.

Let us see if you can figure out the errors in the following sentences. You can check your answers and see the justification in the subsequent discussion:

1. No lesser than eighty persons were killed in the blast.
2. Helen was the more beautiful than all the mortals.
3. She is the oldest in the family.
4. The earthquake shook the eldest building in the town.
5. The anaconda is the most largest of all snakes.
6. Never consider yourself inferior than others.
7. Reading books is more preferable to me than watching television.
8. China's population is greater than any other country.
9. Of Delhi and Jaipur, the last is the more colourful of the two.
10. The three first chapters of the novel are written beautifully.

Use of little, less, and least   *Less* in itself is the comparative of *little*. So, it is not appropriate to use *lesser* in comparisons. Moreover, *less* is used as a comparative before uncountable nouns. For countable nouns therefore, we should use *fewer*. Hence, the first sentence should be corrected as under:

No **fewer** than eighty persons were killed in the blast.

In the second sentence, the comparison is drawn in a wrong way. When an individual is compared to the rest of his/her type, the superlative form of the adjective should be used and words like *other* should be omitted. This is how we can improve this sentence:

> Helen was the **most beautiful** of all mortals.

As suggested earlier, *elder* and *eldest* are used while carrying out comparison within family whereas *older* and *oldest* are used for other comparisons between things and people. Therefore, the third and the fourth sentences should be rewritten as under:

> She is the **eldest** in the family.
> The earthquake shook the **oldest** building in the town.

Further, we are not supposed to use *most* with those adjectives that are already written in a superlative form. Therefore, writing *most largest* is grammatically incorrect. So, the fifth sentence needs to be corrected as shown below:

> The anaconda is the **largest** of all snakes.

As suggested earlier, the words *junior*, *senior*, *inferior*, *superior*, etc. should be followed by *to* and not *than*. So the sixth sentence should actually be written as under:

> Never consider yourself **inferior** to others.

Just as *most* is not required to be put before *largest*, there is no need to use *more* ahead of *preferable* which in itself is the comparative form. So the seventh sentence should be revised as shown below:

> For me, reading books is **preferable** to watching television.

The next sentence again carries out improper comparison. China's population cannot be compared to other countries in their entirety, though its population can be compared to the population of other countries. It should be rewritten as shown below:

> China's population is **greater** than that of any other country in the world.

Again, as suggested earlier, when we refer to two nouns, the second is referred to as *latter* and not as *last*. So, the last but one sentence can be improved as cited below:

> Of Delhi and Jaipur, the **latter** is the more colourful of the two.

The last sentence is completely absurd and we should say the following:

> The **first three** chapters of the novel are written beautifully.

PRACTICE TEST 2.8

Fill in the blanks with appropriate adjectives:

1. _____ rubbish!
2. _____ boys were absent from the class in the morning.
3. I don't like _____ commercial movies.
4. I wish _____ errors do not occur in my writings.
5. This is the _____ sum of this unit.
6. The _____ part of the story comes up when the old man, who was wearing his glasses on his head, starts searching for them everywhere else.
7. _____ son came to see his old father.
8. That _____ man kept on asking _____ questions.
9. He is my _____ brother.
10. The college students put up a _____ show in the auditorium.

PRACTICE TEST 2.9

Fill in the blanks with appropriate adjectives from the list of words given below:

Petrified, boring, talented, huge, ghastly, pampered, creative, stylish, latest, riot-hit, persuasive, sensuous, redundant

1. I was _____ to see a _____ serpent in front of my room.
2. _____ children often get spoilt.
3. I found the movie extremely _____.
4. Painters, writers, singers, and dancers are inherently _____ and highly _____.
5. You have to make your proposal very _____ in order to make it saleable.
6. The Home Minister was apprised of the _____ incidents in the _____ areas.
7. Keats' poetry is remarkably _____.
8. With dead bodies littered around, it was a _____ sight.
9. Fashionable girls tend to dress up in a _____ way.
10. Avoid using _____ words in your speech.

2.2.4 Verbs

The verb is the most essential part of speech in English. You can think of a sentence without a subject or an object but you cannot think of a sentence without a verb. Even the shortest sentence contains a verb. You can make a one-word sentence with a verb, for example in day-to-day life, you use some of the following expressions:

Stop!   Come!   Go!   Sit!

However, you cannot make a one-word sentence with any other part of speech.

Verbs are described as 'action words'. Many verbs give the idea of action, of 'doing' something. For example, words like *write, teach, sing, dance,* and *work* convey some action.

But some verbs do not give the idea of action; they give the idea of existence or a state of 'being'. For example, verbs like *be, appear, exist, seem, feel,* and *belong* convey a state.

A verb always has an explicit or implied subject. For example,

Professor Bhat teaches us English.

Professor Bhat is the explicit subject in this sentence, whereas in the following sentence the subject is implied:

Stop!

You is an implied subject here.

In simple terms, therefore, we can say that *verbs* are words that tell us what a subject **does** or **is**; they describe

1. action (Siddarth **plays** cricket very well.)
2. state (Mohit **looks** very tired.)

There is something very special about verbs in English. Almost all the verbs in English change in form according to subject and tense. For example, the verbs *sing, dance,* and *cry* have following forms:

1. to sing, sing, sings, sang, singing, sung
2. to dance, dance, dances, danced, dancing
3. to cry, cry, cries, cried, crying

Now let us learn more about verbs.

Classification of verbs   Read the following sentences carefully and take a note of the verbs used in them.

1. Mr Verma **is** a doctor.
2. The earth **revolves** round the sun.
3. We **must pay** our taxes in time.

Are the above verbs same in nature? No! In fact, there are many different types of verbs, which are shown in Fig. 2.1 and below.

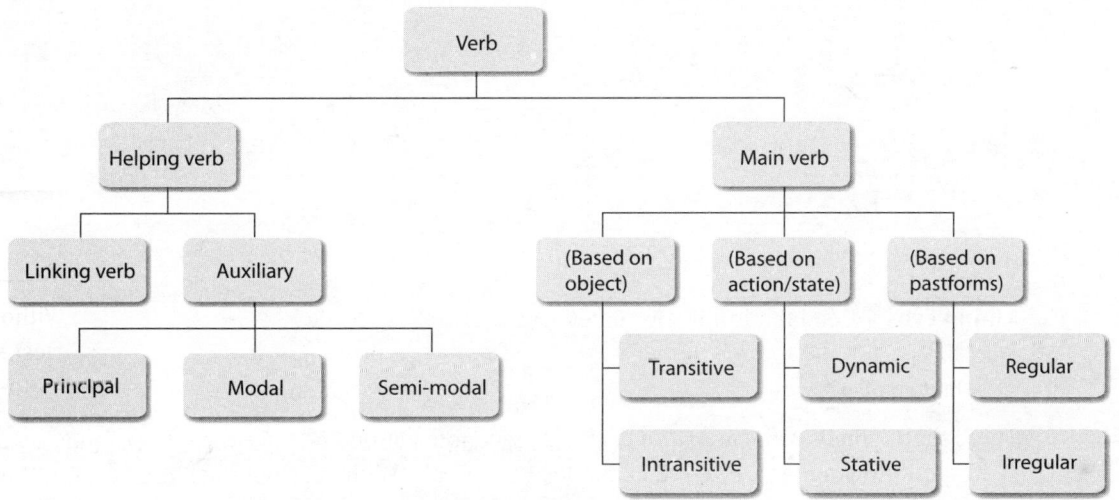

FIG. 2.1   Different Types of Verbs

1. **Helping verbs**: primary/modal
2. **Main verbs**: transitive/intransitive, dynamic/stative, regular/irregular

Linking verbs or copular verbs   Usually, a linking verb shows equality (=) or a change to a different state (→). Linking verbs are mostly used intransitively. Linking verbs such as 'be', 'seem', 'become', and 'look' are quite commonly used in everyday speech. Such verbs are generally followed by complements. For instance, look at the following examples:

1. Manisha is intelligent. (Manisha = intelligent)
2. Dr Gupta is a surgeon. (Dr Gupta = surgeon)
3. Rohit seems tired. (Rohit = tired)
4. Ashok sounds greedy. (Ashok = greedy)
5. The sky became dark. (the sky → dark)
6. The bread has gone stale. (bread → stale)
7. His body turned pale. (body → pale)

Auxiliary verbs   Auxiliary verbs are the verbs which help to form a tense or an expression. They however cannot form a complete sentence on their own and require main verbs to denote the action. The principal auxiliaries are *to be*, *to have*, and *to do*. However, modals and semi-modals also form auxiliaries since they combine the main verb to convey a distinct sense. These will be discussed in detail later under the sections which deal with tenses and modals. Figure 2.2 shows the different types of auxiliary verbs.

Let us read the following examples:

1. I **am** waiting for you. (**to be** auxiliary)
2. My father **has** deposited that money in the bank. (**to have** auxiliary)
3. You **can** do this project in time. (**can** auxiliary)
4. They **used to** visit us quite often (**used to** auxiliary)

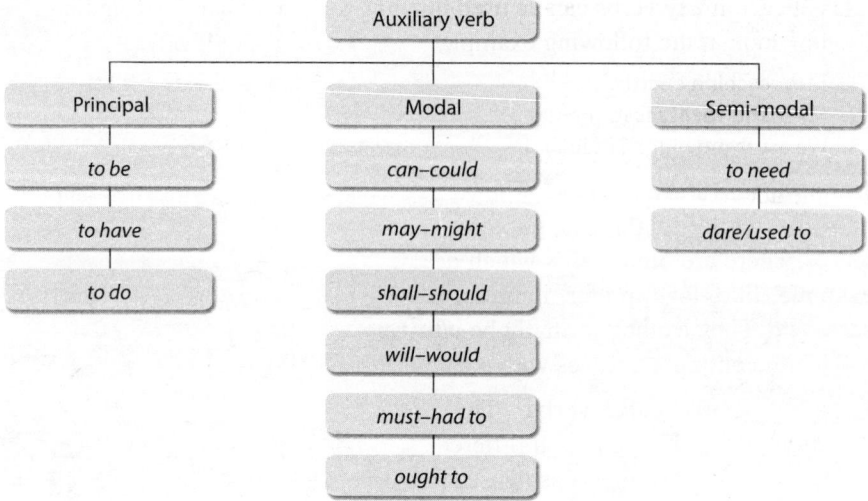

FIG. 2.2   Different Types of Auxiliary Verbs

Main verbs   Read the following sentences:

1. I **write**.
2. Children **play**.
3. They **run**.
4. We **laugh.**

These highlighted words are *main verbs*; they denote action and have meaning of their own.

In the following table, we see examples which have sentences with helping verbs and main verbs. Notice that all the sentences have a main verb. Only some of them have a helping verb.

Subject	Auxiliary Verb	Main Verb	Object/Adverbial
Sania		bought	a cell phone.
We	are	going	to market.
Sanjay	has	written	an essay.
Students	will	enjoy	the fair.

In the following section, we shall know more about main verbs.

Main verbs are also called lexical verbs   Unlike other verbs, main verbs have meaning on their own. There are thousands of main verbs and we can classify them in the following ways:

- Transitive and intransitive
- Dynamic and static
- Regular and irregular

Transitive and intransitive verbs   Read the following sentences carefully:

1. He **read** a novel yesterday.
2. My mother has **planned** a trip to Mumbai.
3. Snigdha **loves** swimming.
4. Maria **wrote** a letter to Peter.

The verbs in the above sentences—**read, planned, loves, wrote**—require objects in order to complete the sentences. Such verbs are called transitive verbs.

A *transitive verb* takes a direct object. For example, *Somebody killed the snake.* An *intransitive verb* does not require a direct object. For example, *He died.*

However, many verbs can be used transitively as well as intransitively.

Now look at the following examples:

1. I am reaching shortly.
2. My friend speaks fast.
3. We are flying high in the sky.
4. The letter reached us last night.
5. They speak English fluently.
6. Children are flying kites today.

Dynamic and stative verbs   The verbs which describe actions are called 'dynamic verbs'. For example, hit, kill, fight, run, go, throw, explode, write, etc. These can be used with continuous tenses. There are other verbs which describe a state or a situation and are called 'stative'. For example, like, love, prefer, impress, hear, see, sound, belong to, consist of, need, resemble, seem, etc. They cannot normally be used with continuous tenses (though some of them can be used with continuous tenses with a change in meaning).

Regular and irregular verbs   This is another classification which you need to understand regarding verbs. The only real difference between regular and irregular verbs is that they have different endings for their past tense and past participle tense forms. For regular verbs, the past tense ending and past participle ending is always the same: -ed. For irregular verbs, the past tense ending and the past participle ending are variable; so it becomes essential to learn them by heart.

Regular verbs   The following examples give the base, past tense, and past participle forms of a few regular verbs:

1. Cook, cooked, cooked
2. Clean, cleaned, cleaned
3. Water, watered, watered
4. Turn, turned, turned
5. Smile, smiled, smiled
6. Wash, washed, washed
7. Smoke, smoked, smoked
8. Work, worked, worked

Irregular verbs   The following examples give the base, past tense, and past participle forms of some irregular verbs:

1. Do, did, done
2. Eat, ate, eaten
3. Drink, drank, drunk
4. Cut, cut, cut
5. Sleep, slept, slept
6. Write, wrote, written
7. Sing, sang, sung
8. Be, was, been
9. Throw, threw, thrown

PRACTICE TEST 2.10

Identify whether the verbs in the following sentences are transitive or intransitive:

1. The children are flying kites in the sky.
2. Planes are flying in the sky.
3. He is a man of letters; he writes quite well.
4. He wrote a letter to his beloved.
5. Always speak the truth.
6. Don't speak too loud.
7. People sometimes have to tell lies.
8. The boss gave us clear instructions.
9. The fat cat sat on the mat.
10. The puppy ate the biscuit.

PRACTICE TEST 2.11

Choose the correct form of the verb in each of the sentences given below:

1. Yesterday afternoon, I lied/lay on the couch in front of the television.
2. After dinner, the mother lay/laid the child in the cradle.

3. The university was found/founded by the Chief Minister.
4. I found/founded a mobile on the road while coming back from college.
5. A beautiful scenery is hanged/hung in their drawing room.
6. The beautiful criminal was hung/hanged for killing her husband.
7. If you can't grow a tree, at least never fall/fell one in your life.
8. He fall/fell from the stairs and broke his leg.
9. We saw/sawed a man coming from the far end of the road.
10. At the far end of the road, a tree was being saw/sawed.

2.2.5 Adverbs

Look at the following sentences and figure out which of the highlighted words function as adverbs and which others serve as adjectives:

1. Zaheer Khan is a **fast** bowler.
2. He bowls **fast**.
3. But he doesn't bowl **very fast**.
4. He bowls **moderately fast**.
5. Jatin is my **fast** friend.
6. He actually is my **very fast** friend.
7. He however doesn't speak **that fast**.
8. Anyhow, he remains an **extremely good** friend.

Remember, an adverb can qualify a verb, an adjective, and another adverb. An adjective, on the other hand, qualifies a noun or a pronoun. So, the word *fast* in the first sentence is an adjective because it qualifies the noun *bowler*. In the second sentence however, the word *fast* qualifies the verb *bowls* and thus functions as an adverb. In the next two sentences, the words *very* and *moderately* function as adverbs as they add to the meaning of another adverb *fast*. In the fifth sentence, the word *fast* again functions as an adjective as it qualifies Jatin—a noun. In the sixth sentence, the word *very* is an adverb as it modifies the adjective *fast*. In the last but one sentence, the word *fast* again qualifies a verb *speak* and hence is an adverb, whereas *that* qualifies *fast* and hence *that* too functions as an adverb in the sentence. Similarly, the word *extremely* in the last sentence functions as an adverb for it qualifies the adjective *good*.

Understanding the distinction between adverbs and adjectives is necessary to avoid errors in their usage.

Look at the following sentences and see how the same words—highlighted here—can function as adjectives and adverbs. Can you tell in which sentence they function as what part of speech?

1. The concert was organized in memory of the **late** artist.
2. He came quite **late** in the night.
3. We didn't have a **long** queue.
4. We didn't have to wait **long**.
5. I went to bed **early**.
6. I had an **early** dinner.
7. They went **straight** into the room.
8. Gavaskar was the master of **straight** drive.

Look at how we can trace them out:

1. The concert was organized in memory of the **late** artist. (Adjective; qualifies the noun **artist**)
2. He came quite **late** at night. (Adverb; qualifies the verb **came**)
3. We didn't have a **long** queue. (Adjective; qualifies the noun **queue**)
4. We didn't have to wait **long**. (Adverb; qualifies the verb **wait**)
5. I went to bed **early**. (Adverb; qualifies the verb **went**)

6. I had an **early** dinner. (Adjective; qualifies the noun **dinner**)
7. They went **straight** into the room. (Adverb; qualifies the verb **went**)
8. Gavaskar was the master of **straight** drive. (Adjective; qualifies the noun **drive**)

At times, it is difficult to distinguish adverbs from prepositions. Look at the following sentences. The following examples suggest how to identify them:

1. The book lies **on** the table. (Preposition; shows the relationship between **the book** and **the table**)
2. Life moves **on**. (Adverb; qualifies the verb **moves**)
3. Is he **in** his room? (Preposition; shows the relationship between **he** and **his room**)
4. Has he come **in**? (Adverb; qualifies the verb **come**)
5. The criminal jumped **off** the train. (Preposition; shows the relationship between **the criminal** and **the train**)
6. The arm of the chair suddenly came **off**. (Adverb; qualifies the verb **come**)
7. Have I seen you **before**? (Adverb; qualifies the verb **seen**)
8. He returned the day **before** yesterday. (Preposition; establishes the relationship between **the day** and **yesterday**)

Before proceeding further, let us see how many types of adverbs are used to add to the meaning of verbs, adjectives, and other adverbs. Broadly speaking, adverbs can be of the following types:

1. *Adverbs of time* (now, then, everyday, yesterday, etc.)
2. *Adverbs of frequency* (twice, often, always, never, ever, etc.)
3. *Adverbs of place* (everywhere, outside, there, here, etc.)
4. *Adverbs of manner* (melodiously, beautifully, stupidly, etc.)
5. *Adverbs of degree* (almost, rather, nearly, etc.)
6. *Adverbs of affirmation or negation* (surely, definitely, certainly, positively, etc.)
7. *Adverbs of reason* (hence, therefore, so, since, because, for, etc.)

PRACTICE TEST 2.12

Identify adverbs/adverbial phrases in the following sentences. Also, define their nature:

1. I have not seen him lately.
2. He therefore could not achieve success.
3. She moved around quite sprightly.
4. The refugees slept fretfully in the tent.
5. Probably, he has gone to the market.
6. The little girl followed the guest everywhere.
7. Don't go that far.
8. The story is not written lucidly.
9. Surely, you are wrong!
10. He drove quite slowly all the way.
11. Don't worry; she is far better now.
12. He is too fat to climb the stairs.
13. We seldom see each other now.
14. Yesterday, I called him late at night.
15. He often comes late these days.

Many adverbs are denoted by adding the suffix –ly to adjectives. For example, look at the following sentences:

1. It is a **hard** exercise.
2. Good books are **hardly** read these days.
3. She seems **happy**.
4. The cricketer answered all the questions **happily**.
5. The bride looks **cute**.
6. The cat looked at me **cutely**.

Use your adverb appropriately   As you can see that the adverbs *hardly*, *happily*, and *cutely* have been formed from the words *hard*, *happy*, and *cute* which function as adjectives and complements in some of the sentences above. Certain adverbs may however have two forms, one with –ly structure and the other without it. It is in these situations that one has to opt for the appropriate forms of adverbs.

Some of the adverbs can give different meanings when used with an –ly particle and when without it.

PRACTICE TEST 2.13

Go ahead and choose the correct adverbs:

1. He cut shortly/short his journey and returned home.
2. Wait! The Guest of Honour is arriving shortly/short.
3. His latest book has been wide/widely appreciated.
4. The window was kept wide/widely open for the lover to make his secret entry.
5. The dandy moved round/roundly the damsel throughout the event.
6. He was round/roundly scolded for being a philanderer.
7. Do you have to talk so loud/loudly?
8. Loud/Loudly, he cleared his throat.
9. Go slow/slowly round this corner.
10. Slow/Slowly, the cat moved towards the kitchen.

Positioning the adverbs in sentences is quite a challenge. Following are some adverbs which change the meaning of a sentence simply by stationing themselves at different places in a sentence. Can you make out the difference in meaning of each of these sentences? Look at the explanation given within parentheses.

1. **Only** an adverb qualifies a verb.
    (It is an adverb and nothing else that can qualify a verb.)
2. An adverb **only** qualifies a verb.
    (An adverb has only one function—to qualify a verb.)
3. I **really** don't know the answer.
    (Truly speaking, I don't know the answer.)
4. I don't **really** know the answer.
    (I am not sure if I know the answer.)
5. I **deliberately** didn't leave the door open.
    (I saw to it consciously that the door was not left open.)
6. I didn't **deliberately** leave the door open.
    (It was an inadvertent mistake that the door was left open; I did not do it consciously.)
7. **Clearly**, they didn't explain things.
    (It is obvious that they did not explain things.)
8. They didn't explain things **clearly**.
    (They probably tried to explain but could not explain things clearly enough.)
9. **Only** she would do a silly thing like that.
    (She is the only one to do a silly thing like that.)
10. She would do **only** a silly thing like that.
    (The only thing she can do is a silly thing like that.)

Since adverbs can easily be manoeuvered to give a different meaning in a given context, it is very important for us to place adverbs carefully in a sentence.

All, however, is not erratic with the ways of adverbs. Mostly, the type of adverb may decide its position in a sentence.

Generally, the adverbs of manner are placed at front/end positions in sentences. Look at the sentences given below:

1. She speaks Russian **fluently**.
2. Try to behave **sensibly**.
3. He moved **swiftly**.
4. The plants grew **rapidly**.
5. I had to choose my words **carefully/with care**.
6. Can't we discuss this **sensibly/in a sensible way**?
7. The policemen inspected the car **officiously/in an officious manner**.
8. **Gently** fry the potato pieces. (Puts emphasis on the manner)
9. I **quickly** ran and got my coat back. (Highlights urgency)

When it comes to adverbials of place and time, such as *here, there, everywhere, now, then, yet, today, next Sunday, at school, in the park, last week, for three days, last year, etc.*, they are placed either after the verb or after the object in a sentence.

See how you would really form the following sentences:

1. Keep **there** the book.	(Incorrect)
2. Keep the book **there**.	(Correct)
3. **Last week** she met him.	(Not so appropriate)
4. She met him **last week**.	(Quite appropriate)
5. Yesterday she sang **melodiously** in the concert.	(Wrong order of adverbs in a sentence—adverbs of time, manner, place)
6. She sang **melodiously** in the concert yesterday.	(Correct order of adverbs in a sentence—adverbs of manner, place, time)

Adverbs of frequency, for example *always, never, after, rarely, usually, generally*, etc. are usually placed between the subject and the verb.

1. He **never** saw me.	(Correct)
2. He saw me **never**.	(Incorrect)
3. I have told him **often** to come early.	(Incorrect)
4. I have **often** told him to come early.	(Correct)
5. We have **usually** lunch at twelve.	(Incorrect)
6. We **usually** have lunch at twelve.	(Correct)

At times, the type of the verb may decide the position of an adverb in a sentence:

1. He speaks **always** the truth.	(Incorrect)
2. He **always** speaks the truth.	(Correct)
3. He **always** is at home in the evening.	(Incorrect)
4. He is **always** at home in the evening.	(Correct)
5. He **never** is late for work.	(Incorrect)
6. He is **never** late for work.	(Correct)
7. He tells **never** lies.	(Incorrect)
8. He **never** tells lies.	(Correct)

The auxiliaries *have to* and *used to* take the adverb before the verb:

1. He has to go to office **often** on foot.   (Incorrect)
2. He **often** has to go to office on foot.   (Correct)
3. He used to **always** come with me to work.   (Incorrect)
4. He **always** used to come with me to work.   (Correct)

Remember to place the adverbs *enough* after and *too* before the words they modify. Also avoid using *too* + adjective for positive connotations unless they are used with sarcasm. Similarly, avoid using *intelligent* + *enough* for negative connotations.

Look at the following sentences:

1. This box is **big enough** to accommodate all your shirts.   (Correct)
2. This box is **too big** to accommodate all your shirts.   (Flawed)
3. This box is **too small** to accommodate all your shirts.   (Correct)
4. He is **too intelligent** to understand the problems.   (Incorrect)
5. He is **intelligent enough** to understand the problems.   (Correct)
6. He is **too slow** to win the race.   (Correct)
7. He is **too fat** to climb the stairs.   (Correct)
8. Tea is **too good**.   (Flawed)
9. Tea is **really good**.   (Correct)

PRACTICE TEST 2.14

Read the following sentences and see if the adverbs are rightly placed. Rewrite the sentences if required.

1. He looks often sad and gloomy these days.
2. Doctors have reported that now one can have cancer also due to depression.
3. She is intelligent enough to marry a fool like you.
4. He has been to Kashmir never before.
5. The committee has been already informed about the incidence.
6. They were seen together going to the party.
7. He brilliantly bats at number three position.
8. The spirit knocks often the door in the night.
9. He always is punctual in his routine.
10. We wash on Sundays our cars.

2.2.6 Prepositions

Prepositions are words placed before a noun or a pronoun to show the relation or connection with the remaining part(s) of the sentence. Look at the highlighted words in the following sentences:

1. Majestic, the super boar, loves travelling **by** air.
2. Pussy, the cute cat, sat **on** the table.
3. Champion, the pampered dog, sat **in** the car.
4. Petty, the tiny mouse, hid **under** the chair.
5. Satan, the wily serpent, slid **underneath** the carpet.
6. Silky, the little squirrel, scurried **behind** the sofa.

See how the words *by, on, in, under, underneath*, and *behind* relate the different nouns *Majestic, Pussy, Champion, Petty, Satan*, and *Silky* to the other set of nouns such as *air, table, car, chair, carpet*, and *sofa*. Now, let us see how by changing only such small words, the whole meaning undergoes a change:

1. Majestic, the super boar, loves travelling **in** the air.
2. Pussy, the cute cat, sat **beside** the table.
3. Champion, the pampered dog, sat **on** the car.
4. Petty, the tiny mouse, hid **behind** the chair.
5. Satan, the wily serpent, slid **inside** the carpet.
6. Silky, the little squirrel, scurried **underneath** the sofa.

Obviously, now all these animal characters with funny names to their credit seem to be doing something different from whatever they seemed to have done earlier. And how is the difference denoted? It is only by changing words like *on, by, in, behind, inside*, and *underneath* in the above sentences that the change in the meaning is communicated. Such words are known as prepositions.

Using prepositions correctly is not always as easy as it sounds and many a time, errors are caused in sentences due to the wrong choice of prepositions. Since prepositions relate more to collocations than to rules, it would be more appropriate to learn them with the help of examples. So, let us see if you can make out the errors that are caused by wrong choice of prepositions in the following sentences:

1. See you in Christmas.
2. Applications must reach the Registrar's office on 31st May.
3. You must be home before twelve o'clock.
4. See you on the theatre.
5. I will discuss this issue on tomorrow.
6. On last Sunday, we went on a picnic.
7. When I listened him, I found him quite boring.
8. What are you doing? I am searching my mobile.
9. For many years, Foxy lived at Delhi; now she is at Jaipur.
10. The Gujarat earthquake registered 8.1 in the Richter scale.

Generally we say *at Christmas, at Deepawali, at Holi*, etc. Therefore, the preposition *in* in the first sentence is completely wrong. Similarly, if the applications can reach the office till 31st May, the preposition should be *by* and not *on*. We use preposition *by* in case something is to be done by the denoted time, day, date, week, month, or year. Therefore, the third sentence should also use *by* instead of *before*.

Further, when we promise to meet someone, normally the preposition chosen to denote the place of meeting is *at*. So the fourth sentence should have the preposition *at* in place of *on*. In the next two sentences, no preposition is required. On the other hand, prepositions are required in the sentences that follow as the word *listen* is followed by the preposition *to* and *search* is followed by *for* to communicate the meaning required to be conveyed here.

Moreover, when people live in big cities, they live *in* them and not *at* them. *At* is chosen for small places not for big cities, such as Delhi and Jaipur; hence error in the penultimate sentence.

Finally, the intensity of earthquake is measured *on* the Richter scale and not *in* it. This is how you should write these sentences:

1. See you **at** Christmas.
2. Applications must reach the Registrar's office **by** 31st May.

3. You must be home **by** twelve o'clock.
4. See you **at** the theatre.
5. I will discuss this issue **tomorrow**.
6. Last Sunday, we went **on** a picnic.
7. When I listened **to** him, I found him quite boring.
8. What are you doing? I am searching **for** my mobile.
9. For many years, Foxy lived **in** Delhi; now she is **in** Jaipur.
10. The Gujarat earthquake registered 8.1 **on** the Richter scale.

Look further how to choose the correct prepositions in the following expressions:

Choices	Correct Usage
• With open mouth, the visitors looked **on/at** the painting.	• With open mouth, the visitors looked **at** the painting.
• The patient is being attended **to/on** by the nurse.	• The patient is being attended **to** by the nurse.
• No one likes to be stared **at/on** while eating.	• No one likes to be stared **at** while eating.
• When the fat man slipped **on/with** a banana peel, the boys standing around laughed **on/at** him.	• When the fat man slipped **on** a banana peel, the boys standing around laughed **at** him.
• The officer lives **at/in** a large bungalow.	• The officer lives **in** a large bungalow.
• Oh! How painful! I have never heard **about/of** such a tragedy.	• Oh! How painful! I have never heard **of** such a tragedy.
• When we saw him last, he was sitting **at/in** that chair.	• When we saw him last, he was sitting **in** that chair.
• Is he going to come **on/by** train?	• Is he going to come **by** train?
• What do you think **about/of** global warming?	• What do you think **about** global warming?
• I am sorry; I can't agree **with/to** your proposal.	• I am sorry; I can't agree **to** your proposal.

As already suggested, not rules but grammatical structures and collocations decide the choice and placement of prepositions. At times, only a particular preposition follows certain verbs.

Look at the following sentences to identify the collocation between the verbs and the prepositions they normally take:

Choices	Correct Usage
• He is endowed **by/with** wonderful creative talent.	• He is endowed **with** wonderful creative talent.
• On a foreign tour, players have to adapt **with/to** the changed climatic conditions.	• On a foreign tour, players have to adapt **to** the changed climatic conditions.
• I am really grateful **to/for** you **to/for** all your support.	• I am really grateful **to** you **for** all your support.

(Contd)

Choices	Correct Usage
• New meanings can always be derived **from/ on** good writings.	• New meanings can always be derived **from** good writings.
• He was quite poorly judged **with/by** his teachers.	• He was quite poorly judged **by** his teachers.
• The Congress has recently been profited **with/ by** infightings in the BJP.	• The Congress has recently been profited **by** infightings in the BJP.
• Though a rebel to the core before marriage, he is now confined **with/to** his wife and kids.	• Though a rebel to the core before marriage, he now is confined **to** his wife and kids.
• The Prime Minister was apprised **with/of** the latest incidents in the riot-hit areas.	• The Prime Minister was apprised **of** the latest incidents in the riot-hit areas.
• The Speaker's timely intervention prevented the members **from/of** coming to blows with each other.	• The Speaker's timely intervention prevented the members **from** coming to blows with each other.
• He just can't help it; he is addicted **with/to** wine.	• He just can't help it; he is addicted **to** wine.

Prepositions are chosen not only by certain verbs but also by other grammatical structures. Observe carefully the prepositions and the grammatical structures that precede them:

Choices	Correct Usage
• Agra is famous **about/for** the Taj Mahal.	• Agra is famous **for** the Taj Mahal.
• Once a rarity, newspaper is accessible **to/with** almost everyone around us.	• Once a rarity, newspaper is accessible **to** almost everyone around us.
• The old man was hard **on/of** hearing.	• The old man was hard **of** hearing.
• Keats' poetry is remarkable **in/for** its sensuousness.	• Keats' poetry is remarkable **for** its sensuousness.
• When we saw him, he was beaming **in/with** enthusiasm and confidence.	• When we saw him, he was beaming **with** enthusiasm and confidence.
• Good scholars are men **with/of** deep learning.	• Good scholars are men **of** deep learning.
• The news of accident made us all worried **about/ with** his safety.	• The news of accident made us all worried **about** his safety.
• Marlowe was a contemporary **of/with** Shakespeare.	• Marlowe was a contemporary **of** Shakespeare.
• My father was fond **for/of** old songs.	• My father was fond **of** old songs.
• Don't worry; I am aware **about/of** these things.	• Don't worry; I am aware **of** these things.

Now, look at the sentences given below. Observe carefully both the incorrect and the correct usages of prepositions:

Incorrect Usage	Correct Usage
• Sir, I take exception **with** this decision.	• Sir, I take exception **to** this decision.
• Every sane man prefers silence **than** the raucous music of modern times.	• Every sane man prefers silence **to** the raucous music of modern times.
• He was always helpful **with** us.	• He was always helpful **to** us.
• How can you afford to live **with** that meagre a salary?	• How can you afford to live **on** that meagre a salary?
• We are sorry **about** the technical glitch you are experiencing at the moment.	• We are sorry **for** the technical glitch you are experiencing at the moment.
• Are you really interested **with** that snazzy girl?	• Are you really interested **in** that snazzy girl?
• However rational, most of us are afraid **from** ghosts.	• However rational, most of us are afraid **of** ghosts.
• Congratulations! Your boss was full **with** praise for you.	• Congratulations! Your boss was full **of** praise for you.
• Parents normally are proud **with** their sons' achievements.	• Parents normally are proud **of** their sons' achievements.
• Will you look **at** our dog in our absence?	• Will you look **after** our dog in our absence?

Sometimes, we add a preposition with certain expressions which do not actually require any preposition. Interestingly enough, sometimes when the preposition is required, we omit it. Look how such errors can be avoided:

Incorrect Usage	Correct Usage
• As he grew in age, he started resembling **to** his father.	• As he grew in age, he started resembling his father.
• I am planning to write him.	• I am planning to write **to** him.
• Resolutely she kept quiet and refused to answer **to** me.	• Resolutely she kept quiet and refused to answer me.
• Suddenly, the tiger attacks **on** the hunter.	• Suddenly, the tiger attacks the hunter.
• What are you looking so stupidly?	• What are you looking **at** so stupidly?
• Something odd struck us as we approached **to** the house.	• Something odd struck us as we approached the house.

(Contd)

(Contd)

Incorrect Usage	Correct Usage
• The soldier fell in love with the beautiful nurse who attended him.	• The soldier fell in love with the beautiful nurse who attended **to** him.
• As we entered **into** the room, we heard a loud explosion outside.	• As we entered the room, we heard a loud explosion outside.
• I have given him everything he has asked.	• I have given him everything he has asked **for**.
• Have you paid all that you have bought?	• Have you paid **for** all that you have bought?

Learn to use prepositions correctly   One way to learn to use prepositions is to frame questions with appropriate prepositions. By looking at the prepositions both in the question and the answer form, we tend to know their ways in a better way. So, let us try this method to learn something more about prepositions. Go ahead and frame questions for all these statements written below:

1. Ice cream is made of milk.
2. No, I am not interested in poetry.
3. I am waiting for a friend of mine.
4. I am looking for my glasses.
5. We wish to speak to Mr Eliot.
6. We have been talking about the book we recently read.
7. We have been living in a spacious house.
8. We are listening to the news.
9. Yes, I have written to all the members of the society.
10. I am looking at the beautiful wrist watch he sent on my birthday.

Here are some of the possible questions which can evoke the answers given in the above set of sentences. Carefully notice the prepositions with which these questions are framed:

1. What is ice cream made of?
2. Are you interested in poetry?
3. Who are you waiting for?
4. What are you looking for?
5. Who do you wish to speak to?
6. What have you been talking about?
7. What kind of a house have you been living in?
8. What are you listening to?
9. Have you written to all the members of the society?
10. What are you looking at?

PRACTICE TEST 2.15

Now, choose the right prepositions in each of the following sentences:

1. When we watch a tragedy, we are overcome in/with emotions.
2. Having been caught using unfair means, he was debarred from/with sitting for/in the examinations of/for three years.
3. Despite all the rumours, we are quite confident about/of securing a win.
4. Many members abstained with/from casting their votes.
5. The captain attributed the victory to/on his team.
6. You need to apologize for/to her immediately.
7. The poem refers with/to the mythical allusions.
8. He was disgusted at/with the idea for/of having to change his child's diapers on/in his wife's absence.
9. If you are ignorant of/about everything, you are likely to fail in life.
10. He sounded particularly obliged to/for his family members.

PRACTICE TEST 2.16

Read the following sentences and give the missing prepositions in the blanks:

1. Armed _____ guns, the terrorists attacked the passengers.
2. We are confident _____ winning the match.
3. Surprisingly, he is excluded _____ the squad.
4. When you enter that house, beware _____ dogs.
5. People with income less than two lakh rupees per annum should be exempted _____ having to pay income tax.
6. One must know what to keep and what to dispense _____.
7. Regardless _____ the fact that we are one nation, people fight over petty issues.
8. Somehow, we need to cope _____ this problem in a better way.
9. You always seem prepared _____ an argument.
10. The suggestions you have given are not in consonance _____ our requirement.

PRACTICE TEST 2.17

Read the following extracts and fill in the blanks with appropriate prepositions:

1. The national selection panel will meet _____ Tuesday _____ pick the Indian cricket squad _____ the home Test series _____ Sri Lanka, starting _____ Ahmedabad _____ November 16.
2. Diabetes is a chronic disorder, the diagnosis _____ which is accompanied _____ considerable physical and mental stress. While physical stress, such as loss _____ weight, weakness, and frequent infections, can be taken care _____ _____ insulin and oral hypoglycaemics, psychological stress is difficult to handle.
3. A questionnaire was sent _____ a large group of people _____ a psychologist _____ Canada, asking what they felt about their clothes when appearing _____ an interview.

2.2.7 Connectives

Connectives are words such as *and, but, after, because, though, as, wherein, whereupon, for, unless, lest, while, whereas*, etc. These are also called conjunctions. Some of these connectives are known as coordinating conjunctions and some others are called subordinating conjunctions. To distinguish these two, look at the following sentences:

1. **As** he was not well, he could not come to the meeting.
   (The conjunction **as** connects the subordinate clause **As he was not well** to the main clause **he could not come to the meeting**; hence a subordinating conjunction.)
2. He was not well **and** he could not come to the meeting.
   (The conjunction **and** connects two independent clauses **he was not well** and **he could not come to the meeting**; hence a coordinating conjunction.)
3. **Unless** you solve sums, you cannot feel confident in Mathematics.
   (The conjunction **unless** connects a subordinating clause **unless you solve sums** to the main clause **you cannot feel confident in Mathematics**; hence a subordinating conjunction.)
4. She was not beautiful **but** she looked attractive in that dress.
   (The conjunction **but** connects two independent clauses **she was not beautiful** and **she looked -attractive in that dress**; hence a coordinating conjunction.)

A main/principal/independent clause can independently convey the meaning, while a dependent/subordinating clause has to depend on the main clause for its meaning.

Main coordinating conjunctions   **And, but, or, also, either…or, neither…nor, etc.**
Main subordinating conjunctions   **Though, although, as, when, unless, while, because, etc.**

The most important thing regarding conjunctions, however, is to use them correctly. Look at the following sentences and learn how to avoid the faulty handling of conjunctions:

Incorrect Usage	Correct Usage
• Hardly I had entered the room, when the phone rang.	• **Hardly had** I entered the room, **when** the phone rang.
• He neither appeared prepared for the interview or confident while speaking.	• He appeared **neither** prepared for the interview **nor** confident while speaking.
• You must work hard lest you should not fail.	• You must work hard **lest you should** fail.
• Not only he is stupid but stubborn as well.	• He is **not only** stupid **but also** stubborn. (OR) **Not only** is he stupid **but also** stubborn.
• Such rituals are seldom or ever observed in America.	• Such rituals are **seldom** or **never** observed in America.
• She not only makes errors, but she does not also admit them.	• She **not only** makes errors, **but also** does not admit them.
• No sooner we had boarded the train, it started to move.	• **No sooner had** we boarded the train **than** it started to move.
• He is a great scholar; he always speaks like an expert.	• He is a great scholar; he always speaks **as** an expert.
• We can either speak our mind or can keep quiet in such situations.	• We can **either** speak our mind **or** keep quiet in such situations.
• He is in a fix; he cannot either leave his job nor can do it well.	• He is in a fix; he can **neither** leave his job **nor** do it well.

Here are a few important points to remember:
- The conjunctions *hardly had, scarcely had, barely had* come as such and are followed by *when*.
- *Neither* is followed by *nor*, not *or*.
- *Lest* is followed by *should* and not by *should not*.
- *Not only* is followed by *but also*.
- *Seldom* is followed by *never* for emphasis; not *ever*.
- *No sooner had* is followed by *than*.
- *Like* is used in comparisons; *as* is correct in the context.

Given below are some more sentences. Find out the error in each of them and rewrite them correctly:

1. Unless you do not work hard, you cannot succeed in life.
2. I don't want to work with him because neither he is knowledgeable nor cooperative.
3. No sooner had the bus stopped, when the passengers started scrambling for a place in it.

4. He is idiotic while his wife is quite intelligent.
5. Although they offered me to join the company, but I decided not to join it.
6. Walk fast, lest you should not miss the train.
7. Desdemona drops her handkerchief wherein Iago misuses it.
8. Contrary to the government's claims notwithstanding, the situation in the flood-hit areas continues to be dismal.
9. My friend wears shorts, even so in winter.
10. They allowed me to continue the diploma even if I had failed in the first two papers.

The conjunction *unless* is negative in meaning and hence should not be followed by another negative element in the sentence. Hence, the first sentence is incorrect. The second sentence flouts the parallel structure required for a coordinating conjunctions *neither...nor*.

In the third sentence, *no sooner* is wrongly followed by *when* and it should be followed by *than*. In the fourth sentence, *while* is wrongly used in place of *whereas*.

Difference between while and whereas: The conjunction *while* suggests a simultaneous action, whereas the conjunction *whereas* is used to show a contrast.

The fifth sentence is erroneous as *although* need not be followed by *but*. In the sixth sentence *lest* is again wrongly followed by a negative.

The conjunction *whereupon*, which means immediately afterwards, should come in place of *wherein* which means 'by which' and hence does not make sense in the seventh sentence.

Contrary to and *notwithstanding* both suggest something like *despite* and only one is required, whereas the eighth sentence uses both of them, hence the error. In the ninth sentence *even* is wrongly replaced by *even so* which means 'all the same'. In the context, *even* needs to be used to emphasize the fact that the speaker wears shorts even in winter.

In the last sentence, *even if* should be replaced by *even though* as *even if* is only conditional whereas *even though* is concessional and hence makes better sense.

Actually the difference between *even if* and *even though* should be clearly maintained. Look at the following sentences to find out how they differ in meaning:

1. Even if the movie is boring, they are going to watch it.
2. Even though the movie is boring, they are going to watch it.

In the first sentence, it is implied that the movie may or may not be boring. *Even if* in this is conditional as it suggests the idea that it will not matter even if the movie turns out to be boring for they are going to watch it. In the second sentence, *even though* implies that it is obvious that the movie is boring. The use of *even though* is concessional and the speaker concedes the idea that though the movie is boring, they are going to watch it.

Therefore, you can revise all the sentences as suggested below:

1. **Unless** you work hard, you cannot succeed in life.
2. I don't want to work with him because he is **neither** knowledgeable **nor** cooperative.
3. **No sooner** had the bus stopped, **than** the passengers started scrambling for a place in it.
4. He is idiotic, **whereas** his wife is quite intelligent.
5. **Although** they offered me to join the company, I decided not to join it.
6. Walk fast, **lest** you should miss the train.
7. Desdemona drops her handkerchief **whereupon** Iago misuses it.
8. **Contrary to** the government's claims, the situation in the flood-hit areas continues to be dismal.

Or

Notwithstanding the government's claims, the situation in the flood-hit areas continues to be dismal.

9. My friend wears shorts, **even** in winter.
10. They allowed me to continue the diploma **even though** I had failed in the first two papers.

Look at another set of sentences and suggest how they can be rewritten in accurate grammatical structures:

1. Despite of the fact that he has five daughters, he is not worried.
2. Both his learning as well as his expression have been widely recognized.
3. We smelt something burning even if we entered the room.
4. Like his brother, he is not all that stupid.
5. Whereas coming back from office, I saw a huge snake on the road.
6. Because you came late, you could not watch the beginning of the movie.
7. We could not catch the train even as we reached the station in time.
8. James made up his mind what to say even though he walked to the office building.
9. Scarcely I had finished my food, than I started feeling giddiness.
10. Seeing that since it is quite dark outside, why don't you stay with us tonight?

Despite and in spite of   The first sentence is wrong as the conjunction *despite* is not followed by *of* though *in spite* is. So, it is correct to say *in spite of* but *despite* is correct, not *despite of*.

The second sentence also is incorrectly written as *both* is followed by *and*, and not by *as well as*.

Even if and even as   In the third sentence, a conditional *even if* is wrongly used in place of *even as*. As the sentence requires to show a simultaneous action, *even as* which is equivalent to *while* is required and not *even if*, which is conditional in its meaning and hence does not serve any purpose in the context.

In the fourth sentence, *unlike* should replace *like* as the contrast is required to be highlighted, whereas *like* is used to denote similarity.

The fifth sentence also goes haywire because in place of *while*—used to denote simultaneous action—*whereas* is chosen, which is comparative and hence serves no purpose in this context. Normally we do not start a sentence with the conjunction *because* and instead prefer other conjunctions such as *since* and *for* which are similar in meaning. Hence, the sixth sentence also needs to be rewritten.

Again in the seventh sentence, the concessional conjunction *even though* is required, whereas *even as*, which suggests *while* and is chosen for a simultaneous action, is wrongly used in the sentence. Since the focus in this sentence is on that fact that the persons in the context could not catch the train despite the fact that they had reached the station in time, we need a concessional conjunction *even though* and the conditional *even as* hardly makes any sense in it.

Even though and even as   In the eighth sentence, the concessional conjunction *even though* should be replaced by *even as* which is similar to *while* and shows a simultaneous action. Since the idea is that James made up his mind about what to say while he moved towards the office building, it is *even as* and not *even though* which is required.

Scarcely had is similar to *hardly had* and is followed by *when*. The ninth sentence wrongly uses *scarcely I had* whereas the correct way to write it is *scarcely had I* and it should be followed by *when* and not *than* as is wrongly used there. The last sentence suffers from the problem of plenty as both *seeing that* and *since* are similar in meaning and only one would suffice to communicate the meaning.

See how in their revised form, these sentences lucidly express the idea intended to be conveyed in the first place:

1. **Despite** the fact that he has five daughters, he is not worried. (OR) **In spite of** the fact that he has five daughters, he is not worried.
2. **Both** his learning **and** his expression have been widely recognized.
3. We smelt something burning **even as** we entered the room.
4. **Unlike** his brother, he is not all that stupid.
5. **While** coming back from office, I saw a huge snake on the road.
6. **As** you came late, you could not watch the beginning of the movie.
7. We could not catch the train **even though** we reached the station in time.
8. James made up his mind about what to say **even as** he walked to the office building.
9. **Scarcely had** I finished my food, **when** I started feeling giddiness.
10. **Seeing that** it is quite dark outside, why don't you stay with us tonight?

PRACTICE TEST 2.18

Fill in the blanks with appropriate conjunctions:

1. They are _____ rich _____ generous.
2. _____ you read good books, you can't improve your vocabulary.
3. _____ he comes now, we can't really listen to his speech.
4. _____ his in-depth knowledge, he can't teach well.
5. I really can't trust you _____ what you say is quite appealing.
6. All his assertions _____, he could not make an impact on his audience.
7. _____ returning from hospital, the patient developed complications on the way.
8. Why do you always behave _____ your father does?
9. He has written a letter to the Prime Minister _____ he has suggested how the common man must be encouraged to combat global warming.
10. He _____ spoke rudely to us _____ slammed the door shut in a very aggressive way.

PRACTICE TEST 2.19

Join each pair of the following sentences by using a suitable conjunction:

1. On his birthday, we went to the market. | We bought a nice gift for him.
2. He is a great actor. | He is not a good human being.
3. She was not well. | She attended her office.
4. Father may be in the office. | Father may be on the way to office.
5. They fought for their country. | They laid down their lives for their country.
6. Good people are simple. | Good people are not simpleton.
7. He watched the movie. | His sister finished her homework.
8. The man was seriously injured. | His wife died on the spot.
9. Man got all the wealth. | Man could not get happiness.
10. You must keep your mouth shut. | You must get out of here.

Fill in the blanks with appropriate conjunctions:

1. _____ I informed her, she did not bother to enquire about your health.
2. They did not try to win _____ they lost the match.
3. _____ you say so, I really cannot trust you.
4. Work hard _____ you should fail.
5. Man _____ wins _____ loses the race of life.
6. You must do it _____ I say so.
7. The salesmen could not convince the customers _____ they were not articulate enough.
8. _____ he comes now, we cannot go _____ watch the match.
9. Not that I loved Caesar less, _____ that I loved Rome more.
10. _____ a borrower, _____ a lender be.

Apart from the basic parts of speech we have learnt so far, there are some other essential elements in a sentence. Let us learn them as well in some detail:

2.3 ARTICLES

The words *a*, *an*, and *the* are called articles. *A* and *an* are known as *indefinite articles*, whereas *the* is called the *definite article*. The indefinite articles *a* and *an* are used before singular countable nouns. Now the question is when to use *a* and when to opt for *an*.

2.3.1 Use of A and An

Actually the decision to choose between *a* and *an* rests on the initial sound of the word that follows them.

As far as the general rule is concerned, we use *a* before words which begin with a consonant sound. *An*, on the other hand, is used before those words which begin with a vowel sound.

The choice, however, is not as easy as it seems. At times, we get words which apparently start with a vowel (a, e, i, o, u) but give the sound of a consonant. Such words should be preceded by *a* and not *an*. Similarly, the initial consonant sound, particularly that of *h*, may be muted in some words. Such words will be preceded by *an* if the sound that follows the unsounded *h* happens to be a vowel sound.

This is how you can make correct choices between *a* and *an*:

1. He is pursuing his M.Phil. from **a university** in Delhi.
   (The word **university** starts with a consonant sound /j/ and therefore must be preceded by **a** and not **an**; hence **a university**.)
2. I saw him **an hour** ago or so.
   (The consonant **h** is muted in the word **hour** and the sound that follows **h** is that of a vowel; hence, **an hour**.)
3. Sir, it is **an honour** to be called here as **a chief guest**.
   (The consonant **h** in the word **honour** is unsounded and the sound that follows **h** in the word honour is that of a vowel; hence, **an honour**.)

4. It's not easy to be **a mother**.

   (The word **mother** starts with a consonant sound /*m*/; hence, **a mother**.)

5. His mother was **an American** though his father was **a European**.

   (The initial sound in the word **American** is that of a vowel; hence, **an American**, whereas the initial sound in the word **European** starts with a consonant sound /*j*/ and therefore we have to write **a European**.)

6. Being the Secretary of the BCCI, he is holding **an honorary** position.

   (The consonant **h** in the word **honorary** is silent; hence, **an honorary**.)

7. The bride was given a necklace as **an heirloom** by her mother-in-law.

   (The consonant sound /*h*/ in the word **heirloom** is silent; hence, **an heirloom**.)

8. There is **a hotel** at the end of this road.

   (The initial consonant sound /*h*/ in the word **hotel** is sounded; hence, **a hotel**.)

9. Be careful when you mount **a horse**; it's a moody creature.

   (The initial consonant sound /*h*/ in the word **horse** is sounded; hence, **a horse**.)

10. Certainly you will be paid **an honorarium**, but only on hourly basis.

    (The initial consonant sound /*h*/ in **honorarium** is silent; hence, **an honorarium**.)

11. When I saw her, she was wearing **a uniform**.

    (Though the word **uniform** starts with a vowel *u*, it emits only a consonant sound /*j*/ initially; hence, **a uniform**.)

12. At college, he was **a union leader**.

    (The initial sound in the word **union** is that of a consonant; hence, **a union leader**.)

13. This city has **a unique quality**.

    (The initial sound in the word **unique** is that of a consonant; hence, **a unique quality**.)

14. **A united family** is thousand times better than **a divided one**.

    (The initial sound in the word **united** is that of a consonant; hence, **a united family**.)

2.3.2 Use of The

Compared to *a* and *an*, the use of article *the* is often more subtle and at times quite confusing. In order to help you understand when to use the definite article *the* and when to avoid it and when actually to replace it with *a* or *an*, let us see some of the grammatical notions that go about choosing or discarding the definite article *the*:

Don't use *the* before the names of substances if they are used in a general sense.

1. **Gold** is a precious metal. (not the gold)
2. **Bread** is made from **flour**. (not the bread … the flour)
3. **Lead** is a very heavy metal. (not the lead)

But use *the* if the reference is to a particular kind.

1. **The gold** mined here is of poor quality.
2. They were grateful for **the bread** we gave them.
3. Thieves stole **the lead** from the roof.

Don't use *the* before the names of meals if you refer to them in a general sense.

1. I usually have **dinner** at 9.00 p.m. (not the dinner)
2. Do you take **breakfast** every morning? (not the breakfast)

But use *the* when the meal refers to a social function or the food itself.

1. **The dinner** will be held in the lawns.
2. We really enjoyed **the lunch** she offered us.

Sometimes, possessive pronouns *my*, *your*, *his*, *her*, *their*, etc. are used in order to emphasize the idea in a personal way.

1. I was having **my dinner** when I heard the doorbell ring.
2. We were taking **our lunch** when we heard a loud explosion outside.

Don't use *the* before plural nouns when they are used in a universal sense.

1. **Mangoes** are grown all across the country.
2. **Good readers** are required for good books.
3. **Festivals** can rejuvenate people all over again.
4. **Floods** can cause havoc in one's life.

But use *the* if the reference is made to a particular type.

1. **The mangoes** grown in Uttar Pradesh are sweeter than those in Haryana.
2. **The readers** were not really pleased with the books.
3. **The festivals** celebrated in India have a special fervour.
4. **The recent floods** have caused havoc in Andhra Pradesh.

Actually, we can generalize an idea both through the singular and the plural forms of nouns. When we use the plural form of a noun to show the universal application of the idea, we drop *the*. We, however, use *the* if we choose a singular noun to express the universal idea. Look at the examples given below:

1. **The cow** is a very docile animal.
2. **Cows** are docile animals.
3. **The tiger** and **the cat** belong to the same family of animals.
4. **Tigers** and **cats** belong to the same family of animals.
5. **The lion** is a brave animal.
6. **Lions** are brave animals.

We, however, do not use *the* while using words such as *man* and *woman*.

1. **Man** is a complex species.
2. **Woman** generally wins where man fails.

Don't use *the* with the names of countries unless they suggest that they are made of several small units, states, or parts. For example:

1. **India** is a great country.
2. **Italy** loves football.
3. **The United States** is the strongest democracy in the world.
4. **The West Indies** were once the champions in cricket.

Use *the* if the name is preceded by words such as *kingdom*, *republic*, *federation*, etc.

1. **The Republic of South Africa** finally made its mark in cricket.

Don't use *the* with the names of games.

1. I play **tennis**.
2. **Football** is a game that requires great running ability.

But use *the* if the game referred to is used in a particular context.

1. **The tennis** played these days is more of a fun.
2. **The football** that Maradona played was superb.

Don't use *the* with times of day and night when prepositions *at, by, after*, and *before* precede them.

1. **At sunrise**, he got up to realize that his friend was not there. (not **at the sunrise**)
2. **By noon**, I am likely to finish this assignment. (not **by the noon**)
3. **After night fell**, he went to bed with a sense of loss. (not **after the night fell**)
4. **Before morning came**, movement on the road started. (not **before the morning came**)

But *the* should be used with times of day and night when other prepositions such as *during, in,* etc. precede them or when they refer to a particular event.

1. On hot days, people generally sleep **during the day**.
2. He is likely to come to us **in the afternoon**.
3. Owls keep awake **in the night** and remain asleep **during the day**.

Don't use *the* before proper nouns but use *the* if they are used as adjectives and the reference is to the characteristics of the person.

Compare the following sets of sentences:

1. **Shakespeare** is probably the greatest writer of all times. (not the Shakespeare)
2. Kalidas is **the Shakespeare** of India.
3. **Switzerland** is one of the most beautiful places on the earth. (not the Switzerland)
4. Kashmir is **the Switzerland** of the East.
5. **Botham** was a wonderful all-rounder. (not the Botham)
6. Flintoff was **the Botham** of the 2005 Ashes.

The can be used before nouns to denote the inhabitants of a country but it should not be used before the languages they speak.

1. **The English (people)** live in England and speak **English (language)**.
2. **Spanish (language)** is spoken in Spain.
3. **The French (people)** love **French (language)** not just in France but everywhere else.

Use *the* before the names of mountain ranges, seas, oceans, rivers, deserts, forests, dams, falls, group of islands, etc.

1. You can almost easily lose yourself in **the Himalayas**.
2. The plane fell into **the Pacific Ocean**.
3. **The Bhakra dam** built on the Sutlej is quite old.
4. **The Sahara desert** is the largest desert on earth.

Don't use *the* before the names of single peaks and islands.

1. **Everest** is the highest peak on the earth. (not the Everest)
2. **Sri Lanka** is a beautiful island. (not the Sri Lanka)

The is used before the names of trains and ships.

1. He is coming by **the Frontier Mail**.
2. **The Titanic** disappeared into the sea.
3. **The Queen Elizabeth** is a famous British liner.

The, however, is not used when the reference is made to a vehicle as a means of travel, particularly if it follows the preposition *by*.

1. He is coming **by train**.
2. Thanks largely to poor public transport; more and more people prefer travelling **by car** these days.
3. We normally go to work **by bus**.

However, when these means of transport are preceded by other prepositions such as on, in, etc., *The* is generally used with them.

1. You can sleep **in the car**, but it won't be all that comfortable.
2. The consignment kept **on the bus** fell on the road.
3. You should prefer travelling **by train** but avoid sitting **on the train** in case you find it very crowded.

Avoid *the* before words such as *hospital*, *school*, *college*, *office*, *church*, etc. if the reference is to the purpose for which the building exists. For example, as a student you *go to college* and not *go to the college*. Similarly, an employee goes *to office* and not *to the office*. If the visit is occasional and for some other purpose, *the* is used to specify the building. Look at the following sentences to understand the concept in detail:

1. On Sundays, Christians go to **church**. (not **the church**)
2. I am going to **the church** to see its artistic design. (**The** is required because the visit is made for a different reason.)
3. You call me up later on; I'm getting late for **office**.
4. He walked straight into **the office** and started looking for the cashier.
5. Tomorrow when you go to **college**, don't forget to carry your books.
6. I am going to **the college** to collect my character certificate.
7. For operation and post-operative care, I had to stay in **hospital** for about a month or so.
8. I am going to **the hospital** to enquire about the health of one of my relatives.

Use *the* with those adjectives which are to be used as nouns denoting an entire class or type. In such a case, they are seen as plural nouns and are followed by the plural verb.

1. **The rich** always exploit the poor.
2. **The contented** never grumble.

Use *the* in sentences where a proper noun is immediately followed by an adjective.

1. We still remember **Ashoka the Great**.
2. Listeners are always enchanted by **Lata, the musical**.
3. Not books, but libraries have been written on **Shakespeare, the immortal**.

Use *the* also in such sentences when the structure is reversed.

1. **The great Ashoka** then marched ahead on the road to expiation and salvation.
2. **The musical Lata** was as enchanting as ever.
3. **The immortal Shakespeare** depicted life so comprehensively that after he retired, there was nothing left for other dramatists to show in their work.

Use *the* before superlative degree of adjectives.

1. He is **the wisest** man I have ever met.
2. It was **the hottest day** of the season.
3. **The most beautiful** of them came up to me and asked if I could help her.

Avoid *the* with comparative degree of adjectives in normal structures.

1. Michael is **elder** to me but Mitchell is the eldest.
2. Jane is **taller** than John.
3. He seems **better** today.

However, *the* is sometimes used also with comparative degree of adjectives, particularly when they are repeated in the same sentence and are not followed by *than*.

1. **The more** you read, **the better** is your expression.
2. **The more** I have, **the more** I'll want.

The is used with adjectives employed to signify different nouns used in singular form.

1. **The red** and **the white** rose in the garden …
2. **The first** and **the second** chapter of the book …

But omit *the* if the nouns following these adjectives are in the plural.

1. The red and **white roses** …
2. The first and **second chapters** of the book …

Also omit *the* with the remaining nouns and use it only with the first noun, if two or more nouns are used for the same person.

1. Arundhathi Roy, the writer and **activist** is coming to Jaipur next week.
2. He is the father and **guardian** of this boy.
3. He is the cashier and **accountant** in the office.

However, use *the* if the separate nouns refer to different persons in the context.

1. Arundhathi Roy and Medha Patekar, **the writer** and **the activist**, are coming to Jaipur next week.
2. **The father** and **the guardian** of the boy have been informed by the school authorities.
3. **The cashier** and **the accountant** of the firm are charged with corruption.

Use *the* with those common nouns which function as abstract nouns in a particular context.

1. At last, **the father** in him was stirred.
2. Finally, **the mother** in her was moved.

Use *the* with ordinal numbers.

1. He was **the first man** to arrive on the scene.
2. **The sixth chapter** of the book is stylistically brilliant.

Use *the* before the names of unique things.

1. **The sun** was beating relentlessly on us.
2. You can see countless stars in **the sky**.

Use *the* with names of musical instruments.

1. He can play **the flute**.
2. She loved playing **the piano**.

Sometimes, *the* precedes the names of certain books.

1. **The Paradise Lost** is an immortal epic.
2. **The Iliad** is still read with zeal and enthusiasm.

However, *the* is omitted from the names of books if they have to be preceded by the name of the author.

1. Milton's **Paradise Lost** is an immortal epic.
2. Homer's **Iliad** is a great work of art.

Use *the* if proper nouns are to be told with nouns in plural form.

1. Last night, I went to **the Beckers'**.
2. **The Guptas** are a famous dynasty in Indian history.

PRACTICE TEST 2.21

Read the sentences given below and use *the* wherever required. Cross out places where an article is not required:

1. They lost their way in _____ Sahara desert.
2. _____ Titanic was considered to be an unsinkable vessel.
3. He is travelling by _____ bus.
4. Let's go to _____ bank to deposit the cash.
5. In _____ Austria, people speak _____ German.
6. If you are seriously ill, you will have to go to _____ hospital.
7. Mr Smith and his family generally go to _____ church on Sunday morning.
8. The inspector went to _____ church to inspect the damage done by the agitators.
9. My wife has gone to _____ hospital to visit a sick friend of hers.
10. They stood there in awe and admired _____ church.
11. He committed a crime and was sent to _____ prison.
12. Let's redecorate _____ institute.
13. I feel tired for I went to _____ bed late last night.
14. Let's lift _____ bed and put it out in the sun.
15. Don't worry, I can make _____ bed.

PRACTICE TEST 2.22

Choose the correct option from *a/an/the* for the following sentences. Cross out places where an article is not required:

1. There was _____ king.
2. _____ king was very magnanimous.
3. We love _____ movies.
4. Don't worry; I will finish _____ work by _____ evening.
5. _____ second girl in _____ first row is my sister.
6. He got up at _____ dawn and went to _____ bed at _____ sunrise.
7. He really loves _____ wine of France.
8. I take _____ dinner at about 8.00 p.m.
9. _____ young and _____ old, _____ high and _____ low, _____ all loved Charlie Chaplin.
10. At last, _____ woman in her was moved.
11. Yesterday, I read _____ *Othello*, _____ great tragedy by Shakespeare, _____ genius.
12. _____ visitors truly admired _____ sunset.
13. _____ 22nd June is _____ hottest day of _____ year.
14. Dr Manmohan Singh, _____ Prime Minister of India, is _____ noted economist.
15. _____ *Hindu* is a prestigious national daily.

Determiners   At times, words such as *this, that, those, these, each, some, a, an, the, one, two, all, any*, etc. are also referred to as determiners. A determiner is used to define and limit the meaning of a noun that follows it. Take a look at the following examples to understand how determiners are used to define nouns:

1. Some boys turned up to listen to the teacher.
2. An idea proposed to her is an idea lost.
3. Any news from the crew?
4. That mistake cost him his life.
5. An apology is better than a threat.

6. Give me some rice.
7. All girls jumped up and down in excitement.

8. These tricks won't work on me.
9. Each boy was given a prize.

2.4 MODALS

Look at the following sentences:

1. It may rain today.
2. It might rain today.
3. It will rain today.
4. I can lift this box.
5. I could lift this box.
6. I must lift this box.
7. I will lift this box.

8. You should work hard.
9. You must work hard.
10. You ought to work hard.
11. Will you listen to me?
12. Would you listen to me?
13. Can you listen to me?
14. Could you listen to me?

Can you make out the most important words in the above sentences—those which can change the meaning? Of course, they are the words such as *will, can, could, may, might, must, should, ought to*, etc. which actually distinguish one sentence from the other. By all means, the structural difference between *I can lift this box* and *I will lift this box* is minimal, but not the meaning. The difference in the meanings of these two sentences is so crucial that if you reach a shop and say *I can lift this box*, the shopkeeper may offer you a job and when you say *I will lift this box*, he may call the police!

Similarly, you say *it may rain today* when you see clouds in the sky and hence a good chance, whereas *it might rain today* expresses a dim possibility of rain. *It will rain today*, comparatively, suggests some kind of certainty as you see thick, dark clouds hovering around pretty low in the sky. Likewise, *I can lift this box* suggests your capacity or capability in the present and if you say *I could lift this box*, it would mean that you have lost the strength to lift a box as heavy as the one referred to here. *I must lift this box*, on the other hand, suggests some compulsion or obligation that you need to fulfil. Saying *I will lift this box* can suggest several things, such as determination or intention.

Moving further, let us see how *you should work hard, you must work hard*, and *you ought to work hard* can be seen differently from one another. *You should work hard* is some kind of expectation that usually those close or elder to us have from us. *You must work hard* is not just an expectation; it is almost an order which needs to be followed. *You ought to work hard* is, on the other hand, a very strong suggestion or advice that one feels like giving to those who usually seem negligent of the importance of such virtues.

Similarly, *will you listen to me* is suggestive of the speaker's irritation or impertinence. *Would you listen to me* is far more mellow in tone and tenor. It is suggestive of a polite idea the speaker intends to pass on to the listener. *Can you listen to me* has nothing to do with politeness or the lack of it. It is an innocuous enquiry and usually asked by those who are speaking to you over the phone or by the speakers who are addressing some sizable audience. The modal **can** here relates to the capability or the lack of it on the part of the listener. It may be used as a way to seek permission which is somewhat informal and of course not all that polite, and therefore, not suggested to be used while seeking permission. Comparatively, *could you listen to me* is quite polite in the present or refers to some capacity or ability of the listener in the past.

We hope by now you must have understood that the words such as *can, could, may, might, will, would, shall, should, must* and *ought to* are called modals and how they differ in their tone and tenor. However, since they are often confused with one another, it would be fruitful to remind ourselves about how these modals are to be used to communicate different ideas.

2.4.1 May and Might; Can and Could

Since our credo is learning by doing, let us sort these modals out through exercises. Given below is an exercise based on how the modal verbs or modals—*may* and *might*, *can* and *could* are used. See the sentences below and choose either of these modals—*may*, *might*, *can*, or *could* to fill in the blanks in each of these sentences:

1. When I was young, I _____ swim very well.
2. _____ you live long!
3. You _____ leave now. _____ I, Sir?
4. _____ you solve this sum for me?
5. Though it's quite late, it _____ just be fine even if we reach now.
6. _____ you help me cross the road?
7. _____ I borrow your pen?
8. You _____ pay a little more attention to your studies!
9. He _____ be relieved with immediate effect.
10. _____ you make out what he said?

This is how you can insert the right modal at the right place:

1. When I was young, I could swim very well.
   (**Could** for reference to the capability in the past)
2. May you live long!
   (**May** for wishes)
3. You may leave now. May I, Sir?
   (**May** for seeking and giving permission. **Can** also possible in informal speech)
4. Could/Can you solve this sum for me?
   (**Could** for polite request; **Can** for ability)
5. Though it's quite late, it might just be fine even if we reach now.
   (**Might** for expressing slight hope)
6. Could you help me cross the road?
   (**Could** for polite request)
7. May/Might I borrow your pen?
   (**May** generally for seeking permission; **Might** somewhat diffident)
8. You might pay a little more attention to your studies!
   (**Might** also used for expressing dissatisfaction with someone's behaviour)
9. He may be relieved with immediate effect.
   (**May** for official orders permitting things to take place)
10. Could you make out what he said?
    (**Could** for judging listener's ability in the past)

PRACTICE TEST 2.23

Now, can you choose the correct modals from the options in the following sentences? Try using correct modals out of the options mentioned:

1. I can/will/might see a lot of dark clouds in the sky; it may/might/will rain today.
2. Earlier you can/may/could eat platefuls of rice; why won't/couldn't/can't you now eat even this much?
3. May/Can/Could I have your attention please!
4. In just thirty seconds, this washing machine can/will/may rinse fifty clothes.
5. May/Will/Can you have some more tea?

6. Though he is likely to win the championship, he will/can/might just not make it after all.
7. Will/Can/May it suit you well?
8. Might/May/Can it be possible for you to wait for me at the bus stand?
9. He will/can/may leave his job any time now.
10. Ashima has called up to ask if you will/may/can spare some time for her now.

2.4.2 Shall and Should; Will and Would

Look at the following sentences and try to figure out which of the modals out of the given options will suit the purpose according to the context:

1. Don't worry; I will/shall carry the books for you.
2. Would/Should/Will that I were there to enjoy the party!
3. The philosopher was so punctual that people will/should/would set their watch by his routine.
4. Will/Shall/Should I open the door for you?
5. Be sincere lest you would/should/may fail miserably in life.
6. She told me that she will/should/would turn twenty-six next month.
7. You will/would/should listen to everyone but do what your heart says.
8. Should/Would/Will you lend us your car for a day, please?
9. Should/Would/Shall it rain, how are you going to protect yourself?
10. I wish you should/would/will keep your mouth shut for a while.

| It might rain | It may rain | It will rain |

Might, May, and Will

This is how you can see the correct options and learn why one modal needs to be preferred to the others:

1. Don't worry; I **will** carry the books for you.
   (**Will** is used to express volition and intention.)
2. **Would** that I were there to enjoy the party!
   (For all hypothetical and improbable situations or wishes, **would** is chosen.)
3. The philosopher was so punctual that people **would** set their watches by his routine.
   (**Would** is chosen to show habits of the past.)
4. **Shall** I open the door for you?
   (**Shall I** is used to know the willingness of the person addressed.)

5. Be sincere **lest you should** fail miserably in life.
   (**Lest**—meaning so that it does not happen—is followed by **should**.)
6. She told me that she **would** turn twenty-six next month.
   (In indirect narration, **will** becomes **would**.)
7. You **should** listen to everyone but do what your heart says.
   (**Should** is chosen to give suggestions and advice.)
8. **Would** you lend us your car for a day, please?
   (When used for making requests, **would** makes them sound more polite; **will** is also possible but won't be all that polite.)
9. **Should** it rain, how are you going to protect yourself?
   (**Should** is used to express some assumption or supposition. In such a case, **should** normally starts a sentence. At times, **if it should...** is also used for a beginning.)
10. I wish you **would** keep your mouth shut for a while.
    (**Would** is used for expressing a strong desire or wish; **would** is again preferred in a *that clause* to suggest a wish or some kind of irritation.)

PRACTICE TEST 2.24

Choose the modal that fits the bill in the following sentences:

1. Will/Would you mind opening the door for me?
2. You should/would be more polite while talking to your juniors in office.
3. May/Shall we have something different for dinner today?
4. You may/might pay some attention to the way you keep your room.
5. Though I am not well, I think I would/should attend the meeting.
6. When I was of your age, I should/could run without stopping.
7. Can/Will you have some more rice?
8. Can/May this be true?
9. With rapt attention, students should/would listen to the mesmerizing words of their magical teacher.
10. You shall/should not enter that girl's house again.

2.4.3 Must, Should, and Ought to

The modals *must*, *ought to*, and *should* are used to express concepts of obligation, necessity, duty, advice, suggestion, command, expectation, etc. For numerous expressions, they can replace one another. Even then, each of these modals has a distinct purpose of its own. Look how we can make judicious choices while using them:

Choices	Correct Usage
• Vehicles **should not/must not** be parked here.	• Vehicles **must not** be parked here. (**Must** is mostly used for commands, orders, and instructions.)
• You **must be/ought to be** ashamed of what you have done.	• You **ought to be** ashamed of what you have done. (**Ought to** normally suggests some obligation that was not carried out.)
• You **must not/should not** give your boss much room for suspicion.	• You **should not** give your boss much room for suspicion. (**Should** is very commonly used for giving suggestions and advice.)

(Contd)

(Contd)

Choices	Correct Usage
• You **ought to have been/must have been** more careful before selecting such a girl as your life partner.	• You **ought to have been** more careful before selecting such a girl as your life partner. (**Ought** to is used for desirability and failed expectation.)
• You **must not/should not** have laughed at the fat man slipping on the banana like that.	• You **should not** have laughed at the fat man slipping on the banana like that. (**Should** not is used for reminding people about their duties and obligations.)
• You **should be/must be** back before dark.	• You **must be** back before dark. (**Must** is used for commands, necessity, and instructions.)
• **Must/Should** you miss the train, how would you go?	• **Should** you miss the train, how would you go? (As suggested earlier, **should** is used for suppositions.)
• You **should have/must have** listened to your friend's woes more seriously.	• You **should have** listened to your friend's woes more seriously. (**Should** is used for moral obligations and duties.)
• **Should/Must** you go so soon?	• **Must** you go so soon? (**Must** is used for showing necessity.)
• She doesn't often get late; something **should have/must have** gone wrong.	• She doesn't often get late; something **must have** gone wrong. (**Must** is used for expressing possibilities and speculations.)

2.4.4 Be, Do, Have, Need (to), Used to, and Dare

At times certain words such as *used to, do, have, need,* and *dare* express notions similar to those expressed by other modals. Of these, *need* and *dare* are known as semi-modals. Look at the following sentences to understand some usages of these expressions:

1. How dare you speak to me like that!
   (**Dare** is used to express the speaker's intensity, consternation, or repugnance at someone behaving in an undesirable way.)
2. In ancient times, people used to be less anxious.
   (**Used to** is used to indicate the discontinued habits of the past.)
3. You need not worry about her; she will be fine.
   (**Need not** suggests that something is not necessary, and hence, appropriate here.)
4. I must leave; I have to receive the chief guest at the airport.
   (**Have to** is used to suggest compulsion or obligation.)
5. We are to be married in a couple of days' time.
   (**Am to** and **are to** are used to denote planning.)
6. When we were kids we had to be back home by ten o'clock in the night.
   (**Have** and **had to** are used to show compulsion.)
7. Do keep quiet, for God's sake!
   (**Do** is used to suggest added emphasis, force, or persuasion in such expressions.)
8. You dare not come near me! (AND)
   You need not come near me!
   (**Dare not** would suggest a challenge and **need not** would indicate a refusal.)
9. We do miss you every day.
   (**Do** adds to the sense of emotion conveyed.)
10. When I was a child we used to have a Fiat.
    (**Used to** is used for the discontinued habits of the past.)

PRACTICE TEST 2.25

Fill in the blanks with appropriate modals:
1. I _____ leave; I am getting late.
2. _____ that I were selected Miss India!
3. What's the point in crying over the spilt milk? You _____ have listened to us earlier!
4. I _____ leave for America next month.
5. Leave it; I _____ do this for you.
6. Your father _____ be close to seventy now?
7. _____ his soul live in eternal peace!
8. You _____ improve your speech.
9. I _____ be a theist but now I don't believe in God.
10. _____ you speak English fluently?

PRACTICE TEST 2.26

Choose appropriate modals for the following expressions:
1. Living in such shanties should be/must be so difficult!
2. Mustn't you have/Couldn't you have informed us about it in time?
3. Would/May you suggest me some more names?
4. He could/used to be a fun loving guy before marriage.
5. He said I should/might come at any time.
6. You must be/ought to be caring for others.
7. You should/must reach office in time.
8. Will I/Shall I drive the vehicle while you take rest?
9. The keynote speaker has to/is to begin his speech any time now.
10. Even as an old man, he would/should walk for miles together.

Interjections   Interjections are words such as *Oh*, *Ah*, *Hurrah*, *Alas*, *Wow*, etc., which are used to express some strong emotions or sudden bursts of feelings. Generally, interjections are placed in the beginning of a statement, giving it the emotional intensity it needs to convey. Take a look at the following examples to understand how interjections are used in our everyday expressions:

1. **Oh!** I have lost my ticket.
2. **Hurray!** We have won the match.
3. **Ah!** The sight is mesmerizing.
4. **Wow!** What a great innings he has played.
5. **Alas!** He is no more.

2.5 SENTENCES AND THEIR TYPES

A sentence is a group of words that makes complete sense. It is different from a word, phrase, or clause. Look at the following sentence:

The boy sang a song in the class.

In this sentence, we have as many as eight words which are strung together to give complete sense. The whole structure would be called a sentence. It has a subject—**The boy**—and a predicate—**sang a song in the class**.

Alternatively, look at the expression **in the class**. Does it make any sense? Yes. But does it make complete sense? No. It does make some sense but it is not complete. Moreover, it does not have a subject or a predicate of its own.

Therefore, a group of words that makes partial sense and does not have a subject and predicate of its own is called a phrase.

Moving further, look at the following sentence:

When we reached home, it was midnight.

This sentence can be broken into two parts:

1. when we reached home
2. it was midnight

Both these parts have a subject and a predicate of their own. Of these, the second part—**it was midnight**—makes complete sense. It has a subject (*it*), a verb (*was*), and a complement (*midnight*). The first part—**when we reached home**—too seems to make complete sense, but it is not as complete as the other part—**it was midnight**—is. Even then, it has a subject (*we*), a verb (*reached*), an object (*home*), and an adverb (*when*). So both these parts are **clauses**.

Therefore, a group of words that forms a part of a sentence and may have a subject and a predicate of its own is called a clause.

2.5.1 Types of Sentences (Based on Sense)

Before venturing further into sentence and its types, let us see how sentences can be divided on the bases of sense and structure. Going by the sense, sentences can be of five types (Fig. 2.3). Read the following sentences:

1. Cassius does his work on time.
2. Catherine does not do her work on time.
3. Does Brutus do his work on time?
4. Antony, do your work on time.
5. You too, Brutus! Then Caesar must die!

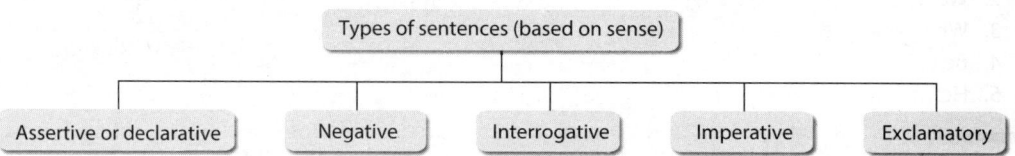

FIG. 2.3 Types of Sentences (Based on Sense)

The first sentence makes a simple statement; such sentences are known as **assertive or declarative sentences**. The second sentence makes a negative statement; such sentences are called **negative sentences**. The third sentence puts a question; such sentences are known as **interrogative sentences**. The fourth sentence gives a command; the sentences which express commands, requests, orders, entreaties, etc. are known as **imperative sentences**. The fifth sentence expresses a strong feeling. The sentences which express strong feelings are known as **exclamatory sentences**.

2.5.2 Types of Sentences (Based on Structure)

At the structural level, we can divide sentences into three categories (Fig. 2.4). In order to get acquainted with them, look at the following sentences:

1. India won the match.
2. We tried hard but we could not win the match.
3. Although we tried hard, we could not win the match.

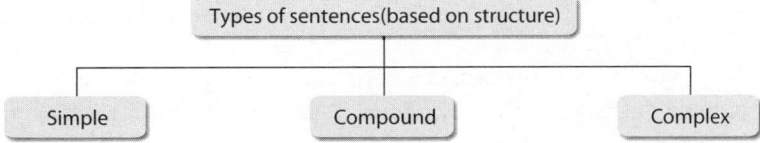

FIG. 2.4   Types of Sentences (Based on Structure)

The first sentence has a simple structure. It has a subject—*India*—and a predicate—*won the match*. Such a sentence is called a **simple sentence**.

The second sentence has a compound structure. It has two independent clauses—*We tried hard* and *we could not win the match*.

Both these sentences are joined by the coordinating conjunction *but*. Such a sentence is a called **compound sentence**.

The third sentence has a complex structure. It has a main clause *we could not win the match* and a subordinate clause *although we tried hard*. Such a sentence is known as a **complex sentence**.

In other words, a simple sentence comprises a single verb. A compound sentence comprises two or more verbs. Both these clauses in a compound sentence are *independent*. A complex sentence, on the other hand, is a combination of one *main* and one or more *subordinate clauses*.

PRACTICE TEST 2.27

Read the following sentences and state whether they are declarative, exclamatory, imperative, interrogative, or negative:

1. Help us, please.
2. We don't like such things.
3. What a shame!
4. Be quiet.
5. He was stabbed in the party.
6. All is well, that ends well.
7. Have you finished your work?
8. How stupid!
9. Go there.
10. Has the chief guest arrived?

PRACTICE TEST 2.28

State whether the following sentences are compound or complex:

1. All are equal but some are more equal than others.
2. You must go or I shall slap you.
3. When we reached back, it was quite dark.
4. Show me the place where he was killed.
5. They wanted to know who stood first in the calligraphy competition.
6. Although she has five sons, she can trust none of them for her old age.
7. I did not attend the party because I was not feeling well.
8. Unless you are a Milton, people won't appreciate your poetry.
9. We tried our best but could not win the match.
10. A guest is unwelcome when he stays far too long.

RECAPITULATION

✓ Noun is a word used to refer to the name of a person, place, or thing. There are four types of nouns: common, abstract, collective, and proper noun. On the basis of numbers, there are singular and plural nouns.

✓ Pronoun is a word that replaces a noun. These are personal, impersonal, demonstrative, distributive, indefinite, relative, reflexive, and emphatic pronouns. They have different cases, such as subjective, objective, and possessive, in which they can be used.

✓ A word used to add to the meaning of a noun or a pronoun is an adjective. There are different types of adjectives such as adjectives of quality, adjectives of number—definite numeral and indefinite numeral adjectives—distributive, demonstrative, interrogative, and emphasizing adjectives.

✓ The part of the sentence that shows an action is called a verb. Verbs are broadly classified into transitive and intransitive verbs. They are also classified as linking verbs, main verbs, auxiliary verbs, and regular and irregular verbs.

✓ Adverb is a word that qualifies a verb, an adjective, or another adverb. The different types of adverbs are adverbs of time, frequency, place, manner, degree, affirmation or negation, and reason.

✓ Prepositions are the words placed before a noun or a pronoun to show the relation or connection with the remaining part(s) of a sentence. Usually, by changing a preposition, we can change the entire meaning of a sentence.

✓ Connectives are words such as and, but, after, because, though, as, wherein, whereupon, for, unless, lest, while, whereas, etc. Some of these connectives are known as coordinating conjunctions and some others are called subordinating conjunctions.

✓ The words *a*, *an*, and *the* are called articles. *A* and *an* are known as indefinite articles, whereas *the* is called the definite article.

✓ The words such as can, could, may, might, will, would, shall, should, must, and ought to are called modals and they differ in their tone and tenor. Just by changing modals, the entire meaning of a sentence can be changed.

✓ A sentence is a group of words that makes complete sense. There are different types of sentences such as assertive or declarative, negative, interrogative, imperative, and exclamatory. Based on their structure we can divide sentences into simple, compound, and complex.

— WISEWELL QUIPS —

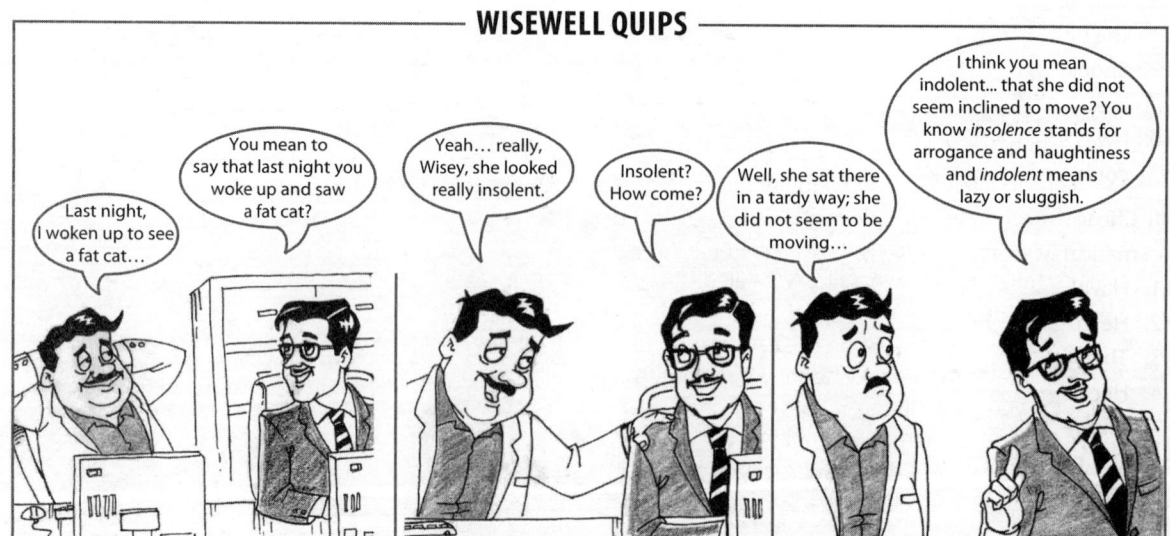

EXERCISES

I. Fill in the blanks with appropriate prepositions:
1. Don't worry; we are prepared _____ anything.
2. She is not capable _____ facing such a trial.
3. The whole country is replete _____ filth.
4. We must abstain _____ bad habits.
5. Even today, we don't have access _____ information in our country.
6. It is believed that he was falsely implicated _____ the case.
7. Though a playback singer, she is well versed _____ classical music.
8. As a country, we are accustomed _____ corruption and dishonesty.
9. He is still ignorant _____ reality.
10. We pine _____ what is not there.

II. Choose the correct modals in the following sentences:
1. He said we can/could/would attend the party.
2. You should not/need not/must not turn on the light; it is already quite bright.
3. Will/Can/Shall I lift this box for you?
4. May/Can/Shall we go home now, Sir?
5. He could/might/used to come on time earlier.
6. Doing that all alone shall be/would be/must be difficult for you.
7. We should/will/must try our best next time; we assure you.
8. Children could/must/should not watch television all the time.
9. Should/Could/Would you see Amrita, ask her to call me.
10. I was scared that if I told her the truth, she must/could/might not trust me ever again.

III. Chose the correct nouns to make the sentence grammatically correct:
1. Have you got all the informations/information?
2. He doesn't know how to give advice/advices.
3. That cost me thirty thousands/thousand rupees.
4. Her sister-in-laws/sisters-in-law made her life miserable.
5. On my way back, I bought four dozens/dozen bananas.
6. A large number of people came to pay their last respects/respect to the departed leader.
7. After his death, his children squandered the entire assets/asset in a matter of months.
8. Rajan could not come to school today; he is down with measle/measles.
9. In Shakespearean world, even handkerchieves/handkerchiefs can spell a tragedy.
10. The good/goods of the family lies in keeping its dark secrets hidden.

IV. Fill in the blanks with correct pronouns:
1. Who/Whom are you speaking to?
2. Let us/we take care of that on our own.
3. Yesterday, she and I/me were taking a walk.
4. Now that he is dead, we can divide the money between you and I/me.
5. After all, the car was my/mine.
6. Its/It's a big problem; don't take it easy.
7. Which/What do you take me to be?
8. Final decision, after all, is your/yours.
9. The poor girl hanged hers/herself.
10. I met her daughter who/whom is a doctor in Delhi.

V. Fill in the blanks with appropriate conjunctions:
1. Give up smoking _____ face the consequences.
2. _____ he pretends to be against hypocrisy, he himself is a hypocrite.
3. _____ pleasure _____ popularity can actually redeem your soul.
4. I saw her _____ returning from office.
5. He is literate _____ not educated.
6. My friend is prodigal, _____ his wife is a skinflint.
7. The drunk man spoiled _____ the party _____ the mood of the people.
8. _____ I entered the room _____ the shriek was heard.
9. _____ the train stop _____ the thief jumped out of it.
10. Vasla is _____ charming _____ intelligent.

VI. Use the given adverbs in their appropriate position:
1. We see each other nowadays seldom.
2. They are late for never work.
3. What you tell is not enough good.
4. The story has begun just.
5. He hasn't done anything wrong really.
6. We take usually our tea in the garden.

7. He cracks witty jokes often.

8. You have to mind always your language in such situations.

9. She has been informed about the incident already.

10. Have you seen ever anything like that?

VII. Rewrite the following sentences by using proper degrees of comparison of adjectives:

1. No lesser than thirty people died in the accident.

2. He is more better today than he was yesterday.

3. Stella is the more beautiful girl in the entire locality.

4. Steve is the oldest in the family.

5. Of the three points discussed, we need to concentrate on the latest.

6. The streets of Ludhiana are dirtier than Jaipur.

7. What I need is few days' rest.

8. Little smile can do wonders to your face value.

9. Few students who were there in the class were intelligent.

10. I have a little doubt that he will succeed.

VIII. Use the following verbs transitively as well as intransitively:

1. Stop
2. Play
3. Fly
4. Fell
5. Speak
6. Decide
7. Move
8. Try
9. Sink
10. Read

IX. Use articles a, an, or the wherever required in the following sentences:

1. Give me _____ some rice; I really don't need _____ whole of it.

2. Both _____ friends fall for _____ charms of _____ solitary girl in _____ movie.

3. In _____ afternoon, my grandfather often goes to _____ library.

4. _____ Trojan War has _____ historical significance.

5. _____ Browns now live in _____ France. Earlier they were in _____ USA. They came to _____ India and stayed here for _____ couple of years. From _____ India they went to _____ United Kingdom and then moved to _____ France.

6. _____ man who is mortal does not just aspire for _____ bread he intends to earn.

7. _____ books are essential in _____ student's life.

8. _____ Golden Temple of _____ Amritsar is _____ very famous religious monument in _____ Punjab.

9. _____ Mutiny of 1857 was quite instrumental in _____ India's efforts for achieving _____ freedom.

10. I spoke to _____ Principal of _____ college in _____ morning.

ANSWER KEY

Practice Tests

2.1
1. Jaipur—Proper noun; City—Common noun
2. Kathak—Proper noun; Dance—Common noun
3. Mohammad Rafi—Proper noun; Singer—Common noun
4. Nokia—Proper noun; Mobile—Common noun
5. Delhi—Proper noun; Capital—Common noun; India—Proper noun
6. Pilani—Proper noun; Town—Common Noun
7. Oranges—Common noun; Reliance Fresh—Proper noun
8. The Tribune—Proper noun; Newspaper—Common noun
9. The God of Small Things—Proper noun; Arundhathi Roy—Proper noun
10. Steve Waugh—Proper noun; Captain—Common noun

2.2
1. Choose = Choice
2. Judge = Judgement
3. Broad = Broadness
4. Sane = Sanity
5. Starve = Starvation
6. Captain = Captaincy
7. Good = Goodness
8. Proud = Pride
9. Woman = Womanhood
10. Quick = Quickness
11. Dark = Darkness
12. Hero = Heroism
13. Hate = Hatred
14. Think = Thought
15. Bond = Bondage
16. Vacant = Vacancy

2.3
1. Girl = Countable
2. Wisdom = Uncountable
3. Idea = Countable
4. Imagination = Uncountable/Countable
5. Style = Uncountable/Countable
6. Composure = Uncountable
7. Kite = Countable
8. Intuition = Uncountable/Countable
9. Paper = Countable
10. Tub = Countable
11. Pass = Countable
12. Title = Countable
13. Chair = Countable
14. Magazine = Countable
15. Seminar = Countable
16. Crime = Countable
17. Integrity = Uncountable
18. Movie = Countable
19. Cricket = Countable if used for Insects; Uncountable if refers to the game
20. Speech = Countable

2.4
1. The girl was a stunner; she had blonde hair.
2. He loves listening to quality music.
3. The speaker was appreciated with loud applause from the audience.
4. Cutlery has become quite stylish of late.
5. Business cannot grow in times of recession.
6. The company is planning to purchase more equipment.
7. Time has come when we need to be serious about climate change.
8. Housewives always have lots of household work to do.
9. There is no point in visiting Rajasthan during summer; since it is hot during that period, sightseeing cannot be enjoyed.
10. In the laughter show the audience had lots of fun.

2.5
1. India and America have signed quite a few memoranda.
2. What is the criterion for selection in this organization?
3. Our country is facing several types of crises.
4. The phenomenon of Indian engineers and doctors going abroad has to change.
5. The parentheses shown within the text are to be removed.

2.6
1. I; you
2. I/he/she; it
3. I; you; your
4. his; me; him
5. your; it
6. your; this/that/it
7. He; it; me
8. his; her
9. He; his; him/me/her
10. He; he; he; it OR She; she; she; it

2.7
1. ourselves
2. them
3. you/him/her
4. his
5. me
6. their
7. It
8. their
9. its
10. It

2.8
1. What
2. Five (or some other number)
3. such
4. Such
5. most difficult/easiest
6. most interesting
7. Neither
8. stupid, silly
9. elder/younger
10. stupendous/remarkable

2.9
1. I was **petrified** to see a **huge** serpent in front of my room.
2. **Pampered** children often get spoilt.
3. I found the movie extremely **boring**.
4. Painters, writers, singers, and dancers are inherently **creative** and highly **talented**.
5. You have to make your proposal very **persuasive** in order to make it saleable.
6. The Home Minister was apprised of the **latest** incidents in the **riot-hit** areas.
7. Keats' poetry is remarkably **sensuous**.
8. With dead bodies littered around, it was a **ghastly** sight.
9. Fashionable girls tend to dress-up in a **stylish** way.
10. Avoid using **redundant** words in your speech.

2.10
1. Transitive
2. Intransitive
3. Intransitive
4. Transitive
5. Transitive
6. Intransitive
7. Transitive
8. Transitive
9. Intransitive
10. Transitive

2.11
1. Yesterday afternoon, I **lay** on the couch in front of the television.
(Past form of the verb lie/lay/lain)

2. After dinner, the mother **laid** the child in the cradle.
   (Past form of the verb lay/laid/laid)
3. The university was **founded** by the Chief Minister.
   (Past form of the verb found/founded/founded)
4. I **found** a mobile on the road while coming back from college.
   (Past form of the verb find/found/found)
5. A beautiful scenery is **hung** in their drawing room.
   (Past participle form of the verb hang/hung/hung)
6. The beautiful criminal was **hanged** for killing her husband.
   (Past participle form of the verb hang/hanged/hanged)
7. If you can't grow a tree, at least never **fell** one in your life.
   (Base form of the verb fell/felled/felled)
8. He **fell** from the stairs and broke his leg.
   (Past form of the verb fall/fell/fallen)
9. We **saw** a man coming from the far end of the road.
   (Past form of the verb see/saw/seen)
10. At the far end of the road, a tree was being **sawed**.
    (Past participle form of the verb saw/sawed/sawed)

2.12
1. I have not seen him lately. (lately; adverb of time)
2. He therefore could not achieve success. (therefore; adverb of reason)
3. She moved around quite sprightly. (around; adverb of place and quite and sprightly; adverbs of manner)
4. The refugees slept fretfully in the tent. (fretfully; adverb of manner)
5. Probably, he has gone to the market. (probably; adverb of certainty)
6. The little girl followed the guest everywhere. (everywhere; adverb of place)
7. Don't go that far. (that far; adverb of degree)
8. The story is not written lucidly. (lucidly; adverb of manner)
9. Surely, you are wrong. (surely; adverb of certainty)
10. He drove quite slowly all the way. (slowly; adverb of manner)
11. Don't worry; she is far better now. (far better; adverb of degree, now; adverb of time)
12. He is too fat to climb the stairs. (too; adverb of degree)
13. We seldom see each other now. (seldom; adverb of frequency, now; adverb of time)
14. Yesterday, I called him late at night. (yesterday, late; adverb of time)
15. He often comes late these days. (often; adverb of frequency; late, these days; adverb of time)

2.13
1. He cut **short** his journey and returned home.
2. Wait! The Guest of Honour is arriving **shortly**.
3. His latest book has been **widely** appreciated.
4. The window was kept **wide** open for the lover to make his secret entry.
5. The dandy moved **round** the damsel throughout the event.
6. He was **roundly** scolded for being a philanderer.
7. Do you have to talk so **loud**?
8. **Loudly**, he cleared his throat.
9. Go **slow** round this corner.
10. **Slowly**, the cat moved towards the kitchen.

2.14
1. He **often** looks sad and gloomy these days.
2. Doctors have **now** reported that one can have cancer also due to depression.
3. She is intelligent **enough** not to marry a fool like you.
4. He has **never** been to Kashmir before.
5. The committee has **already** been informed about the incidence.
6. They were seen going **together** to the party.
7. He bats **brilliantly** at number three position.
8. The spirit **often** knocks the door in the night.
9. He is **always** punctual in his routine.
10. We wash our cars **on Sundays**.

2.15
1. When we watch a tragedy, we are overcome **with** emotions.
2. Having been caught using unfair means, he was debarred **from** sitting **in** the examinations **for** three years.
3. Despite all the rumours, we are quite confident **of** securing a win.

4. Many members abstained **from** casting their votes.
5. The captain attributed the victory **to** his team.
6. You need to apologize **to** her immediately.
7. The poem refers **to** the mythical allusions.
8. He was disgusted **at** the idea of having to change his child's diapers **in** his wife's absence.
9. If you are ignorant **of** everything, you are likely to fail **in** life.
10. He sounded particularly obliged **to** his family members.

2.16 1. with
2. of
3. from
4. of
5. from
6. with
7. of
8. with
9. for
10. with

2.17 1. on, to, for, against, in, on.
2. of, by, of, of, by.
3. to, by, from, for.

2.18 1. neither… nor
2. Unless
3. Even if
4. Despite
5. even though
6. notwithstanding
7. While
8. as
9. wherein
10. not only…but also

2.19 1. On his birthday, we went to the market **and** bought a nice gift for him.
2. He is a great actor **but** not a good human being.
3. **Though** she was not well, she attended her office.
4. Father may be **either** in the office **or** on his way (to office).
5. They fought for their country **and** laid down their lives for it.
6. Good people are simple, **but** never simpleton.
7. He watched the movie, **while** his sister finished her homework.
8. The man was seriously injured, **whereas** his wife died on the spot.
9. Man got all the wealth **but** not happiness.
10. You must keep your mouth shut **or** get out of here.

2.20 1. **Although** I informed her, she did not bother to enquire about your health.
2. They did not try to win; **so** they lost the match.
3. **Though** you say so, I really cannot trust you.
4. Work hard **lest** you should fail.
5. Man **either** wins **or** loses the race of life.
6. You must do it **for** I say so.
7. The salesmen could not convince the customers **because** they were not articulate enough.
8. **Even** if he comes now, we cannot go **and** watch the match.
9. Not that I loved Caesar less, **but** that I loved Rome more.
10. **Neither** a borrower, **nor** a lender be.

2.21 1. the
2. The
3. X
4. the
5. X, X
6. X
7. X
8. the
9. the
10. the
11. X
12. the
13. X
14. the
15. the

2.22 1. a
2. The
3. X
4. the, X
5. The, the
6. X, X, X
7. the
8. X
9. The, the, the, the
10. the
11. X, a, the
12. The, the
13. X, the, the
14. the, a
15. The

2.23 1. can, will
2. could, can't
3. May
4. can
5. Will
6. might
7. Will
8. Might (for diffidence)
9. may
10. can

2.24 1. Would
2. should
3. Shall
4. might
5. should
6. could
7. Will
8. Can
9. would
10. should

2.25 1. have to/must
2. Would
3. should
4. will
5. will
6. must
7. May
8. must
9. used to
10. Can

2.26 1. must be
2. Couldn't you have
3. Would
4. used to be
5. might
6. ought to be
7. must
8. Shall
9. is to
10. would

2.27	1.	Imperative		4.	Imperative		7.	Interrogative		10.	Interrogative
	2.	Negative		5.	Declarative		8.	Exclamatory			
	3.	Exclamatory		6.	Declarative		9.	Imperative			

2.28	1.	Compound		4.	Complex		7.	Complex		10.	Complex
	2.	Compound		5.	Complex		8.	Complex			
	3.	Complex		6.	Complex		9.	Compound			

Exercises

I.
1. Don't worry; we are prepared **for** anything.
2. She is not capable **of** facing such a trial.
3. The whole country is replete **with** filth.
4. We must abstain **from** bad habits.
5. Even today, we don't have access **to** information in our country.
6. It is believed that he was falsely implicated **in** the case.
7. Though a playback singer, she is well versed **in** classical music.
8. As a country, we are accustomed **to** corruption and dishonesty.
9. He is still ignorant **of** reality.
10. We pine **for** what is not there.

II.
1. He said we **could** attend the party.
2. You **need not** turn on the light; it is already quite bright.
3. **Shall** I lift this box for you?
4. **May** we go home now, Sir?
5. He **used to** come on time earlier.
6. Doing that all alone **must be** difficult for you.
7. We **will** try our best next time; we assure you.
8. Children **should** not watch television all the time.
9. **Should** you see Amrita, ask her to call me.
10. I was scared that if I told her the truth, she **might** not trust me ever again.

III.
1. Have you got all the **information**?
2. He doesn't know how to give **advice**.
3. That cost me thirty **thousand** rupees.
4. Her **sisters-in-law** made her life miserable.
5. On my way back, I bought four **dozen** bananas.
6. A large number of people came to pay their last **respects** to the departed leader.
7. After his death, his children squandered the entire **assets** in a matter of months.
8. Rajan could not come to school today; he is down with **measles**.
9. In Shakespearean world, even **handkerchiefs** can spell a tragedy.
10. The **good** of the family lies in keeping its dark secrets hidden.

IV.
1. **Whom** are you speaking to?
2. Let **us** take care of that on our own.
3. Yesterday, she and **I** were taking a walk.
4. Now that he is dead, we can divide the money between you and **me**.
5. After all, the car was **mine**.
6. **It's** a big problem; don't take it easy.
7. **What** do you take me to be?
8. Final decision, after all, is **yours**.
9. The poor girl hanged **herself**.
10. I met her daughter **who** is a doctor in Delhi.

V.
1. Give up smoking **or** face the consequences.
2. **Though** he pretends to be against hypocrisy, he himself is a hypocrite.
3. **Neither** pleasure **nor** popularity can actually redeem your soul.
4. I saw her **while** returning from office.
5. He is literate **but** not educated.
6. My friend is prodigal, **whereas** his wife is a skinflint.
7. The drunk man spoiled **not only** the party **but also** the mood of the people.

8. **Hardly had** I entered the room **when** the shriek was heard.
9. **No sooner** did the train stop **than** the thief jumped out of it.
10. Vasla is **both** charming **and** intelligent.

VI. 1. We **seldom** see each other nowadays.
2. They are **never** late for work.
3. What you tell is not good **enough**.
4. The story has **just** begun.
5. He hasn't **really** done anything wrong.
6. We **usually** take our tea in the garden.
7. He **often** cracks witty jokes.
8. You **always** have to mind your language in such situations.
9. She has **already** been informed about the incident.
10. Have you **ever** seen anything like that?

VII. 1. **No fewer** than thirty people died in the accident.
2. He is **better/much better** today than he was yesterday.
3. Stella is the **most beautiful** girl in the entire locality.
4. Steve is the **eldest** in the family.
5. Of the three points discussed, we need to concentrate on **the last**.
6. The streets of Ludhiana are **dirtier** than those of Jaipur.
7. What I need is **a few** days' rest.
8. **A little** smile can do wonders to your face value.
9. **The few** students who were there in the class were intelligent.
10. I have **little** doubt that he will succeed.

VIII.

Stop (Intransitive):	The play was stopped half way through.
Stop (Transitive):	Stop being a fool!
Play (Intransitive):	While playing in the garden, he saw a snake.
Play (Transitive):	When he plays cricket, he forgets about everything else.
Fly (Intransitive):	The plane is flying in the sky.
Fly (Transitive):	He flies planes in the sky.
Fell (Intransitive):	He fell on the ground and hurt himself.
Fell (Transitive):	If you can't grow a tree, at least don't fell one.
Speak (Intransitive):	Don't speak so loud.
Speak (Transitive):	Always speak the truth.
Decide (Intransitive):	When are you going to decide?
Decide (Transitive):	Decide the matter quickly and act.
Move (Intransitive):	Move to your left please.
Move (Transitive):	He moved the proposal and it was accepted.
Try (Intransitive):	He is trying hard to succeed.
Try (Transitive):	Try this once more.
Sink (Intransitive):	He sank rapidly into the sea.
Sink (Transitive):	One stone is enough to sink a ship.
Read (Intransitive):	He is reading in his study.
Read (Transitive):	These days I am reading Bacon's essays.

IX. 1. X, the
2. the, the, the, the
3. the, X
4. The, a
5. The, X, the, X, a, X, the, X
6. X, the
7. X, a
8. The, X, a, X
9. The, X, X
10. the, the, the

Applied Grammar and Usage

Learning Objectives After reading this chapter, you will be able to

- learn in detail about the subject–verb concord and the correct use of tenses
- learn how to identify a clause, differentiate between a dependent clause and an independent clause, and learn about the different types of clauses
- learn how to change an active voice into passive and direct speech into indirect
- learn how to frame tag questions
- understand the correct usage and application of various punctuation marks

3.1 INTRODUCTION

In Chapter 2, we learnt the fundamental concepts of grammar. Having known the basic grammatical components such as noun, pronoun, adjective, verb, adverb, preposition, conjunction, modal, etc., it is time to move further and understand some other related concepts and see their application and usage in expressions that we use in our everyday life.

3.2 SUBJECT–VERB CONCORD

Let us look at the following conversation between Anshita and her mother:

Anshita: Mom, where is my reading glass?
Mom: Well, it is there on the table.
Anshita: And my shorts? Where is my shorts?
Mom: Can't you see, it's there on the bed.
Anshita: Can you find out for me?
Mom: No, you have to search it yourself.
Anshita: Why, are you not vacant?
Mom: Yes. I am not free.
Anshita: What are you doing?
Mom: I am reading the newspaper.
Anshita: Any interesting newses?

What do you think about the level of accuracy with which the daughter–mother duo have exchanged the above information regarding the whereabouts of the *shorts* and the *glasses*? Of course, one cannot appreciate their linguistic competence since out of the eleven sentences constructed, as many as nine are grammatically incorrect. The very first sentence is incorrect because there is nothing like *reading glass*. In fact, there are some nouns which are always used in the plural form. For example, you always say, 'Where are my reading glasses/trousers/shorts/goggles/scissors/tweezers?' since the nouns that are seen as pairs are always seen as plural and they should take only a plural verb.

You must have heard someone saying, 'Yesterday, I fell on the road and broke my glasses.' Now, even if just one glass of your spectacles is broken, it would be wrong to say, 'Yesterday, I fell on the road and broke my glass.' In fact, the word *glass* can create some ambiguity in the sentence as it may refer to some sort of a looking glass or a tumbler used for drinking water. Similarly, it would not be appropriate to say, 'Where is my shorts?' It is so because the word *shorts* also takes a plural verb. So, the correct way to enquire about your shorts is 'Where are my shorts?' and not 'Where is my shorts?'

Some of the words are used as plurals such as **jeans, pants, scissors, squeezers, tweezers, trousers, belongings, outskirts, goods, congratulations, clothes, particulars, etc**. and they invariably take plural verbs.

Therefore, the words *shorts* and *glasses* in this conversation should be referred to in plural and only the plural pronouns such as *they*, *them*, *their*, etc. should replace them.

Similarly, there are other nouns, such as *information, hair, fish, airplane, news*, which are usually seen as singular nouns and they take a singular verb. Therefore, it is incorrect to say, 'Any interesting newses?' as even in the plural sense *news* remains *news* and does not become *newses*. Now read the dialogue between Anshita and her mother again and see how exactly they should have exchanged the information:

Anshita:	Mom, where are my reading glasses?
Mom:	Well, they are there on the table.
Anshita:	And my shorts? Where are my shorts?
Mom:	Can't you see? They are there on the bed.
Anshita:	Can you find them out for me?
Mom:	No, you have to search for them yourself.
Anshita:	Why, are you not free?
Mom:	No. I am not free.
Anshita:	What are you doing?
Mom:	I am reading the newspaper.
Anshita:	Is there any interesting news?

Can you now identify what went wrong initially? You will see that most of these errors were related to the incorrect subject–verb agreement. We should always choose the verbs that suit the subjects. If the subject of a verb is singular in nature, we cannot choose a verb which is plural and vice versa.

Now, look at the following table and observe the correct subject–verb concord in the sentences:

Choices	Correct Usage
• Diabetes **is/are** a silent killer.	• Diabetes **is** a silent killer.
• I like Physics. It **is/are** an interesting subject.	• I like Physics. It **is** an interesting subject.

(Contd)

Choices	Correct Usage
• Statistics **prove/proves** that our population is still growing at an alarming rate.	• Statistics **prove** that our population is still growing at an alarming rate.
• The jury gave **its/their** verdict in an unbiased manner.	• The jury gave **its** verdict in an unbiased manner.
• Politics today **has/have** degenerated awfully.	• Politics today **has** degenerated awfully.
• For achieving success, we sometimes employ means which **is/are** unfair.	• For achieving success, we sometimes employ means which **are** unfair.
• What's the matter with this institute? Why **is/are** its premises always locked?	• What's the matter with this institute? Why **are** its premises always locked?
• Where **are/is** my **clothes/cloth**?	• Where **are** my **clothes**?
• That day, I saw as many as twenty **aircraft/airplanes** which **was/were** performing in the parade.	• That day, I saw as many as twenty **aircraft** which **were** performing in the parade.
• My **advices/advice are/is** always important. Don't ignore it/them.	• My **advice is** always important. Don't ignore it.
• The **information/informations** sent by you **is/are** not really sufficient.	• The **information** sent by you **is** not really sufficient.

Do you understand the logic behind the options chosen? Let us make this clear to you. The first example mentions the name of a disease, which is *diabetes*, and it should be seen as a singular noun and should automatically take 'is' and not a plural verb 'are'. Similarly, some other diseases such as *rabies*, *measles*, *mumps*, *rickets*, etc. too take the singular verbs *is/was* and it would be wrong to use the plural verbs *are/were* with these names.

Sometimes, the names of the diseases are named after the medical scientists who discover them. In such a case, an apostrophe s ('s) is added after the name of the discoverer and the noun continues to take a singular verb. Look at the following examples:

1. His grandmother died yesterday. She had been suffering from Parkinson's. It is truly an awful disease.
2. At one point of time, it was rumoured that the great actor was suffering from the Alzheimer's disease. Later on however, no traces of it were found in his performance.

In both these examples, the nouns *Parkinson's* and *Alzheimer's* are followed by singular pronoun 'it' because they are the names of the diseases, named after their discoverers, and are treated as singular nouns.

Likewise, names of subjects such as *electronics*, *physics*, *mathematics*, *economics*, and *statistics* are treated as singular nouns and take singular verbs. Similarly, the word *politics* too would take singular verb because it essentially is one of those nouns which appear to be plural but have singular connotations.

Incorrect Usage	Correct Usage
• Statistics **are** a very scoring subject	• Statistics **is** a very scoring subject.
• Politics **are** a dirty game	• Politics **is** a dirty game.

But the words *statistics* and *economics* take a plural verb when they do not stand for the names of subjects.

See the following examples:

1. Statistics (figures) clearly **reveal** that more and more people are now moving to cities from villages in India.
2. The economics (economic policies) of the third world countries **have** always baffled most economists.

 When the words *statistics* and *economics* mean *data* and *economic policies* respectively, they are treated as plural nouns.

Going further, read the following sentences:

1. India beats Australia by an innings and fifty-two runs.
2. Billiards is the game of the rich.
3. The news is good; he has come back safe.

In the above examples, the words *innings*, *billiards*, and *news*, though end with 's' in spelling, are essentially seen as singular nouns.

Similarly, the collective nouns such as *committee, team, herd, fleet, jury, council*, etc. usually take singular verbs because they are seen as a single, indivisible lot.

Look at the following sentences where a collective noun is followed by a singular verb:

1. Our team **has** won the first prize.
2. The herd **was** seen passing through the tunnel.
3. A committee **is** constituted to look into the problem.

In each of these sentences, the collective nouns *team, herd*, and *committee* are followed by the singular verbs *has, was*, and *is*.

On the other hand, some of the nouns, such as *means, premises*, and *clothes*, are generally written in plural forms and usually take a plural verb. Unlike these, there are some other nouns which appear in the plural form but are used both in the singular and plural sense. Read the following sentences carefully and find out the nouns which are used both in the singular and the plural sense:

1. Man is a very complex species.
2. Some of the species that used to exist on the earth have now become extinct.
3. When they toured India, they lost both the one-day and the test series.
4. Though we have lost this series, we haven't lost much.

See, the words *series* and *species* do not change their form and are used similarly to convey the singular as well as the plural sense. Some other nouns which have the singular and the plural forms alike are *spacecraft, swine, aircraft, deer, sheep, cod*, etc.

However, there are some other nouns which are not usually written in plural forms but still take plural verbs. Look at the following expressions:

1. The people of this country are quite poor.
2. Whose cattle are these?
3. Vermin are quite dangerous.

At times, the collective noun *people* is used as a common noun and means 'nation'. In such situations, it is perfectly all right to treat it as any other common noun, for example, *boy, girl, country, nation*, etc. and use it likewise. The following examples illustrate the point:

1. All the **peoples** (nations) of the world need to unite to fight against terrorism.
2. The **people** (group of persons) of this country need to unite and fight against terrorism.

Some other nouns are written in the singular form and are followed by singular verbs. See the following examples:

1. I need your **advice** (not advices). Your advice **is** (not are) highly solicited.
2. Give me some **information**. (not informations). It **is** (not these are) urgently required.

Some other nouns, however, are used only in the plural sense and also appear in the plural form. Look at the following examples:

1. In the entire **annals** of history, there is nobody like Ashoka, the great.
2. In the name of **assets**, he just has a small house.
3. The **obsequies** are yet to be performed.
4. **Thanks** are due to the organizers.

 The words *annals, assets, obsequies,* and *thanks* are normally used in the plural form.

Some other nouns that are used only in plural forms are *alms, riches, environs, tidings, proceeds* (only the noun form and not the singular base form of the verb), *nuptials, eaves, chattels,* etc.

Unlike them, the nouns of measurement are normally used only in singular, particularly when they are used after numerals. Take a look at the following examples:

1. The billion dollar (not dollars) question is—would he win the race?
2. I bought five dozen (not dozens) bananas.
3. Send a draft for five thousand (not thousands) rupees.

Some other nouns in this category are *pair, gross, hundred, million, score,* etc. However, when these nouns are not followed by numerals, it is all right to refer to them in their plural forms. For instance:

1. Thousands of people turned up to attend the funeral.
2. Billions are lost in protecting the world from terrorism every year.
3. Dozens of people are injured in the accident.

When a particular phrase refers to measurements, quantities, and amounts taken as a whole, it takes a singular verb. For example, take a look at the following sentences:

Incorrect Usage	Correct Usage
• The twenty thousand rupees you gave me **are** lost.	• The twenty thousand rupees you gave me **is** lost.
• Twenty thousand rupees **are** a meagre salary these days.	• Twenty thousand rupees **is** a meagre salary these days.

Further, a singular verb is chosen when some percentage (%) precedes a noun that is plural in nature. Look at the following examples:

1. 20% of ₹2000 **is** not a big amount.
2. Even today, 30% of the country's population **lives** below poverty line.

In both these sentences, percentage is followed by a seemingly plural noun but takes only a singular verb. This norm, however, does not apply when percentage is followed by other plural nouns. Look at the sentences given below:

1. Around 50% of the country's people **are** still superstitious.
2. Nearly 40% of students in this institute **have** problem with spoken English.
3. Despite the rise of feminism, nearly 80% Indian women **are** still discriminated against their male counterparts.

Sometimes, for choosing the subject of a verb, two separate nouns are mentioned, but if they express a single idea, only singular verb is chosen, as shown in the following examples:

1. Bread and butter **is** the breakfast for many in the country.
2. Gandhiji's aim and objective in life **was** to help the deprived classes.
3. Slow and steady **wins** the race.

For all the above expressions, we should choose a singular verb.

However, when the separate nouns do not essentially express a single idea, they need to be treated as plural nouns and should take plural verbs. See the following examples:

1. Becker and Edberg are participating in the tournament. (Two different players)
2. Thorpe and Thorpe have written *Objective English*. (Two different authors)
3. BITS, Pilani, and IIT, Roorkee, are organizing the event. (Two different institutes)

 The expressions such as *a majority of/majority of/the majority of/a number of/a lot of/plenty of/ all of*, etc. are generally followed by plural nouns and take plural verbs.

For example:

1. A lot of people **have** decided to abstain from casting their votes.
2. When meetings **are** conducted, a number of ideas are shared.
3. Plenty of steps **are** required to be taken before we can be sure of our security system.
4. A lot of exercises **are** given in this book for practice.

However, when expressions such as *plenty of/most of/a lot of/a great deal of*, etc. are not followed by a plural noun, the verbs chosen are singular. Look at the following examples:

1. Despite all efforts by the government, a great deal of water is still wasted.
2. Lot of bloodshed is seen during wars.
3. Plenty of room (opportunity) is available for all those who want to succeed in life.
4. Most of the discussion was addressed to the task of bringing equality between sexes.
5. Plenty of food goes waste everyday.

 The expressions *everybody, everyone, everything, each and every*, etc. are followed by singular verbs. For example:

1. Everything was destroyed in the attack.
2. Though nobody died, everybody was injured.
3. Each and every student needs to be regular in the class.
4. Everyone was listening to the speaker with avid interest.

 When two separate singular nouns are denoted through coordinating conjunctions such as *either...or* and *neither...nor*, the verb chosen is singular. However, when one of these nouns is plural, the verb chosen is also plural and it is placed closer to the plural noun. Therefore we should write this as

1. Either Geeta or her **sister has** done this. (Correct)
2. Either Geeta or her **sisters have** done this. (Correct)

Remember, choosing a plural verb for a singular noun or putting a singular noun closer to a plural verb is considered flawed.

For example:

1. Either Geeta or her sisters has done this. (Incorrect)
2. Either her sisters or Geeta have done this. (Incorrect)

Keeping in mind this rule, observe the correct subject–verb concord in the following sentences:

1. Neither the monk nor the pupils **know** about the mystery of life.
2. Either the poet or the thinker **is** likely to chair the session.
3. Neither the teacher nor the students **are** expected to come today.
4. Either you or your team members **are** to blame for the failure of the project.
5. Neither the government nor the people **are** serious about the issue.
6. Neither man's culture nor his family **is** to blame for his actions.
7. Either John or his friends **have** informed the police.

Further, expressions such as *neither of the…, either of the…, one of the…, none of the…,* which are generally followed by plural nouns, normally take singular verbs. Look at the following examples:

1. One of the employees was fired.
2. Neither of the students has reported for admission.
3. Each of the Williams sisters is a terrific tennis player.
4. Either of the boys has broken the lock.
5. None of the issues is important to us.

These beginnings are different from those which start with *most of…, many of…, some of…, almost all…,* etc. These too are followed by plural nouns and take plural verbs. See the following examples:

1. Many of the teachers have left the university.
2. Most of the girls were screaming at the scary sight.
3. Some of the members have expressed a desire to quit.
4. Almost all the spectators were thrilled to witness the intensity of the contest.

At times, it becomes difficult to decide whether the subject of the verb is singular or plural. Owing to this complexity, sentences are wrongly structured since a singular noun is yoked together with a plural verb and vice versa. Look at the following sentences to understand the point:

1. The minister along with the bodyguards were killed. (Incorrect)
2. The minister along with the bodyguards was killed. (Correct)
3. The ministers along with the bodyguards was killed. (Incorrect)
4. The ministers along with the bodyguards were killed. (Correct)

Out of the given structures, the first and the third are grammatically wrong. In all these sentences, the real subject is *the minister* or *the ministers*. In such sentences, whatever comes after the conjunction *along with* should not really affect the verb. Look at the second sentence, where a singular subject *the minister* takes a singular verb 'was'. Similarly, the fourth sentence also correctly chooses the plural verb 'were' for the plural subject *the ministers*.

Always remember the following rules:

> **Singular Noun + along with + Singular/Plural Noun takes Singular Verb**
> **Plural Noun + along with + Singular/Plural Noun takes Plural Verb**

While dealing with such linguistic constructions, you can bear in mind one simple instruction. Don't consider the nouns that follow expressions such as *along with, besides, in addition to, with,* etc. Decide whether to choose a singular or a plural verb on the basis of the noun that precedes such expressions. Look at the examples given below to understand the point clearly:

1. **The minister** besides the bodyguards **was** killed.

2. **The ministers** besides the bodyguards **were** killed.

3. **The minister** with the bodyguards **was** killed.
4. **The ministers** with the bodyguards **were** killed.
5. **The minister** in addition to the bodyguards **was** killed.
6. **The ministers** in addition to the bodyguards **were** killed.

See, in all these sentences, whatever precedes the expressions *besides*, *with*, *in addition to* decides the verb. Wherever the singular noun *minister* comes, the verb chosen is also singular, that is, 'was'. The plural noun *ministers*, on the other hand, invariably takes the plural verb 'were'.

A real conundrum confronts us sometimes when the expression *as well as* connects two nouns. Look at the following expressions and choose the right verbs for them:

1. The minister as well as the bodyguards was/were killed.
2. The cashier as well as the accountant has/have come.
3. The Prime Minister as well as other important dignitaries has/have visited our institute.
4. Shakespeare as well as his contemporaries is/are still regarded the greatest dramatists in the entire range of English literature.

Essentially, the expression *as well as* is not different from *besides*, *with*, *in addition to*, *along with*, etc. as all these are parenthetical in nature and the real subject normally precedes them. The first two examples, therefore, should take singular verbs 'was' and 'has' as whatever follows 'as well as' should not affect the verb. The remaining two examples should also follow the suit. However, if we choose a singular verb in the last two sentences, the whole sense is lost. This is a classic example of how *as well as* is wrongly used in place of *and*. Actually, Indian English does not make much distinction between *as well as* and *and*. However, it is suggested that the distinction is maintained as *and* is not parenthetical in nature, whereas *as well as* is. Therefore, it is advisable to use *and* when the idea is to introduce a plural subject (this + that) and choose *as well as* when whatever follows it is only parenthetical and does not really affect the verb.

Whenever expressions such as *besides*, *along with*, *as well as*, *in addition to*, *with*, etc. conjoin two subjects, the verb needs to be chosen according to the first subject.

Look how by choosing *and* in place of *as well as* we can suggest the idea in a lucid way:

1. The minister and the bodyguards **were** killed.
2. The cashier and the accountant **have** come.
3. The Prime Minister and other important dignitaries **have** visited our Institute.
4. Shakespeare and his contemporaries **are** still regarded the greatest dramatists in the entire range of English literature.

We can also draw distinction between a singular noun and a plural one by looking at the number of times the definite article 'the' is used in the sentence. If 'the' precedes both the nouns, the subject of the sentence and also the verb should be plural. Look at the expressions given below:

1. The cashier and accountant **has** come.
2. The poet and philosopher **is** dead.
3. The leader and orator **has** arrived.

Though these may appear to be ungrammatical constructions, the sentences are actually correct. It is so because when we wish to refer to two different attributes of the same person, we do not repeat the definite article 'the'. Since the different attributes mentioned here refer to the same person in each of these sentences, a singular verb is more appropriate. Look at the alternative structures in which not one person, but two different people that are mentioned:

1. The cashier and the accountant **have** come.
2. The poet and the philosopher **are** dead.
3. The leader and the orator **have** arrived.

The subject–verb concord goes haywire because at times it becomes difficult to decipher whether the subject is singular or plural, particularly in sentences in which the subject comprises several nouns and adjectives. Look at the following sentences to understand the idea:

1. Many difficult issues related to our society is/are required to be put into proper perspective.
   Many difficult issues related to our society **are** required to be put into proper perspective.
2. The only ones who came to attend the party was/were his kith and kin.
   The only ones, who came to attend the party, **were** his kith and kin.
3. The boy who is wearing brown jeans is/are my brother.
   The boy, who is wearing brown jeans, **is** my brother.
4. The fact that in our party itself there are many dissidents who attempt/attempts to destabilize the government is/are quite disheartening.
   The fact that in our party itself there are many dissidents, who **attempt** to destabilize the government, is quite disheartening.
5. The only way you could listen to his endless stories was/were to shut your ears.
   The only way you could listen to his endless stories **was** to shut your ears.

In all such sentences, the nature of the subject should make us choose the verb. If the subject is singular, it should be followed by a singular verb. Otherwise, it should take a plural verb.

PRACTICE TEST 3.1

Choose the correct subject–verb combinations in the following sentences:

1. It is obvious that everything is/are fine.
2. The majority of nurses in our country is/are women.
3. Either James or his brothers has/have written the mail.
4. The jury has/have decided to hear the case at an early date.
5. That they are fools is/are known to everyone.
6. A lot of Indians have/has raised the issue of colour discrimination.
7. None of the reports are/is worth consideration.
8. One of the boys was/were hurt in the class.
9. Quite a few students find/finds the grammar classes quite boring.
10. The foreign delegation comprising many experts is/are likely to visit our university next week.

PRACTICE TEST 3.2

Given below is a small passage. Identify whether the subject in each of the sentences below is singular or plural. Choose the verb accordingly:

Today, all the teachers is/are enjoying a free day at college. You know the annual function is drawing closer and students is/are preparing for various activities which is/are being planned by them. The increased participation of students in different types of extra-curricular activities has/have led to a substantial decrease in attendance in the classroom. In many classes you see only a handful of students who too is/are sitting in a lacklustre mood. Only a few of them is/are not really yawning. That remind/reminds me of a very alarming proposition. The student community these days is/are hardly interested in doing anything related to learning. What generally interest/interests them is/are fun and frolic, and not studies.

3.3 USING TENSES

Read the following sentences:

1. I take lessons in English language skills.
2. I took lessons in English language skills last year.
3. I shall take lessons in English language skills next month.

What do you notice in the above sentences? Yes, in sentence 1, *take* refers to the present. In sentence 2, *took* refers to the past. And, in sentence 3, *shall take* refers to the future. Thus, a verb may refer to any of the following:

- Present time
- Past time
- Future time

A verb that refers to a time in the present is said to be in present tense. For example:

1. I often write a letter to my friend.
2. He loves his younger brother.
3. We listen to classical music.
4. They go to Jaipur during holidays.
5. The baby cries bitterly.

A verb that refers to a time in the past is said to be in past tense. For example:

1. Rita washed her sweater.
2. They prepared fried rice for dinner.
3. We celebrated Diwali joyfully.
4. She bought a car yesterday.

A verb phrase that refers to a time in the future is said to be in future tense. For example:

1. I shall visit my friend today.
2. We shall discuss our future plans.
3. My friends will take the CAT in December.
4. They will watch a movie today.

PRACTICE TEST 3.3

Read the following sentences and identify the tense they carry:

1. Tukka cleaned her room today.
2. I shall visit their house sometime in the next week.
3. They waited for me for an hour.
4. Ruth speaks very fast.
5. I shall wish him a very happy birthday.
6. Roshan gives me a sweet smile everyday.
7. They will send the invitation cards next month.
8. He gained the goodwill of his customers.
9. The authorities will discuss the terms and conditions with their employees in the meeting.
10. I love singing old Hindi songs.

This exercise might not have been really challenging for you mainly because they all are related to simple time frames in the present, past, and future. However, the varied human actions do not necessarily relate to simple time frames and hence we are required to learn in detail all the possible ways in which the different time frames related to the present, the past, and the future should be employed while we speak and write English.

Therefore, let us learn all the tenses in English in detail. To begin with, take a look at the following set of sentences:

I get up at five o'clock in the morning. I go for a walk with my father. After returning from the walk, I study for half an hour and get ready for my college. My father drops me there in his car. I get back from college at 4 p.m.

When do you think we write like this? We do it when we want to express the actions that happen in a general or routine manner. This means that the action may not be happening at the time of speaking. The statements given above merely suggest a routine activity usually followed. The tense used in these sentences is known as **present indefinite tense** or **simple present tense**. While using simple present tense, we use the following structure:

> **Subject + Base Form of Verb + s/es + Object**

See the following examples:

1. She likes soft toys. (III Person Singular + Base Form of Verb + s + Object)
2. We fight with each other. (I Person Plural + Base Form of Verb + Object)

By using the above structure of sentences you can express the following:

- The actions that are done as habits in everyday life
- General facts
- Universal truths or facts

Read the following sentences carefully and see how this structure helps us express our actions in the present:

1. The earth revolves round the sun. (**Universal truth**)
2. Mr Arora watches a movie on Sunday. (**Habitual action**)
3. I usually get up at five in the morning. (**Habitual action**)
4. He often plays in the evening. (**Habitual action**)
5. Horses run faster than donkeys. (**General fact**)
6. We go to some hill station during summer holidays. (**Habitual action**)

Table 3.1 should help you observe the structure of the simple present tense/present indefinite tense.

Table 3.1   Sentence Structure in Simple Present Tense

Subject	Verb	Object/Adverb/Complement
I	do not play (do not + base form of verb with I, we, you, they, plural subject)	football.
Do we {Interrogative (Do/Does) + Subject}	sing (Base form of verb)	well?
She	cooks (Base form of verb + s/es with he, she, it, singular subject)	food.
Doesn't he {Interrogative + Negative (Don't/ Doesn't) + Subject}	obey (Base form of verb)	his parents?
My teacher	teaches (Base form of verb + s/ es with singular noun)	English.

PRACTICE TEST 3.4

Fill in the blanks with an appropriate verb in simple present tense:

1. My mother _____ (buy) vegetables from the Reliance Fresh store.
2. Normally, he _____ (walk) very fast.
3. She _____ (rebuke) her children for playing computer games.
4. We _____ (not find) the solution of this problem on the Internet.
5. They _____ (not trust) their employees.
6. Girls _____ (not like) wrestling.
7. The chief librarian usually _____ (purchase) new books and journals for the library in July.
8. Snigdha _____ (dance) skillfully.
9. Siddharth _____ (take) coffee after dinner.
10. It _____ (rain) frequently in London.

Read the following sentences and see how the actions are expressed.

1. Priya **is learning** French from the American Institute of Foreign Languages.
2. We **are celebrating** the fourth marriage anniversary of my brother today.
3. He **is toying** with the idea of creating a new robot with emotions.
4. The cost of living **is increasing** day by day.

What do you observe in these sentences? In these sentences, the action is not a routine activity; in fact, it is going on even at the time of speaking. The action in all these sentencs seems to be in progression. This is called **present continuous** or **present progressive tense**. The structure for this tense is as follows:

> **Subject + is/are/am + (Base Form of Verb + ing) + Object**

Singular subject takes *is*; plural subject takes *are*; and the subject *I* takes *am*.

Subject		Verb
I person singular (I)	→	am
I person plural (we)	→	are
II person (you)	→	are
III person singular (he, she, it)	→	is
III person plural (they)	→	are

Table 3.2 should help you understand the structure of the present continuous or present progressive tense.

Table 3.2 Sentence Structure in Present Progressive Tense

Subject	Verb	Object
I	am writing	a letter.
She	is watching	a movie.

(Contd)

Table 3.2 (Contd)

Subject	Verb	Object
We	are discussing	a problem.
Boys	are laughing at	the joke.

PRACTICE TEST 3.5

Fill in the blanks with appropriate verbs. Look for the clues that should help you choose between simple present and present progressive tenses:

1. I _____ (work) hard for my GRE test these days.
2. We _____ (face) an acute problem in power supply these days.
3. Since Mr Smith is going to France, he _____ (learn) French these days.
4. Due to recession, many companies _____ (downsize) their operations.
5. Look! Anne _____ (relish) orange juice and French fries in the sun.
6. Our institute _____ (plan) to bring a change in its promotion policy.
7. This time, our government _____ (not cut) down the prices of petroleum and gas.
8. The patient _____ (wear) a blanket because he is feeling very cold.
9. The watchman normally _____ (bring) tea and snacks for the hostel inmates from outside.
10. That great man always _____ (donate) a lot of money for poor children.

In the above exercise the expressions such as *these days*, *always*, *normally*, etc. helped us choose the tenses correctly.

But we may not be lucky all the time and in the want of obvious clues, the simple present/present indefinite and present continuous/present progressive tenses sometimes wrongly replace each other. Owing to their specific features, they should be selected carefully. The following section helps you understand where to use simple present tense and where to opt for present progressive tense.

3.3.1 Simple Present or Present Progressive

Look at the following sentences:

1. The cake is smelling sweet.
2. I am thinking that the idea given by you is quite good.
3. I am loving it.
4. He is owning a very big bungalow in the heart of the city.
5. Somehow he is always disliking my suggestions.
6. When I went to the party last night, I was not recognizing my uncle.
7. Sorry, I am not understanding your point.
8. These days, I am not having any vehicle.
9. He is not believing in God.
10. He is never trusting his neighbours.

It is not difficult to observe that all the above sentences are written incorrectly. Let us see how we can use them correctly:

1. The cake **smells** sweet.
2. I **think** that the idea given by you is quite good.
3. I **love** it.
4. He **owns** a very big bungalow in the heart of the city.

5. Somehow he always **dislikes** my suggestions.
6. When I **went** to the party last night, I could not recognize my uncle.
7. Sorry, I **do not understand** your point.
8. These days, I **don't have** any vehicles.
9. He **does** not believe in God.
10. He never **trusts** his neighbours.

In fact, in all these sentences, the verbs of perception, thoughts, emotions, senses, or possession are used. None of these verbs are usually expressed in continuous form as they are suggestive of more stable emotions, thoughts, perspectives, beliefs and feelings.

Look at more such verbs which normally do not appear in the progressive tense:

Incorrect Usage	Correct Usage
• I **am not agreeing** to your point of view.	• I don't **agree** to your point of view.
• The committee **is consisting** of three members.	• The committee **consists** of three members.
• He **is appearing** to be sad today.	• He **appears** to be sad today.
• Normally, I **am preferring** tea to coffee.	• Normally, I **prefer** tea to coffee.
• He is **seeming** to be all right now.	• He **seems** all right now.
• I **am feeling** for the poor.	• I **feel** for the poor.
• We **are not meaning** this.	• We **don't mean** this.
• **Are you minding** moving a little?	• **Do you mind** moving a little?
• We **are hoping** to see you sometime.	• We **hope** to see you sometime.
• We all **are wishing** to be happy in life.	• We all **wish** to be happy in life.

So, be cautious and use a simple present tense when the verbs to be chosen are any of the following:

1. **Verbs of emotion**, for example, wish, desire, like, love, hate, want, refuse, etc.
2. **Verbs of thought**, for example, think, believe, agree, understand, mean, mind, know, etc.
3. **Verbs of senses**, for example, see, hear, taste, feel, smell, touch, etc.
4. **Verbs of perception**, for example, recognize, notice, perceive, imagine, remember, etc.
5. **Verbs of appearance**, for example, appear, seem, look, etc.
6. **Verbs of possession**, for example, own, possess, belong, contain, have, consist, etc.

PRACTICE TEST 3.6

Pick the correct options in the following sentences:

1. Why **are you**/**do you** always come late?
2. He is **likely**/**likes** to visit our school during vacation.
3. "Look, how glorious **is appearing**/**appears** the Sun today!"
4. He **is going to resume**/**resumes** his innings on a nervous 99 after lunch.
5. Why **are you**/**do you** inform me about that now?
6. Sometimes it **is raining**/**rains** here quite torrentially.
7. He **is writing**/**writes** his poems in Urdu.
8. We **are waiting**/**wait** for the signal to clear.

9. Normally, the temperature **is ranging/ranges** between 10 and 20 degrees in this region.
10. He **is conducting/conducts** a meeting right now in his chamber.

3.3.2 Present Perfect or Simple Past

Moving further, now read the following sentences carefully:

1. He **gave** me a watch.
2. My friend **asked** for help.
3. The tourist **went** around the city.
4. His uncle **telephoned** him to bring a laptop for his daughter.

What have you observed? In these sentences, the action is completed in the past. For expressing such actions, we use simple past/past indefinite tense and use the following structure:

> **Subject + Past Form of Verb + Object/Complement/Adjunct**

Now go through the following set of sentences and observe how the simple past tense is used:

1. Parimal **decided** to leave his job.
2. She **did not understand** the problem.
3. Sukumar **disclosed** the secret.
4. **Did** you **enjoy** your trip?

PRACTICE TEST 3.7

Complete the following sentences by using the structure for simple past tense:

1. Earlier, I _____ (cannot contact) you.
2. _____ the officer _____ (inform) you about that?
3. My children _____ (play) badminton while I _____ (sleep).
4. _____ your company _____ (plan) to start something in Kashmir earlier this year?
5. Last month, my wife _____ (visit) New York for attending a meeting.
6. His father _____ (drive) us back home very fast.
7. Mr Batra _____ (deposit) five thousand rupees in his son's account.
8. They _____ (open) new branches in all the cities but surprisingly (close) each of these one by one.
9. Since she was not well, she _____ (consult) a doctor.
10. The teachers at the conference _____ (identify) the innovative methods of teaching.

Moving on to the next tense, let's now read the following set of sentences:

1. India **has contributed** immensely in the field of science and technology.
2. The gardener **has planted** a few more shrubs and trees this month.
3. Sunil Gavaskar **has redefined** cricket in our country.
4. The journalist **has brought** the unsung heroes to light.

All the above sentences use the present perfect tense as each of these follows the following structure:

> **Subject + has/have + Past Participle Form of Verb + Object**

The present perfect form of the verb is preferred when the focus is on the completion of action and the immediacy of the past. Attempting the practice tests that follow should help you understand when to use the present perfect form of the verb.

PRACTICE TEST 3.8

Complete the following sentences with appropriate verbs in present perfect tense:

1. I am not feeling hungry, I _____ (take) heavy breakfast.
2. It is nice to see you again; we _____ (not meet) each other for such a long time!
3. Recently, the state government _____ (pass) a bill for free college education for girls.
4. Though she loves going on long drives, she _____ (not learn) how to drive.
5. _____ you ever _____ (think) how stressful life _____ (become) these days.
6. I know I _____ (leave) no stone unturned to pass the CAT exam.
7. The music band _____ (display) a stupendous show this time.
8. The film director _____ (develop) the plot of the story and used the camera and lights skillfully.
9. He is not aware of the legal implications as he _____ (not purchase) any land before.
10. The policy _____ (come) as a result of the scathing attack from the media.

Since the present perfect tense is at times confused with simple past tense, a careful use is desirable. Remember that we could go wrong not just in choosing between the simple present and the present progressive but also while picking between the present perfect and the simple past tense.

PRACTICE TEST 3.9

In the following sentences, identify the places where the present perfect and the simple past tense are either mixed or chosen not so judiciously:

1. Sir, I **have passed** my B.Tech in 2003.
2. We **finished** our assignment just now.
3. Don't worry; I already **informed** her about that.
4. We **have seen** a tiger five years ago.
5. I **have spoken** to him last night.
6. Wait; I **did not** yet **finish** my work.
7. I **have read** that story in Class IX.
8. We **have spent** a very enjoyable evening yesterday.
9. She **has submitted** her project this morning.
10. Last year, we **have conducted** several workshops for our teachers.

None of these sentences chooses the correct tense. Whenever a particular particle of time such as *yesterday, this morning, the year 2002, last month, couple of hours back,* etc. appears, we normally choose simple past tense. The expressions such as *just now, recently, already, yet,* etc., on the other hand, should make us choose the present perfect tense.

PRACTICE TEST 3.10

Look at the sentences given below and choose either present perfect tense or simple past tense:

1. Modern scientific discoveries challenged/have challenged many established notions.
2. India won/has won the one-day series against Australia. In the last match played in Delhi yesterday, team India defeated/have defeated the Aussies by 37 runs to clinch the series.
3. Prices increased/have increased alarmingly in the past six months.

4. She left/has left yesterday itself.
5. She left/has left just now.
6. It's fine. I received/have received the mail.
7. Keeping in view the recession times, many companies started/have started downsizing.
8. We decided/have decided to continue the training, even though it was hardly of any use.
9. In his directorial career, Raj Kapoor made/has made several memorable movies.
10. Doctors linked/have linked many diseases, such as diabetes and cancer, to anxiety and unhappiness in life.

PRACTICE TEST 3.11

Choose the correct tense in the following sentences:

1. Call the doctor, I think he broke/has broken his toe.
2. Now that you came/have come, why don't you stay with us for a while?
3. They questioned/have questioned the actor at the airport before they let/have let him go.
4. We spent/have spent the entire weekend worrying about his whereabouts.
5. Research showed/has shown that exercises help people remain cheerful.
6. Shakespeare wrote/has written his last play possibly in 1613.
7. Don't worry, I informed/have informed his parents about his illness.
8. Earlier he made/has made some comedies but for some time now he produced/has produced only serious movies.
9. I am sure I read/have read about that episode in some magazine.
10. You grew/have grown so big ever since I saw you last.

3.3.3 Simple Past or Past Perfect

Now read the following set of sentences.

1. I **had finished** my lunch before the kids came from school.
2. The students **had checked** their pre-comprehensive marks before their final exams.
3. After we **had identified** the areas of improvement, we commenced the training programme.
4. Mr Jones **had gone through** the manuscript several times before sending it to the publisher.

In the above sentences, both the actions happen in the past time but one action takes place earlier than the other. To demarcate the earlier action from the succeeding one in the past, you are required to use the past perfect tense as shown in the above examples. This tense is formed by using **had** with the **past participle form** of the verb. So the basic structure of this tense is as follows:

Subject + had + Past Participle Form of Verb + Object/Complement/Adjunct

Now let us understand where we should use simple past tense and where to use past perfect tense.

PRACTICE TEST 3.12

Check if the past perfect and simple past tense in the following sentences are used correctly:

1. The British had ruled over India for nearly two hundred years before they had handed it over to us.
2. We had walked all the way to the station only to know that the train had already left.

3. When the police had interrogated the criminal, they had put him behind the bars.
4. My neighbour had told me that his father died.
5. By the time the doctor had reached, the patient had died.
6. We thought of raising your pay, but had not been able to get enough funds.
7. After he finalized the deal, he had signed it.
8. The movie already began when we had entered the hall.
9. He planned to retire at the age of sixty-five, but he had died much before that.
10. I just entered the building, when the bomb had gone off suddenly.

Of course, none of these sentences is correct. Choosing between the simple past and past perfect tense is at times difficult. To avoid confusion, choose past perfect (had + past participle form of the verb) for the action that precedes the other action in the past. Choose past simple for the action that follows. In other words, choose **past perfect** for the **earlier** action and **simple past** for the **later action**.

Thus, it is important to distinguish two actions of the past if one of them takes place before the other. However, when two actions of the past happen successively and also when we want to emphasize that the second event is the result of the first, we choose simple past tense in both the clauses. Look at the following sentences:

1. The thief **broke** into the shop (not had broken into the shop) and stole jewellery.
2. Rushdie **shot** to fame (not had shot to fame) after he wrote *Midnight's Children*.
3. The cow **entered** (not had entered) the farm and walked over the plants.
4. When fog **cleared** (not had cleared), we resumed our journey.
5. When we **reached** the station (not had reached the station), we found them waiting for us.

3.3.4 Simple Future or Future Progressive

Let us read the following sentences:

1. My brother **will come** tomorrow.
2. We **shall abide** by the rules of the institute.
3. They **will go** on strike if there is hike in petroleum prices.
4. I **shall visit** my friend since he is not keeping fine these days.

What do you observe? These sentences express an action or a situation that will occur in the future. This tense is formed by using **will/shall** with the **base form** of the verb. So, the basic structure for **simple future/future indefinite tense** is as follows:

Subject + will/shall + Base Form of Verb + Object/Complement/Adjunct

Going back to the past tense, look at the following sentences:

1. We **were waiting** for the class to be over.
2. The children **were playing** cricket in the garden.
3. They **were patrolling** the territory when the bomb went off.
4. When my friend called me, I **was trying** to figure out the cross-word puzzle.

In all the above sentences, the structure that is followed can be grammatically summed up as follows:

> **Subject + was/were + Base Form of Verb + ing + Object/Complement/Adjunct**

The tense that follows this structure is called **past continuous/past progressive tense** and is chosen to represent a progressive action in the past.

Now let us go through the following set of sentences:

1. Praveen **will be going** to Mumbai for cancer treatment.
2. We **shall be completing** our degree next year.
3. My uncle **will be joining** me for dinner tonight.

Future progressive tense is also chosen to express the actions that will take place in future. The basic structure for this tense is as follows:

> **Subject + will be/shall be + Base Form of Verb + ing + Object/Complement/Adjunct**

3.3.5 Future Perfect

Future perfect tense describes an action that will occur in future before some other action. This tense is formed by using **will have** with the **past participle form** of the verb.

1. By the time summer approaches, I **shall have finished** my exams.
2. When their son turns 16, they **will have decided** where they will build a house.

As you can make out from the above sentences, the basic structure of the future perfect tense is as follows:

> **Subject + will have/shall have + Past Participle Form of Verb + Object/Complement/Adjunct**

3.3.6 Present Perfect Continuous

Let us read the following set of sentences:

1. We **have been waiting** for you all day.
2. She is tired. She **has been working** all day.
3. They **have been studying** since 5 o'clock.

What do you observe? In these sentences, actions start sometime in the past and they continue in the present. This is called **present perfect continuous tense**. Since present perfect continuous tense refers to an action that started in the past, but has continued in the present, it needs to be chosen judiciously.

Read the following sentences:

1. We are waiting for you for the last two hours.
2. He is making a fool of us for such a long time!
3. We are not seeing them for the last couple of days.

None of these sentences is appropriate because all these sentences choose present continuous tense while denoting an action that has started in the past and is still continuing at the time of speaking. Hence, always use present perfect continuous tense in such situations.

Observe how the revised version of these sentences seems appropriate:

1. We **have been waiting** for you for the last two hours.
2. He **has been making** a fool of us for such a long time!
3. We **have not been seeing** them for the last couple of days.

Here are a few more examples :

1. It **has been raining** since morning.
2. Though I **have been reading** the novel all afternoon, I have not finished it.
3. I **have been sleeping** in this bed since morning.

The basic structure of present perfect continuous tense is as follows:

> **Subject + has been/have been + Base Form of Verb + ing + Object/Complement/Adjunct**

Remember, mostly there is a mention of time in this tense and it is denoted through either of the following:

(a) **For**—a length of time   (b) **Since**—a point in time

Read the following sentences carefully:

1. I have been working at BITS, Pilani, **since 1994**. (a point in time)
2. Deevan Knitwares has been exporting to China and Korea **since its inception**. (a point in time)
3. It has been snowing in Shimla **for one week**. (a length of time)
4. My brother has been dealing in tyres for the **past twenty-five years**. (a length of time)

3.3.7 Past Perfect Continuous

Read the following sentences:

1. We **had been talking** since morning.
2. I **had been walking** on the road for two hours.
3. The girls **had been dancing** on stage for three hours tirelessly.

The past perfect continuous tense is used to express actions that started and continued over a period of time in the past with some part of the action having been completed.

See the following examples:

1. We found them waiting at 9 a.m. The doctor arrived at 11 a.m. When the doctor arrived, they **had been waiting** for two hours.
2. Suddenly, my car broke down. I was not surprised. It **had** not **been running** well for a long time.

The past perfect continuous tense is generally formed this way:

> **Subject + had been + Base Form of Verb + ing + Object/Complement/Adjunct**

Let us see a few more examples:

1. Milkha Singh was very tired. He **had been running**.
2. I could smell cigarettes. Somebody **had been smoking** here.

3. They went to watch a movie. They **had been working** very hard for their final exams for the past one and a half months.

4. **Had** the pilot **been trying** to give some signal of helplessness before the crash?

3.3.8 Future Perfect Continuous

Read the following sentences carefully:

1. I **will have been working** here for three years by the end of next month.
2. Shane Bond **will have been playing** international cricket for nine years by next March.
3. Siddharth **will have been studying** for tomorrow's exam for eleven hours by seven o'clock next morning.

The Future Perfect Continuous is used when we wish to emphasize on an action that will continue to happen up to a certain time in future.

The basic sentence structure for future perfect continuous tense is as follows:

> **Subject + will/shall + have been + Base Form of Verb + ing + Object/Complement/Adjunct**

See the following examples:

1. She **will have been raising** all speculations by the time she becomes the dean of her unit.
2. By the time she finishes this semester, Susheela **will have been studying** about parasites for four years.
3. By next Thursday, I **will have been working** on this project for three years.

In brief, let us revise what we have learnt so far with the help of the verb *write* (Table 3.3).

Table 3.3 Structure of Verb Phrases

Tense	Simple	Continuous	Perfect	Perfect Continuous
Present	Write(s)	Am/Is/Are writing	Have/Has written	Have been/Has been writing
Past	Wrote	Was/Were writing	Had written	Had been writing
Future	Will/Shall write	Will/Shall be writing	Will have/Shall have written	Will have been/Shall have been writing

Finally, to recapitulate what we have learnt in this section, let us remind ourselves how a sentence can be written in different tense forms (Table 3.4).

Table 3.4 An Action Expressed in Different Tenses

Examples	Tense
I read novels during holidays.	Present indefinite tense
I am reading a novel.	Present continuous tense
I have recently read that novel.	Present perfect tense
I have been reading novels since my childhood.	Present perfect continuous tense

(Contd)

Table 3.4 (Contd)

Examples	Tense
I read that novel last week.	Past indefinite tense
I was reading a novel at that time.	Past continuous tense
I had read that novel before I lent it to a friend of mine.	Past perfect tense
I had been reading that novel before it was stolen.	Past perfect continuous tense
I'll read that novel during my holidays.	Future indefinite tense
I will be reading that novel in a couple of days.	Future continuous tense
I will have read this novel by then.	Future perfect tense
I will have been reading this novel for a month or so.	Future perfect continuous tense

PRACTICE TEST 3.13

Fill in the blanks with the correct form of the verb given within brackets:

1. The Speaker of the House _____ (finish) her term in May next year.
2. The sociologist _____ (examine) the effects that racial discrimination has on society.
3. The explorer _____ (explain) the latest discovery regarding pyramids in Egypt in his research.
4. The leader _____ (vanish) from the city, when protests began against him on the streets.
5. Dr Jahangir _____ (present) his ongoing research on sexist language next week.
6. The researchers _____ (travel) several countries in order to collect more significant data.
7. Women _____ (get) the right to vote in presidential elections of the United States in 1921.
8. By the time the troops arrive, the army on the boarder _____ (spend) several weeks waiting.
9. The CEO _____ (consider) a transfer to Mumbai where profits would be larger.
10. Before annual examination, the students _____ (participate) in many extracurricular activities.

PRACTICE TEST 3.14

Select the appropriate verb from the choices given below and choose their correct forms so as to fit in the blank:
(grow, have, decide, watch, work, paint, rain, write, want, keep, put)

1. I did my homework when I _____ television.
2. Since it _____, we cannot go to beach.
3. Yesterday, I _____ breakfast at 7.30 a.m.
4. Where's my wallet? It was on the table. Probably, somebody _____ it somewhere else.
5. We _____ (not) to bother them. So we stopped asking them stupid questions.
6. After finishing his work he _____ to go out.
7. Don't disturb me I _____ an essay.
8. I'm very tired today. I _____ all day.
9. Unemployment _____ at an alarming rate for the past two years because of recent economic depression.
10. This room was white. Now it is blue. He _____ it blue.

3.4 MOODS OF VERBS

Don't you feel that verbs can also be moody! After all, they express actions. Hence, they have all the swings of temperament! Verbs do have different ways and manners of expressions which we refer to as their moods.

There are three moods of verbs in English. To understand them, look at the following expressions:

1. He speaks wonderful English.
2. Learn from him.
3. If I were him, I would not have been all that fluent.
4. I wish I knew how to speak that well.

The first sentence makes a statement. So we say that the verb is in an **indicative mood**. The verb chooses to be in this mood to make a statement of facts, ask a question, or express a supposition which is taken for granted. Look at the different expressions in which the verb appears to be in the indicative mood:

1. The movie was quite impressive.	(Statement)
2. Have you finished your meal?	(Question)
3. I am a great admirer of Brian Lara.	(Statement)
4. If he is the project leader, I shall blame him for the failure of the project leader.	(Supposition taken for granted; the speaker knows that person referred to as *he* is the project leader)
5. Am I audible to you?	(Question)
6. Mr Smith taught us English.	(Statement)
7. If it rains, I shall stay back.	(Supposition taken for granted)
8. He gets up by six in the morning.	(Statement)

The other most commonly observed mood of verb is **imperative mood**. The verb chooses to be in this mood to express a *command, request, order, caution, exhortation, prayer, entreaty*, etc. Whenever the verb acquires this mood, the subject of verb (you) is omitted. Look at the following sentences and see the different expressions of the verbs in their imperative mood:

1. Please listen to me.	(Request)
2. Don't leave us, please!	(Entreaty)
3. Avoid chewing tobacco.	(Suggestion/Caution)
4. Don't park your vehicle here.	(Order)
5. O God! Give us our daily bread.	(Prayer)
6. Come here.	(Command/Request)
7. Be careful in life.	(Caution)

The third and the most unusual mood of verb is **subjunctive mood**. It is in this mood that the verb chooses to express itself in peculiar grammatical structures. Look at the following sentences and see how unusually they are structured:

1. We recommend that the director be removed.
   (Not *we recommend that the Director should be removed*)
2. I wish I knew her name.
   (Not for the present situation; replaces *I wish I know her name*)
3. He talks to me as though I were his servant.
   (Not *he talks to me as though I am his servant*)
4. I would rather you kept your mouth shut.
   (Not *I would rather you keep your mouth shut*)
5. It is high time we did something about corruption.
   (Not *it is high time we do or should do something about corruption*)

6. If I were to be the captain, I would not make that error.
   (Not *if I was the captain, I would not make that error*)
7. I wish I were a superstar.
   (Not *I wish I am/was a superstar*)
8. The landlord demanded that the boys vacate the house.
   (Not *the landlord demanded that the boys should vacate the house*)

The subjunctive mood of verb is expressive of formal expressions, which is used more frequently in American English. The subjunctive mood of the verb is often chosen in a that-clause following words such as *demand, insist, recommend, suggest, vital, propose*, etc. In such clauses, present subjunctive mood is chosen, for instance,

1. The workers got together and demanded that the union leader be released immediately.
2. The company insisted that the employee withdraw his claims.

The past subjunctive is chosen to express the following:

(i) After the verb *wish* to express a desire contrary to the reality, for instance,

1. I wish I were the Prime Minister of India.
2. She wishes she were a lot more educated.

(ii) After expressions such as *if, as if, as though* to express some improbability, for instance,

1. The rookie behaves as though he were the greatest bowler around.
2. If I were you, I would accept that offer.
3. She tells me as if this were the only thing for me to do in life.

(iii) After expression *it is time/it is high time + subject*, for instance,

1. It Is time we moved back.
2. It is high time, we listened to our children.

(iv) To express a preference in expressions with would rather, for instance,

1. They would rather you kept out of it.
2. She would rather you stayed back home.

PRACTICE TEST 3.15

Tell whether the verbs in the following sentences are in indicative, imperative, or subjunctive mood:

1. I move that Ms Slipslop be appointed Managing Director.
2. Steve craves for Stella as though she were his beloved.
3. Don't speak to me like that!
4. I wish I knew how to drive a car.
5. The sun rises in the east.
6. Save some money.
7. It is high time we left the party.
8. Please don't be so rude!
9. If I were you, I would never speak to her like that.
10. Shut up!

PRACTICE TEST 3.16

Change the following sentences using subjunctive mood of the verb:

1. It is insisted that we should achieve these targets by the year end.
2. If I had been you, I would not make such an error.
3. You talk to me as if I know nothing about the matter.
4. I wish I was not in love with her.
5. It is time that we should teach them a lesson.
6. We would rather prefer that you issue the letter today itself.

7. It is recommended that we should plug the unplugged holes.

8. If I was the monarch of the world, I would never promote corruption.

9. He orders me as though I am his assistant.

10. They would rather you listen to them first.

Read the following sentences:

1. The committee cut the budget.

2. The budget was cut by the committee.

Of course, you can make out that the second sentence is the passive form of the first sentence. As you must know, we make the passive voice of a sentence by interchanging the position of the subject and the object in the sentence. In the above sentence, *the committee* is in the subject position, whereas *the budget* is in the object position. The case is reversed in the second sentence.

What else is done? The agent of the action has been shown with the help of the preposition *by*. Further, the simple past *cut* becomes *was cut* in the passive form.

Let us first see how to convert the active voice into the passive in Table 3.5.

Table 3.5   Converting Active Voice into Passive Voice

Active Form	Tense Employed	Original Structure	Passive Form	Recast Structure	Focuses on
Geeta writes a letter	Present indefinite/ simple present tense	Base form of the verb + s/es with III Person singular	**A letter is written by Geeta**	Is/Am/Are + Past participle form of the verb	**A letter is written**
Geeta is writing a letter	Present progressive/ Continuous tense	Is/Am/Are + Base form of the verb + ing	**A letter is being written by Geeta**	Is/Am/Are + Being + Past participle form of the verb	**A letter is being written**
Geeta has written a letter	Present perfect tense	Has/Have + Past participle form of the verb	**A letter has been written by Geeta**	Has been + Past participle form of the verb	**A letter has been written**
Geeta wrote a letter	Past indefinite/ Simple past tense	Past form of the verb	**A letter was written by Geeta**	Was/Were + Past participle form of the verb	**A letter was written**
Geeta was writing a letter	Past continuous/ Past progressive tense	Was/Were + Base form of the verb + ing	**A letter was being written by Geeta**	Was/Were + Being + Past participle form of the verb	**A letter was being written**

(Contd)

Table 3.5 (Contd)

Active Form	Tense Employed	Original Structure	Passive Form	Recast Structure	Focuses on
Geeta had written a letter	Past perfect tense	Had + Past participle form of the verb	A letter had been written by Geeta	Had been + Past participle form of the verb	A letter had been written
Geeta will write a letter	Simple future/Future indefinite tense	Shall/Will + Base form of the verb	A letter will be written by Geeta	Will be + Past participle form of the verb	A letter will be written
Geeta will have written a letter	Future perfect tense	Will have/ Shall have + Past participle form of the verb	A letter will have been written by Geeta	Will have been + Past participle form of the verb	A letter will have been written

Apart from the other structural changes, an important change in passive from active *is the shift of emphasis*. The last column in the above table illustrates how the focus in the passive shifts *from the doer or agent* of the action *to the action* itself. That is why, in many passive forms, we see the agent of the action being dropped altogether.

Going back to the opening sentence, can you now figure out the difference between these two versions of the same sentence? Look at them once more:

1. **The committee** cut the budget.
   (The focus is on who cut the budget.)

2. **The budget** was cut by the committee.
   (The focus is on what was done.)

Actually, the focus in the passive is generally on the action part. Therefore, in most cases, it can be written without mentioning the agent.

Since this aspect is not usually stressed, we end up using active voice where the passive would be most suitable.

PRACTICE TEST 3.17

Look at the sentences given below and suggest whether it is an active or a passive that is required to suit the occasion:

1. We are sorry to inform you that **we cannot do anything** about your problem.
   (Sounds like a real bad news; prefer passive voice in case the active sounds too blunt)
2. **Accidents kill** thousands of people every year.
   (Normally what happens should be more important in such statements; the focus in this sentence, however, is what/who kills thousands of people)
3. **The newscasters telecast** the news at 8.00 p.m.
   (It is not important who telecast them but the action)
4. **You have not** yet informed me.
   (Sounds blunt and accusatory)
5. **They held** the gathering near the temple.
   (The focus should be on the action and not on the agent)

6. **The nurse has given** the medicine to the patient.
(The fact that the medicine has been given should be highlighted and not who has done that)
7. **The agitators handed** a memorandum to the district collector.
(The action should be stressed in such statements and not the agent)
8. **As we started walking, we saw** a small cab coming from the other direction.
(What was seen should be more important than who saw it)

Note: By suggesting you to prefer using passive to active voice in such situations, we do not intend to suggest that the active versions of these sentences are incorrect. However, the passive voice puts the focus on the right place in the above sentences, hence, recommended by us.

PRACTICE TEST 3.18

Read the following passage and rewrite it choosing passive voice wherever required:

They did not invite me to the party. Still, I was determined to attend the same. So, I reached the place uninvited. What I saw was something really forgettable. What was happening around was strange. Waiters were supplying wine openly to the guests. The guests were consuming that unashamedly. That was not enough. I could also see the young girls dancing with their suitors. The attendants were coming again and again and filling their wine glasses with more wine. Later on, I observed these devotees of god Bacchus inserting their fangs into the non-vegetarian dishes which the other set of waiters were supplying at a frantic pace.

In the above passage, the narrator chooses active voice throughout the description of his/her experience. While seeing the revised version in the Answer Key that mostly chooses a passive voice, you can make out that passive voice suits the occasion much better than active voice.

Further, passive voice is also preferred when we tend to describe a procedure or method.

PRACTICE TEST 3.19

Rewrite the following passage correctly:

Ladies and Gentlemen, I have immense pleasure in demonstrating to you all the functioning of our washing machine. Well, this is how it functions—we first put all the soiled clothes into it. Then we sprinkle the detergent on them. After that, we dip them in a bucketful of water. We cover the lid and turn the machine on. Once we do it, we don't have to worry about what it does with our clothes. You will see that the machine automatically washes them, rinses them, and dries them.

In the above speech, the speaker chooses active voice most of the time. Do you think it suits the occasion?

Passive voice is also preferred when the focus is on a generalized notion. Read the following expressions:

1. **People regard Shakespeare** to be the greatest playwright of all times.
2. **People require water** for everything.

Don't you feel that both these expressions will sound better in the passive voice? Of course, unless otherwise the focus is entirely on **people**. In normal circumstances, you feel like saying the same thing this way:

1. **Shakespeare is regarded** as the greatest playwright of all times.
2. **Water is required** for everything.

In the above sentences, we not only use the passive form of voice but also eliminate the agent when the focus is rested firmly on the action and the agent is irrelevant. Read the following expressions and decide which option you would choose:

1. You should apply by 31st December. (OR)
   Applications must be sent by 31st December.
2. We are providing this service as per our contract. (OR)
   This service is being provided as per our contract.
3. Someone stole my mobile. (OR)
   My mobile was stolen.

Of course, the options in the passive voice seem far more relevant and appropriate.

The passive is usually chosen also for writing sentences with expressions *get/got, it is said/ reported/thought/believed, said to be*, etc. Look at the sentences given below:

1. They got married last June.
   (Not **they married each other last June.**)
2. I got selected for the armed services twice.
   (Not **the armed forces selected me twice.**)
3. These forts are believed to be built in the twelfth century.
   (Not **they are believed to have built these forts in the twelfth century.**)
4. It is said that animals can quite well make out what we want from them.
   (Not **people say that animals can quite well make out what we want from them.**)
5. The prisoner is known to have assaulted the inmate earlier too.
   (Not **we know that the prisoner has assaulted the inmate earlier too.**)

Moving further, let us see another aspect of active and passive voice. We trust you know that some verbs can have two passives. See the following examples:

1. She gave me a book. (Active)
   I was given a book by her. (Passive)
   A book was given to me by her. (Passive)
2. The academy gave the scientist the prestigious award. (Active)
   The scientist was given the prestigious award by the academy. (Passive)
   The prestigious award was given to the scientist by the academy. (Passive)

On the other hand, certain expressions with verbs *lack, resemble, suit, let*, etc. cannot be converted into passive. See the following examples:

1. I have three jackets in the wardrobe.
   (Not **three jackets are had by me in the wardrobe.**)
2. The dress does not suit you.
   (Not **you are not suited by the dress.**)
3. He resembles his father.
   (Not **his father is resembled by him.**)
4. Our new boss lets us have a little chat in the office.
   (Not **we are let to have a little chat in the office.**)

But see the following example:

We are allowed to have a little chat in the office.

(The verb **allow** can have a passive form but not the verb **let**.)

PRACTICE TEST 3.20

Rewrite the following sentences using passive voice:

1. We advise the patients of swine flu to wear a mask.
2. We can prove that Darwin's theory has some chinks in the armour.
3. They will organize the function in the auditorium.
4. You have not yet reported to me.
5. The attendants change bed sheets everyday in this hotel.
6. They pay me monthly.
7. The producers are likely to offer a staggering signing amount to the star.
8. For many centuries we did not know that plants can breathe.
9. Translators the world over have translated Premchand's stories into different languages.
10. People mostly consider learning a language to be a difficult task.
11. In the meeting, the members agreed that new norms be brought into effect.
12. The minister announced that he will give pension to all the old people in the state.
13. They shifted a large number of victims to the relief camps.
14. Dr Samuel Johnson compiled the first English dictionary.
15. The government has announced that it will introduce the new act in the next parliamentary session.
16. The captain desired that they should play the game in the positive spirit.
17. The glass broke as you walked over it.
18. I always service my car at this service station.
19. Earlier people believed that AIDS spreads by touch.
20. Fans expected our team to win, but we lost.

PRACTICE TEST 3.21

Read the following sentences carefully. Change the voice wherever it is required:

1. This mistake should not be repeated.
2. The new fast bowler lacks the ability to turn the bowl.
3. It is alleged that he killed his wife.
4. It looks simple but interesting.
5. You cannot do much about it.
6. The government nationalized banks in 1970s.
7. Read the instructions carefully before you sign the document.
8. Today, he seems to be in a good mood.
9. He has become more arrogant ever since he has brought a brand new car.
10. We will conduct the workshop at the main centre.

3.6 DIRECT/INDIRECT SPEECH

To begin with, let us see what Julia's teacher announces in her class:

Girls, on 15 December, we are going to organize a speech competition. Those of you who are interested in delivering a speech may give their names to me. Girls, you can choose any topic for your speech. You can give your speech on a great person, great event, great achievement, great movement, or any other thing that interests you. You will be required to speak at least for five minutes and the maximum time you can take will be seven minutes. Those of you who come first, second, and third would be given attractive prizes.

Now, Julia has to inform her mother about what the teacher told her. Read her version carefully and find out where she goes wrong. This is what she told her mother:

> Mom, **our teacher today said** that on 15 December, **we** are going to organize a speech competition. **Those of you** who are interested in delivering a speech may give **their names to me**. **You can** choose any topic for **your speech**. She also told **you can** give **your speech** on a great person, great event, great achievement, great movement, or any other thing that interests **you**. **You will be** required to speak at least for five minutes and the maximum time **you can take** will be seven minutes. **Those of you** who come first, second, and third would be given attractive prizes.

What do you think of Julia's version of the announcement? She did try well but at certain places, she went wrong. The expressions highlighted above are to be changed. Let us start from the very beginning. In indirect speech, *told* replaces *said* if the reporting verb has an object. Since the announcement was made to the students by the teacher, Julia should have chosen *our teacher told us* instead of our *teacher said*.

Further, in indirect speech, the pronouns used are changed according to the subject of the verb in the reporting speech; *we* should be replaced by *they* or *the school*. Similarly, some other changes which should have been employed are as follows:

Those of you	should be replaced by	**those of us**
Their names to me	should be replaced by	**our names to her**
You can	should be replaced by	**we can**
Your speech	should be replaced by	**our speech**
You	should be replaced by	**us**
You will be	should be replaced by	**we will be**
You can take	should be replaced by	**we can take**
Those of you	should be replaced by	**those of us**

This is how actually Julia should have informed her mother about the announcement made by her teacher:

> Mom, our teacher today told us that on 15 December, the school is going to organize a speech competition. Those of us who are interested in delivering a speech may give our names to her. We can choose any topic for our speech. She also told us that we can give our speech on a great person, great event, great achievement, great movement, or any other thing that interests us. We would be required to speak at least for five minutes and the maximum time we can take would be seven minutes. Those of us who come first, second, and third would be given attractive prizes.

Hence, while using indirect speech, we change the tense, person, and time according to the subject. To learn further, let us see how the changes are affected in indirect narration:

Direct narration

1. Mildred said, 'I am busy today.'
2. The teacher said, 'God is everywhere.'
3. 'You have come first in your class,' my friend told me.
4. 'I am feeling hungry,' the little boy said to his mother.
5. Jack said to Rose, 'You look nice in this dress.'

6. Wilber said to Agatha, 'I am leaving tomorrow.'
7. 'You shall come back by eleven,' my father told me.
8. 'It may rain this evening,' said Jane to Jasper.
9. Kate said to me, 'I know the way.'
10. Scott's mother told him, 'You must not forget your tiffin in your class.'

Indirect narration

1. Mildred said that she was busy that day.
   (**I am** becomes **she was** and **today** becomes **that day**.)
2. The teacher said that God is everywhere.
   (Universal statements are not changed in indirect structure/narration.)
3. My friend told me that I had come first in my class.
   (**You** becomes **I**; **your** becomes **my**. **Have** becomes **Had**. **Have** can be retained if the statement is still relevant.)
4. The little boy told his mother that he was feeling hungry.
   (**I am** becomes **he was**; **said to** becomes **told**.)
5. Jack told Rose that she looked nice in that dress.
   (**Said to** becomes told; **you look** becomes **she looked**; **this** becomes **that**.)
6. Wilber told Agatha that he was leaving the next day.
   (**Said to** becomes **told**; **I am** becomes **he was**; **tomorrow** becomes **the next day**.)
7. My father told me that I should come back by eleven. (OR) My father asked me to come back by eleven. (OR) My father advised me to come back by eleven.
   (Instead of quoting the exact words we can report the meaning in our own words. Hence, **asked/ cautioned/suggested/advised** are also possible.)
8. Jane told Jasper that it might rain that evening.
   (**Said to** becomes **told**; **may** becomes **might**; and **this evening** becomes **that evening**.)
9. Kate told me that she knew the way.
   (**Said to** becomes **told**; **I** becomes **she**; **know** becomes **knew**.)
10. Scott's mother asked/warned him not to forget his tiffin in his class.
    (The sense of the sentence can make us choose **warned** and **asked** in place of **told**.)
    (OR)
    Scott's mother told him that he must not forget his tiffin in his class.
    (**Said to** becomes **told**; **you** becomes **he**; **your** becomes **his**.)

Remember that some verbs should be followed by an indirect object. For instance,

1. The principal **promised us** a holiday.
   (Not 'The principal promised a holiday.')
2. The chairman **informed us/everyone/members** that the meeting would start at three.
   (Not 'The chairman informed that the meeting would start at three.')
3. His teacher **told him** that he should be serious in the class.
   (Not 'His teacher told that he should be serious in the class.')

Remember that if the direct speech is in the simple past, it is changed to past perfect. For instance,

I said to my friend, 'I **saw the Taj** when I was eight years old.'

would be changed into

I told my friend that I **had seen the Taj** when I was eight years old.

The use of past perfect in place of simple past indicates that the monument was seen further in the past. However, when it becomes clear that something happened long before, past perfect is not required. Also when the reference is to something improbable or unreal, simple past is chosen.

See the following examples:

1. 'I felt awful,' she told me.
   She told me that she felt awful.
2. She said to the police, 'If I knew, I'd tell you.'
   She told the police that if she knew, she would tell them.
3. 'I wish I had a girlfriend!' Stephen said.
   Stephen wished he had a girlfriend.

If the reported speech puts a question, question verbs such as **ask**, **enquire**, **wonder**, **want to know**, etc. can be used. If the reported speech starts with a yes/no question, **if** or **whether** is used in the indirect narration.

See the following examples:

1. 'Is some more time left?' spoke the student.
   The student asked if some more time was left.
2. 'Who is there?' she spoke in a surprised voice.
   She was wondering who was there.
3. He asked, 'Lisa, do you have any answer?'
   He enquired/wondered if Lisa had any answer.

To put a request or order into indirect speech, the verbs **tell** and **ask** are commonly chosen. For example,

1. 'Please go out of the room,' the teacher said to the students.
   The teacher asked the students to go out of the room.
2. 'Would you move a little to your left,' he said to his fellow passenger.
   He asked/requested his fellow passenger to move a little to his left.
3. 'Just keep quiet,' the teacher told us.
   The teacher asked us to keep quiet.

Other modal verbs, such as **must, should, have to**, etc., are also used to express the intended emotion.

See the following examples:

1. 'Please let me go,' the little child said to the rogue.
   The little child pleaded with the rogue to let him go.
2. 'Please stand in a queue,' the manager said to the candidates.
   The manager asked the candidates to stand in a queue.
3. 'Do you mind if I smoke?' her boss said.
   Her boss asked if he could smoke. (OR) Her boss sought her permission before smoking in her presence/at her place.

Since the focus is on the emotion or idea expressed, we can certainly use an alternative pattern while using the indirect speech. Look at the following examples:

1. 'I am sorry,' he said.
   He apologized.
2. 'I am not going to help you,' he told me.
   He refused to help me.
3. 'Well done for playing so well,' the coach said to his players.
   The coach complimented the players on playing well.
4. 'You have to reach office in time!' the director told the officials.
   The director insisted that the officials reach office in time. (OR) The director asked his officials to reach office in time.

5. 'Be careful,' the road is meandering.

He warned/cautioned us that the road was meandering. (OR) He warned about the meandering road that was ahead.

6. 'I am sorry, I lost your notebook,' he said.

He regretted losing my notebook.

Thus, it becomes really important for us to correctly recapture the mood, tone, and tenor in indirect speech.

PRACTICE TEST 3.22

Given below is an exercise to find out whether you can guess the emotion/idea appropriately. Hence, choose the verb that you feel would adequately express the idea in the context:

1. He told/warned us not to be rude with him next time.
2. The plumber warned/informed us that there was a leakage in the pipeline.
3. My doctor advised/wanted me to cut down on sweets.
4. The clerk refused/advised to help us.
5. The teacher requested/asked us to be silent.
6. She cautioned/reminded me that it was our marriage anniversary that day.
7. He criticized/complained us for being sloppy in our attitude.
8. He reassured/reminded me that we would get a seat.
9. Roberts worried/wondered why he should always be neglected for promotion.
10. The captain warned/thanked the players for winning him the crucial match.

PRACTICE TEST 3.23

Change the following expressions into indirect speech:

1. 'Why do you always vex me like that?'
2. 'May we leave now?'
3. 'I'm afraid we can't do anything about that.'
4. 'Don't worry; I'll help you.'
5. 'We are not going to speak to the journalists.'
6. 'Be careful; he is a clever fellow.'
7. 'Why can't I get the same appreciation?'
8. 'Thank you very much.'
9. 'Let's watch a movie.'
10. 'You really need to see a doctor.'

Note: There may be more than one way of converting these sentences into an indirect speech.

▮ 3.7 CLAUSE AND ITS TYPES

In simple words, a clause is a group of words that forms part of a sentence and has a subject and a predicate of its own. Read the following expression:

Take this or leave it.

In this sentence, we have two different sentences:

1. Take this.

2. Leave it.

Now, both these sentences are combined with a coordinating conjunction *or*. Hence both—*take this* and *leave it*—are coordinate clauses. They both can stand on their own and are thus **independent clauses**. Now, look at this sentence:

Take an umbrella because it is going to rain.

In this sentence we have two clauses—**take an umbrella** and **because it is going to rain**. Out of these two clauses, *take an umbrella* is an **independent clause**. It can stand on its own. The other clause—*because it is going to rain*—has to depend on the other clause. Thus, it is a **subordinate clause**.

Therefore, clauses can be divided into two broad categories:

1. Independent or main or principal clause
2. Dependent or subordinate clause

Dependent clauses can be of various types. Read the following sentences and see how the dependent/subordinate clause can be distinguished from the main clause:

1. When I read her letter, I realized my mistake.
2. Since you say so, I must look into the matter.
3. She lives as Americans do.
4. I could not buy the ticket as I had no money.
5. You may go wherever you like.
6. Though he is rich, he has no friends.
7. As soon as the thief heard the noise, he jumped over the fence.
8. If I see him, I will inform him.
9. I don't know what you are trying to prove.
10. This is the place where the battle of Panipat was fought.

This is how you can separate the main clause from the subordinate clause (highlighted in bold):

1. **When I read her letter**, I realized my mistake.
2. **Since you say so**, I must look into the matter.
3. She lives **as Americans do**.
4. I could not buy the ticket **as I had no money**.
5. You may go **wherever you like**.
6. **Though he is rich**, he has no friends.
7. **As soon as the thief heard the noise**, he jumped over the fence.
8. **If I see him**, I will inform him.
9. I don't know **what you are trying to prove**.
10. This is the place **where the battle of Panipat was fought**.

Dependent clauses can function as nouns, adjectives, and adverbs. Therefore, we can name them noun clause, adjective clause, and adverb clause. Now let us know about such clauses a little more so that we may sharpen our writing skills in a systematic manner.

3.7.1 Noun Clauses

Read the following sentences and identify what kind of function the subordinate clauses perform:

1. The child wondered **if his parents bought him what he wanted for Diwali**. (Object of the verb *wondered*)
2. **What I want for dinner** is a pizza. (Subject of the predicate)
3. The stranger told us **how he escaped the militants' clutches**. (Object of the verb *told*)
4. Vacation is **what I need most**. (Complement of the subject *vacation*)
5. Give it to **whoever requires it the most**. (Object of the preposition *to*)

So you can notice that all the above subordinate clauses are functioning as subjects, objects, or complements. All these clauses are noun clauses.

 A noun clause can replace any noun in a sentence, by functioning as a subject, object, or complement.

3.7.2 Adjective Clauses

Let us read the following set of sentences and see what kind of function the subordinate clauses perform here:

1. I listened to the song that you told me about.
2. The novel that won the Pulitzer Prize had not sold well when it was first published.

In the first sentence, the clause 'that you told me about' functions as the modifier of the noun phrase 'the song'. Similarly, in the second sentence, the clause 'that won the Pulitzer Prize' can't stand by itself and modifies the noun phrase 'the novel'. Such dependent and modifying clauses are known as adjective clauses.

 An adjective clause mostly modifies a noun or a pronoun that appears in the subject or object position of a sentence.

See a few more similar examples:

1. The building that they built on Juhu Beach in Mumbai is worth seven thousand dollars.
2. The ceremony, which several celebrities attended, received intense coverage.

Now read the sentences given below:

1. I'll do the laundry when I'm out of clothes.
2. Radha brushed her long hair while she waited for her husband to come.

In the above two sentences, we observe that both the subordinate clauses are an answer to the 'when' part of the action. Since they function as adverbs, they are called adverb clauses.

 Like all adverbials, adverb clauses express when, where, why, and how something occurs. A dependent clause is an adverb clause if it can be replaced by an adverb.

We hope by now you must be in a position to distinguish the subordinate clause from the main clause. Go ahead and identify the subordinate clauses in the following sentences:

1. We asked whomever we saw for a reaction to the play.
2. We asked whoever called us to call back later.
2. They gave the money to whoever presented the winning ticket.
4. While we saw fumes in the air, we drove away as quickly as we could.
5. The group of tourists decided to have lunch in the village because the van needed repairs.

This is how you can separate the main clauses from the subordinate ones. The clauses highlighted below are the subordinate clauses:

1. We asked **whomever we saw** for a reaction to the play.
2. We asked **whoever called us** to call back later.
3. They gave the money to **whoever presented the winning ticket**.
4. **While we saw fumes in the air**, we drove away as quickly as we could.
5. The group of tourists decided to have lunch in the **village because the van needed repairs**.

3.7.3 Adverbial Clauses

An adverbial clause is a subordinate clause that acts as an adverb in a sentence. It may denote time, place, purpose, condition, concession, cause, reason, or result. See the following sentences:

1. My daughter was playing with her friends **while I was busy in the kitchen**. (Adverbial clause of time)
2. **Since it was raining heavily**, I could not attend my friend's wedding. (Adverbial clause of reason)
3. **When the boss entered the conference hall**, all stood up and clapped. (Adverbial clause of time)

4. **Although what you say might be right**, I can't change my decision. (Adverbial clause of concession)
5. **If you go to Delhi**, bring some new books for me. (Adverbial clause of condition)
6. Sit **wherever you like**. (Adverbial clause of place)
7. Take medicine **so that you get well soon**. (Adverbial clause of purpose)

3.7.4 Relative Clauses

A relative clause is a subordinate clause that begins with a question word (e.g., who, which, where) or that. You can use it to modify a noun or a pronoun. A relative clause helps to identify or at times gives more information about a noun or a pronoun in a sentence.

Read the following set of sentences:

1. Students who have critical listening and thinking skills often achieve good academic results.
2. There is a new novel that highlights the pangs of the recent Mumbai terrorist attack.
3. A university is a place where students can pursue advanced studies in specific disciplines.
4. This is the banquet hall in which my niece's marriage will take place next week.

Words such as **who**, **that**, and **when** are often referred to as relative pronouns when they are used to introduce relative clauses.

They are used for the following:

Who	→	for people
Which	→	for things
That	→	for both people and things.
Whom	→	as the object of a relative clause (in more formal English), though it is increasingly common to replace it with **who**.
Whose	→	to indicate possession, as a determiner before nouns.

See the following examples:

1. Can you tell me the name of the person who first landed on the moon?
2. All the students whose registration was done last month will continue attending the classes.
3. In this conference I will be meeting Professor David Crystal, a renowned Professor of Linguistics whom I wanted to meet for a long time.

Types of relative clauses   There are two types of relative clauses:

- Defining (or restrictive)
- Non-defining (or non-restrictive)

Defining/Restrictive clauses   We use a defining (or restrictive) relative clause to 'identify' or 'restrict the reference of' a noun. We do not separate it from the rest of the sentence by commas (in text) or pauses (in speech).

Let us read the following sentences:

1. The student who scores highest marks in this essay competition will get a cash prize of five thousand rupees.
2. The computer games and movies that involve fighting leave a negative impact on young children.

Non-defining/Non-restrictive clauses   We use a non-defining (or non-restrictive) relative clause to provide additional information about the noun, whose identity or reference is already established.

1. Albert Einstein, **who put forward the theory of relativity**, is considered by many as the most intelligent person in human history.
2. The CIEFL, **which provides opportunities for advanced research in teaching of English**, is located in Hyderabad.

 You should not use the relative pronoun *that* in non-defining relative clauses.

Reduction of relative clauses You can sometimes reduce a relative clause to create a more concise style.

Read the examples given below:

1. The training session (that was) arranged for middle level managers has been postponed for fortnight.
2. The girl (who/whom) you met at the party last night is my cousin.

PRACTICE TEST 3.24

Combine the following sentences by using a coordinating or subordinating conjunction:

1. You invited me.	I came.
2. Children played.	Mother slept in the afternoon.
3. Walk fast.	You should miss the train.
4. Give me this purse.	I shall snatch it away from you.
5. We found the lock broken.	We reached back home.
6. They tried to win the race.	They lost the race.
7. You are a genius.	You cannot solve this puzzle.
8. Keep quiet.	Get lost.
9. He is a complete idiot.	His brother is a wizard.
10. I can do this.	I won't do this.

PRACTICE TEST 3.25

Choose the adverbial clauses in the following sentences and identify their types:

1. Before you leave, tell me the whole story.
2. Since you say so, I must believe it.
3. If I want it, I'll let you know.
4. Search for it, where you kept it.
5. Don't turn on the television, until you finish your homework.
6. Though the exercise seems difficult, it has its own advantage.
7. They were asked to wait till they fainted in the sun.
8. Had I seen him, I would have informed him.
9. Some of us live so that we may accumulate more wealth.
10. Since you were not there, I left the letter with your sister.

PRACTICE TEST 3.26

Complete the following sentences with a clause:

1. Since we wanted to watch a movie, _____.
2. I went to Delhi so that _____.
3. We discussed our problems with our teacher _____.

4. _____, the teacher had already started teaching the lesson.
5. They tried their best _____.
6. The man _____ was my uncle.
7. Give me some money, otherwise _____.
8. The officer asked me not to go home unless _____.
9. Had they worked hard, _____.
10. Pay your taxes in time or _____.

3.8 USING NON-FINITES

Read the following sentences:

1. I **play** in the evening.
2. I **have been playing** since morning.
3. I **was playing** yesterday.
4. I **will play** tomorrow again.
5. In the morning, I **played** for more than two hours.
6. I **had played** for half an hour or so before the bell rang.

In all the above sentences, the verb *play* has been contextualized with the help of the auxiliary verbs *have been*, *was*, *will*, *had*, etc. These auxiliaries are chosen according to the tense that is intended to be communicated. That is why, when the purpose is to suggest a continuation of an action over a period of time in the present, *have been* playing has been used.

For a continuous action in the past, *was playing* has been chosen and in order to show future, *will play* is written. And for a routine action, simply *play* is chosen. Similarly, for some action that took place in the past, *played* is used and finally to refer to the earlier action (of playing) in the past, in comparison to the later one (the ringing of the bell) *had played* is used. Therefore, we see the verb *play* being used in different forms—*play*, *have been playing*, *was playing*, *will play*, *played*, and *had played*. Thus, *play* is a finite verb. A finite verb changes according to the tense that it has to denote. Now, to see the other side of the coin, look at the following sentences:

1. I plan to play in the evening.
2. I have been planning to play since morning.
3. I was planning to play yesterday morning.
4. I will plan to play tomorrow again.
5. In the morning, I planned to play with my friend.
6. I had planned to play before the bell rang.

Do you observe how the verb *to play* remains as such throughout all the above sentences? It has not changed its construction regardless of the tense employed, whereas the other verb *plan* in these sentences keeps on changing as it becomes *plan*, *was planning*, *have been planning*, *will plan*, *had planned*, etc. In this sentence, *plan* is a finite verb, whereas *play* is used as a to-infinitive. Thus the verb that gets changed according to the tense and time is a finite verb and the one that does not get affected is a non-finite verb.

It is not only the tense and time but also person or subject of the verb that forces the change in the form of a finite verb. Look at the following examples:

1. I want to play.
2. You want to play.
3. Ann wants to play.
4. Hughes wants to play.
5. Clarke and Lee want to play.

See, how *to play* stays as such in all these sentences. It simply disregards whether the subject of the sentences is *I*, *You*, *Ann*, *Hughes*, or *Clarke and Lee*. The other verb *want*, however, keeps changing according to the subject that propels it. For the first person singular pronoun *I*, the second person singular/plural pronoun *you* and the third person plural nouns *Clarke and Lee* it remains *want*, whereas for the third person singular nouns *Ann* and *Hughes* it gets changed into *wants*. No such changes are noticed in the verb *to play* which remains the same throughout. Thus, the non-finite verb does not change according to the tense, time, or person, whereas the finite verb changes.

There are three types of non-finite verbs—**infinitive**, **gerund**, and **participle**. Let us see the following set of sentences to understand them in detail:

1. My friend, John, wants **to swim**.
2. Not just that, John simply loves **swimming**.
3. **Swimming** is his absolute passion.
4. His girlfriend, Jane, however dislikes **swimming**.
5. Because of this, Jane never learnt how **to swim**.
6. So, whenever Jane finds him **swimming**, she goes nuts.
7. This does not affect John who just keeps **swimming**.
8. Even yesterday, Jane found John in the **swimming** pool.
9. When she came in the morning, John was **swimming** and when she left in the afternoon, he was still **swimming**.
10. It became too much when she found him **swimming** again in the evening.
11. Angry and distraught, Jane decided to walk out on John, leaving him in the sweet company of his not so sweet hobby—**swimming**.

Probably, the expression *to swim* in the first and fifth sentences is not a problem now. It is similar to the expression *to play* that occurred in the examples that preceded this set of sentences. So *to swim*, like *to play*, is a **to-infinitive**. It has a *to* and the base form of the verb. So the expressions, such as *to swim*, *to dance*, *to speak*, *to kill*, and so on and so forth are examples of **to-infinitives** which consist of *to* and the base form of the verbs. In the other type of infinitive, only the base form of the verb is used. Such infinitives are known as plain/bare infinitives.

For example:

They made me **laugh**.

In all the remaining sentences, the word *swimming* occurs.
Let us see all the remaining sentences one by one:

1. Not just that, John simply loves **swimming**.

The verb *loves* in this sentence answers *what John loves*, i.e., *swimming*. If you see it carefully, this sentence has a structure which is similar to the structure of the sentences given below:

1. John loves her. 2. John loves coffee. 3. John loves this house.

Now, what are these words—*her, coffee, this house*? These are nouns and pronouns. Hence, in the sentence *John loves swimming*, the word *swimming* replaces other nouns or pronouns such as *her, coffee, this house*, etc. Seen thus, the class of the word *swimming* appears to be similar to that of a noun or a pronoun. However, *swimming* is not exactly like *her, coffee*, and *this house* because it embodies some action *swim* also, which other words such as *her, coffee*, and *this house* do not embody. Therefore, *swimming* in this sentence is a gerund which though appears to be a verb but is used as a noun.

Further, since a noun or a pronoun can take the subject and the object positions in a sentence, the word *swimming* in both the second and the third sentences takes exactly these positions. So, we write the following:

1. John loves swimming.
   {John (subject) loves (verb) swimming (object)}

2. Swimming is his absolute passion.
   {Swimming (subject) is (verb) his absolute passion (subject complement)}

Similarly, in the fourth sentence observe that the word *swimming* becomes the object as it answers what Jane, the subject, dislikes:

His girlfriend, Jane, however, dislikes swimming.
{(His girlfriend, Jane (subject), however dislikes (verb) swimming (object)}

In the sixth sentence, however, the word *swimming* is used in an entirely different way. Look at it:

So, whenever Jane finds him swimming, she goes nuts.

The word *swimming* here seems to qualify the object *him* doing something. The word *swimming* in this sentence has the qualities both of a verb, as it denotes some action, and that of an adjective, for it also acts as a modifier. This is what a *participle* is. Look some more examples of participles:

1. **Thinking** all was safe, she decided to take risk.
2. **Driven** by passion and desire, Dr Faustus falls headlong into the hell.
3. **Studded** with armour, he approached the battlefield.
4. **Pining** for more, man keeps **hankering** after material possessions.

Look carefully to observe that all the words here—*thinking, driven, studded,* and *pining*—suggest some action but at the same time, they also qualify the subjects—*she, Dr Faustus, he,* and *man*, respectively. So, the words *thinking, driven, studded,* and *pining* are participles. *Thinking* and *pining* are present participles and *driven* and *studded* are past participles. *Hankering* in the last sentence, however, functions as a gerund.

In the next sentence, the word *swimming* is again a gerund. In the eighth sentence the word *swimming* qualifies *pool* and hence is a **participle adjective** that is formed from verbs with –ing and –ed or –en endings; for example, interesting story, tired face, broken toy, etc. and act as adjectives. In the ninth sentence, *swimming* is used twice and both the times, it is a **pure verb** in the progressive tense. In the tenth sentence, swimming is used as a participle and in the last sentence, it is a gerund.

Believing that you now can make out the difference between a gerund, a participle, and a to-infinitive, we put you through the task of making such distinction. Go ahead and identify the non-finites, namely gerunds, participles, and to-infinitives in the following sentences:

1. Believing people seldom doubt.
2. This is not the time to fight with each other.
3. Singing gives me immense pleasure.
4. Crooning an old tune, she picked up the phone.
5. We are terribly sorry to hear this.
6. Beating the door wildly, the girl screamed for help.
7. He is fond of playing cricket.
8. Jumping over the fence, the thief escaped.
9. He saw a clown standing on his head.
10. His dejected look won't make him a good public speaker.

See how we can make out these three non-finites:

1. **Believing** people seldom doubt.
   (Those people who believe qualifies people besides showing action, hence present participle)
2. This is not the time **to fight** with each other.
   (to + the base form of the verb, hence to-inifinitive.)

3. **Singing** gives me immense pleasure.
   (What gives me pleasure—singing—a verbal noun, hence gerund)
4. **Crooning** an old tune, she picked up the phone.
   (She picked up the phone doing what—crooning; hence *crooning* qualifies *she* besides denoting action; hence a verbal adjective; therefore, present participle)
5. We are terribly sorry **to hear** this.
   (to + base form of the verb, hence *to hear* is a to-infinitive)
6. **Beating** the door wildly, the girl screamed for help.
   (*Beating* qualifies the person besides denoting action hence adjective verb—participle)
7. He is fond of **playing** cricket.
   (He is fond of what—*Playing*. Hence, *playing* is an object + verb = noun + verb = gerund)
8. **Jumping** over the fence, the thief escaped.
   (Adds to the meaning of the action by qualifying the person doing it; hence, a participle)
9. He saw a clown **standing** on his head.
   (The clown was seen doing what—standing on his head. Hence, a participle)
10. His **dejected** look won't make him a good public speaker.
    (What type of look—dejected; hence participle adjective)

PRACTICE TEST 3.27

Figure out the gerunds, participles, to-infinitives, and verbs in progressive tense in the following expressions:

1. Making a lame excuse, he sounded quite cheap.
2. Children love making mud castles.
3. Seeing is believing.
4. Seeing, he believed.
5. Waving their hats and handkerchiefs, the crowd cheered the king.
6. Asking questions is easier than answering them.
7. Making food is not difficult.
8. You are making a fool of yourself.
9. Making tea, she answered the call on her cell.
10. He died in the waiting room.
11. Waiting is often painful.
12. He died waiting for the train.
13. King Lear decides to divide his kingdom.
14. The thief entered through the broken widow.
15. We live to succeed.

At times, we make errors in using participles. Read the following sentences and decide if participles are correctly chosen:

1. **Crying** over a broken toy, the mother consoled her child.
2. **Blinded** by the fog, the bike rammed into a tree.
3. **Sitting** in the park, a snake bit him.
4. **Driven** by greed, valuables were pinched.
5. **Upon landing** at the airport, the company sent a Toyota to pick her up.
6. **Being corpulent and overweight**, the hill could not be mounted.
7. **Having flooded** the street, she could not reach home.
8. **Going** up the hill, the old ramparts became visible.
9. **Entering** the room, the darkness was quite blinding.
10. **Being built** on the rock, the experts predicted that the building would not settle.

Can you figure out what goes wrong with these sentences? Actually, the participle beginnings such as **crying over the broken toy, blinded by the fog, sitting in the park, driven by greed, upon landing, being corpulent and overweight, having flooded, going up the hill, entering the room**, and **being built on the rock** are all modifying in nature. They all add to the meaning of the word that they modify.

Now, as a norm, the modified and the modifier need to be placed close to each other to avoid ambiguity in expression. For example, the participle beginning—*crying over a broken toy*—is followed by *the mother*. Now, is it the *mother* who is *crying over the broken toy* or her *child*? Of course, the child.

Similarly in the third sentence, it appears as though it were a *snake* that was *sitting in the park*! Look at the sixth sentence. It seems that the *hill is corpulent and overweight* and hence it *cannot be mounted*! Thus, the wrong positioning of the participle beginning makes these sentences amusingly ambiguous. Therefore, remember to keep the participles close to the words they are required to modify. The above sentences can be recast in a meaningful manner as follows:

1. The mother **consoled her child who was crying over a broken toy**.
2. **As the biker was blinded by the fog**, the bike rammed into a tree.
3. A snake bit him **while he was sitting in the park**.
4. **Driven by greed, they** pinched the valuables.
5. **Upon her landing at the airport**, the company sent a Toyota to pick her up.
6. **As they were corpulent and overweight**, they could not mount the hill.
7. **As the street was flooded, she could** not reach home.
8. **As we went up the hill**, the **old ramparts** became visible.
9. **As we entered the room, the darkness** was quite blinding.
10. The experts predicted that the **building would not settle because it was being built on the rock**.

Wrong placement of participle is not the only type of error that we commit while dealing with non-finites. At times, gerunds and to-infinitives are wrongly used in place of each other making the entire grammatical structure go haywire. Look at the following sentences and decide whether they are grammatically correct:

1. The school authorities failed in informing the parents of the sick child.
2. Avoid to copy others!
3. When he realized that smoking causes cancer, he stopped to smoke.
4. The rich refuse accepting the poor in their fold.
5. To err is human; still, we should avoid to make the same error in life.
6. When you finish to read the book, would you lend me that for a couple of days?
7. We hope achieving the target by the year end.
8. If you have some doubt, don't hesitate calling me.
9. They have managed completing their task in time.
10. Though he has to stay in a city, he has always missed to be in his village.

Can you make out what went wrong? Gerunds and to-infinitives have been wrongly chosen in these constructions. In fact, there are some verbs, such as *agree, manage, aim, demand, fail, offer, decline, want, wish, prepare, tend*, etc., which are normally followed by a *to-infinitive*, that is, to + verb's base form.

On the other hand, there are certain other verbs, such as *avoid, dread, deny, feel like, imagine, consider, suggest, consider*, etc., which are usually followed by a *gerund*, that is, –ing with the base form of the verb. The above sentences are rewritten correctly as follows:

1. The school authorities **failed to inform** the parents of the sick child.
2. Avoid **copying** others!
3. When he realized that smoking causes cancer, he **stopped smoking**.
4. The rich **refuse to accept** the poor in their fold.
5. To err is human; still, we should **avoid making** the same error in life.

6. When you **finish reading** the book, would you lend me that for a couple of days?
7. We hope **to achieve** the target by the year end.
8. If you have some doubt, don't **hesitate to** call me.
9. They have managed **to complete** their task in time.
10. Though he has to stay in a city, he has always **missed being** in his village.

Even when some of the non-finite verbs such as *come, try, regret, remember, stop,* etc. take both **to-infinitive** and **–ing forms**, they don't mean the same thing. Look how the meaning communicated with to-infinitive is different from the one communicated through –ing construction in the following sentences:

1. He **stopped to take** a cup of tea on the way.
   (He stopped so that he could take a cup of tea.)
2. He **stopped taking** a cup of tea in the morning.
   (The practice of taking a cup of tea in the morning was stopped.)
3. The child **came running** with a kite in his hand.
   (It describes the manner in which the child came, i.e., running.)
4. Every child **has to come** to realize that he/she (or it) cannot afford to be a child any longer.
   (It highlights the gradual change that takes place in every child's life.)
5. **Remember to** take your umbrella if you go out; it is going to rain today.
   (The action of remembering precedes the other action.)

Avoid being Nonsensical with Non-finites!

6. I don't **remember having seen** him all these days.
   (The other action precedes the action of remembering.)
7. **Stop telling** me that you are always right!
   (It focuses on what needs to be discontinued.)
8. **He stopped to have a closer** look at the report before resuming further.
   (It highlights an action completed in the past.)
9. I **regret having spoken** to you so rudely at that moment.
   (Not happy about what the speaker had done in the past.)
10. We **regret to inform** you that despite our best efforts, we are unable to accommodate your request.
    (Not happy about what the speaker regrets doing now.)

PRACTICE TEST 3.28

Find out whether the expressions written in bold in the following sentences act as gerunds, participles, participle adjectives, or verbs:

1. **Challenging** the enemy, he threw his handkerchief on the floor.
2. **Challenging** someone's beliefs is not always pleasant.
3. He was chucked out for he was **challenging** the authorities.
4. **Speaking** over the phone, he sounded like a moor.
5. Don't disturb, I am **speaking** to my boss over the phone.

6. He avoids **speaking** to me directly.
7. We don't need inarticulate fools, we require **speaking** people.
8. **Reading** is a wonderful habit.
9. **Reading** newspaper in the morning is a matter of routine for many.
10. The parents were shocked to find the child **reading** an erotica.
11. Last night, I broke my **reading** glasses.
12. Men are often seen **reading** at the breakfast table while women are seen **talking** incessantly.
13. **Seeing** a snake in the room, she shouted at the top of her voice.
14. You cannot avoid **seeing** the truth.
15. Why are you **telling** me this?
16. He gave his enemy a **telling** blow.
17. **Crying** inconsolably, children take their **broken** toys to their elders.
18. **Pampered** children are not those who are told a story by their grandmothers.
19. The **told** story needs to be retold.
20. **Exuding** confidence, he did not try to conceal his **vaulting** ambition.

PRACTICE TEST 3.29

Rewrite the following sentences placing the participles correctly:

1. We saw walking on a rope a little girl on the way.
2. Buried for more than two days, the rescue team disinterred five dead bodies.
3. Talking to the stars were seen lots of journalists on the occasion.
4. Having arrived late for the movie, tickets could not be got.
5. Holding their kids at the party were seen quite a few young mothers.
6. Angry and distraught, the papers were signed by the members in a huff.
7. Walking down the street, a solitary cab was sighted.
8. Wanted an intelligent governess for a child, twenty-five-year-old.
9. Playing in the lawn, the bumblebee stung the child.
10. Amusing all the way, the tourists were shown the fort by the guide.

PRACTICE TEST 3.30

Rewrite the following sentences using appropriate gerunds, participles, or to-infinitives:

1. Stop to talk to me in such a rude manner!
2. I want going through the whole exercise all over again.
3. The members resented the idea to invite her to the meeting.
4. I don't feel like to tell you that I am not happy with your performance.
5. Avoid to be late all the time!
6. He admitted to be there at the time of the crime.
7. The speaker failed making a proper impression on the audience.
8. We all detest to face difficult situations.
9. To our surprise, she declined helping us even in those circumstances.
10. How can you imagine to do such a thing!

3.9 TAG QUESTIONS

Tag questions are questions that are attached to a statement in order to provide emphasis on it. Though tag questions are questions, they are rhetorical in nature and in asking them, the speaker really does not seek an answer but just intends to stress the idea suggested in the statement.

Though tag questions are mostly used in conversational language, it is nevertheless important for us to use them correctly. Look at the following statements and see if the tag questions appended to them are correctly used:

1. It is a wonderful idea, **is it**?
2. That was a great shot, **wasn't it**?
3. Jane is very brave, **isn't it**?
4. We are Indians, **isn't it**?
5. Jack looks ill, **isn't he**?
6. We all need money, **isn't it**?
7. You are not listening to me, **aren't you**?
8. Reading grammar is quite boring, **isn't that**?
9. Americans are quite rich, **is it not**?
10. You are not a fool, **isn't**?

None of the tag questions is correctly appended to any of these statements. The basic rule for putting a tag question is that *a positive statement takes a negative tag question and a negative statement takes a positive tag question*. When we use it, we put a comma (,) after the statement and use small letters in a contracted form, followed by a question mark. Now, look at the first sentence.

The statement is positive; so it requires a negative tag question *isn't it* and not a positive tag question *Is it*. So, it should be written as follows:

1. It is a wonderful idea, **isn't it**?

Further, as a general rule, whatever forms the subject of a statement is repeated as it is in the tag question. The second sentence, however, does not follow this sequence.

There is no need to write *it* in the tag question when the subject of the statement is *that*. So, it should be rewritten as follows:

2. That was a great shot, **wasn't that**?

The third statement also does not carry an appropriate tag question. See, in this statement, it is *Jane* who is the subject of the sentence. Since *Jane* is a feminine noun, it should be replaced by a feminine personal pronoun, *she*, and not *it*. So the sentence should be written as follows:

3. Jane is very brave, **isn't she**?

Similarly, the fourth sentence also needs to be revised. Since the subject of the sentence is *we*, the impersonal pronoun *it* cannot replace it in the tag question. Therefore, the sentence should be written as:

4. We are Indians, **aren't we**?

The fifth sentence chooses the correct pronoun, *he*, for the masculine noun *Jack*. However, it uses an inappropriate tense for the tag question. Since the main statement is in the simple present tense, the tag must also be written in the same tense and hence *does* and not *is* should be written here:

5. Jack looks ill, **doesn't he**?

For the remaining sentences in question, carefully observe the correct way of adding the tag questions to the statements:

6. We all need money, **don't we**?
7. You are not listening to me, **are you**?
8. Reading grammar is quite boring, **isn't it**?
9. Americans are quite rich, **aren't they**?
10. You are not a fool, **are you**?

Going further, look at the following statements and find out whether the tag questions that follow them are correctly used:

1. Few want to help others these days, **don't they**?
2. We could hardly concentrate, **couldn't we**?
3. There is little hope of his recovery, **isn't there**?
4. We seldom meet each other these days, **don't we**?
5. We could barely do anything, **couldn't we**?
6. Few ideas were exchanged in the meeting, **weren't they**?
7. We could scarcely make out what he said, **couldn't we**?
8. None of them were interested, **weren't they**?
9. We could remember nothing that was interesting about it, **couldn't we**?
10. Few students turned up for the class, **didn't they**?

All the tag questions appended to the statements given above are inappropriate. Actually, all the statements listed above communicate a negative idea. Remember, words such as *few* and *little* mean hardly anything, which is equivalent to nothing or none. Similarly, words such as *hardly*, *scarcely*, *barely*, and *seldom* also suggest a negative idea and hence the tag questions appended to all these statements should be positive.

In short, remember to *use a positive tag for a statement which connotes a negative thought even though it may seem to use a positive verb*. This is how we can revise these sentences:

1. Few want to help others these days, **do they**?
2. We could hardly concentrate, **could we**?
3. There is little hope of his recovery, **is there**?
4. We seldom meet each other these days, **do we**?
5. We could barely do anything, **could we**?
6. Few ideas were exchanged in the meeting, **were they**?
7. We could scarcely make out what he said, **could we**?
8. None of them were interested, **were they**?
9. We could remember nothing that was interesting about it, **could we**?
10. Few students turned up for the class, **did they**?

Adding a tag question to a statement may not always be as easy as it seems. Given below are some statements which are followed by certain tag questions. Read them carefully and find out if they are appropriate:

1. Please stay with us tonight, **isn't it**?
2. Let's play cricket, **isn't it**?
3. One must not lose patience, **must he**?
4. Everyone enjoyed the party, **didn't he**?
5. Some of the girls danced really well, **isn't it**?
6. Keep quiet, **OK**?
7. We have our lunch at one o'clock in the afternoon, **isn't it**?
8. Pass me the salt, **can you**?
9. I am a fool, **isn't it**?
10. Everybody has come, **has he**?

Imperative statements that connote a request, order, command, suggestion, etc. usually take *will you* as tag questions. *Let's* is followed by *shall we*. Further, when the verb *have* is used in the sense of *take* or *eat*, the tag question will be according to the tense, for example, in the present tense it should take *don't we*.

Moreover, expressions such as *everybody, everyone, someone,* and *somebody* are followed by a plural tag question whereas *one* is followed by *one* and not *he*. This is how we can use the tag questions in the sentences appropriately:

1. Please stay with us tonight, **will you/won't you**?
2. Let's play cricket, **shall we**?
3. One must not lose patience, **must one**?
4. Everyone enjoyed the party, **didn't they**?
5. Some of the girls danced really well, **didn't they**?
6. Keep quiet, **will you**?
7. We have our lunch at one o'clock in the afternoon, **don't we**?
8. Pass me the salt, **will you**?
9. I am a fool, **aren't I**?
10. Everybody has come, **haven't they**?

PRACTICE TEST 3.31

Use correct tag questions in the following statements:

1. Everybody was excited about the idea.
2. We aren't all that stupid.
3. One can't have one's pie and eat it too.
4. The snake was quite huge.
5. Drive carefully on the road.
6. Let's enjoy a movie tonight.
7. I have three brothers.
8. He could hardly see anything in the darkness.
9. It is not a brilliant idea after all.
10. Someone is going to come.
11. I usually have tea after dinner.
12. You are such a complete joker.
13. None of them paid any attention to us.
14. Just keep quiet.
15. We aren't really enjoying our lives.
16. Few passers-by respond to the mishaps on the road these days.
17. He seldom calls on us these days.
18. That boy is quite athletic.
19. Some of you are going to flunk maths.
20. I always listen to you carefully.
21. Helen was a beautiful woman.
22. I think you are just thirteen.
23. He would just not listen to us.
24. I am such an idiot.
25. Life is strange.

3.10 PUNCTUATION MARKS

Punctuation marks are visual indicators used in a written or printed text in order to separate sentences or a part of a sentence from another. They are used to make an idea readable. Look at the following example:

1. I can't do it he said speaking at the top of his voice she listened to it and said ok go to hell.

Would you like to read something of this sort? Of course, it would be quite a challenge to read a thing like that. Can you make out what is wrong with the above expression? The sentence does not have any commas, inverted commas, full stop, or exclamation marks, in brief, the signs that are known as **punctuation marks**.

In this section, we shall learn how to use punctuation marks correctly. Given below are the important punctuation marks which are normally used in written English:

1. Full stop (.)
2. Comma (,)
3. Dash (—)
4. Hyphen (-)
5. Semicolon (;)
6. Double inverted commas (" ")
7. Single inverted commas (' ')
8. Colon (:)
9. Apostrophe (')
10. Parentheses ( )

11. Sign of interrogation/Question mark (?)
12. Exclamation mark (!)
13. Capital letters

Let us see all these punctuation marks in detail.

3.10.1 Full Stop

The **full stop** (.) is used to mark the end of an affirmative, negative, or imperative sentence. For example:

1. Marie Curie was a great scientist.
2. She did not know the way to the market.
3. Listen to me.

It is also used in abbreviations, such as the following:

1. He is an M.B.B.S. doctor.
2. She works for I.D.B.I.
3. Our teacher is pursuing his Ph.D.

Not long ago, it was customary to write *Mr.* and *Mrs.* in English, that is, with a full stop at the end of these words. In current usage, however, we write these abbreviations as *Mr* and *Mrs*, that is, without a full stop since they are now seen as full spellings.

3.10.2 Comma

Just as a full stop marks the end of a sentence, a **comma** (,) suggests a pause in the writing. Following are the main uses of a comma:

It indicates omission of a word, especially a verb.

1. You can do that; I, never.
2. Her mother was an English; her father, an American.
3. She got her prize; I, my punishment.

It separates the co-ordinate clause(s) in a compound sentence.

1. I came, I saw, I conquered.
2. Father is in the office, mother is in the kitchen.
3. Men may come and men may go, but I go on for ever.

It separates the subject and the long preceding phrase that characterizes it.

1. Harassed and distraught right from the early days of her marriage, she decided to embark on a journey of her own.
2. Contrary to the notion that it consumes a lot of your time and gives you only a little, reading gives you an opportunity to look not just beyond but inside you as well.

It separates the same parts of speech used in the same sentence.

1. She was tall, slim, and beautiful.
2. Books, chairs, tables, desks, and settees could be seen in the lawn.
3. Unwilling to go to school, the child whined, cried, groaned, and protested whichever way he could.

It separates the parenthetical ideas from the core ideas in a sentence.

1. Your suggestion, however, is quite tempting.
2. The villain, having trapped the hero, gave a malicious smile.
3. No such efforts, therefore, are going to yield fruit.

It marks a non-defining clause. It is used to contribute to the original idea in a parenthetical way and can be omitted without doing any harm to the core meaning of the sentence.

1. My friend, who is a journalist, doesn't think so.
2. The poet, the one who always defied the system, decided not to comply with the king's orders.
3. His book, the one that he wrote at the fag end of his career, is likely to create a stir.

It is used to separate two or more nouns in apposition.

1. Shakespeare, the greatest dramatist of all times, was born in 1564.
2. Indira Gandhi, the only woman Prime Minister of India so far, was a great politician.
3. Sam, my uncle, is returning from England.

It is used to address people.

1. Sir, I am indebted to you.
2. How are you, my dear?
3. Come here, little girl.

It marks off direct quotations from the rest of the sentence.

1. Mother said to her children, 'Have your food.'
2. 'What do you think,' said he, 'is the cause of your failure?'
3. 'Come in and tell what happened,' the fat man said and moved away from the door.

It is used to separate an adverbial clause from the principal clause.

1. When we came back, we found the doors open.
2. If you do not work hard, you cannot succeed.
3. Though it is none of my business, I cannot force myself into aloofness on this issue.

It is used before and after words, phrases, or clauses that are introduced to the main thought in a parenthetical way.

1. This, in no way, is my problem.
2. The poor little children, when they first saw him, thought he was an angel.
3. Your story, in all probability, is fairy tale.

3.10.3 Dash

A **dash** (—) indicated by a long horizontal line is often used in place of a colon or parenthesis. Here are its uses:

It is used to emphasize the idea anticipated in the sentence.

1. Finally, we got what we had all along desired—our first television set at home.
2. He is what you expect him to be—the greatest fool on earth!
3. They told us whatever was to be told—nothing could be done.

At times, much like a comma, a dash is used to separate an expression from the rest of the sentence.

1. He is—after all—his mother's son.
2. In the end—to be precise—I would say that all that shines is not gold.
3. We are—generally speaking—people of short memories.

It is also used after the colon to indicate something that follows.

1. These are some of the views—
2. He says—*frailty thy name is woman*!

It is also used to indicate an abrupt change of idea.

1. Had he not boarded the plane—but what is the use of thinking like that?
2. Once you reach here—but wait, you are coming, aren't you?
3. Only if we were a little more educated—but how would that have changed our lives?

3.10.4 Hyphen

A **hyphen** (-) is a shorter line than a dash. The uses of a hyphen are as follows:
It is used to join two or more words in a compound word.

1. She was truly tormented by her daughter-in-law.
2. The commander-in-chief refused to sanction any leave to the sergeant.
3. The ex-director of the company is paying a visit this afternoon.
4. These days, in the name of scholars, you will see jack-of-all-trades but master-of-none.

3.10.5 Semicolon

A **semicolon** (;) stands for a longer pause than a comma. Following are the uses of a semicolon:
It is used to separate clauses.

1. Reading maketh a full man; conference a ready man; and writing an exact man.
2. Man proposes; God disposes.
3. Not that I loved Caesar less; but that I loved Rome more.
4. It is easy to be difficult; but difficult to be easy.

It is used to express different ideas without writing a new sentence.

1. Today, we don't do anything regarding global warming; tomorrow there is nothing we can do about it.
2. In the morning, he fought with his wife; in the afternoon, he reconciled with her.
3. One man kept her in good humour; the other kept her in the need of the first.

3.10.6 Colon

A **colon** (:) is used to list examples and enumeration.

1. Following are the examples of parts of speech: noun, pronoun, adjective, adverb, etc.
2. These are the points to be kept in mind: …
3. The team consists of eleven players: Sachin Tendulkar, Rahul Dravid, V.V.S. Laxman,…

3.10.7 Single and Double Inverted Commas

Double inverted commas (" ") are used to quote the exact words of the person being quoted.

1. He said, "You are my friend."
2. "One cannot fool oneself for a long time," cleaning his glasses he spoke, "but you have tried to keep yourself foolish enough for a pretty long time."
3. Eliot begins by saying, "April is the cruelest month."

Single inverted commas (' ') are used to cite a quotation within another quotation.

1. "There is no point in keeping a pulled face", said he, "even if you are perturbed by the 'to be or not to be' conundrum."
2. "What sort of movie was that—so loud and so pompous?" felt she, "they seem to have forgotten that 'art lies in concealing art'."

Many publications, including this book, follow the exact opposite of the rules mentioned in the examples, as per their individual house-style guidelines.

3.10.8 Apostrophe

The **apostrophe** (') is used to indicate the possessive of a noun. If the noun is singular, the apostrophe is followed by an **s**; if the noun is plural, the **s** is followed by the apostrophe, except when the plural does not end in **s**, as in the case of a few irregular nouns, for example, children's:

1. The children's books are lying there.
2. Brown's house is next to ours.
3. Waugh's captaincy is still appreciated.
4. The girl's purse was lost. (the purse of a girl)
5. The girls' purses were stolen. (purses of many girls)

- It is also used to show words in a contracted form.

1. I feel it's time to move out of the house. (*it's* stands for *it is*)
2. It's been ages since we met her. (*it's* stands for *it has*)
3. Let's go and watch some play. (*let's* stands for *let us*)
4. You're just a complete fool. (*you're* stands for *you are*)
5. Won't you come inside? (*won't* stands for *will not*)
6. Don't you dare speak to him like that! (*don't* stands for *do not*)

- It is sometimes also used to show the letters and figures in the plural form to avoid confusion.

1. In 1970's was seen the first wave of Parallel Hindi Cinema.
2. Articulate your s's and sh's properly.
3. Round off all the 0.25's and 0.50's in the final total.

3.10.9 Parentheses

Parentheses ( ) are used by writers to indicate an afterthought by introducing some words, a phrase, or a clause:

1. The great man (this is how he is seen to be in the area) is reported to have killed his wife.
2. The development (so it seems) was achieved by turning the poor out of their huts.

3.10.10 Sign of Interrogation/Question Mark

The **sign of interrogation/question mark** (?) is used after a direct question or a tag question that is appended to a statement:

1. Do you understand what I say?
2. Shall we take some rest?
3. You are stupid, aren't you?

Remember that a question mark is not used after an indirect question.

1. I am not sure what to do in life.
2. The inspector could not make out if she was telling a lie.
3. They asked their children whether they are doing good parenting.

3.10.11 Exclamation Mark

The **exclamation mark** (!) is used in phrases and sentences that express sudden, strong emotion or a wish:

1. May you live long!
2. What a terrible sight!
3. O Hamlet, speak no more!
4. Oh, you fool! Listen to me first and then decide.

3.10.12 Capital Letters

Capital letters are used for various purposes. To begin with, we start a sentence with a capital letter:

1. We can't do anything about it.
2. No problem.
3. Has he come?

They are used to begin a sentence inside inverted commas.

1. It is said, 'To err is human.'
2. Shakespeare says, 'One may smile and smile, and still be a villain.'

- They normally begin a proper noun and the adjectives we form from it.

1. Pinter is known for his theatre language popularly known as Pinteresque idiom.
2. Italy is a place of intellect; the Italian thinker Machiavelli is still well known.
3. Though *Maqbool* does try to recreate Shakespeare's classic *Macbeth*, the sweep of imagination and grandeur of spectacle is hardly Shakespearean.

- They are used to refer to a person's title or degree.

1. Pandit Nehru was the first prime minister of India.
2. Sir V.S. Naipaul is visiting India next year.
3. Dr R.P. Pareek, an MD, is an expert in diabetics.

- They are used to refer to the names of festivals.

1. Christmas falls on 25th December.
2. Diwali is the single most important festival in our family.

- They are used to refer to the names of days, weeks, months, and events.

1. On Sundays, we generally get up quite late.
2. North India is quite cold in January.
3. The Trojan War has acquired mythical significance in our collective unconscious.

- They are used to mark the important words in a title. Normally, the head words, such as nouns, pronouns, adjectives, verbs, and adverbs, are written in capital letters, whereas conjunctions, prepositions, and articles are written in small letters.

1. 'Ode on a Grecian Urn' is a great poem by John Keats.
2. The title of my book, *Language as Stratagem in Pinter's Plays*, has won me laurels on many occasions.
3. His sister is pursuing her research and is currently busy in the writing of a project entitled *Rediscovering Indian Diaspora: A Study of Postmodern Fiction in English with Special Reference to the Works of Jhumpa Lahiri, Rohinton Mistry, and Jaishree Mishra*.

- They are chosen to refer to the word *God* and the pronouns replacing it.

1. God is great.
2. No one knows His ways.
3. Don't worry; He knows that you are innocent.

- They are used in words of exclamations.

1. Oh! You are back.
2. This is the solution, Eh!
3. Ugh! I forgot to call you.

- The personal pronoun *I* is always written in capital letter.

1. I can't see you.
2. 'It is doubtful,' I said to her.
3. I will not say I have won the battle unless I am convinced that I have done it.

PRACTICE TEST 3.32

Punctuate the following choosing appropriate punctuation marks at appropriate places:

1. That he was alone and wanted to be alone was a matter of concern for all
2. If you want to be healthy in life do this get up early work hard lead a natural life
3. When I heard a knock at the door I turned around
4. It is however not all that important to speak all the time
5. When he was young Shakespeare who went on to become the greatest writer of all times married a woman eight years his senior
6. God made women beautiful so that men may love them and men foolish so that they may return the favour
7. His attitude to say the least was really horrible
8. America England and France got together and went after Germany
9. Milton the great Puritan poet went blind at the age of forty-four
10. He told his wife 'Learn to live with my silence'

PRACTICE TEST 3.33

Punctuate the following using suitable punctuation marks:

1. Have you finished your meal
2. Alas he is no more
3. That's what said he
4. What's wrong with you asked he
5. I am not sure which way to decide
6. What a superb shot
7. Oh I have forgotten the keys
8. You can always win said he but you need
9. Shelley says we pine for what is not there
10. They asked me whether I knew the answer

PRACTICE TEST 3.34

Punctuate the following using capital letters wherever required:

1. in *macbeth* shakespeare starts the action on an ironical note.
2. if you go to jaipur, don't forget to visit the amber fort, the jaigarh fort, the hawa mahal and the city palace.
3. do you think that i am responsible for everything that goes wrong?
4. i always read *the times of india*; it is a good newspaper.
5. the english live in england and speak english.
6. the commentators are sir ian botham and sunil gavaskar.
7. through his the rape of the lock, alexander pope intends to bring to the fore the foibles and follies of the society.
8. oh it's time we left.
9. shakespeare says foul is fair, fair is foul.
10. when you turn left, you will find an atm of the axis bank.

 RECAPITULATION

✓ If the subject of a sentence is singular, it should be followed by a singular verb. Otherwise, it should take a plural verb.

✓ When diseases such as rabies, measles, mumps, rickets, etc. come as subjects, the sentence should take singular verbs *is/was* and it would be wrong to use plural verbs *are/were*. The names of subjects such as *Electronics, Physics, Mathematics, Economics,* and *Statistics* are treated as singular nouns and they take singular verbs. The words *innings, billiards,* and *news,* though end with 's' in spelling, are essentially seen as singular nouns.

✓ The expressions such as *a majority of, majority of, the majority of, a number of, a lot of, plenty of, all of,* etc. are followed by plural nouns and take plural verbs.

✓ Whenever expressions such as *as well as, besides, along with, in addition to, with,* etc. conjoin two subjects, the verb is chosen according to the first subject.

✓ For expressing universal truth, habitual action, and general fact, we use subject + base form of verb + s/es + object whereas when action is in progression, subject + is/are/am + (I form of verb + ing) + object.

✓ The verbs of perception, emotions, senses, or possession are usually expressed in continuous form as they normally are suggestive of more stable emotions and feelings.

✓ Whenever a particular point/period of time such as *yesterday, this morning, the year 2002, last month, couple of hours back,* etc. appears, we normally choose simple past tense. The expressions such as *just now, recently, already, yet,* etc. on the other hand should make us choose the present perfect tense.

✓ Choosing between past simple and past perfect tense is at times difficult. To avoid confusion, choose past perfect for the earlier action and simple past for the later action.

✓ Verbs do have different modes and manners of expressions, which we refer to as their moods. Verbs have indicative, subjunctive, and imperative moods.

✓ An important change in the passive form from the active is the shift of the emphasis. The focus in the passive shifts from the doer or agent of the action to the action itself. That is why, in many passive forms, we see the agent of the action being dropped altogether.

✓ While using indirect speech, we change the tense, person, and time according to the subject. If the reported speech puts a question, the question verbs such as *ask, enquire, wonder, want to know,* etc. can be chosen. If the reported speech starts with a yes/no question, *if* or *whether* is used in the indirect narration. To put a request or order into indirect speech, verbs *tell* and *ask* are commonly chosen.

✓ A clause is a sentence within a sentence. Clauses are divided into two broad categories: independent or main or principle clause and dependent or subordinate clause. Dependent clauses can function as nouns, adjectives, and adverbs.

✓ The non-finite verb does not change according to the tense, time, or person, whereas the finite verb changes. There are three types of non-finite verbs—to-infinitive, gerund, and participle.

✓ Though tag questions are questions, they really are rhetorical in nature and in asking them, the speaker does not seek an answer but just intends to stress the idea suggested in the statement.

✓ Punctuation marks are used to make an idea readable. Commas, inverted commas, full stop, exclamation marks, etc. are known as punctuation marks.

EXERCISES

I. Choose the correct subject–verb combinations in the following sentences:

1. Neither the parents nor their offspring is/are to blame for the lack of communication.
2. The number of institutes offering engineering courses in the country has/have suddenly skyrocketed.
3. None of your suggestions is/are required here.
4. Each and every student was/were informed about the programme.
5. Notwithstanding the issues raised by the opposition parties, the government is/are quite keen to go ahead with its policy on educational reform.
6. Diabetes and rabies is/are dangerous diseases, though in a different way.
7. Statistics reveal/reveals that corruption is increasing in our country.
8. The crowd was/were not appreciative of the speaker's views.

9. Ever since the convict gave them a slip, his whereabouts has been/have been a mystery for the entire police force.

10. The love and care of both the parents is/are essential for the healthy development of a child.

II. Rewrite the following sentences correctly:

1. I had finished reading God of Small Things; so you can now borrow that book for some time.

2. Sorry, my mother cannot come and see you at the moment; she sleeps at the moment.

3. I talked to my colleague before I had informed the boss about the problem.

4. I have not known how to lock this suitcase; when the sales girl read out to me the instructions I was not listening to her.

5. When I worked as a Manager in the bank, we were getting quite a few enquiries from the RBI.

6. We can't help you at this stage; we are having our exams next week.

7. So far, he is doing his work very diligently.

8. I had called him for the movie but said he had watched it already.

9. Can you bring one LCD projector for us when you will go to Delhi?

10. The moment I had reached the venue, I had informed the organizers about my arrival.

III. Choose the correct tense in the following sentences:

1. Why didn't/haven't you inform/informed me about that earlier?

2. Mr Marshall passed away/has passed away in the night.

3. On hearing the news, they all drove to/have driven to their wards staying in hostels.

4. No problem; we have had/we had our dinner.

5. I know a girl who has spoken/spoke five languages.

6. The little child was extremely delighted with herself. She had found/found what she was looking for.

7. After the thief left/had left with the jewellery, she decided/had decided to call the police.

8. The poet had died/died last week. He was/had been suffering from cancer for quite some time.

9. When I entered/had entered the room, I realized/had realized that something was/had been wrong.

10. Once she packed/had packed her briefcase, she stormed/had stormed out of the house.

IV. Fill in the blanks by choosing correct prepositions in the following sentences:

1. We are supposed to send the application _____ 31 March.

2. When I called the hospital and enquired _____ Romika's health, they told me that he had gone home.

3. Christopher Marlow was a great English dramatist, who also was a contemporary _____ Shakespeare, the great.

4. This picture is remarkable _____ its simplicity.

5. Congratulations! Your boss was full _____ appreciation for you.

6. We must abstain _____ unnecessary distractions.

7. What are you looking _____ so stupidly?

8. Sir, I take exception _____ this decision.

9. Good scholars are men _____ deep learning.

10. How can you afford to live _____ such a paltry amount?

V. Inventing necessary details, change the following direct statements into indirect speech. Choose the most appropriate verb for each sentence so as to fit the context:

1. 'I know I have done something wrong.'

2. 'You are selected.'

3. 'Sorry! We couldn't save the patient.'

4. 'The stalkers have been troubling us for a long time.'

5. 'Our dishes are still as popular as ever.'

6. 'Don't worry! I'll take you home safely.'

7. 'I must say that I could not recognize you.'

8. 'Please save my son!'

9. 'He is still quite popular among the masses.'

10. 'Get lost!'

VI. Choose the correct option in each of the sentences given below:

1. Before becoming a musician, he worked (as/like) a waiter in a hotel.

2. Our college will remain closed (on behalf of/on account of) Diwali holidays.

3. (No sooner/hardly) was I in the office (than/then) my boss called me in.

4. (The more/more) you accumulate in life, (the more/more) you need.

5. We often wonder what will happen (while/when) there is no water on earth.

6. (Despite of/in spite of) several chinks in their armour, the police could nab the thief.

7. (Even if/even though) they have alerted the public, storm can unleash substantial damage.

8. Since the speaker has arrived, there is (little/a little) that you need to worry about now.

9. Som's house is bigger than (our/ours).

10. We (need not to/need not) inform you each and every time.

VII. Choose the correct non-finite verbs in the following sentences:

1. I know the answer.

2. They don't know how to behave.

3. I am not a fool.

4. Everybody seemed satisfied.

5. We cannot do anything about it.

6. Let's move from here and sit out in the sun.

7. You hardly ever feel happy these days.

8. One must be very careful these days.

9. Do this for me.

10. Sit down.

VIII. Complete the following sentences by introducing a relative or participle clause in the space provided:

1. The project leader _____ is very friendly.

2. The girl _____ is my younger sister.

3. The book _____ is out of stock.

4. _____, he decided to retire half way through the match.

5. If you meet anyone _____, let me know.

6. We stood paralysed _____.

7. The excessive anxiety and worry _____ is beginning to take a toll on her mother's health.

8. The maestro, _____, is also likely to participate in the series.

9. The devastating earthquake _____ is one of the major calamities of the century.

10. _____ is something Indians still cannot reconcile to.

IX. Punctuate the following passage using appropriate punctuation marks wherever required:

Running a massive enterprise is a tricky business being a woman and being at the helm of such an empire makes the situation all the more difficult to handle says apoorva tandon the ceo of silkways Designers, New Delhi, and a premier fashion designer making men believe in womans calibre is the toughest thing in the world they can make errors and still can sound error free whereas even if an efficient woman errs just once she becomes fallen for ever says ms tandon one of the most talked about fashion designers of the last two decades in the country

X. Rewrite the following sentences keeping in view the accuracy of subject–verb concord:

1. Each of the boys were given a uniform.

2. The officer as well as the driver were injured in the accident.

3. Neither the villagers nor the police knows about the thief.

4. One of the several problems confronting us are the scarcity of water.

5. Either of his two sisters are coming today.

6. Some of the issues related to terrorism is likely to figure in the discussions.

7. That he keeps reading such books and magazines are not unknown to many.

8. Neither the bus nor the train go there.

9. Each and every member were aware of the seriousness of the situation.

10. Now that everyone have come, we can start the proceedings.

ANSWER KEY

Practice Tests

3.1 **1.** It is obvious that everything **is** fine.

2. The majority of nurses in our country **are** women.

3. Either James or his brothers **have** written the mail.

4. The jury **has** decided to hear the case at an early date.

5. That they are fools **is** known to everyone.

6. A lot of Indians **have** raised the issue of colour discrimination.
7. None of the **reports** is worth consideration.
8. One of the boys was hurt in the clash.
9. Quite a few students **find** the grammar classes quite boring.
10. The foreign delegation comprising many experts **is** likely to visit our university next week.

3.2 Today, all the teachers **are** enjoying a free day at college. You know the annual function is drawing closer and students **are** preparing for various activities which **are** being planned by them. The increased participation of students in different types of extra-curricular activities **has** led to a substantial decrease in attendance in the classroom. In many classes you see only a handful of students who too **are** sitting in a lacklustre mood. Only a few of them **are** not really yawning. That **reminds** me of a very alarming proposition. The student community these days **is** hardly interested in doing anything related to learning. What generally interests them **is** fun and frolic, and not studies.

3.3
1. Simple Past	2. Simple Future	3. Simple Past	4. Simple Present
5. Simple Future	6. Simple Present	7. Simple Future	8. Simple Past
9. Simple Future	10. Simple Present		

3.4
1. My mother **buys** vegetables from the Reliance Fresh Store.
2. **I walk** very fast.
3. **She rebukes** her children for playing computer games.
4. **We do not find the solution** of this problem on the Internet.
5. **They do not trust** their employees.
6. **Girls do not like** wrestling.
7. **The chief librarian usually purchases** new books and journals for the library in July.
8. **Snigdha dances** skillfully.
9. **Siddharth takes coffee** after dinner.
10. **It rains** frequently in London.

3.5
1. I **am working** hard for my GRE test these days.
2. We **are facing** an acute problem in power supply these days.
3. Since Mr Smith **is going** to France, he **is learning** French these days.
4. Due to economic recession many companies **are downsizing** their operations.
5. Look! Anne **is relishing** orange juice and French fries in the sun.
6. Our institute **is planning** to bring a change in its promotion policy.
7. Our government **is not cutting** down the prices of petroleum and gas.
8. The patient **is wearing** a blanket because he is feeling very cold.
9. The watchman normally **brings** tea and snacks for the hostel inmates from outside.
10. That great man always **donates** a lot of money for poor children.

3.6
1. **Why do you always come late?**
2. **He is likely to visit our school during vacation.**
3. **"Look, how glorious appears the Sun today!"**
4. **He is going to resume his innings on a nervous 99 after lunch in a short while from now.**
5. **Why are you informing me about that now?**
6. **Sometimes it rains here quite torrentially.**
7. **He writes his poems in Urdu.**
8. **We are waiting for the signal to clear.**
9. **Normally, the temperature ranges between 10 and 20 degrees in this region.**
10. **He is conducting a meeting right now in his chamber.**

3.7
1. Earlier, I **could not** contact you.
2. **Did** the officer **inform** you about that?
3. My children **played** badminton while I **slept**.
4. **Did** your company **plan** to start something in Kashmir earlier this year?
5. Last month, my wife **visited** New York for attending a meeting.
6. His father **drove** us back home very fast.
7. Mr Batra **deposited** five thousand rupees in his son's account.
8. They **opened** new branches in all the cities but surprisingly **closed** each of these one by one.
9. Since she was not well, she **consulted** the doctor.
10. The teachers at the conference **identified** the innovative methods of teaching.

3.8
1. I am not feeling hungry, I **have taken** heavy breakfast.
2. It is nice to see you again; we **have not met** each other for such a long time!
3. Recently, the state government **has passed** a bill for free college education for girls.
4. Though she loves going on long drives, she **has not learnt** how to drive.
5. **Have** you ever **thought** how stressful life has become these days?
6. I know I **have left** no stone unturned to pass the CAT exam.
7. The music band **has displayed** a stupendous show this time.
8. The film director **has developed** the plot of the story and used the camera and lights skillfully.
9. He is not aware of the legal implications as he **has not purchased** any land before.
10. The policy **has come** as the result of scathing attack from the media.

3.9
1. Sir, I **passed** my B.Tech in 2003.
2. We **have finished** our assignment just now.
3. Don't worry; I **have** already **informed** her about that.
4. We **saw** a tiger five years ago.
5. I **spoke** to him last night.
6. Wait; I **have not** yet **finished** my work.
7. I **read** that story in Class IX.
8. We **spent** a very enjoyable evening yesterday.
9. She **submitted** her project this morning.
10. Last year, we **conducted** several workshops for our teachers.

3.10
1. Modern scientific discoveries **have challenged** many established notions.
(Present perfect is preferred for something happened in the past and continues to impact the present)
2. India **has won** the one-day series against Australia.
(Present perfect to introduce something that has happened in the recent past; something that holds its significance in the present; what has happened is important and not when it happened)
In the last match played in Delhi yesterday, team India defeated the Aussies by 37 runs to clinch the series.
(Past indefinite for something wherein we intend to indicate when something happened at a specific time in the past.)
3. Prices **have increased** alarmingly in the past six months.
(Present perfect for what has happened; event more important than the time of its occurrence)
4. She **left** yesterday itself.
(Past indefinite for something that happened in the past and is completely separated from the present)
5. She **has left** just now.
(The Adverbial *just now* connects the past and the present, hence present perfect tense)
6. It's fine. I **have received** the mail.
(Present perfect for emphasizing the completion of the action; what has happened is important and not when)
7. Keeping in view the recession times, many companies **have started** downsizing.
(Present perfect for something that continues to impact the present)
8. We **decided** to continue the training, even though it was hardly of any use.
(Simple past for some event referring to a finished activity in the past)
9. In his directorial career, Raj Kapoor **made** several memorable movies.
(Simple past for some activity done in the past; not continued in the present)
10. Doctors **have linked** many diseases, such as diabetes and cancer, to anxiety and unhappiness in life.
(Present perfect for something has been recently invented, reported or discovered)

3.11
1. Call the doctor; I think he **has broken** his toe.
2. Now that you **have come**, why don't you stay with us for a while?
3. They **questioned** the actor at the airport before they **let** him go.
4. We **spent** the entire weekend worrying about his whereabouts.
5. Research **has shown** that exercises help people remain cheerful.
6. Shakespeare **wrote** his last play possibly in 1613.
7. Don't worry; I **have informed** his parents about his illness.
8. Earlier he **made** some comedies but for some time now he **has produced** only serious movies.
9. I am sure I **have read** about that episode in some magazine.
10. You **have grown** so big ever since I saw you last.

3.12
1. The British **had ruled** (the action that precedes) over India for nearly two hundred years before they **handed** (the action that follows) it over to us.
2. We **walked** (later action) all the way to the station only to know that the train **had already left** (earlier action).

3. When the police **had interrogated** (earlier action) the criminal, they **put** (later action) him behind the bars.
4. My neighbour **told** (later action) me that his father **had died** (earlier action).
5. By the time the doctor **reached** (later action), the patient **had died** (earlier action).
6. We **had thought** of raising your pay (earlier action), but **could not get** (later action) enough funds.
7. After he **had finalized** the deal (earlier action), he **signed** it (later action).
8. The movie **had already begun** (earlier action) when we **entered** the hall (later action).
9. He **had planned** to retire at the age of sixty-five (earlier action), but he **died** much before that (later action).
10. I **had just entered** the building (earlier action), when the bomb **went off** suddenly (later action).

3.13
1. will have finished
2. has examined
3. has explained
4. vanished
5. will be presenting
6. have travelled
7. got
8. will have spent
9. has been considering
10. had been participating/will have participated

3.14
1. was watching
2. is raining
3. had
4. put
5. did not want
6. decided
7. am writing
8. have been working
9. has been growing
10. has painted

3.15
1. Subjunctive
2. Subjunctive
3. Imperative
4. Subjunctive
5. Indicative
6. Imperative
7. Subjunctive
8. Imperative
9. Subjunctive
10. Imperative

3.16
1. It is insisted that we achieve these targets by the year end.
   (Or) It is important that these targets be achieved by the year end.
2. If I were you, I would not make such an error.
3. You talk to me as if I knew nothing about the matter.
4. I wish I were not in love with her.
5. It is time we taught them a lesson.
6. We would rather you issued the letter today itself.
7. It is recommended that the unplugged holes be plugged.
8. If I were the monarch of the world, I would never promote corruption.
9. He orders me as though I were his assistant.
10. They would rather you listened to them first.

3.17
1. We are sorry to inform you **that nothing much can be done about your problem**.
   (The passive highlights the lack of control over the circumstances and hence communicates the situation to the reader in a better way)
2. **Thousands of people are killed** in accidents every year.
   (Highlights the tragic reality. In such expressions, what happens is more important than who causes it)
3. **The news was telecast** at 8.00 p.m.
   (The focus is rightly placed on the action that took place)
4. **I have not yet been informed**.
   (Correctly implies the wait/disappointment/impatience without sounding blunt and full of accusations)
5. **The gathering was held** near the temple.
   (The focus properly shifts to the **action conducted**)
6. **The patient has been given the medicine.**
   (The passive correctly highlights that what was important, has been done)
7. **A memorandum was handed** to the district collector.
   (The action is properly stressed through the passive)
8. As we started walking, **a small cab was seen** coming from the other direction.
   (What was seen has been given more importance than who saw it)

3.18
They did not invite me to the party. Still, I was determined to attend the same. So, I reached the place uninvited. What I saw was something really forgettable. What was happening around was strange. **Wine was being openly served and consumed** by the guests. That was not enough. **Young girls were/could be seen dancing** with their suitors. Their glasses of **wine were being filled** again and again. Later on, these devotees of god Bacchus **were observed** inserting their fangs into the non-vegetarian dishes **which were being supplied** at a frantic pace.

3.19
Ladies and Gentlemen, I have immense pleasure in demonstrating to you all the functioning of our washing machine. Well, this is how it functions—first of all, **the soiled clothes are put into it**. Then the **detergent is sprinkled on them**. After that, **they are dipped** in a bucketful of water. The **lid is covered** and the **machine**

is turned on. **Having done that**, we don't have to worry about what it does with our clothes. You will see that automatically the **clothes are washed, rinsed, and dried…**

3.20 1. The patients of swine flu **are advised** to wear a mask.
2. It **can be proved** that Darwin's theory has some chinks in the armour.
3. The function **will be organized** in the auditorium.
4. **I have not yet been reported**.
5. **Bed sheets are changed** everyday in this hotel.
6. **I am paid monthly**.
7. The **star is likely to be offered** a staggering signing amount.
8. For many centuries **it was not known** that plants can breathe.
9. Premchand's **stories have been translated** into different languages the world over.
10. Learning a **language is mostly considered** to be a difficult task.
11. In the meeting, **it was agreed that** new norms be brought into effect.
12. The minister announced that all the old people in the state **would be given** pension.
13. A large number of **victims were shifted** to the relief camps.
14. The first English **dictionary was compiled** by Dr Samuel Johnson.
15. The government has announced that the new **act will be introduced** in the next parliamentary session.
16. The captain desired that the **game should be played** in the positive spirit.
17. The **glass was broken** as you walked over it.
18. I always **get my car serviced** at this service station.
19. Earlier **it was believed** that AIDS spreads by touch.
20. **We were expected** to win, but we lost.

3.21 1. No change
2. No change
3. No change
4. No change
5. In order to sound impersonal, replace it with: Nothing much can be done about it.
6. Banks were nationalized in 1970s.
7. Read the instructions carefully before you sign the document.
   In order to sound impersonal, replace it with: The instructions must be read carefully before signing the document.
8. No change
9. No change
10. To put focus on the action, it may be reworded as: The workshop will be conducted at the main centre.

3.22 1. He **warned** us not to be rude with him next time.
2. The plumber **informed** us that there was a leakage in the pipeline.
3. My doctor **advised** me to cut down on sweets.
4. The clerk **refused** to help us.
5. The teacher **asked** us to be silent.
6. She **reminded** me that it was our marriage anniversary that day.
7. He **criticized** us for being sloppy in our attitude.
8. He **reassured** me that we would get a seat.
9. Roberts **wondered** why he should always be neglected for promotion.
10. The captain **thanked** the players for winning him the crucial match.

3.23 1. She complained about being vexed like that.
2. They sought permission before leaving.
3. The man expressed regret/felt sorry for not being able to help us.
4. He promised to help me.
5. The duo decided not to/refused to speak to the journalists.
6. They warned us to be careful for the man was a clever fellow.
7. She complained about not getting the same appreciation.
8. They thanked us.
9. He suggested watching a movie.
10. He advised me to see a doctor.

3.24 1. I came because you invited me.
2. Children played while mother slept in the afternoon.
3. Walk fast lest you should miss the train.

4. Give me this purse otherwise I shall snatch it away from you.
5. When we reached back home, we found the lock broken.
6. They tried to win the race but lost.
7. Unless you are a genius, you cannot solve this puzzle.
8. Keep quiet or get lost.
9. He is a complete idiot, whereas his brother is a wizard.
10. Though I can, I won't do this.

3.25 1. Before you leave (Adverbial Clause of Time)
2. Since you say so (Adverbial Clause of Cause or Reason)
3. If I want it (Adverbial Clause of Condition)
4. where you kept it (Adverbial Clause of Place)
5. until you finish your homework (Adverbial Clause of Time)
6. Though the exercise seems difficult (Adverbial Clause of Concession)
7. till they fainted in the sun (Adverbial Clause of Time)
8. Had I seen him (Adverbial Clause of Condition)
9. so that we may accumulate more wealth (Adverbial Clause of Purpose)
10. Since you were not there (Adverbial Clause of Cause or Reason)

3.26 1. Since we wanted to watch a movie, we left the office early.
2. I went to Delhi so that I could visit a few libraries and collect material for my research.
3. We discussed our problems with our teacher while Sahil kept playing game on his cell.
4. When we reached our classroom, the teacher had already started teaching the lesson.
5. They tried their best but failed miserably.
6. The man whom you met yesterday was my uncle.
7. Give me some money, otherwise I will not be able to buy medicine for my mother.
8. The officer asked me not to go home unless I had finished the report.
9. Had they worked hard, they would have cleared the exam.
10. Pay your taxes in time or you will be behind bars.

3.27 1. **Making** a lame excuse, he sounded quite cheap. (Participle)
2. Children love **making** mud castles. (Gerund)
3. **Seeing** is **believing**. (*Gerund, Gerund*)
4. **Seeing**, he believed. (Participle)
5. **Waving** their hats and handkerchiefs, the crowd cheered the king. (*Participle*)
6. **Asking** questions is easier than **answering** them. (Gerund, Gerund)
7. **Making** food is not difficult. (*Gerund*)
8. You are **making** a fool of yourself. (Verb)
9. **Making** tea, she answered the call on her cell. (*Participle*)
10. He died in the **waiting** room. (Participle Adjective)
11. **Waiting** is often painful. (*Gerund*)
12. He died **waiting** for the train. (Participle)
13. King Lear decides **to divide** his kingdom. (to-infinitive)
14. The thief entered through the **broken** widow. (Participle Adjective)
15. We live **to succeed**. (to-infinitive)

3.28 1. Participle 2. Gerund 3. Verb 4. Participle
5. Verb 6. Gerund 7. Participle Adjective 8. Gerund
9. Gerund 10. Participle 11. Participle Adjective 12. Participle, Participle
13. Participle 14. Gerund 15. Verb 16. Participle Adjective
17. Participle, Participle Adjective 18. Participle Adjective 19. Participle Adjective
20. Participle, Participle Adjective

3.29 1. On the way, we saw a little girl walking on a rope.
2. The rescue team disinterred five dead bodies buried for more than two days.
3. Lots of journalists were seen talking to the stars on the occasion.
4. As we arrived late for the movie, we could not get the tickets.
5. Quite a few young mothers were seen holding their kids at the party. (OR) At the party, quite a few young mothers were seen holding their kids.

6. Angry and distraught, the members signed the papers in a huff.
7. As we walked down the street, a solitary cab was sighted.
8. Wanted a twenty-five-year-old intelligent governess for a child.
9. The bumblebee stung the child while he was playing in the garden.
10. Amusing all the way, the guide showed the tourist the fort.

3.30
1. Stop talking to me in such a rude manner!
2. I want to go through the whole exercise all over again.
3. The members resented the idea of inviting her to the meeting.
4. I don't feel like telling you that I am not happy with your performance.
5. Avoid being late all the time!
6. He admitted being there at the time of the crime.
7. The speaker failed to make a proper impression on the audience.
8. We all detest facing difficult situations.
9. To our surprise, she declined to help us even in those circumstances.
10. How can you imagine doing such a thing!

3.31
1. Everybody was excited about the idea, weren't they?
2. We aren't all that stupid, are we?
3. One can't have his pie and eat it too, can one?
4. The snake was quite huge, wasn't it?
5. Drive carefully on the road, will you/won't you?
6. Let's enjoy a movie tonight, shall we?
7. I have three brothers, don't I?
8. He could hardly see anything in the darkness, could he?
9. It is not a brilliant idea after all, is it?
10. Someone is going to come, aren't they?
11. I usually have tea after dinner, don't I?
12. You are such a complete joker, aren't you?
13. None of them paid any attention to us, did they?
14. Just keep quiet, will you?
15. We aren't really enjoying our lives, are we?
16. Few passers-by respond to the mishaps on the road these days, do they?
17. He seldom calls on us these days, does he?
18. That boy is quite athletic, isn't he?
19. Some of you are going to flunk maths, aren't you?
20. I always listen to you carefully, don't I?
21. Helen was a beautiful woman, wasn't she?
22. I think you are just thirteen, aren't you?
23. He would just not listen to us, would he?
24. I am such an idiot, aren't I?
25. Life is strange, isn't it?

3.32
1. That he was alone and wanted to be alone was a matter of concern for all.
2. If you want to be healthy in life, do this: get up early; work hard; lead a natural life.
3. When I heard a knock at the door, I turned around.
4. It is, however, not all that important to speak all the time.
5. When he was young, Shakespeare, who went on to become the greatest writer of all times, married a woman eight years his senior.
6. God made women beautiful so that men may love them; and men foolish so that they may return the favour.
7. His attitude, to say the least, was really horrible.
8. America, England, and France got together and went after Germany.
9. Milton, the great Puritan poet, went blind at the age of forty-four.
10. He told his wife, 'Learn to live with my silence.'

3.33
1. Have you finished your meal?
2. Alas! He is no more.
3. 'That's what,' said he.
4. 'What's wrong with you?' asked he.
5. I am not sure which way to decide.
6. What a superb shot!
7. Oh! I have forgotten the keys.
8. 'You can always win,' said he, 'but you need to work hard for it.'
9. Shelley says, 'We pine for what is not there.'
10. They asked me whether I knew the answer.

3.34 **1.** In *Macbeth*, Shakespeare starts the action on an ironical note.
 2. If you go to Jaipur, don't forget to visit the Amber Fort, the Jaigarh Fort, the Hawa Mahal and the City Palace.
 3. Do you think that I am responsible for everything that goes wrong?
 4. I always read *The Times of India*; it is a good newspaper.
 5. The English live in England and speak English.
 6. The commentators are Sir Ian Botham and Sunil Gavaskar.
 7. Through his *The Rape of the Lock*, Alexander Pope intends to bring to the fore the foibles and follies of the society.
 8. Oh! It's time we left.
 9. Shakespeare says, 'Foul is fair, fair is foul.'
 10. When you turn left, you will find an ATM of the Axis Bank.

Exercises

I. **1.** Neither the parents nor their offspring are to blame for the lack of communication.
 2. The number of institutes offering engineering courses in the country has suddenly skyrocketed.
 3. None of your suggestions is required here.
 4. Each and every student was informed about the programme.
 5. Notwithstanding the issues raised by the opposition parties, the government is quite keen to go ahead with its policy on educational reform.
 6. Diabetes and rabies are dangerous diseases, though in a different way.
 7. Statistics reveal that corruption is increasing in our country.
 8. The crowd was not appreciative of the speaker's views.
 9. Ever since the convict gave them a slip, his whereabouts have been a mystery for the entire police force.
 10. The love and care of both the parents is essential for the healthy development of a child.

II. **1.** I **have** finished reading God of Small Things; so you now can borrow that book for some time.
 2. Sorry, my mother cannot come and see you at the moment; she is **sleeping** at the moment.
 3. I **had talked** to my colleague before I **informed** the boss about the problem.
 4. I **do not know** how to lock this suitcase; when the sales girl read out to me the instructions, I was not listening to her.
 5. When I worked as a Manager in the bank, **we used to get** quite a few enquiries from the RBI.
 6. We can't help you at this stage; we **have** our exams next week.
 7. So far, he **has been doing** his work very diligently.
 8. I **called** him for the movie but said he had watched it already.
 9. Can you bring one LCD projector for us when **you go** to Delhi?
 10. The moment I had reached the venue, I informed the organizers about my arrival.

III. **1.** Why **didn't you inform** me about that earlier?
 2. Mr Marshall **passed away** in the night.
 3. On hearing the news, they all **drove** to their wards staying in hostels.
 4. No problem; we **have had** our dinner.
 5. I know a girl who **spoke** five languages.
 6. The little child was extremely delighted with herself. She **had found** what she was looking for.
 7. After the thief **had left** with the jewellery, she **decided** to call the police.
 8. The poet **died** last week. He **had been suffering** from cancer for quite some time.
 9. When I **entered** the room, I **realized** that something **was** wrong.
 10. Once she **had packed** her briefcase, she **stormed** out of the house.

IV. **1.** We are supposed to send the application **by** 31st March.
 2. When I called the hospital and enquired **about** Romika's health, they told me that he had gone home.
 3. Christopher Marlow was a great English dramatist who also was a contemporary **of** _____ Shakespeare, the great.
 4. This picture is remarkable **for** its simplicity.
 5. Congratulations! Your boss was full **of** appreciation for you.
 6. We must abstain **from** unnecessary distractions.
 7. What are you looking **at** so stupidly?
 8. Sir, I take exception **to** this decision.
 9. Good scholars are men **of** deep learning.
 10. How can you afford to live **on** such a paltry amount?

V.
1. He **confessed** that he had done something wrong.
2. The chairman of the board **announced** that I was selected.
3. The doctors **regretted** that they could not save the patient.
4. The girls **complained** that the stalkers had been troubling them for a long time.
5. The owner of the hotel **claimed** that their dishes were still as popular as ever.
6. The driver **told** us not to worry assuring that he would take us home safely.
7. She **admitted** that she could not recognize me.
8. The mother **pleaded** with the doctor to save her son.
9. The report **suggests** that he is still quite popular among the masses.
10. The boss **asked** her to get lost.

VI.
1. Before becoming a musician, he worked **as** a waiter in a hotel.
2. Our college will remain closed **on account** of Diwali holidays.
3. No **sooner** was I in the office, **than** my boss called me in.
4. **The more** you accumulate in life, **the more** you need.
5. We often wonder what will happen **when** there is no water on earth.
6. **In spite of** several chinks in their armour, the police could nab the thief.
7. **Even though** they have alerted the public, storm can unleash substantial damage.
8. Since the speaker has arrived, there is **little** that you need to worry about now.
9. Som's house is bigger than **ours**.
10. We **need not** inform you each and every time.

VII.
1. I know the answer, **don't I?**
2. They don't know how to behave, **do they?**
3. I am not a fool, **am I?**
4. Everybody seemed satisfied, **didn't they?**
5. We cannot do anything about it, **can we?**
6. Let's move from here and sit out in the sun, **shall we?**
7. You hardly ever feel happy these days, **do you?**
8. One must be very careful these days, **mustn't one?**
9. Do this for me, **will/won't you?**
10. Sit down, **will/won't you?**

VIII.
1. The project leader **with whom we work** is very friendly.
2. The girl **sitting in that car** is my younger sister.
3. The book **that you are looking for** is out of stock.
4. **Sensing that he cannot conquer his opponent**, he decided to retire half way through the match.
5. If you meet anyone **who can help us with the domestic work**, let me know.
6. We stood paralysed **seeing that huge snake coming out of the debris**.
7. The excessive worry **that she is prone to depression** is beginning to take a toll on her mother's health.
8. The maestro, **who has scored more than thirteen thousand runs in test cricket**, is also likely to participate in the series.
9. The devastating earthquake **that rocked Haiti recently** is one of the major calamities of the century.
10. **Seeing their grown-up girls in the company of boys** is something Indians still cannot reconcile to.

IX. Running a massive enterprise is a tricky business. 'Being a woman and being at the helm of such an empire makes the situation all the more difficult to handle,' says Apoorva Tandon, the CEO of Silkways Designers, New Delhi, and a premier fashion designer. 'Making men believe in a woman's calibre is the toughest thing in the world; they can make errors and still can sound error free, whereas even if an efficient woman errs just once, she becomes 'fallen' for ever,' says Ms Tandon, one of the most talked about designers of the last two decades in the country.

X.
1. Each of the boys was given a uniform.
2. The officer as well as the driver was injured in the accident.
3. Neither the villagers nor the police know about the thief.
4. One of the several problems confronting us is the scarcity of water.
5. Either of his two sisters is coming today.
6. Some of the issues related to terrorism are likely to figure in the discussions.
7. That he keeps reading such books and magazines is not unknown to many.
8. Neither the bus nor the train goes there.
9. Each and every member was aware of the seriousness of the situation.
10. Now that everyone has come, we can start the proceedings.

CHAPTER 4

Common Errors and Misappropriations

Learning Objectives After reading this chapter, you will be able to

- spot common errors
- avoid errors usually committed

In the preceding chapters on grammar, you learnt the basic rules required for writing and speaking correct English. However, there are many common errors which are normally committed by most of us. In this chapter, you will be introduced to many such errors and you will learn how to avoid making such mistakes in your speech and writing.

To begin with, look at the following set of sentences:

1. The murderer killed her, sleeping in cold blood.
2. Have you met our professor in biochemistry?
3. Yes, he just passed away me when I was entering the college gate.
4. Students, if you have any doubt, meet me behind the class.
5. While playing in the garden, the scorpion bit the child.
6. We are trying to change this house for the last three years.
7. Yesterday, the interview of Amitabh Bachchan was telecasted.
8. He is not at home; he has gone out in the morning.
9. She is one of the best student in our class.
10. Both the sister were seen at the party.

Can you identify the errors in the above sentences? Let us try to see what goes wrong and where.

In the first sentence, the modifier *in cold blood* should modify the verb *killed*, but it is misplaced and it seems to modify the verb *sleeping*. Hence, the error is of a *faulty modified–modifier arrangement*.

The second sentence uses a wrong preposition, as in place of the preposition *of*, the preposition *in* is chosen.

The third sentence is really funny, as in place of the phrasal verb *pass by*, the phrasal verb *pass away*—which means *to die*—is used.

The fourth sentence also wrongly uses the preposition *behind* in place of *after*.

In the fifth sentence, once again the modifier *while playing in the garden* is wrongly placed. It should follow the main clause.

The sixth sentence wrongly uses the present progressive (continuous) tense *we are trying* for an action which has been going on for some period of time and is still continuing. To denote it, the present perfect continuous tense has to be used.

The seventh sentence has two errors. The verb *telecast* does not become *telecasted* in the passive. Moreover, to denote the possession by a living being, we normally use *'s*.

In the eighth sentence, the phrase *in the morning* should not be preceded by the present perfect tense *has gone out*. In such a case, past indefinite tense should be used.

In the ninth sentence, *she is one of the* should be followed by the plural form *students* and not by the singular *student*.

In the tenth sentence, 'both' suggests that there are two sisters and hence the plural form *sisters* has to be used.

This is how the sentences cited above are to be structured:

1. While sleeping, she was murdered in cold blood.
2. Have you met our professor of bio-chemistry?
3. Yes, he just passed by me as I entered the college gate.
4. Students, if you have any doubt, meet me after the class.
5. The scorpion bit the child while he/she was playing in the garden.
6. We have been trying to change this house for the last three years.
7. Yesterday, Amitabh Bachchan's interview was telecast.
8. He is not at home; he went out in the morning.
9. She is one of the best students in our class.
10. Both the sisters were seen at the party.

PRACTICE TEST 4.1

Try to identify the best way to express the intended idea in the following:

1. (a) This is soup which I made with rice and barley.
   (b) This is the soup that I made with rice and barley.
   (c) The soup I had made with rice and barley.
   (d) The soup which I made used rice and barley.

2. (a) I have heard these news right in the morning.
   (b) I heard these news in the morning.
   (c) I had heard this news only in the morning.
   (d) I heard this news in the morning.

3. (a) My younger brother didn't want to listen to my advices.
   (b) My brother younger didn't want to listen to my advice.
   (c) My younger brother hadn't want to hear to my advice.
   (d) My younger brother had not wanted to listen to my advices.

4. (a) Please credit this amount to my name.
   (b) Please credit this amount in my name.
   (c) Please credit this amount to my account.
   (d) Please credit this amount for my name.

5. (a) I like her childish face.
   (b) I like her childlike face.
   (c) I like her face like child.
   (d) I like her face like a child.

6. (a) The convict stood in front of the judge.
   (b) The convict stood against the judge.
   (c) The convict stood beside the judge.
   (d) The convict stood before the judge.

7. (a) We have many faculties in our institute.
   (b) We are having many faculties in our institute.
   (c) We have many a faculties in our institute.
   (d) We have many faculty members in our institute.

8. (a) Being build on rock, the scientists knew that the lab would not settle.
   (b) As the lab was being built on rock, the scientists knew that it would not settle.

(c) Because the lab was being built on rock, the scientists knew that it would not settle.

(d) Because it was being built on rock, the scientists knew that the lab would not settle.

9. (a) Pinter has left Portia so that he can paint in Paris.

(b) Pinter has left Portia to paint in Paris.

(c) Pinter has left Portia so that we can go and paint in Paris.

(d) Pinter has left Portia in Paris so that they can paint there.

10. (a) While playing in the garden, a wasp string him.

(b) While he was playing in the garden, a wasp stung him.

(c) He was playing in the garden while a wasp stung him.

(d) While being strung by a wasp, he played in the garden.

PRACTICE TEST 4.2

Choose the expression that is closest to standard English usage among the given choices in the following sentences:

1. (a) Each boy and each girl were given a book.
   (b) Each boy and each girl was given book.
   (c) Each boy and each girl was given a book.
   (d) Each boy and girl was given a book.

2. (a) The news was broadcast at 2.00 p.m.
   (b) The news were broadcast at 2.00 p.m.
   (c) The news was broadcasted at 2.00 p.m.
   (d) The news had been broadcast at 2.00p.m.

3. (a) I have passed my B.Tech in 2007.
   (b) I had passed my B.Tech in 2007.
   (c) I did pass my B.Tech in 2007.
   (d) I passed my B.Tech in 2007.

4. (a) He shook the hand of his beloved.
   (b) He shook hands with his beloved.
   (c) He shook hands of his beloved.
   (d) He shook hand with his beloved.

5. (a) One half of the women thinks high heels make their feet look smaller.
   (b) One half of the women thinks high heels makes their feet look smaller.
   (c) One half of the women think high heels make their feet look smaller.
   (d) One half of the women think high heel makes their feet look smaller.

6. (a) On my way, I saw a girl of five years old.
   (b) On my way, I have seen a girl of five years.
   (c) On my way, I saw a five-year-old girl.
   (d) On my way, I had seen a five years' girl.

7. (a) Yesterday, there was an accident in the corner of the street.
   (b) Yesterday, there was an accident at the corner of the street.
   (c) Yesterday, there was an accident at the corner off street.
   (d) Yesterday, there was an accident on the corner of the street.

8. (a) He picked up a quarrel on his way.
   (b) He picked a quarrel on his way.
   (c) On his way, he picked at a quarrel.
   (d) On the way, he picked a quarrel.

9. (a) God forbade Adam not to eat the apple.
   (b) God forbade Adam to eat the apple.
   (c) God had forbidden Adam to eat the apple.
   (d) God had forbidden Adam not to eat the apple.

10. (a) If she has been crying, she must been very upset.
    (b) If she was crying, she must have been very upset.
    (c) If she was crying, she must have very upset.
    (d) If she was crying, she have been very upset.

PRACTICE TEST 4.3

Choose the best option from the following choices:

1. (a) I can't cope up with this problem.
   (b) I can't cope at this problem.
   (c) I can't cope with this problem.
   (d) I can't cope of with this problem.

2. (a) He was elected the Chairman.
   (b) He was elected for the Chairman.
   (c) He was elected for a Chairman.
   (d) He was elected as the Chairman.

3. (a) In this poem, the poet describes about his childhood.
   (b) In this poem, the poet describes his childhood.
   (c) In this poem, the poets describe his childhood.
   (d) In this poem, the poet described childhood.

4. (a) He was proved to be wrong.
   (b) He was proved in the wrong.
   (c) He proved wrong.
   (d) He was proved wrong.

5. (a) He resembles to his father.
   (b) He resembles his father.
   (c) He resembled with his father.
   (d) He resembled by his father.

6. (a) He shot himself dead after bidding his wife goodbye with a gun.
   (b) He himself shot his wife dead after bidding her goodbye with a gun.
   (c) He shot his wife himself with a gun after bidding his wife goodbye.
   (d) He shot himself dead with a gun after bidding his wife goodbye.

7. (a) Let us discuss about the matter.
   (b) Let us discuss on the matter.
   (c) Let us discuss the matter.
   (d) Let us discuss matter.

8. (a) My dog is better than Ramesh.
   (b) My dog is better than that of Ramesh.
   (c) My dog is better than Ramesh's.
   (d) My dog is better than that of Ramesh's.

9. (a) I'll report back in about a week or so.
   (b) I'll report at about a week or so.
   (c) I'll report in about week or so.
   (d) I'll report in about a week or so.

10. (a) I consider her to be a brilliant student.
    (b) I consider her a brilliant student.
    (c) I consider her brilliant student.
    (d) I consider her to be brilliant.

PRACTICE TEST 4.4

Choose the option that best conforms to standard English usage from the given options in the following sentences:

1. (a) I am sure whether or not to pay for the ticket.
   (b) I am not sure whether to pay or not for the ticket.
   (c) I am not sure whether to pay for the ticket.
   (d) I am not sure whether to pay for the ticket or not.

2. (a) The manager ordered for his dismissal.
   (b) The manager ordered his dismissal.
   (c) The manager has ordered for his dismissals.
   (d) The manager has ordered to his dismissal.

3. (a) I am awaiting for your reply.
   (b) Your reply is being waited.
   (c) I am waiting on your reply.
   (d) Your reply is being awaited.

4. (a) If Isha hadn't come home yet, she must still be waiting for us in the coffee shop.
   (b) If Isha hasn't come home yet, she must still be waiting for us in the coffee shop.
   (c) If Isha didn't come home yet, she must still be waiting for us in the coffee shop.
   (d) If Isha hasn't come home yet, she will be still waiting for us in the coffee shop.

5. (a) She is contesting in the Pilibhit Lok Sabha seat.

(b) She is contesting the Pilibhit Lok Sabha seat.

(c) She is contesting Pilibhit Lok Sabha seat.

(d) She is contesting with the Pilibhit Lok Sabha seat.

6. (a) He asked whether I would like to accompany him.

(b) I would like to accompany him whether he asked.

(c) He asked me that I would like to accompany him.

(d) He told me whether I would like to accompany him.

7. (a) His services at the hospital were dispensed.

(b) His services at the hospital were dispensed with.

(c) His service was dispensed with.

(d) His service was dispensed.

8. (a) You must at once start, otherwise you will be late to catch the train.

(b) You must start at once, otherwise you will be late to catch the train.

(c) You at once must start, otherwise you will be late to catch the train.

(d) You must start at once, otherwise you will late to catch the train.

9. (a) The poors don't have a house to live.

(b) The poors don't have a house to live in.

(c) Poor don't have a house to live in.

(d) The poor don't have a house to live in.

10. (a) After graduation, I will apply to few multinational companies for a job as a graphic designer.

(b) After graduation, I will apply to a few multinational companies for a job as a graphic designer.

(c) After graduation, I will apply for a few multinational companies for a job as a graphic designer.

(d) After graduation, I will apply to a few multinational companies for a designer job as a graphic.

PRACTICE TEST 4.5

Read the following sentences and pick the best alternative from the given options:

1. (a) We are indebted highly to Mr S Majumdar, Corporate HR Head of our company for his valuable suggestions.

(b) We are highly indebted to Mr S Majumdar, Corporate HR Head of our company for his valuable suggestions.

(c) We are highly indebted to Corporate HR Head of our company, Mr S Majumdar, for his valuable suggestions.

(d) We are highly indebted to Mr S Majumdar, Corporate HR Head of our company because of his valuable suggestions.

2. (a) It is regarded sacred.

(b) It is regarded to be sacred.

(c) It is regarded as sacred.

(d) It is regarded sacreds.

3. (a) Both of them did not care for money.

(b) Neither of them cared for money.

(c) None of them cared of money.

(d) Both of them did not care with money.

4. (a) Either give me an advice or keep quiet.

(b) Either give me an advise or keep quiet.

(c) Either advise me or keep quiet.

(d) Either give me some advise or keep quiet.

5. (a) It is high time we do something in this regard.

(b) It is high time we did something in this regard.

(c) It is high time we should do something in this regard.

(d) It is high time we have to do something in this regard.

6. (a) Many a times, it is the money that decides the issue.

(b) Many a time, it is money that decides the issue.

(c) Many a times, it is the money that decides the issue.

(d) Many a time, it is the money that decides the issue.

7. (a) They made me learn English at school.
   (b) They made me learn English at the school.
   (c) They made me to learn English at school.
   (d) They made me to learn English at the school.

8. (a) One of our achievements has been the prize that our students have won recently.
   (b) One of our achievements have been the prize that our students have won recently.
   (c) One of our achievement has been the prize that our students won recently.
   (d) One of our achievement has been the prize that our students had won recently.

9. (a) The manager of our department gave me some useful information.

   (b) The manager of our department gave useful informations.
   (c) The manager of our department gave a useful information.
   (d) The manager of our department gave an useful information.

10. (a) Two thousand rupees are a meagre salary these days.
    (b) Two thousand rupees is a meagre salary these days.
    (c) Two thousand rupee is a meagre salary these days.
    (d) Two thousands rupees is meagre salary these days.

PRACTICE TEST 4.6

In the following sentences, the same idea has been expressed in different ways. Choose the one that conforms to standard English usage for each of the sentences.

1. (a) The poet and philosopher are dead.
   (b) The poet and the philosopher is dead.
   (c) The poet and philosopher is dead.
   (d) Poet and philosopher is dead.

2. (a) The captain as well as their soldiers are dead.
   (b) The captain as well as his soldier are dead.
   (c) The captains as well as his soldiers is dead.
   (d) The captain as well as his soldiers is dead.

3. (a) Neither he nor his secretary were present.
   (b) Neither he nor his secretaries was present.
   (c) Neither his secretary not he were present.
   (d) Neither he nor his secretary was present.

4. (a) Either of the two sisters are at fault.
   (b) Either of the two sisters is at fault.
   (c) Either of the two sister are at fault.
   (d) Either of the two sister is at fault.

5. (a) The economics of the government has always baffled me.
   (b) The economic of the government has always baffled me.
   (c) The economics of the government has baffled me always.
   (d) The economics of the government have always baffled me.

6. (a) Mother Teresa's aim and objective in life was to provide relief to the poor.
   (b) Mother Teresa's aims and objectives in life was to provide relief to the poor.
   (c) Mother Teresa's aim and objective in life was to provide relief to poor.
   (d) Mother Teresa's aim and objective in life were to provide relief to the poor.

7. (a) Chair's legs are broken.
   (b) The chair's legs are broken.
   (c) The legs of the chair are broken.
   (d) Legs of chair are broken.

8. (a) The star and orator have arrived.
   (b) The star and the orator have arrived.
   (c) The star and orator have been arrived.
   (d) Star and orator have arrived.

9. (a) Sorry, I am not understanding your point.
   (b) Sorry, I don't understand your point.
   (c) Sorry, I am cannot understand your point.
   (d) Sorry, I cannot be able to understand your point.

10. (a) I am loving it.
    (b) I am in love with it.
    (c) I love it.
    (d) I have loved it.

PRACTICE TEST 4.7

Choose the best possible alternative from the given options:

1. (a) You did not yet informed me.
   (b) I haven't yet been informed.
   (c) I am not yet been informed.
   (d) I have yet not been informed by you.

2. (a) When I saw her, she had been wearing a gaudy dress.
   (b) When I had been seeing her, she was wearing a gaudy dress.
   (c) When I saw her, she wore a gaudy dress.
   (d) When I saw her, she was wearing a gaudy dress.

3. (a) Had I seen him, I would have informed him.
   (b) Had I seen him, I will inform him.
   (c) Had I seen him, I would inform him.
   (d) Had I seen him, I shall inform him.

4. (a) I am a fool, amn't I?
   (b) I am a fool, isn't it?
   (c) I am a fool, aren't I?
   (d) I am a fool, is it?

5. (a) The trainees have been going through a three-day orientation programme.
   (b) The trainees have been going through a three days orientation programme.
   (c) The trainees are going through three-day orientation programme.
   (d) The trainees are going through a three days orientation programme.

6. (a) He comes late often to the office.
   (b) He comes often late to office.
   (c) He often comes late to office.
   (d) Often he comes late to the office.

7. (a) Recently, scientists invented a device to predict earthquakes.
   (b) Recently, scientists have invented a device for predicting earthquakes.
   (c) Recently, scientists had invented a device to predict earthquakes.
   (d) Recently, scientists are inventing a device to predict earthquakes.

8. (a) When I reached office, the meeting had already begun.
   (b) When I had reached office, the meeting had already begun.
   (c) When I reach office, the meeting had already begun.
   (d) When I had reached office, the meeting did have begun.

9. (a) I am not believing what my boss says.
   (b) I don't believe what my boss says.
   (c) I have not believed what my boss says.
   (d) I may not believe what my boss says.

10. (a) Though Mr Verma is rude, he is not malicious.
    (b) Though Mr Verma is rude, but he is not malicious.
    (c) Though Mr Verma is rude, yet he is not a malicious.
    (d) Though Mr Verma is rude, still will not be malicious.

PRACTICE TEST 4.8

Read the following sentences and choose the best way to express the idea in each of the sentences given below:

1. (a) Either of the sister is coming today.
   (b) Either of the sisters are coming today.
   (c) Either of the sisters is coming today.
   (d) Either of the sisters could have been coming today.

2. (a) Had we bowled well, we would won the match.
   (b) Had we bowled well, we would have won the match.
   (c) Had we bowled well, we will win the match.
   (d) Had we bowled well, we shall win the match.

3. (a) The officer along with his secretaries have come.
   (b) The officer along with his secretaries has come.
   (c) The officer along with his secretaries is come.
   (d) The officer along with his secretaries are come.

4. (a) These days, I am reading a detective story.
   (b) These days, I have been reading a detective story.

(c) These days, I had read a detective story.

(d) These days, I read a detective story.

5. (a) When I was young, I can walk ten miles.

(b) When I was young, I shall walk ten miles.

(c) When I was young, I could walk ten miles.

(d) When I was young, I could have been walked ten miles.

6. (a) Prepare well lest you should not fail the test.

(b) Prepare well lest you should fail the test.

(c) Prepare well lest you will not fail the test.

(d) Prepare well lest you will fail the test.

7. (a) Hardly had I entered the room, than the mobile started ringing.

(b) Hardly had I entered the room, when the mobile had started ringing.

(c) Hardly had I entered the room, when the mobile will have started ringing.

(d) Hardly had I entered the room, when the mobile started ringing.

8. (a) When the master came home, the dog wagged it's tail.

(b) When the master comes home, the dog wagged its tail.

(c) When the master had come home, the dog wagged its' tail.

(d) When the master came home, the dog wagged its tail.

9. (a) John and I often study together.

(b) John and me often study together.

(c) I and John often study together.

(d) Me and John often study together.

10. (a) The farmer reaped the crop and sold it in the market.

(b) The farmer had reaped the crop and had sold it in the market.

(c) The farmer reaped the crop and had sold them in the market.

(d) The farmer had reaped the crop and sold it in the market.

PRACTICE TEST 4.9

Read the following sentences and choose the best possible way to express each of them:

1. (a) He likes taking tea after coming from office.

(b) He likes to take tea after coming from office.

(c) He likes taking the tea after coming from office.

(d) He likes to take tea after coming from the office.

2. (a) Everyone has come, has he?

(b) Everyone has come, isn't it?

(c) Everyone has come, haven't they?

(d) Everyone has come, have they?

3. (a) Before he entered the room, he remembered what his teacher says.

(b) Before he entered the room, he remembered what his teacher had said.

(c) Before he entered the room, he had remembered what his teacher said.

(d) Before he entered the room, he remembered what his teacher said.

4. (a) As the talk was not interested, I dozed off in the room.

(b) As the talk was not interesting, I had dozed off in the room.

(c) As the talk was not interesting, I dozed off in the room.

(d) As the talk had not been interested, I dozed off in the room.

5. (a) The teacher told me that she was not happy with my performance.

(b) The teacher told that she was not happy with my performance.

(c) The teacher said that she was not happy with my performance.

(d) The teacher told that she had not been happy with my performance.

6. (a) He read the magazine for a while before he cast it away.

(b) He had read the magazine for a while before he cast it aside.

(c) He read the magazine for a while before he had cast it aside.

(d) He read the magazine for a while before he cast it aside.

7. (a) As regards the payment, please note that it is in the process.

(b) As regarding the payment, please note that it is in the process.

(c) As regards the payment, please not that it has been in the process.

(d) As regard the payment, please note that it is in the process.

8. (a) A rebel finds it difficult to conform into the system.

(b) A rebel finds it difficult to conform with the system.

(c) A rebel finds it difficult to confirm to the system.

(d) A rebel finds it difficult to conform to the system.

9. (a) Science and technology has brought to us many comforts of life.

(b) Science and technology have brought to us many comforts in life.

(c) Science and technology has brought with many comforts of life.

(d) Science and technology has brought us with many comforts of life.

10. (a) In earlier time, people used to live in caves.

(b) In earlier times, people used to live in the caves.

(c) In earlier times, people used to live in caves.

(d) In earlier times, the people used to live in caves.

PRACTICE TEST 4.10

Find out the best possible alternative from the choices given below for each of the following ideas:

1. (a) The higher you go, cooler it gets.
   (b) The higher you go, the cooler it gets.
   (c) Higher you go, cooler it gets.
   (d) Higher you go, the cooler it gets.

2. (a) My brother has applied for lectureship in this college.
   (b) My brother has applied for lecturership in this college.
   (c) My brother had been applied for lectureship in this college.
   (d) My brother applies for lecturership in this college.

3. (a) Many a times, you will find your mobile causing you undesirable anxiety.
   (b) Many a times, you shall find your mobile causing you undesirable anxiety.
   (c) Many a time, you will find your mobile causing you undesirable anxiety.
   (d) Many a times, you will find your mobile causing you undesirable anxiety.

4. (a) It is time we left the party.
   (b) It is time we leave the party.
   (c) It is time we left in the party.
   (d) It is times we left the party.

5. (a) Both his brothers are working in this bank for the last seven years.
   (b) Both his brother have been working in this bank for the last seven years.
   (c) Both his brothers will be working in this bank for the last seven years.
   (d) Both his brothers have been working in this bank for the last seven years.

6. (a) Since 2009, the prices are only soaring higher and higher.
   (b) Since 2009, the prices have only been soaring higher and higher.
   (c) Since 2009, the prices has only been soaring higher and higher.
   (d) Since 2009, the prices have been only soaring higher and higher.

7. (a) If you don't find the book anywhere, buy the new one for me.
   (b) If you won't find the book anywhere, buy the new one for me.
   (c) If you are not finding the book anywhere, buy the new one for me.
   (d) If you would not find the book anywhere, buy the new one for me.

8. (a) While I listened to him, I realized how difficult the situation is.
   (b) While I listened to him, I realized how difficult the situation was.

(c) While listening to him, I realized how difficult the situation was.

(d) While listening to him, I realized how difficult the situation is.

9. (a) If I were you, I would not waste a single moment.

(b) If I am you, I would not waste a single moment.

(c) If I was you, I would not waste a single moment.

(d) If I had been you, I would not waste a single moment.

10. (a) He is one of the poets who has won the Pulitzer award.

(b) He is one of the poets who have won the Pulitzer award.

(c) He is one of the poet who has won the Pulitzer award.

(d) He is one of the poets who have won the Pulitzer awards.

PRACTICE TEST 4.11

Read the following sentences and choose the structure that expresses the idea in the best possible manner:

1. (a) The girl who sat besides my friend is his sister.

(b) The girl who sat beside my friend is his sister.

(c) The girl who sat beside my friend has been his sister.

(d) The girl besides friend was his sister.

2. (a) After he had written the mail, he sent it to his partner.

(b) After he wrote the mail, he sends it to his partner.

(c) After he has written the mail, he sent it to his partner.

(d) After he wrote the mail, he had sent it to his partner.

3. (a) He is too clever to solve this problem.

(b) He is clever enough not to solve this problem.

(c) He is not too clever to solve this problem.

(d) He is clever enough to solve this problem.

4. (a) There are so many as 1000 small activities involved in these whole process.

(b) There are as many as 1000 small activities involved in whole process.

(c) There are as many as 1000 small activities involved in this whole process.

(d) There are as many as 1000 small activities in these whole process involved.

5. (a) Even as he stops taking sugar, he will not lose weight.

(b) Even though he stops taking sugar, he will not lose weight.

(c) Even while he stops taking sugar, he will not lose weight.

(d) Even if he stops taking sugar, he will not lose weight.

6. (a) Let me read the report and decide the matter.

(b) Let I read the report and decide the matter.

(c) Let me read the report and decide about the matter.

(d) Let I read the report and decide about the matter.

7. (a) The peacock is one of the most beautiful bird in the world.

(b) The peacock is one of the beautiful bird in the world.

(c) The peacock is one of the most beautiful birds into the world.

(d) The peacock is one of the most beautiful birds in the world.

8. (a) At his arrival, the spectators greeted the superstar.

(b) On his arrival, the spectators greeted the superstar.

(c) With his arrival, the spectators greeted the superstar.

(d) In his arrival, the spectators greeted the superstar.

9. (a) Thirty security personnel are reported to have been injured in the blast.

(b) Thirty security personnels are reported to have been injured in the blast.

(c) Thirty security personnel have reported to be injured in the blast.

(d) Thirty security personnels have reported to have been injured in the blast.

10. (a) Hope for the best, but be prepared for worst.

(b) Hope of the best, but be prepared for the worst.

(c) Hope for the best, but be prepared for the worst.

(d) Hope for best, but be prepared for worst.

PRACTICE TEST 4.12

Choose the correct options in the following sentences:

1. (a) Finally, it is the brave who wins the final frontier.
   (b) Finally, it is brave who wins the final frontier.
   (c) Finally, it is the braves who win the final frontier.
   (d) Finally, it is the brave who win the final frontier.

2. (a) In the family of seven, she is eldest.
   (b) In the family of seven, she is oldest.
   (c) In the family of seven, she is the eldest.
   (d) In the family of sevens, she is the eldest.

3. (a) The mob shouted at the top of their voice.
   (b) The mob shouted at the tops of their voices.
   (c) The mob shouted at the top of his voice.
   (d) The mob shouted at the top of her voices.

4. (a) When I reached, I found him to read the book.
   (b) When I had reached, I found him reading the book.
   (c) When I reached, I found him reading the book.
   (d) When I reached, I found him to read the book.

5. (a) The candidates waited in the lobby impatiently for their turn.
   (b) The candidates waited impatiently for their turn in the lobby.
   (c) The candidates waited impatiently for the lobby in their turn.
   (d) The candidates impatiently waited for their turns in the lobby.

6. (a) Don't worry; he is used to going through pain in life.
   (b) Don't worry; he is used to go through pain in life.
   (c) Don't worry; he used to through pain in life.

(d) Don't worry; he is used to going through pain in lives.

7. (a) Unless you don't practise hard, you can't win the match.
   (b) Unless you practise hard, you can't win the match.
   (c) Unless you will practise hard, you cannot win the match.
   (d) Unless you can't practise hard, you can't win the match.

8. (a) When I thought about the incident, it brought a smile to my face.
   (b) When I thought about the incidence, it brought a smile on my face.
   (c) When I thought about the incident, a smile was brought about my face.
   (d) When I thought for the incidence, it brought me a smile on my face.

9. (a) Both the husband and the wife is to blame for the discord in the family.
   (b) Both the husband and wife are to blame for the discord in the family.
   (c) Both husband and wife is to blame for the discord in the family.
   (d) Both the husbands and the wives are to blame for the discords in the family.

10. (a) When the child fell asleep, the mother lay it in the cradle.
    (b) When the child fell asleep, the mother laid in the cradle.
    (c) When the child fell asleep, the mother lied it in the cradle.
    (d) When the child fell asleep, the mother laid it in the cradle.

PRACTICE TEST 4.13

Read the following sentences and choose the option that best expresses the idea intended:

1. (a) After lunch, he dosed off sitting in the chair.
   (b) After lunch, he dozed off sitting on the chair.
   (c) After lunch, he dozed of sitting in the chair.
   (d) After lunch, he dozed off sitting in the chair.

2. (a) The poet was felicitated for his achievements.
   (b) The poet was facilitated for his achievements.
   (c) The poet was felicitated with his achievements.
   (d) The poet was felicitated from his achievements.

3. (a) I would appreciate if you share these information with your colleagues and associates.
   (b) I would appreciate if you could share these information to your colleagues and associates.
   (c) I would appreciate if you could share this information about your colleagues and associates.
   (d) I would appreciate if you could share this information with your colleagues and associates.

4. (a) I am watching a movie which was released yesterday.
   (b) I am seeing a movie which was released yesterday.
   (c) I am looking at a movie which was released yesterday.
   (d) I am looking a movie which was released yesterday.

5. (a) On hearing this joke, my grandmother would burst into peals of laughter.
   (b) On hearing this joke, my grandmother would burst into a peel of laughter.
   (c) On hearing this joke, my grandmother would burst into peals of laughters.
   (d) On hearing this joke, my grandmother would burst into peels of laughter.

6. (a) You are invited cordially to submit manuscripts for the coming issues.
   (b) You are cordially invited to submit manuscripts for the coming issues.
   (c) For the coming issues you are cordially invited to submit manuscripts.
   (d) You have cordially invited to submit manuscripts for the coming issues.

7. (a) It is high time we should do something about corruption.
   (b) It is highest time we do something about corruption.
   (c) It is high time we did something about corruption.
   (d) It is high time and we will do something about corruption.

8. (a) Having been fed by the kind master, the obliged dog wagged its tail vigorously.
   (b) Having been fed by the kind master, the obliged dog wagged it's tail vigorously.
   (c) Having been fed by the kind master, the obliged dog waged its tail vigorously.
   (d) Having been fed by the kind master, the obliged dog wagged its tale vigorously.

9. (a) In the complimentary close you may write 'your's sincerely' even in business letters.
   (b) In the complementary close you may write 'yours' sincerely' even in business letters.
   (c) In the complimentary close you may write' yours sincerely' even in business letters.
   (d) In the complementary close you may right 'yours' sincerely' even in business letters.

10. (a) While laying in the bed, I was suddenly reminded of her.
    (b) While living in the bed, I was suddenly reminded of her.
    (c) While lying on the bed, I was suddenly reminded of her.
    (d) While lay in the bed, I was suddenly reminded of her.

PRACTICE TEST 4.14

Choose the expression that best conveys the intended idea in the following sentences:

1. (a) While I lay on the couch yesterday afternoon, I heard a loud explosion outside.
   (b) While I laid on the couch yesterday afternoon, I heard a loud explosion outside.
   (c) While I lie on the couch yesterday afternoon, I heard a loud explosion outside.
   (d) While I lain on the couch yesterday afternoon, I heard a loud explosion outside.

2. (a) The foundation stone of this institute was lay by the education minister in 1996.
   (b) The foundation stone of this institute was lain by the education minister in 1996.
   (c) The foundation stone of this institute has been laid by the education minister in 1996.
   (d) The foundation stone of this institute was laid by the education minister in 1996.

3. (a) Unmindful of the grave environmental challenges, many tribal communities keep on falling the trees around them.
   (b) Unmindful of the grave environmental challenges, many tribal communities keep on felling the trees around them.
   (c) Unmindful of the grave environmental challenges, many tribal communities keep on feeling the trees around them.
   (d) Unmindful of the grave environmental challenges, many tribal communities keep on feeding the trees around them.

4. (a) This hospital was founded to provide medical facilities to the poor.
   (b) This hospital was found to provide medical facilities to the poor.
   (c) This hospital was founded to provide medical facilities to poor.
   (d) This hospital was found to provide medical facilities to the poors.

5. (a) Its a pity that our education system does not help a child grow into a creative and imaginative individual.
   (b) Its' a pity that our education system does not help a child grow into a creative and imaginative individual.
   (c) It's a pity that our education system does not help a child grow into a creative and imaginative individual.
   (d) It a pity that our education system does not help a child grow into a creative and imaginative individual.

6. (a) Overawed by his superiors, he always behaved fawningly in their presence, especially so lest something adverse might be inferred about him.
   (b) Overawed by his superiors, he always behaved fawningly in their presence, especially so lest something adverse might not be inferred about him.
   (c) Overawed by his superiors, he never behaved fawningly in their presence, especially so lest something adverse is inferred about him.
   (d) Overawed by his superiors, he always behaved fawningly in their presence, especially so lest something adverse should not be inferred about him.

7. (a) A confirmed hypocrite and knowing nothing about life, he still tries to impress everyone around him by speaking as an expert.
   (b) A confirmed hypocrite and knowing nothing about life, he still tries to impress everyone around him by speaking like an expert.
   (c) A confirmed hypocrite and knowing nothing about life, he still tries to impress everyone around him by speaking with an expert.
   (d) A confirmed hypocrite and knowing nothing about life, he still tries to impress everyone around him by speaking for an expert.

8. (a) No sooner when he entered the room, he heard the woman scream wildly.
   (b) No sooner he had entered the room, when he heard the woman scream wildly.
   (c) No sooner that he entered the room, then he heard the woman scream wildly.
   (d) No sooner had he entered the room, than he heard the woman scream wildly.

9. (a) Hardly had the speaker started his address when the audience dozed off.
    (b) Hardly had the speaker started his address that the audience dozed off.
    (c) Hardly had the speaker started his address than the audience dozed off.
    (d) Hardly had the speaker started his address but the audience dozed off.

10. (a) It is not only difficult to catch a snake but it is dangerous as well.
    (b) It is not only difficult to catch a snake but it is dangerous in fact.
    (c) It is not only difficult to catch a snake but also dangerous.
    (d) It is not only difficult but also dangerous to catch a snake.

PRACTICE TEST 4.15

Choose the best option in the following sentences:

1. (a) The tiny, timid girl spoke in so low a voice that she was almost inaudible.
    (b) The tiny, timid girl spoke in such a low voice that she was almost inaudible.
    (c) The tiny, timid girl spoke in so much low voice that she was almost inaudible.
    (d) The tiny, timid girl spoke in such low a voice that she was almost inaudible.

2. (a) The book is either written by Shakespeare or one of his contemporaries.
    (b) The book is written either by Shakespeare or one of his contemporary.
    (c) The book is written by Shakespeare either or by one of his contemporary.
    (d) The book is written either by Shakespeare or by one of his contemporaries.

3. (a) Either the actors or the director are to be blamed for the failure of the show.
    (b) Either the actors or the director is to be blamed for the failure of the show.
    (c) Either the director or the actors are to blame for the failure of the show.
    (d) Either the directors or the actor is to be blamed for the failure of the show.

4. (a) Due to information technology, people across the globe are exposed to each other's cultures as never before.
    (b) Due to information technology, people across the globe are exposed to each other's cultures like ever before.
    (c) Due to information technology, people across the globe are exposed into each other's cultures like never before.
    (d) Due to information technology, people across the globe are exposed to each other's culture like never before.

5. (a) How can he be promoted? After all, he is junior to me.
    (b) How can he be promoted? After all, he is junior than me.
    (c) How can he be promoted? After all, he is junior than I.
    (d) How can he be promoted? After all, he is junior than I am.

6. (a) He stopped smoking once he realized that it can cause cancer.
    (b) He stopped to smoke once he realized that it can cause cancer.
    (c) He stopped for smoking once he realized that it can cause cancer.
    (d) He stopped the smoking once he realized that it can cause cancer.

7. (a) The doctor prohibited him to take sugar once he was diagnosed for diabetes.
    (b) The doctor prohibited him from taking sugar once he was diagnosed for diabetes.
    (c) The doctor prohibited him from taking sugar once he was diagnosed with diabetes.
    (d) The doctor prohibited him not to take sugar once he was diagnosed with diabetes.

8. (a) Of the two most exciting tennis players, you seem to like the most glamorous one.
    (b) Of the two most exciting tennis players, you seem to like the more glamorous one.
    (c) Of the two most exciting tennis players, you more seem to like the glamorous one.

(d) Of the two most exciting tennis player, you seem to like the more glamorous one.

9. (a) If I had gone with my friends to Pahalgam, Icould take skiing lessons.

   (b) If I had gone with my friends to Pahalgam, I could have taken skiing lessons.

   (c) If I would have gone with my friends to Pahalgam, I could have taken skiing lessons.

   (d) If I had gone with my friends to Pahalgam, I would take skiing lessons.

10. (a) The students of today needs to be trained about cultural diversity as this will facilitate constructive learning outcomes.

   (b) The students of today need to be trained about cultural diversity as this will facilitate constructive learning outcomes.

   (c) The students of today need to be trained for cultural diversity as this will facilitate constructive learning outcomes.

   (d) The students of today need to be trained about cultural diversity since this will facilitate constructive learning outcomes.

PRACTICE TEST 4.16

Read the following sentences and choose the option that best conveys the intended idea:

1. (a) My parents have been living in this house since 2011.

   (b) My parents has been living in this house since 2011.

   (c) My parents are living in this house since 2011.

   (d) My parents are living this in this house from 2011.

2. (a) Say whatever, I don't agree with your proposal.

   (b) Say whatever, I won't agree to your proposal.

   (c) Say whatever, I won't agree with your proposal.

   (d) Say whatever, I cannot agree with your proposal.

3. (a) The clowns' actions were so risible that I could not help myself from laughing.

   (b) The clowns' actions were so risible that I could not help laughing.

   (c) The clowns' actions were so risible that I could not help to laugh.

   (d) The clowns actions were so risible that I could not help laughing.

4. (a) She is intelligent enough not to marry a fool like you.

   (b) She is not intelligent enough to marry a fool like you.

   (c) She is too intelligent to marry a fool like you.

   (d) She is too intelligent not to marry a fool like you.

5. (a) Swati is fat enough to climb the stairs.

   (b) Swati is not thin enough to climb the stairs.

   (c) Swati is too fat to climb the stairs.

   (d) Swati is too fat not to climb the stairs.

6. (a) Have you read *Waiting for Godot*? Who had written it?

   (b) Have you read *Waiting for Godot*? Who wrote it?

   (c) Have you read *Waiting for Godot*? Who has written it?

   (d) Have you read *Waiting for Godot*? Who is the writer?

7. (a) How long are you using this mobile?

   (b) How long were you using this mobile?

   (c) How long you have been using this mobile?

   (d) How long have you been using this mobile?

8. (a) I can't be able to imagine she is dead; she was such a phenomenon.

   (b) I won't be able to imagine she is dead; she was such a phenomena.

   (c) I can't imagine she is dead; she was such a phenomenon.

   (d) I won't imagine she is dead; she was such a phenomena.

9. (a) You'd better go if you really don't want to annoy your father.
   (b) You'd better to go if you really don't want to annoy your father.
   (c) If you better do not go, you father will not be annoyed.
   (d) You must go so that your father does not feel annoyed.

10. (a) Smith was as good person as you could ever see.
    (b) Smith was as much good a person as you could ever see.
    (c) Smith was as good a person as you could ever see.
    (d) Smith was not as good a person as you could ever see.

PRACTICE TEST 4.17

Read the following sentences and choose the best possible alternative from the choices given below:

1. (a) My mother must have tired because she has been working since morning.
   (b) My mother must be tired because she has been working since morning.
   (c) My mother must have been tired because she has been working since morning.
   (d) My mother must be tired because she had worked since morning.

2. (a) The class was fine. However, the few students turned up.
   (b) The class was fine. However, only few students turned up.
   (c) The class was fine. However, only a few students turned up.
   (d) The class was fine. However, only some few students turned up.

3. (a) The computer isn't working. It should have been damaged during transit.
   (b) The computer isn't working. It must have been damaged in transit.
   (c) The computer isn't working. It must have being damaged while transit.
   (d) The computer doesn't work. It must have damaged during transit.

4. (a) During holidays, kids don't have to go to a school.
   (b) During holidays, kids don't have to go to the school.
   (c) During holidays, kids don't have school.
   (d) During holidays, kids don't have to go to school.

5. (a) I am going to hospital to see my friend who is admitted there.
   (b) I am going to the hospital to see my friend who is admitted there.
   (c) I am going to the hospital to see my friend who is admitted their.
   (d) I am going to a hospital to see my friend who is admitted there.

6. (a) That painting might not has been painted by Picasso, it could be a forgery.
   (b) That painting might not have been painted by Picasso, it could be a forgery.
   (c) That painting didn't not have been painted by Picasso, it could be a forgery.
   (d) That painting could not have been painted by Picasso, it could be a forgery.

7. (a) The Bhatnagars live in the house next door.
   (b) The Bhatnagars' live in the house next door.
   (c) Bhatnagars' live in the house next doors.
   (d) The Bhatnagar's live in the house next door.

8. (a) We haven't met anyones that much strange before.
   (b) We haven't met anyones that strange before.
   (c) We haven't met anyone that strange before.
   (d) We haven't met anyone that much strange before.

9. (a) During the meeting, everyone voiced his opinion.
   (b) During the meeting, everyone voiced their opinion.

(c) During the meeting, everyone voiced her opinion.

(d) During the meeting, everyone voiced opinion.

10. (a) As soon as my car had been repaired, I resumed my journey.

(b) As early as my car had been repaired, I resumed my journey.

(c) As my car had been repaired, I resumed my journey.

(d) As soon as my car was repaired, I resumed my journey.

PRACTICE TEST 4.18

Read the following sentences and choose the expressions that best convey the ideas intended:

1. (a) Unfortunately, few of injured died while we were taking them to the hospital.
   (b) Unfortunately, a few of the injured while we were taking them to the hospital.
   (c) Unfortunately, a few of the injured died while we were taking them to hospital.
   (d) Unfortunately, the few of the injured died while we were taking them to the hospital.

2. (a) Nobody else enjoyed Bishop's sermon, but so did I.
   (b) Nobody else enjoyed the Bishop's sermon, nor did I.
   (c) Nobody else enjoyed the Bishop's sermon, and so did I.
   (d) Nobody else enjoyed the Bishop's sermon, and so I did not.

3. (a) His job requires him to do a lot of travelling.
   (b) His job requires him to do a lot of travellings.
   (c) His job requires him to do the lot of travelling.
   (d) His job requires from him to do a lot of travelling.

4. (a) While on tour, I always prefer tea to coffee.
   (b) While on tour, I am always preferring tea to coffee.
   (c) While on tour, I will always prefer tea than coffee.
   (d) While on tour, I always prefer tea for coffee.

5. (a) When I went to the dentist last week, he asked me to show all my teeths.
   (b) When I went to the dentist last week, he asked me to show him all my tooths.
   (c) When I went to the dentist last week, he asked me to show him all my teeth.

   (d) When I went to the dentist last week, he asked me to show to him all my teeths.

6. (a) The coffee in this coffee shop is best one in town.
   (b) The coffee in this coffee shop is the best in town.
   (c) The coffee in this coffee shop is a best in the town.
   (d) The coffee in this coffee shop best in the town.

7. (a) I had never dreamt that the show would be such success.
   (b) I had never dreamt that the show would be such a success.
   (c) I had never dreamt that the show would have been such a success.
   (d) I had never dreamt that the show would have been such success.

8. (a) She was knitting as beautiful hat that she could do.
   (b) She was knitting as beautiful a hat as she could.
   (c) She was knitting so much beautiful a hat as she could do.
   (d) She was knitting so much beautiful a hat as she could.

9. (a) If I were you, I wouldn't do that.
   (b) If I am you, I will not do that.
   (c) If I had been you, I would not have done that.
   (d) Had I been you, I would not have do that.

10. (a) Have some tea, would you?
    (b) Have some tea, shall you?
    (c) Have some tea, won't you?
    (d) Have some tea, can't you?

PRACTICE TEST 4.19

Read the following sentences and choose the expressions that convey the intended idea in the manner that is closest to standard English usage:

1. (a) We had always to wear uniform in the school.
   (b) We always had to wear uniform at schools.
   (c) We always had to wear uniform in school.
   (d) We always had to wear uniform at the school.

2. (a) For vacations, Sheetal is planning to go to Jodhpur to polish her Jodhpuri up.
   (b) During vacations, Sheetal is planning to go to Jodphur to polish with her Jodhpuri.
   (c) In vacation, Sheetal is planning to go to Jodhpur to polish her up Jodhpuri.
   (d) During vacation, Sheetal is planning to go to Jodhpur to polish up her Jodhpuri.

3. (a) Income tax return is to be filed until 31st July.
   (b) Income tax return is to be filed by 31st July.
   (c) Income tax return is to be filed upto 31st July.
   (d) Income tax return is to be filed till 31st July.

4. (a) It is difficult to know why she is doing it.
   (b) It is difficult to know why is she doing it.
   (c) It is difficult to know however is she doing it.
   (d) It is difficult to know however she is doing it.

5. (a) My daughter had accompanied me in my trip to London.
   (b) My daughter had accompanied me for my trip to London.
   (c) My daughter had accompanied me on my trip to London.
   (d) My daughter had accompanied me with my trip to London.

6. (a) Though Federer played well, but he never looked like beating Nadal.
   (b) Though Federer played well, yet he never looked like beating Nadal.
   (c) Though Federer played well, he never looked like beating Nadal.

   (d) Though Federer had played well, he never looked like beating Nadal.

7. (a) So tired Mom was that she went straight to the bed.
   (b) So tired Mom was that she went straight upto bed.
   (c) So tired Mom was that she went straight on bed.
   (d) So tired was Mom that she went straight to bed.

8. (a) She got poor marks in two of her subjects; therefore she had to give them again to get into college.
   (b) She got poor marks in two of the subjects; therefore she had to take them again to get into college.
   (c) She got poor marks in two of the subjects; therefore she had to reappear them again to get into college.
   (d) She got poor marks in two of the subjects; therefore she had to appear them again to get into college.

9. (a) When I saw that book, I knew it was exactly what I had looked for.
   (b) When I saw that book, I knew it was exactly what I had been looking for.
   (c) When I saw that book, I had known it was exactly what I had been looking for.
   (d) When I saw that book, I knew it was exactly what I have been looking for.

10. (a) Do take your umbrella, it may rain.
    (b) Do take your umbrella, it is going to rain.
    (c) Do take your umbrella, it must rain.
    (d) Do take your umbrella, it might rain.

PRACTICE TEST 4.20

Read the following sentences and choose the expressions that best convey the idea:

1. (a) Ladies and Gentlemen, sorry for the technical glitch; but we start the show shortly.
   (b) Ladies and Gentlemen, sorry for the technical glitch; but we may start the show shortly.
   (c) Ladies and Gentlemen, sorry for the technical glitch; but we have to start the show shortly.

(d) Ladies and Gentlemen, sorry for the technical glitch; but we are going to start the show shortly.

2. (a) I have seen the rapist roaming around the other day.
   (b) I saw the rapist roaming around the other day.
   (c) I had seen the rapist roaming around the other day.
   (d) I saw the rapist roaming round the other day.

3. (a) Sorry, but I am not seeing any point in trying this option.
   (b) Sorry, but I don't be able to see any point trying out this option.
   (c) Sorry, but I am not to see any point trying this option out.
   (d) Sorry, but I don't see any point trying this option.

4. (a) When I went to the party, I was not recognizing my uncle; he was grown so old.
   (b) When I went to the party, I had not recognized my uncle; he had grown so old.
   (c) When I went to the party, I could not recognize my uncle; he had grown so old.
   (d) When I went to the party, I did not recognize my uncle; he had grown so old.

5. (a) Did you ever see the Taj?
   (b) Have you ever seen the Taj?
   (c) Have you ever seen to the Taj?
   (d) Had you ever been seen the Taj?

6. (a) Have you have seen the Taj when you lived in Agra?
   (b) Did you ever see the Taj when you lived in Agra?
   (c) Had you been seeing the Taj when you lived in Agra?

(d) Were you ever been to the Taj when you lived in Agra?

7. (a) When I was proposing her, she was not looking interested.
   (b) When I had proposed her, she did not look interested.
   (c) When I proposed to her, she did not look interested.
   (d) When I propose her, she did not look interested.

8. (a) During the curfew, the streets wore a desolate look.
   (b) During the curfew, the streets were wearing a desolate look.
   (c) During the curfew, the streets had been wearing a desolate look.
   (d) During the curfew, the streets have had worn a desolate look.

9. (a) Don't worries, the boss understand why you haven't come earlier?
   (b) Doesn't worry, the boss understand why you did not come earlier.
   (c) Don't worry, the boss understands why you did not come earlier.
   (d) Don't worry, the boss understand did you not come earlier?

10. (a) Even as a child, Ian Botham conveyed to his parents that he will become a cricketer.
    (b) Even as a child, Ian Botham conveyed to his parents that he would become a cricketer.
    (c) Even as a child, Ian Botham had conveyed to his parents that he would have become a cricketer.
    (d) Even as a child, Ian Botham conveyed to his parents that he should have become a cricketer.

 RECAPITULATION

✓ While speaking and writing, quite a few errors creep into our expressions. These can be avoided by carefully observing and following the grammatical rules learnt in the last two chapters.

ANSWER KEY

Practice Tests

	1.	2.	3.	4.	5.	6.	7.	8.	9.	10.
4.1	(b)	(d)	(a)	(c)	(b)	(d)	(d)	(b)	(a)	(b)
4.2	(c)	(a)	(d)	(d)	(c)	(c)	(b)	(b)	(b)	(b)
4.3	(c)	(a)	(b)	(d)	(b)	(d)	(c)	(b)	(d)	(b)
4.4	(c)	(b)	(d)	(b)	(b)	(a)	(b)	(b)	(d)	(b)
4.5	(b)	(c)	(b)	(c)	(b)	(b)	(a)	(a)	(a)	(b)
4.6	(c)	(d)	(d)	(b)	(d)	(a)	(c)	(b)	(b)	(c)
4.7	(b)	(d)	(a)	(c)	(a)	(c)	(b)	(a)	(b)	(a)
4.8	(c)	(b)	(b)	(a)	(c)	(b)	(d)	(d)	(a)	(a)
4.9	(a)	(c)	(b)	(c)	(a)	(d)	(a)	(d)	(a)	(c)
4.10	(b)	(a)	(c)	(a)	(d)	(b)	(a)	(c)	(a)	(b)
4.11	(b)	(a)	(d)	(c)	(d)	(a)	(d)	(b)	(a)	(c)
4.12	(d)	(c)	(a)	(c)	(b)	(a)	(b)	(a)	(b)	(d)
4.13	(d)	(a)	(d)	(a)	(a)	(b)	(c)	(a)	(c)	(c)
4.14	(a)	(d)	(b)	(a)	(c)	(a)	(b)	(d)	(a)	(d)
4.15	(b)	(d)	(c)	(d)	(a)	(a)	(c)	(b)	(b)	(b)
4.16	(a)	(b)	(b)	(a)	(c)	(b)	(d)	(c)	(a)	(c)
4.17	(b)	(c)	(b)	(d)	(b)	(b)	(a)	(c)	(b)	(d)
4.18	(c)	(b)	(a)	(a)	(c)	(b)	(b)	(b)	(a)	(c)
4.19	(c)	(d)	(b)	(a)	(c)	(c)	(d)	(b)	(b)	(b)
4.20	(d)	(b)	(d)	(c)	(b)	(b)	(c)	(a)	(c)	(b)

Jumbled Sentences

Learning Objectives   After reading this chapter, you will be able to

- understand what a jumbled sentence is
- learn how to unscramble a jumbled sentence and write it using correct grammatical structure
- identify the main clause and the dependent clause of a complex sentence, so as to put them in the correct order
- learn how to construct a well-knit paragraph with the help of given sentences

5.1 INTRODUCTION

By now you know that a sentence has various parts and that these parts have to be organized in a proper sequence to make some sense. What happens at times when you want to say something or write something is that you get various words related to the idea but find it difficult to arrange them in the proper sequence. Your experience tells you that something is wrong somewhere. You need to sort out whatever is wrong with the arrangement. Moreover, to test your ability to frame correct sentences, jumbled words are given in exams and you are expected to rearrange them to construct a meaningful sentence.

Let us read the following sentences carefully:

1. We decorated in white the kitchen last year.
2. Good movies are flops always.
3. All families happy one another resemble.
4. If you have to meet the professor today, you want to take a leave.

Are these sentences correct? No, you would perhaps like to rearrange them in the following manner:

1. We decorated the kitchen in white last year.
2. Good movies are always flops.
3. All happy families resemble one another.
4. If you want to meet the professor today, you have to take a leave.

So you see, proper sentences are made up of phrases. These phrases were jumbled in the first set, which we corrected as shown above. How you should approach the jumbled sentences is explained in the following section.

5.2 STEPS TO APPROACH JUMBLED SENTENCES

Under this section, we will discuss steps to approach jumbled sentences.

5.2.1 For Simple Sentences

- Identify the subject.
- Identify the action (verb).
- Identify the object.
- Identify the complement.
- Make the sentence and see if it makes proper sense.

5.2.2 For Complex Sentences

- Identify the main clause.
- Identify the subordinate clause.
- Put them in the right sequence to make the sentence.

Example 1
About/dreamt/you/I/yesterday

1. Identify the subject.	**I**
2. Identify the action.	**dreamt**
3. Identify the object.	**about you**
4. Identify the adverbial adjunct.	**yesterday**

I dreamt about you yesterday.

Example 2
Sita/has not finished/her/yet/work.

1. Identify the subject.	**Sita**
2. Identify the action.	**has not finished**
3. Identify the object.	**her work**
4. Identify the adverbial adjunct.	**yet**

Sita has not yet finished her work.

Example 3
that/lived in/the hills/I dreamt/I/the Himalayas/of

1. Identify the main clause.	**I dreamt**
2. Identify the subordinate clause.	**that I lived in the hills of the Himalayas**
3. Make the sentence.	

I dreamt that I lived in the hills of the Himalayas.

PRACTICE TEST 5.1

In the following questions, the jumbled parts of sentences are given. Rearrange them to make meaningful sentences:

1. should have/teachers/to the changing beliefs/and practices/an open mind
2. in/situations/thrive/people/extraordinary/difficult

3. and soul/I/my heart/Jenny with/love
4. the college/I/at/waited/for you/gate
5. choose/when you/life/to love/feels possible/anything
6. never/waver/people/of doubt/even if/thoughts/creep in/great
7. fair share/their/through/everyone/goes/of/tribulations
8. due to/also/rail traffic/was/fog/hit hard
9. take care/mother/I/need/of/to/my aging
10. people/killed/were/from/three/in/during/police firing/protests/tribal community
11. in advance/you/the tickets/have/to/the play/book/for
12. imposed/has been/in the/affected areas/curfew/also/of Sonitpur
13. zero-tolerance/the centre/has adopted/all acts of terror/against/a/policy
14. educationist Madan Mohan Malaviya/Bharat Ratna/awarded/posthumously/has been
15. tribute/melody/to/legendary/Rafi/the/playback/paid/singer/Lata/rich/Mangeshkar/queen/

PRACTICE TEST 5.2

Rearrange the following phrases so as to make meaningful sentences:

1. (a) something to eat
   (b) never turned a beggar
   (c) the old lady
   (d) without giving him
   (e) away from her door
2. (a) finished washing
   (b) the tap off
   (c) when you've
   (d) don't forget
   (e) to turn
3. (a) to make
   (b) I advised
   (c) a fool of himself
   (d) my friend
   (e) not
4. (a) so we
   (b) may as well
   (c) a little soup left
   (d) finish it up
   (e) there's only
5. (a) was fitted
   (b) with
   (c) every
   (d) the yacht
   (e) modern comfort
6. (a) among
   (b) the passengers on
   (c) broke out
   (d) board

(e) cholera
7. (a) killed
   (b) of unarmed
   (c) in cold blood
   (d) men were
   (e) hundreds
8. (a) lay down
   (b) for his country
   (c) he said
   (d) his life
   (e) he would willingly
9. (a) we can
   (b) always relay on
   (c) if the electricity supply
   (d) fails
   (e) the batteries
10. (a) there is no need
    (b) to be poorer than you are
    (c) just because he happens
    (d) to look down
    (e) upon him
11. (a) found him
    (b) have always
    (c) discussing
    (d) I
    (e) political and social issues
12. (a) bad news
    (b) how quickly
    (c) gets about

(d) it is amazing

(e) in this town

13. (a) before

   (b) had died

   (c) arrived

   (d) the patient

   (e) the doctor

14. (a) their assignment

   (b) to complete

(c) the teacher

(d) by the following day

(e) told the boys

15. (a) and the staff

   (b) a great show

   (c) the chief guest

   (d) for putting up

   (e) congratulated the boys

PRACTICE TEST 5.3

In the following questions, jumbled parts of sentences are given. Rearrange them to make meaningful sentences:

1. the official/English is/several countries/language of/of the world
2. no/at all/have/children/inhibitions
3. always have/poor people/faith//those/in/responsible/their/well-being/for/who are
4. Individuals/children/own perception/in their/are
5. mind of/have a/girls/and know/what they/their own/want
6. that as parents/we believe/the right/we have/to/and instruct/guide/for everything/our kids
7. spiders/most/begins/as soon as/die/winter
8. has begun/India's first/the Government of India/scholarship programme/multi-faceted child/this year
9. than the traditional/cooks/oven/faster/the microwave
10. to walk/are careful/thread of their web/only/spiders/on the non-sticky
11. will be/to live in/it/place/a great/think/I
12. it's/city mall/to the/ten minutes/only/my flat/from
13. busy and/I found/was a very/that Bengaluru/expensive city/terribly
14. asked/Ram/why he/going to/him/was/Delhi
15. on the cake/finely chopped/icing is spread/and decorated/with/dry fruits

PRACTICE TEST 5.4

In the following questions, jumbled parts of sentences are given. Rearrange them to make meaningful sentences:

1. in Beijing/winter/in most people's/mind's eye/is harsh/bleak, and chilly
2. withered leaves/the lane is paved/that is made of/with a thick yellow carpet/thousands of
3. along a peaceful lane/the noise of traffic/away from the crowds of people/and the stinging light of neon lamps/I sometimes saunter
4. all people are swathed/of clothing/in order to keep out/the severe cold/in many layers
5. the/unexpected/shocked and angry/attack/left/by the crowd/the leader
6. a growing number of people/today/winter sports and/enjoy outdoor/outdoor recreation/activities
7. surroundings/the city/mountains and/is surrounded by/beautiful lush green
8. he/walked into/as/the hotel,/to see/there/he was/his childhood friend/delighted
9. snow is/happens only once a year/a special winter weather occurrence/in most people's views/in New Zealand/which

10. known to mankind/of expression/is perhaps/dance/one of the oldest forms
11. on society/there are/view the impact/different ways to/of media technology
12. and a dull one/the difference/a positive attitude/can make//between/an amazing life
13. the key to/a positive attitude/their dreams/all the people who/might be/want to fulfil
14. work for another/no two people/what works for one person/study the same way/little doubt that/may not/and there is
15. help you/this guide/skills/develop/is designed to/effective study

5.3 UNSCRAMBLING A PARAGRAPH

When you get a jumbled paragraph and are asked to unscramble it, you should do the following:

1. Identify the theme/topic of the paragraph.
2. Know how the discussion may begin.
3. Comprehend the various points being communicated, related to the topic under discussion.
4. Understand what makes the idea complete.

Let us read the following scrambled paragraph and try to learn how to construct an effective paragraph.

> Writing an essay is a difficult thing to do./Brainstorming is a good tool for finding the right idea./It is hard to organize your thoughts./You may have many different ideas about the subject./Organizing your thoughts is an important first step to writing an essay./It is important to pick an idea that is easy to write about.

Step one Try to identify the theme/topic of the paragraph, which in this case is writing an essay.

Step two Figure out how the discussion may begin—writing an essay is a difficult thing to do.

Step three Understand the various points being communicated, related to the topic under discussion, which are as follows:

1. It is hard to organize your thoughts.
2. You may have many different ideas about the subject.
3. It is important to pick an idea that is easy to write about.
4. Brainstorming is a good tool for finding the right idea.

Step four Finally, understand what makes the idea complete—organizing your thoughts is an important first step to writing an essay.

Suggested answer

> Writing an essay is a difficult thing to do. It is hard to organize your thoughts. You may have many different ideas about the subject. It is important to pick an idea that is easy to write about. Brainstorming is a good tool for finding the right idea. Organizing your thoughts is an important first step to writing an essay.

PRACTICE TEST 5.5

In each of the following questions, there are six parts of a sentence or six sentences marked S$_1$, S$_6$, P, Q, R, S. The position of S$_1$ and S$_6$ are fixed. Choose the best alternative from the options P, Q, R, and S:

1. (S$_1$) If all people were
   - (P) drastically different
   - (Q) born with
   - (R) the world would be
   - (S) no distinct characteristics,
   (S$_6$) from what we see today.
   - (a) RSPQ (b) SRQP (c) QSRP (d) PRQS

2. (S$_1$) Then she sailed
   - (P) from door to door
   - (Q) to sea shore
   - (R) the boat back
   - (S) and sold the fish
   (S$_6$) at cheap price.
   - (a) RQPS (b) SPRQ (c) QSPR (d) RQSP

3. (S$_1$) The settlement between the Government of Assam and the leaders of Bodoland Movement,
   - (P) public property and socio-economic destabilization,
   - (Q) that has cost the state
   - (R) rather heavily in terms of human lives,
   - (S) augurs well for the fulfilment of
   (S$_6$) the legitimate aspirations of an articulate group of tribals.
   - (a) PQRS (b) PRQS (c) QRPS (d) QRSP

4. (S$_1$) On one hand, we are proud of being Indians,
   - (P) on the other hand, we behave as if we were still at the dawn of our civilization;
   - (Q) murdering our own brothers and sisters is neither the way to please Ram or Rahim,
   - (R) the citizens of the land where Buddha and Gandhi taught
   - (S) the principle of love and non-violence,
   (S$_6$) nor does it fetch us any prosperity.
   - (a) PQRS (b) QRSP (c) RSQP (d) RSPQ

5. (S$_1$) Science has proved that
   - (P) the rates of
   - (Q) rise drastically
   - (R) due to drinking alcohol, smoking, excess weight
   - (S) miscarriages and birth defects
   (S$_6$) and worries.
   - (a) RQSP (b) PSQR (c) RPQS (d) PQRS

6. (S$_1$) Irrespective of how many
   - (P) or how many failures you cross
   - (Q) mistakes you make
   - (R) the success and happiness
   - (S) you gain through it is fantastic
   (S$_6$) and better than anything.
   - (a) QPRS (b) RQPS (c) SRPQ (d) QSRP

7. (S_1) Smoke oozed up between the planks.
   (P) Passengers were told to be ready to quit the ship.
   (Q) The rising gale fanned the smouldering fire.
   (R) Every one now knew there was a fire on board.
   (S) Flames broke out here and there.
   (S_6) Most people bore the shock bravely.
   (a) QPSR (b) QSRP (c) RSPQ (d) SRQP

8. (S_1) On the basis of experiments with rats,
   (P) health experts say that
   (Q) exercise more, and consume vitamins,
   (R) they will live up to 100 years or more
   (S) if humans eat less,
   (S_6) and be vigorous in their eighties and nineties.
   (a) RQSP (b) PSQR (c) RPQS (d) PQRS

9. (S_1) In theory, India is an example of a perfect democracy.
   (P) The vast majority of people are illiterate and ignorant.
   (Q) But, in practice, the working of democracy does not provide much cause for satisfaction.
   (R) General elections have no doubt been held, at regular intervals.
   (S) But the very first condition of a properly educated electorate is lacking in this country.
   (S_6) it is said that the greatest of all dangers of democracy is that it may turn out to be the rule of ignorance.
   (a) QPRS (b) RQPS (c) SRPQ (d) QSRP

10. (S_1) I urge you
    (P) Fake it if you must and
    (Q) deserve to live an extraordinary life.
    (R) as though you are miraculous and
    (S) to begin living every moment
    (S_6) keep faking it until it's real to you.
    (a) QPSR (b) QSRP (c) RSPQ (d) SRQP

PRACTICE TEST 5.6

In each of the following questions, there are six parts of a sentence or six sentences marked S_1, S_6, P, Q, R, S. The position of S_1 and S_6 are fixed. Choose the best option from the choices P, Q, R, and S:

1. (S_1) Former Australian star Adam Gilchrist's idea
   (P) an Olympic sport has received backing
   (Q) that twenty 20 cricket should bid to become
   (R) who feel it would be a superb vehicle
   (S) from the likes of Sourav Ganguly, Steve Waugh, and Stephen Fleming
   (S_6) to globalize the game.
   (a) PSQR (b) RSQP (c) QPSR (d) SQRP

2. (S_1) If the announcement of an all-party meeting
   (P) it appears to have failed miserably
   (Q) was intended as a balm,
   (R) and the funeral of two men killed in police firing on the previous day

(S) with violence escalating on Tuesday across Jammu

(S$_6$) acting as the rallying point for tens of thousands of protesters.

    (a) PSQR   (b) RSQP   (c) SQRP   (d) QPSR

3. (S$_1$) Helped by a rescue party,

    (P) after spending four nights on the world's second-highest mountain

    (Q) following a climbing disaster

    (R) on Tuesday morning

    (S) an Italian climber hobbled on frostbitten feet down K2,

(S$_6$) that killed 11 others.

    (a) PSQR   (b) RSQP   (c) SRPQ   (d) QPSR

4. (S$_1$) Helen Keller has an ageless quality about her, in keeping with her amazing life story.

    (P) Although warmed by this human reaction, she had no wish to be set aside from the rest of humankind.

    (Q) She is an inspiration to both the blind and the seeing everywhere.

    (R) When she visited Japan after World War II, boys and girls from remote villages ran to her, crying 'Helen Keller'.

    (S) Blind, deaf, and mute from early childhood, she rose above her triple handicap to become one of the best known characters in the modern world.

(S$_6$) She believed that the blind should live and work with their fellows, with full responsibility.

    (a) PSQR   (b) RSQP   (c) SQRP   (d) QPSR

5. (S$_1$) Jawaharlal Nehru was born in Allahabad on 14 November 1889.

    (P) Nehru met Mahatma Gandhi in February 1920.

    (Q) In 1905 he was sent to London to study at a school called Harrow.

    (R) He became the first Prime Minister of independent India on 15 August 1947.

    (S) He married Kamla Kaul in 1915.

(S$_6$) He died on 27 May 1964.

    (a) QSPR   (b) RSQP   (c) SRPQ   (d) QPSR

6. (S$_1$) A woman came into the room with tea and a small platter.

    (P) the woman placed a steaming

    (Q) cup of tea before me and served some snacks

    (R) then she disappeared making no sound

    (S) I slouched in the chair and sipped the strong black tea

(S$_6$) The host came and was shocked to see me.

    (a) QSPR   (b) RSQP   (c) SRPQ   (d) PQRS

7. (S$_1$) The sixties were historically

    (P) and culturally a complex period

    (Q) yet for others it was a period

    (R) For some it was a golden era

    (S) when the old order of morality

(S$_6$) authority, and discipline lost its legitimacy.

    (a) QSPR   (b) PRQS   (c) SRPQ   (d) PQRS

8. (S$_1$) It however does not mean that

    (P) In fact, corruption is ubiquitous

    (Q) Corruption occurs only in democratic countries.

    (R) But it is in a democracy that corruption is rampant

(S) as it is as very much present in other forms of government.

(S$_6$) when compared to other forms of government.

    (a) PSQR   (b) RSQP   (c) QPSR   (d) SQRP

9. (S$_1$) According to the latest media reports, the Indian Entertainment and Media Industry

    (P) and thus, looking at the present scenario of growth of Indian Entertainment,

    (Q) is annually growing at a rate of 12.5%

    (R) anticipating a phenomenal gap

    (S) Media & Marketing industry through production of Films, TV Serials, AD Films etc.,

(S$_6$) in the demand and supply of the right professionals.

    (a) PSQR   (b) RSQP   (c) SRPQ   (d) QPSR

10. (S$_1$) Dr. Patil is a name to reckon with in the field of education.

    (P) the prestigious civilian award, the Padmashree

    (Q) for his contribution to the society

    (R) he has been honoured with

    (S) A philanthropist and educationalist,

(S$_6$) through various educational, social and cultural initiatives.

    (a) PSQR   (b) RSQP   (c) SRPQ   (d) QPSR

RECAPITULATION

✓ A jumbled sentence can be unscrambled by identifying and maintaining correct grammatical order between the different parts of a sentence, such as its subject, verb, object, complement, adverbial adjunct, etc.

✓ In complex sentences, identifying the main and the subordinating clauses helps in unscrambling a jumbled sentence.

✓ In unscrambling a paragraph and maintaining order in it, you need to put in correct sequence the introducers, developers, and terminators besides identifying the central idea expressed through the topic sentence in a paragraph. You will learn about these terms in detail in Chapters 17 and 20.

— **WISEWELL QUIPS** —

ANSWER KEY

Practice Tests

5.1
1. Teachers should have an open mind to the changing beliefs and practices.
2. Extraordinary people thrive in difficult situations.
3. I love Jenny with my heart and soul.
4. I waited for you at the college gate.
5. When you choose to love life, anything feels possible.
6. Great people never waver even if thoughts of doubt creep in.
7. Everyone goes through their fair share of tribulations.
8. Rail traffic was also hit hard due to fog.
9. I need to take care of my aging mother.
10. Three people from tribal community were killed in police firing during protests.
11. You have to book the tickets for the play in advance.
12. Curfew has been also imposed in the affected areas of Sonitpur.
13. The Centre has adopted a zero-tolerance policy against all acts of terror.
14. Educationist Madan Mohan Malaviya has been awarded Bharat Ratna posthumously.
15. Melody Queen Lata Mangeshkar paid rich tribute to the legendary playback singer Rafi.

5.2

1. cbeda	4. ecabd	7. ebdac	10. cbade	13. dbaec
2. debca/cadeb	5. dabce	8. ceadb	11. dbace	14. cebad
3. bdeac	6. ecabd	9. cdabe	12. dbace	15. ceadb

5.3
1. English is the official language of several countries of the world.
2. Children have no inhibitions at all.
3. Poor people always have faith in those who are responsible for their well-being.
4. Children are individuals in their own perception.
5. Girls have a mind of their own and know what they want.
6. We believe that as parents we have the right to guide and instruct our kids for everything.
7. Most spiders die as soon as winter begins.
8. This year, the Government of India has begun India's first multi-faceted child scholarship programme.
9. The microwave cooks faster than the traditional oven.
10. Spiders are careful to walk only on the non-sticky thread of their web.
11. I think it will be a great place to live in.
12. It's only ten minutes from my flat to the city mall.
13. I found that Bengaluru was a very busy and terribly expensive city.
14. Ram asked him why he was going to Delhi.
15. Icing is spread on the cake and decorated with finely chopped dry fruits.

5.4
1. In most people's mind's eye, winter in Beijing is harsh, bleak and chilly.
2. The lane is paved with a thick yellow carpet that is made of thousands of withered leaves.
3. Away from the crowds of people, the noise of traffic and the stinging light of neon lamps, I sometimes saunter along a peaceful lane.
4. In order to keep out the severe cold, all people are swathed in many layers of clothing.
5. The unexpected attack by the crowd left the leader shocked and angry.
6. Today, a growing number of people enjoy outdoor winter sports and outdoor recreation activities.
7. The city is surrounded by mountains and beautiful lush green surroundings.
8. As he walked into the hotel, he was delighted to see his childhood friend there.
9. In most people's views in New Zealand, snow is a special winter weather occurrence which happens only once a year.
10. Dance is perhaps one of the oldest forms of expression known to mankind.
11. There are different ways to view the impact of media technology on society.
12. A positive attitude can make the difference between an amazing life and a dull one.
13. A positive attitude might be the key to all the people who want to fulfil their dreams.
14. No two people study the same way, and there is little doubt that what works for one person may not work for another.
15. This guide is designed to help you develop effective study skills.

5.5

1. c	3. c	5. b	7. d	9. a	**5.6** 1. c	3. c	5. a	7. b	9. d
2. d	4. d	6. a	8. B	10. d	2. d	4. c	6. d	8. c	10. c

Indianisms

After reading this chapter, you will be able to

- understand what is meant by Indianisms in terms of the English language
- understand some of the commonly used expressions in Indian English
- identify the errors commonly made by a large number of Indians while speaking and writing English
- learn how to avoid the commonly employed incorrect Indian English usages and prefer the standard English usage instead

6.1 INTRODUCTION

Since English is not our native language, it is not easy for us to attain the level of accuracy and comfort that comes with the luxury of using a mother tongue. In fact, there are a large number of expressions, words, phrases, and other linguistic structures commonly employed by the Indian speakers and writers of English, which are not in consonance with standard English used by the native speakers and writers of English. We generally fail to differentiate the erroneous expressions from the correct ones. Moreover, since a large number of people talk or write erroneously in a particular way, it gains currency over the years and is never noticed for it being anomalous to standard English usage. However, when it comes to interacting with native speakers, such errors not only confuse the native users of English, but also bring into question our credibility as proficient users of English.

The following section lists some commonly used erroneous words, expressions, phrases, and other linguistic structures and their standard English usage.

Indian English	Standard English Usage
By walk	**On foot**
Many of us say something like 'These days I am going to office *by walk*.' The standard English usage would suggest that we say: 'These days I am going to office on foot'.	

(Contd)

Indian English	Standard English Usage
Dickey Quite commonly, we come across people telling each other: 'Don't go for this car; it has a very small *dickey*'. Now, the word *dickey* in standard English does not exist in this sense. The proper word for this is *boot* or *trunk*. So, the correct expression is 'Don't go for this car, it has a small *boot*'.	**Boot**
Hail from Once *hail from* was a common replacement for *belong to* and *come from*. It is no longer in use and we should instead say, 'I come from/belong to Rajasthan,' rather than telling someone something like 'Don't you know, I hail from Rajasthan.' Such an expression is now obsolete.	**Belong to**
Cut jokes It is very common for us to hear something like 'Siddhu knows how to cut witty jokes.' Now, jokes are not *cut*, they are *cracked* or *told*. Therefore, it is more appropriate to say: 'Siddhu knows how to crack jokes.'	**Crack jokes**
Good name It is customary for most of us to make our question polite in some such way: 'What is your good name?' However, the expression 'good name' seems culture specific as in our country, everyone has two names—one good/formal and the other informal one, something that does not happen all over the world. Therefore, it will suffice to say—'What is your name?'	**Name**
Tight slap 'I will give you a *tight slap* on your cheek, if you say that again.' This is a very common way to admonish someone, especially children, though it seems that the method of punishment is not explained in standard English. A native speaker is more likely to structure it with *hard slap* and *on your cheek*, is likely to be *in the face*. So, it is more likely to be 'I will give you a hard slap on your face.'	**Hard slap**
Marriage Most of us in India firmly believe in adages like 'Marriages are made in heaven'. Supposing, we change it to 'Weddings are made in heaven', won't there be some eyebrows raised in protest? Certainly, there would be quite a few. But there are not any for sure when somebody invites you to the marriage of their son. Now just as *marriage* refers to the relationship between a husband and wife, *wedding* refers to the ceremony that solemnizes this relationship. Therefore, the standard English user is required to stick to *wedding* in place of *marriage* when the reference is to the ceremony. Therefore, instead of saying, 'Please do come to the marriage of my brother', we should say, 'Please do come to my brother's wedding.'	**Wedding**

(Contd)

Indian English	Standard English Usage
Get down Consider a sentence like 'He fell while getting down from the horse.' In standard English usage the expression *getting off* is likely to be preferred to *getting down*. Similarly, 'I am going to get down at the next station,' needs to be modified as 'I am going to get off at the next station'.	**Get off**
On/Off the light before one moves out Very commonly, we come across a sentence like 'Off the light before you move out of the room'. A sentence like this is likely to sound confusing to a native speaker as *off* cannot be used as a verb in standard English. Therefore, we are required to use a proper verb such as *turn off* or *switch off* instead of just *off*. Moreover, the expression *move out* suggests permanently vacating a house, etc. Hence, it would be more appropriate to say, 'Turn off the light before you leave the room.'	**Turn on/off the light before leaving the room**
Co-brother/Co-sister The native speaker of English will have difficulty in figuring out the meaning of a bizarre expression like 'Muthuswamy was bad as a brother, but worse as a co-brother'. The expression *co-brother* is very commonly used in South India whereas its counterpart *co-sister* is quite frequently used in some other parts of the country as well. Co-brother is used to refer to the husband of someone's wife's sister. Similarly, co-sister stands for the wife of someone's husband's brother. In standard English usage such relations are yet to be specified with these connotations and therefore they stick to the standard *brother-in-law* and *sister-in-law* arrangement. Therefore, rather than telling someone, 'Meet my co-brother, he lives in Mumbai', say 'Meet my brother-in-law, he lives in Mumbai'.	**Brother-in-law/Sister-in-law**
Bio-data and jack Quite commonly, you hear someone speaking loudly on the mobile—'Just send me your biodata and I will use the jack to get you this job!' Most native speakers of English are likely to cut a sorry figure when they are required to understand an expression like this. Obviously, the word 'jack' seems quite crude and far too literal to make its way into the proper English structure in this sense. Also, the word *biodata* is not all that well known to a native speaker who sticks either to *curriculum vitae* or *résumé*. So, instead of saying, 'He got the job not because of his biodata but because of the jack', say 'He got the job not on the strength of his résumé but because of the approach.'	**Résumé and approach**

(Contd)

Indian English	Standard English Usage
Purchase Indian users of English do not create a proper distinction between these two words. Therefore, it is not surprising to hear something like 'I am going to purchase two breads.' The native speaker, if he/she happens to hear this sentence, is likely to only smile indulgently. Actually, *purchase* is used for some elaborate and formal type of buying or for buying something very expensive. Therefore, the word *purchase* should give way to *buy* in the above context. Moreover, *two breads* should be replaced either with *two slices of bread* or *two loaves of bread*, as the sense requires. Therefore, the better way to say this would be 'I am going to buy two loaves of bread.'	**Buy**
On somebody's face In Indian English, we get to hear a sentence like 'I told him on his face that it was a dirty joke'. Similarly, an expression like 'When I cracked the joke, it brought a smile on his face' is also not so uncommon. In standard English usage however, *to his/my face* is preferred. Therefore, it is better to say 'When I cracked the joke, it brought a smile to his face.'	**To somebody's face**
Time is over Have you not heard an announcement like 'Stop writing; time is over'? In standard English usage, the common expression is *time is up*. Similarly, when we run short of a grocery item, the correct expression is …*run out of something*, and not that *it is over*. Therefore, the correct expression is 'Stop writing! Time is up!'	**Time is up**
Lesson is delivered By saying 'I have delivered the lesson to the child; now he has to revise it,' some happy tutor could be suggesting that she has finished her job. In standard English *delivered*, apart from 'delivering babies', is used with *speech* and *lecture* and *talk*, and not with 'lesson' which is just *given*. Therefore, the correct expression is 'I have given the lesson to the child; now he has to revise it.'	**Lesson is given**
Cannot engage one's period today In Indian English, we often listen to the teachers saying 'I am sorry, I will not be able to engage my class today.' In standard English, the expression *engage the class* actually connotes the idea that someone is engaging the class as a replacement for someone else. For example, a teacher of Fine Arts in Indian schools may be asked to engage the students of class IX when their regular Science teacher has not turned up. So, following standard English, stick to *taking the class* rather than *engaging the class*.	**Cannot take my class**

(Contd)

Indian English	Standard English Usage
Pin drop silence Many speakers have been appreciated in Indian English as when they spoke '…there was a pin-drop silence.' A native speaker is more likely to stick to something like …*there was perfect silence*, or *complete silence*.	**Perfect/complete silence**
Yesterday evening Most of us experience guilt/embarrassment while saying 'I saw him yesterday evening.' Now, it is fine to say *yesterday morning*, or *yesterday afternoon*, but saying *yesterday evening*, or *yesterday night* is simply far too much, obviously because you cannot refer to a division of time by putting *day* with *evening* and *night*. Therefore, it is appropriate to say—'I saw him last evening/last night.'	**Last evening**
Footpath We never question a statement like 'Road for vehicles while *footpath* for pedestrians.' In standard English, the word that is chosen to denote a 'footpath' is *pavement*.	**Road for vehicles and pavement for pedestrians**
Expire None of us wants to be dead. Likewise when others die we cannot really try to avoid saying that 'Mr Sadajeevan Ram is dead.' It's fine as long as we say, 'Mr Sadajeevn Ram is no more,' or 'Mr Sadajeevan Ram has passed away.' But the moment we start saying, 'Mr Sadajeevan Ram has expired,' it is likely to create only a comic effect no matter how mournful you sound while admitting this. It is so because in standard English usage the term 'expire' is used to suggest the ending of a term or period by which some of the items like medicines and food items are required to be consumed or utilized. Therefore, say 'He has passed away, or 'He is *no more*', rather than saying 'He has *expired*.'	**Pass away**
Demise As suggested earlier, most of us are really keen to find out a rosy replacement or euphemism for unpleasant expressions which seem blatantly crude and impolite. Particularly when related to someone being dead, we are quite keen to suggest that someone has just *passed away*, or *expired*. Similarly, while reporting someone's death, we also use an expression like 'Ever since our father's demise, the business has not prospered as desired.' Now, in standard English *demise* is hardly in practice as it is considered rather obsolete. So, a standard way of saying that is 'Ever since our father's death, the business has not prospered as desired.'	**Death**

(Contd)

Indian English	Standard English Usage
Carrying It is understandable that someone's death needs some euphemism on the part of the speaker so that he/she can express this without being rude. But there appears to be no reason—at least to a native speaker—when we avoid something like 'My sister is pregnant', and start replacing it with 'My sister is carrying'. Now, structurally the verb *carrying* is required to be followed by some object and though it is acceptable to say—'My sister is carrying a baby', with possible connotations that she is already a mother and is carrying her baby, it is certainly more appropriate to say 'My sister is pregnant.'	**Pregnant**
Better half People proudly introduce their wives saying, 'Meet my better half…'. The expression in standard English is more likely to be understood as something said in order to sound humorous. However, that may not be the intention of many of the Indian husbands while introducing their wives formally to others. Therefore, the more fanciful *better half* had better be replaced with something more standard such as 'Meet my wife'.	**Wife**
Mrs A simple enquiry like 'What is your Mrs doing these days?' is not likely to raise any eyebrows when both the speaker and the listener are Indians. It is so because in Indian English parlance, *Mrs* is a direct replacement for *wife*, something that is not the case in standard English, where *Mrs* is strictly regarded as a title and is not used as a name itself. Therefore, it is fine to say—'She is Mrs Chaudhary…', or 'Mrs Bansal, why are you so upset?', etc. But not something like 'Where is your Mrs?' The proper structure in standard English would be 'Where is your wife?'	**Wife/Spouse**  Calling Somebody's Wife *Mrs* is not Humorous
Madam It seems that Indian English consistently tries to avoid a seemingly unintimidating word like *wife*. So, quite often you get to face a question like 'Where is Madam? Why have you not brought her?' Probably it is just an expression of someone's position at home, but the native users of English are certainly likely to find themselves somewhat bemused, for they would naturally expect 'madam' to be followed by some proper noun. Therefore, on formal occasions, 'Where is your Madam' should be replaced by a standard usage, for example, 'Where is your wife?'	**Wife**

(Contd)

Indian English	Standard English Usage
Receive 'I am going to the bus stand to receive my wife.' When a British or American wife is going to listen to such an effusive sense of welcome with which someone is likely to *receive* their wife, she is bound to feel envious of her Indian counterpart. It is so because when you *receive* someone, you have garlands ready and you are giving someone a rousing and formal welcome. Therefore, in a less hyperbolic and realistic situation, you are only going to *meet and bring home* your wife.	**Meet and bring home**
See somebody at five fifteen Some of us are very sloppy when it comes to suggesting what time of the day/night it is. So, we often end up saying something like 'See you at five fifteen!' A native speaker of English, however, is more likely to stick to the formal expressions like 'See you at quarter past five,' or 'The train is going to leave at quarter to six', or 'The meeting will resume at half past ten,' etc.	**See somebody at quarter past five**
OK In Indian English, the expression *OK* serves various different purposes. It stands for the traditional 'all correct' suggesting that the listener has agreed to the speaker's proposition. For instance, 'Shall we dine out today?' may be answered with an *OK*. This suggests the approval of the listener. However, very frequently, particularly when a telephonic conversation is going on, the expression *OK* replaces the more appropriate *Yes*. For example, to the statement 'When I reached the bus stand, I missed my bus', the listener can very innocuously add his/her *OK* as a way of response which actually suggests, 'Carry on, I am listening.'	**Right/tell me**
How do you do? Indian speakers of English do not mind exchanging their pleasantries with this particular expression even if they have met the same person many times earlier. In standard English, 'How do you do?' is used when you meet the person for the first time. In the subsequent meetings, it is more appropriate to say 'How are you?'	**How are you?**
Between ... to Quite often, you hear a receptionist telling you that 'The—office remains closed for lunch between 1.00 to 2.30 p.m.' In standard English usage, the correct form is likely to be either 'between 1.00 and 2.30 p.m.', or 'from 1.00 to 2.30 p.m.'	**Between ... and/from ... to**
Bearer For many of us, *bearer* is a synonym of the word *waiter*. Therefore, it surprises none of us when we come across a sentence like 'The bearers in that hotel are not so polite.' In standard English, however, the word *bearer* does not replace the word *waiter*.	**Waiter**

(Contd)

Indian English	Standard English Usage
August audience Time and again, we listen to speakers going up the dais and beginning their speech by addressing the audience with an apparently unctuous remark such as '...I am so delighted to be addressing an august audience like this...' Of course, the speaker is trying to suggest to the audience that they are quite knowledgeable and respected—just like the respected critics and connoisseurs of the Augustan age. However, before we start using such an expression, we must understand that in modern standard English, the expression is regarded as somewhat hackneyed and clichéd.	**Special/Distinguished gathering**
Colony In Indian English, it is quite common to say something like 'Our colony is truly Indian in letter and spirit! We have people from different professions, languages, and religions residing in the same colony.' In standard English, the word *colony* refers to a locality where people involved in the same profession or with similar backgrounds are grouped together in a neighbourhood with each other. So, we have *the doctor's colony*, *the teacher's colony*, etc. In order to communicate the sense intended above, the word *colony* gives way to *locality* or *neighbourhood* or *residential area*.	**Locality/Residential area**
Hotel The kids are feeling hungry! Why don't you stop at some hotel?' could be a very common expression in Indian English. It is so because in Indian English *hotel* is used for a place where one can just eat and not necessarily stay. In standard English, however, *hotel* is a place where you also stay besides taking your food. Therefore, the more appropriate replacement for *hotel* in this context should be *restaurant*.	**Restaurant**
Wheatish 'Wanted a good natured, educated, well-settled boy for a homely, beautiful, educated girl with a wheatish complexion....' Such advertisements are abundantly seen in advertisements seeking matrimonial alliances. The phrase *wheatish complexion* suggests someone's complexion that is *not dark*. So it would be better to say *on the fairer side* instead of *wheatish*.	**Fair**
Innocently divorced In Indian English coinage, *innocently divorced* is used to describe a bride who gets divorced even before her marriage is properly consummated. The concept of innocence stems from a conventional Indian marriage where sometimes the bride is sent to the husband's house months/years after her marriage. It used be only after the ritual, known as *gauna* in some parts of the	**Divorced**

(Contd)

Indian English	Standard English Usage
country, that the girl was considered married in the real sense and before that her marriage was not really 'consummated'. Before that consummation, she was deigned to be innocent. The phrase *innocently divorced* still lurks much like the practice itself which is still in existence in parts of the country.	

Indian English	Standard English Usage
To eat somebody's salt 'Sir, I have eaten your salt! How can I go against you?'—The moment an Indian speaker speaks these words to a native speaker, ostensibly of course to suggest his/her unfailing commitment to the latter, the native speaker is most likely to blink. It is so because *eating someone's salt* has a quintessential cultural tinge to it which refuses to sound universally comprehensible. The idea is that someone is trying to convince the listener that they *cannot betray them for they owe them a lot.*	**To owe somebody a lot and so cannot betray him/her**  Literal Translations of Indian Proverbs Sound Funny
Lose somebody 'Ladies and gentlemen, I have lost my son. If you see him anywhere, please let me know immediately.' A dramatic appeal like this is most likely to toss the audience comprising native listeners into serious bewilderment. The first part of the information—'… I have lost my son!'—is surely going to evoke feelings of shock and sympathy for the speaker as *losing the son* connotes the death of the son. Therefore, the second part '… If you see him anywhere, please let me know immediately' is certainly going to befuddle the native speaker for he/she knows quite well that *seeing somewhere* someone who is already dead is really impossible.	**Not to be found anywhere; somebody has gone missing!**
To keep firing somebody all the time 'My boss is very rude. Repeatedly we get firing from him.' 'You don't know my boss; he keeps firing us everyday.' It would really be nice if a native user of English keeps away from a learned conversation like this. It is so because nobody can really keep firing someone everyday, for *firing* someone essentially means 'dismissing them from a job/employment'. However, the way it is sometimes used in Indian English suggests that someone is so rude that he/she keeps yelling at you.	**To keep yelling at someone all the time**

(Contd)

Indian English	Standard English Usage
To put up When you are *put up* somewhere, it refers to the hotel where you are lodging during a short stay. However, many of the Indian speakers, in asking 'Where are you put up?', actually try to find out from the listener where he/she lives. A native speaker in seeking such an information is more likely to ask 'Where do you live?'	**To live**
Tell me 'Is this Madam Slambang speaking?' 'Yeah, tell me?' A very commonly heard *tell me* is supposed to be offered in Indian English as a worthy replacement for the more sophisticated and polite '…how can I help you?' A native speaker of English is more likely to take offense while being greeted like that over the phone.	**How can I help you?**
Shifting one's house 'Sorry, we can't come to your place at least for a week or so. You know, we are shifting our house!' Wondering at the gargantuan might of the speaker, the listener—if he/she is a native speaker of English—is more likely to curse themselves for not being able to do what the speaker on the other side can so effortlessly perform. It is so because the native speaker is likely to suggest that they would only be *moving to a new house/place* something nowhere comparable to *shifting the house*, for if you are really able to *shift your house*, you have moved it or changed its position from one place or position to another.	**Changing one's house**
Come to one's senses 'You don't know how dreadful it was! I lay there unconscious for seven weeks. And when I came to my senses, everything was over.' It looks like someone's tragic tale. The tragedy though is not just human but also linguistic. It is so because in this context *came to my senses* replaces *regained consciousness*. In standard English, however, when someone *comes to their senses*, he/she starts behaving sensibly and the corollary unmistakably is that he/she had not been doing that at least for some time.	**Regain consciousness**
Itself/Only 'The marriage is going to take place at our home itself/only.' The announcement made in Indian English is likely to sound quite odd to the native speaker of English. In the first place, he/she would like *marriage* to be replaced by *wedding* and then there is no *itself* or *only* required, written or spoken, to imply the sense of extra insistence. A native speaker is more likely to introduce an inversion in the structure of the sentence to secure emphasis and is probably going to speak or write something like 'It is at our home that the wedding would take place.'	**No word for emphasis or insistence**

(Contd)

Indian English	Standard English Usage
Too good	**Very good**
'The movie is too good! You must go and watch it.' This expression suggests that Indian English does not differentiate the good from the bad. To suggest that something is *too good* is an example of bad collocations because *too* normally emphasizes something bad in standard English in the same may *very*, *really*, etc. are chosen to emphasize something good. Therefore, the idea can be better expressed by saying 'The movie is really good! You must go and watch it.'	
Even	**So do/did I**
I'I liked the movie very much.' 'Even I.' This seemingly innocuous support to your friend's likings can offend him/her if he/she is aware of the nuances of *even I*. In the context cited above, the use of *even I* suggests '…it is fine you liked it, even I (what to talk of you, even someone as good or great as I) too liked it!' You now understand how offending it might be to your listener. Therefore, in standard English usage, the response is more likely to be 'So did I. 'Look at the following sentences and see the standard English replies:' We can win the championship.' 'So can we.' 'I never enjoyed grammar.' 'Nor did I.'	
Issue	**Children**
'How many issues do you have?' 'Many. But none that we can resolve.' In the question, *issues* stands for *children* and in the reply it refers to *differences*. It is so because in standard English, *issue* is rarely used to mean *child* unless the question relates to the legal inheritance. Therefore, it is much more appropriate to keep it simple—'How many children do you have?'—especially when you interact with a native speaker of English.	
Timepass	**Average**
'How was the movie?' 'Just timepass!' The expression *timepass* is used quite frequently in Indian English in order to suggest that something was ordinary and not as good as it was expected to be. There may not be much in the same vein for a native user where the phrase timepass is not used for such connotations.	
Give an exam	**Take an exam**
'Stop watching television! You have to give an exam tomorrow.' A whole lot of English speaking Indian mothers have admonished their kids thus, in a language that continues to dodge them for it refuses to be	

(Contd)

Indian English	Standard English Usage
translated according to the needs of their mother tongue. In standard English usage, it is the examiner who *gives* the exam/test and the student can only *take* it. Therefore, one of the more appropriate expressions would be 'You have an exam to take tomorrow.'	

Guest/Visitor — **Visitor**

'When I was about to start, suddenly some guests came.' If the listener happens to be a native user of English, he/she is more likely to blink than make out as to what kept them waiting for you. It is so because in standard English, *guests* are always invited and the one that 'drops in when you are about to start' is actually a *visitor*.

Female — **Woman**

'Why can't the females do all that a man does?' Even if you keep supporting the doctrine of *feminism*, you cannot agree to the proposition of allowing all the *females* do all that a man does. It is so because if you do, we would see a tigress driving a metro, a cow delivering a keynote address at a conference and a mare sermonizing in a church! Since all these are 'females' and not necessarily *women*, and it would be a free for all affair if we tend to use English, a foreign language, so casually. The word *female* in standard English usage is mostly used as a qualifier and denotes some noun in expressions such as female staff, female dancer, etc. Therefore, the equality between men and women can only be demanded by asking the question more appropriately as 'Why can't women do all that a man does?'

Did She Mean 'Women'?

Not mind — **Sure**

'Shall we celebrate Jatin's birthday at a five star hotel?'
'I don't mind.'
A native speaker of English is bound to fathom some sort of reluctance or lack of enthusiasm in the respondent's reply. The expression 'I don't mind' is used to suggest that you do not have any objection to what others have proposed. Unmindful of such connotations, many Indian speakers of English are seen using 'I don't mind' to suggest something that they themselves are very keen to do. By the way, there are many other ways by which you can express your approval, such as 'Really! That would be great',
'Oh yes! Shall we?', 'Yes, by all means', 'Sure', etc.

(Contd)

Indian English	Standard English Usage
Mixy 'Why don't we go for a new mixy? It creates such a din even if I use it for a minute.' A conversation like this is fully comprehensible provided the dialogue takes places between two users of Indian English, because here the *mixer* conveniently steps into its Indianised version *mixy*.	**Mixer**
Do One Thing Many a time we get to hear this as a beginner, particularly when the speaker is giving some instruction, and at times, many instructions and not just one! Therefore, quite often you get to hear a suggestion such as "Do one thing, turn off your device, pack it in a neat *cartoon*, put your name on the top of it, mention the date on which your dear chapatti maker breathed its last, and then put it in a dustbin!"	**There exists no such thing as "Do one thing" in Standard English. Moreover, it is not cartoon, but carton in this context.**
Kindly Revert Back A very commonly observed expression in Indian English is "Kindly revert back as soon as possible." We use it repeatedly in our emails. To begin with, *revert* actually suggests going back or returning to a former state. Therefore, in the given context, it cannot stand for 'reply' or 'respond.' Moreover, 'revert' should not be followed by 'back' whenever used in the appropriate context.	**Please reply...**
One-eyed man 'We can't blame Chandru; he is, after all the only one-eyed man among blinds.' Another example of a literal translation of some idea/expression into English where something as graceful as *a figure among ciphers* is amusingly reduced to *one-eyed man among blinds*.	**A figure among ciphers**
Heater 'On the heater. It's so cold today.' A typical expression articulated in Indian English. As discussed earlier, the word *turn* is omitted and *on* is made to act as verb. Moreover, it is not certain whether the heater that is referred to is really some heater and not actually a *blower*. It is so because many Indians do not see much of a difference between a *heater* and a *blower*. A good user of English though is expected to make a distinction between these two objects.	**Blower**
Bed tea 'The British taught us bad habits! Look they taught us to take bed tea.' A patriotic speech may sound like that. However, though the British may be accused of having taught us to *take tea while in bed*, we cannot really blame them for teaching us to take *bed tea*. It is so because for a native speaker of English, it is generally an *early morning tea*.	**Early morning tea (in bed)**

WISEWELL QUIPS

EXERCISES

I. Rewrite the following sentences using standard English usage:

1. You missed something! The speech was too good.
2. My father was born in Hyderabad only.
3. The institute will remain closed between 2.00 p.m. to 3.00 p.m.
4. We had to prepone the meeting because the chairman is leaving for Mumbai next week.
5. In order to help the environment, he has started going to office by walking these days.
6. All the females here are requested to assemble in the conference hall.
7. My co-brother has started drinking a lot these days.
8. Seen after such a long time! How do you do?
9. We are sorry to announce that the shopkeeper has expired.
10. Why don't you off the lights when you go somewhere?

II. Replace the highlighted expressions in the following sentences with those that conform to standard English usage:

1. Stop writing! Time is **over**.
2. I don't like her. **Even I don't like her**.
3. Why don't you employ some trained **bearers**?
4. We are expecting thirty more **visitors** to join us on lunch.
5. Are you planning to have any more **issues**?
6. We took our lunch **in a hotel** on the way.
7. The meeting will resume at **seven thirty** in the evening.
8. It was **a timepass** movie.
9. Prepare well; tomorrow you have a test to **give**.
10. Those indulging in **eve teasing** must be severely punished.

ANSWER KEY

Exercises

I.
1. You missed something! The speech was really good.
2. It was in Hyderabad that my father was born.
3. The institute will remain closed between 2.00 p.m. and 3.00 p.m.
4. We had to advance the meeting because the chairman is leaving for Mumbai next week.
5. In order to help the environment, he has started going to office on foot these days.
6. All the women here are requested to assemble in the conference hall.
7. My brother-in-law has started drinking a lot these days.
8. Seeing after such a long time! How are you?
9. We are sorry to announce that the shopkeeper has passed away.
10. Why don't you turn off the lights when you go somewhere?

II.
1. Stop writing! Time is **up**.
2. I don't like her. **Nor do I**.
3. Why don't you employ some trained **waiters**?
4. We are expecting thirty more **guests** to join us for lunch.
5. Are you planning to have any more **children**?
6. We took our lunch **at a restaurant** on the way.
7. The meeting will resume **half past seven** in the evening.
8. It was **an ordinary** movie.
9. Prepare well; tomorrow you have a test to **take**.
10. Those indulging in **sexual harassment** must be severely punished.

Basics of Phonetics

Learning Objectives   After reading this chapter, you will be able to

- identify the various reasons for incorrect pronunciation
- acquaint yourself with the standard variety of British, American, and Indian English
- learn how to correctly produce vowel and consonant sounds
- understand what a syllable is and how words are divided into different syllables
- get to know the rules of word stress and intonation in the English language
- learn the use of contrastive stress in sentences

7.1 INTRODUCTION

We are all aware that some speakers of the English language attract us with their good command of language. Among other things, it is their pronunciation that creates the right kind of impact on us as listeners. It is absolutely essential on the part of every speaker of English, as is the case with any other language, to speak with the right pronunciation. Since we are not native speakers of English, there exists a very serious problem with regard to the pronunciation of the Indian speaker's English.

Because of a variety of English spoken in different parts of the world, there is no purity of either language or pronunciation. Therefore, we often come across alternate pronunciations and mispronunciations. However, no matter how common the incorrect pronunciation is, you always need to strive to acquire correct pronunciation.

7.2 REASONS FOR INCORRECT PRONUNCIATION

Let us first understand the two major reasons which lead us to acquire incorrect pronunciation:

1. In a native/first language situation, from a very early stage children learn to respond to sounds and tones which their elders habitually use while talking to them. In due course, children start learning English in English speaking countries; they tend to speak in the mother tongue accent. But in our country, where English is used as a second language, children listen to wrong sounds and tones spoken by their teachers/grown ups in their environment and tend

to pick up faulty pronunciation. This happens mainly due to their lack of sufficient exposure to the right variety of the language.

2. Moreover, we tend to speak English as we speak our mother tongue; therefore, we tend to commit mistakes due to its influence.

7.3 RECEIVED PRONUNCIATION (RP)

You must be wondering what standard English is, considering this language is spoken the world over. Your confusion is justified, as English is spoken as a first or second language by a very large number of people throughout the world. In some countries such as the UK, the USA, Canada, and Australia, English is the native or first language. In other countries such as India, Pakistan, Sri Lanka, Bangladesh, etc., English is spoken as a non-native or second language. As there is such a wide range of variation in pronunciation and accent, it is essential for us to follow a standard. One native regional accent that has gained social prestige is the Received Pronunciation of English (RP for short). It is the pronunciation of the people of south-east England and is used by educated English speakers. It is now equated with the correct pronunciation of English.

Before we start learning the sounds of English, let us first clear a great misconception that exists among us about the sounds of English.

7.4 MISCONCEPTION ABOUT SOUNDS

In our school days, we were told that there are twenty six sounds in the English language as there are twenty six letters in it. Even today, the same thing is being taught in a large number of schools. Because of this misconception, most of the Indian students fail to get acquainted with the right English pronunciation and accent. There are actually *forty four* sounds in the English language even though there are only twenty six letters in it.

For instance, pronounce the following words:

1. Cat   2. Keen   3. Occasion   4. Chemistry

What do you observe? Here, the /k/ sound is used for *c* in cat, *k* in keen, *cc* in occasion, and *ch* in chemistry. In English, different alphabets can give the same sound. It is also possible that the same alphabets in English give different sounds in words. For example:

chemist and **ch**aracter   **ch** gives /k/
                        whereas
chest and **ch**eese   **ch** gives /tʃ/

We hope you have now understood that alphabet is different from sound. Therefore, let us stop erring at least on this front, and master these forty four sounds straightaway.

7.5 TRANSCRIPTION

Since there are standard sounds in English, these are put together with standard symbols. These symbols are called the *International Phonetic Alphabet* symbols or in short IPA. You should familiarize yourself with these symbols as you will find that they are used in dictionaries for indicating the correct pronunciation of words. These symbols are given in the ensuing section of the chapter. Using standard symbols for standard sounds is known as *transcription*. We shall learn how to transcribe words and sentences later.

7.6 SOUNDS

There are two types of sounds, which are

1. vowels   2. consonants

Now, most of us remember our teachers telling us that there are five vowels—a, e, i, o, and u—in English. But indeed there are as many as twenty vowel sounds in English. Out of these twenty vowel sounds, twelve are *pure* vowels and eight of them are *diphthongal glides*. It means that out of these twenty vowel sounds, eight are a combination of two vowel sounds. In phonetics, when there is a glide from one vowel sound to another vowel sound, it is called a *diphthong*. Apart from these twenty vowels there are twenty four consonant sounds as shown in Fig. 7.1.

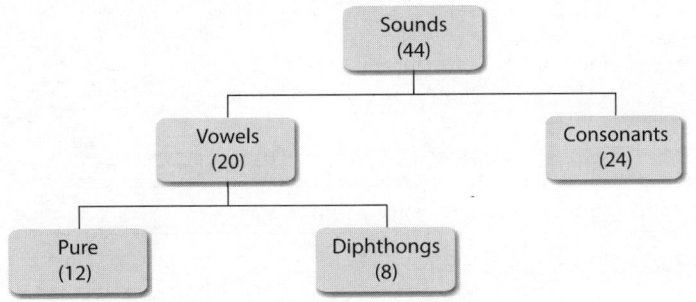

Fig 7.1   Classification of Sounds in English

So, let us learn how to correctly pronounce all the twenty vowels of English, and the symbols they are transcribed with.

7.6.1 Vowels

A vowel is a sound in spoken language, such as the English 'Ah!' [ɑ] or oh! [o], pronounced with an open vocal tract so that there is no build-up of air pressure at any point above the glottis. This contrasts with consonants, such as English 'Sh!' [ʃ], where there is a constriction or closure at some point along the vocal tract.

Pure vowels   A vowel sound whose quality does not change over the duration of the vowel is called a pure vowel. There are 12 pure vowel sounds in English which are given below with examples.

Sounds	Position	Examples
/iː/	**Initial**	each, eat
	Medial	these, seed, seen, cream, dream, shield, peach, thief, piece, deceive, seize, complete, replete, feel, peep, beat, heat, sheep, need, peel, leave, deal
	Final	pea, key, see, knee, plea
/ɪ/	**Initial**	it, in, is, intelligent, index, individual, induct, inch
	Medial	silk, thick, fill, slip, sip, hit, bit, begin, ticket, silk, cliff, city, build, money, busy, hills, live

(Contd)

Sounds	Position	Examples
	Final	duty, beauty, lonely, promptly, quickly
/e/	Initial	enter, exit, empire, entire, any
	Medial	bed, dead, head, many, said, fell, let, test, met, tell, bury, friend, leisure, melt, rest, set, wet, breath, feather, bend, men, led, red, wet, pet
	Final	**Does not occur in the final position**
/æ/	Initial	axe, actor, apple, at, an
	Medial	bank, man, sad, fan, mass, rank, sad, tax, cattle, back, mango, gradual, sand, stand, battle, cash, bag, back
	Final	**Does not occur in the final position**
/ʌ/	Initial	utter, under, understand, undo, umbrella, unable, unborn
	Medial	bus, dull, dust, gun, hunt, munch, much, pump, run, son, come, done, month, double, enough, trouble, young, blood, does, butter, country, couple, study, cup, bun
	Final	**Does not occur in the final position**
/ɑ:/	Initial	art, answer, aunt
	Medial	card, farm, hard, large, march, fast, task, master, pass, dance, branch, path, bath, staff, calm, half, laugh, bath, drama, last, rather, clerk, heart
	Final	car, jar, mar, bar, tar, far, mortar
/ɒ/	Initial	office, oxygen, object, October, odd
	Medial	hot, bottle, dog, fond, lock, not, pot, solve, borrow, quality, want, because, shone, gone, off, God, knowledge, socks, robbed, cost, top
	Final	**Does not occur in the final position**
/ɔ:/	Initial	all, awkward, ought, audition, August, audible
	Medial	ball, call, hall, corn, morning, north, pour, nor, water, door, thought, cause, fault, chalk, board, warm
	Final	saw, raw, claw, paw, more, sore
/ʊ/	Initial	**Does not occur in the initial position**
	Medial	book, cook, took, wood, cushion, push, could, would, should, woman, foot, good, look, hook, crook, bullet, wool
	Final	**Does not occur in the final position**
/u:/	Initial	oodles, ooze, oops
	Medial	rule, approve, groove, suit, lose, foolish, stupid, move, goose, music, beautiful, pupil, June, soon, group, wound, fruit, juice, tooth, choose
	Final	sue, new, you, shoe, two
/ɜ:/	Initial	early, earn, urge, urgent, earnest, earth
	Medial	third, murder, surface, turn, nurse, purple, bird, stern, circle, dirt, thirst, burn, hurt, heard, learn, search, journey, flirt, skirt

(Contd)

Sounds	Position	Examples	
	Final	Does not occur in the final position	
/ə/	Initial	about, effort, ago, allow, ahead	
	Medial	sentence, liberty, condition, factory, society, famous, gentleman, human, substance	
	Final	motor, colour, doctor, beggar, collar, dollar, finger	

Diphthongs

Sounds	Examples
/eɪ/	bait, mail, fail, train, age, may, say, pray, jail, aim, straight, eight, grey, weight, great, waste, date, paste, rate, drain, trail, hail, vain, pain
/aɪ/	ice, fine, pipe, nice, write, rival, silence, tidy, type, cry, dry, fly, reply, satisfy, die, flies, high, might, right, tight, child, kind, buy, island, height
/ɔɪ/	spoil, loin, boil, choice, noise, oil, point, annoy, boy, toil, toy, foil, employ, join, soil
/əʊ/	roll, blow, home, bone, nose, rope, both, open, go, no, so, social, bold, most, post, know, narrow, window, boat, soap, shoulder
/aʊ/	sound, mouth, out, round, allow, cow, town, now, down, mouse, doubt, house, trounce, trousers, about, bound, around, crowd, town
/ɪə/	here, appear, period, mere, cheer, tear(noun), jeer, queer, career, dear, near, fear, deer, serious, zero, clear, idea, real, fierce, dear
/eə/	their, wear, there, air, chair, fair, pair, hair, bare, care, share, various, bear, tear(verb), prayer, dare, rare
/ʊə/	cruel, poor, sure, tour, actual, pure, fuel, tour, virtuous

7.6.2 Consonants

As discussed at the beginning of the chapter, there are as many as forty four sounds in English, and we have already known twenty vowel sounds. We are obviously thus left with twenty four consonant sounds. A consonant is a speech sound that is articulated with complete or partial closure of the vocal tract. Examples are /p/, pronounced with the lips; /t/, pronounced with the front of the tongue; /k/, pronounced with the back of the tongue; /h/, pronounced in the throat; /f/ and /s/, pronounced by forcing air through a narrow channel; and /m/ and /n/, which have air flowing through the nose. Some of the words using these and other consonant sounds in English are listed below:

Sounds	Examples	Sounds	Examples
/p/	pit, spill, keep	/f/	fat, feather, half
/b/	bit, imbibe, jumble	/v/	vat, wave, velvet
/t/	tin, after, what	/θ/	thin, thank, wrath
/d/	din, lead, order	/ð/	then, feather, breathe
/k/	cut, character, leak	/s/	sap, sound, pistol, cross

(Contd)

Sounds	Examples	Sounds	Examples
/g/	**g**ut, **g**irl, ea**g**er	/z/	**z**ap, **z**ing, ma**z**e, **g**a**z**e
/tʃ/	**ch**eap, na**t**ure, wa**tch**	/ʃ/	**sh**e, na**t**ion, **sh**out
/dʒ/	**j**eep, **j**ealous, **j**ud**g**e	/ʒ/	mea**s**ure, plea**s**ure, trea**s**ure
/m/	**m**ap, re**m**ind, **m**i**m**e	/h/	**h**arm, **h**ouse
/n/	**n**ap, **n**ear, li**n**e, Christia**n**	/r/	**r**un, **r**uin, c**r**aze
/ŋ/	ba**ng**, so**ng**, bri**ng**, si**ng**ing	/w/	**w**e, **w**icked, **w**atch
/l/	**l**eft, re**l**ate, detai**l**	/j/	**y**es, **y**ell, **y**esterday

7.6.3 Consonant Cluster (CC)

A *consonant cluster* in a word is a group of two or more consonants with no vowels in between them. The consonant sounds that come before vowel sound are called **onset,** the ones that come after vowel sound **coda,** and the consonant cluster between the vowel sounds is known as **medial**.

Examples

black, bread, trick, twin, flat, splash, spring, strong, scream – the highlighted consonant sounds are *Onset*

length, sixths, bursts, glimpse, astray – the highlighted consonant sounds are *coda*

handspring, sightscreen, abridge, drastic, perquisite – the highlighted consonant sounds are *medial*

PRACTICE TEST 7.1

Identify the sounds represented by the underlined letter(s) in the following words.
   Example: o**cc**asion = /k/

1. Cur**iou**s
2. B**oa**t
3. Dish**o**nest
4. Trag**e**dy
5. Bird**s**
6. Tru**n**k
7. App**ro**ve
8. Awa**re**
9. Exp**l**oit
10. Dear**th**

PRACTICE TEST 7.2

Give two examples in orthography for each of the following:

1. Words with two syllables
2. Words ending in /dʒ/
3. Words ending in /g/
4. Words beginning with /ð/

PRACTICE TEST 7.3

What phonetic symbol/symbols would be used for the underlined letter(s) in each of the following words?

1. fa**th**om
2. voca**bu**lary
3. noi**se**s
4. val**ve**s
5. li**b**erty
6. rai**se**d
7. d**ia**mond
8. occu**rre**nce

7.7 PROBLEMS OF INDIAN ENGLISH

In Indian English, /z/ and /s/ are not used correctly and this leads to confusion between pairs such as the ones given below:

/z/	/s/
eyes	ice
falls	false
fears	fierce
his	hiss
knees	niece

Some Hindi speakers interchange the sounds /s/ and /ʃ/ and that leads to confusion in pairs such as the following:

/ʃ/	/s/
shave	save
she	see
sheet	seat
shine	sign

In order to discriminate between these two sounds, we need to repeatedly speak out tongue-twisters such as the one given below:

She sells sea shells on the sea shore.

Further, most Indian speakers are unaware of the difference between the consonant sounds /v/ and /w/. That is why, the following pairs are often confused:

/w/	/v/
while	vile
west	vest
why	vie

In fact, both these sounds are used indiscriminately by most of us, and careful practice to achieve a distinction between them is mandatory.

7.8 SYLLABLE

A *syllable* is a basic unit of spoken language, which consists of an uninterrupted sound that can be used to make up words. In other words, a syllable always has one vowel sound. So a word has as many syllables as there are vowel sounds. While denoting the syllabic structure of a word, 'C' is used for consonant sounds and 'V' is used for vowel sounds.

Let us look at the structure of a few words:

Train	/treɪn/
Cold	/kəʊld /

For example, 'man', 'code', 'eye', 'lead', 'strength', and 'sixths' are a few of many English words that have one syllable each.

A-go, ho-tel, free-dom, and a-gree are examples of words with two syllables.

Syl-la-ble, dic-tion-ary, and re-la-tion are examples of words with three syllables.

7.8.1 Rules for Counting Syllables

To find the number of syllables in a word, use the following steps:

1. Count the vowel sounds in the word.
2. Subtract any silent vowel, such as the silent /ə/ at the end of a word.
3. Count diphthongs as only one vowel sound.
4. The number of vowel sounds is the same as the number of syllables.

This means that the number of syllables that you hear when you pronounce a word is the same as the number of vowel sounds heard. For example:

1. Words such as *came, rule, blue,* and *name* have two vowels, but the *e* is silent. Therefore, there is only one vowel sound and thereby one syllable in each of these words .
2. The word *outside* has four vowels, but the *e* is silent and the *ou* is a diphthong which counts as only one sound, so this word has only two vowel sounds, and therefore two syllables in this word. Look at its syllabic structure:

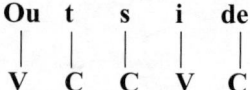

7.8.2 Dividing Words into Syllables

Given below are some of the words with their syllabic division, phonemic transcription, and syllabic structure:

Sl. No.	Word	Transcription	Syllabic Structure
1.	Clap	/klæp/	CCVC
2.	Hope	/həʊp/	CVC
3.	Late	/leɪt/	CVC
4.	Fauna	/ˈfɔː-nə/	CV-CV
5.	Scream	/skriːm/	CCCVC
6.	Remember	/rɪ-ˈmem-bə/	CV-CVC-CV
7.	Telephone	/ˈte-lɪ-fəʊn/	CV-CV-CVC
8.	Despite	/dɪs-ˈpaɪt/	CVC-CVC
9.	Quality	/ˈkwɒ-lə-tɪ/	CCV-CV-CV
10.	Potato	/pə-ˈteɪ-təʊ/	CV-CV-CV
11.	Policeman	/pə-ˈliːs-mən/	CV-CVC-CVC
12.	College	/ˈkɒ-lɪdʒ/	CV-CVC
13.	University	/ˌjuː-nɪ-ˈvɜː-sə-tɪ/	CV-CV-CV-CV-CV
14.	Considering	/kən-ˈsɪ-də-rɪŋ/	CVC-CV-CV-CVC
15.	Ability	/ə-ˈbɪ-lə-tɪ/	V-CV-CV-CV
16.	Passenger	/ˈpæ-sɪn-dʒə/	CV-CVC-CV

(Contd)

Sl. No.	Word	Transcription	Syllabic Structure
17.	Imagine	/ɪ-'mæ-dʒɪn/	V-CV-CVC
18.	Languages	/'læŋ-wɪ-dʒɪz/	CVCC-CV-CVC
19.	Atlantic	/ət-'læn-tɪk/	VC-CVC-CVC
20.	Psychologists	/saɪ-'kɒ-lə-dʒɪsts/	CV-CV-CV-CVCCC
21.	Apologize	/ə-'pɒ-lə-dʒaɪz/	V-CV-CV-CVC
22.	Listlessness	/'lɪst-lɪs-nɪs/	CVCC-CVC-CVC
23.	Disdainfully	/dɪs-'deɪn-fʊ-lɪ/	CVC-CVC-CV-CV
24.	Disembarkment	/ˌdɪs-ɪm-'baːk-mənt/	CVC-VC-CVC-CVCC
25.	Iconoclastic	/aɪ-ˌkɒ-nəʊ-'klæs-tɪk/	VC-CV-CV-CCVC-CVC

PRACTICE TEST 7.4

Show the division of syllables in the following words:

1. offer
2. sudden
3. different
4. September
5. January
6. children
7. college
8. disappear
9. accident
10. proper
11. neighbour
12. confident
13. introduction
14. faithfully
15. dentist

PRACTICE TEST 7.5

Show the division of syllables in the following words:

1. window
2. sympathy
3. perhaps
4. information
5. activity
6. telephone
7. management
8. electrician
9. disappearance
10. concentrate

PRACTICE TEST 7.6

Identify the words with one syllable from the list of words:

1. fulfil
2. awesome
3. space
4. phonetics
5. book
6. grass
7. roman
8. home
9. your
10. plant
11. conquer
12. folder
13. prepare
14. said
15. you

7.9 WORD STRESS

Are you now prepared to transcribe the words as per the RP sounds of English? Not yet. In fact, before we move on to learn how to transcribe the words, we need to know where to stress a word. Did you know that English is an accent-based language, and that in a word not all the syllables are pronounced with equal emphasis? For example, in the word *ability* it is -*bi* and not '*a*' that is heard prominently. If you look up in the dictionary for this word, it would be seen something like /ə'bɪlətɪ/. Notice the little mark "'" after /ə/ and before -/bɪlətɪ/. This is known as *word stress*. Now this stress changes the way a word is to be pronounced. For example, if you look up the word *convict*, it would be shown to have been transcribed as /ˈkɒnvɪkt/ and /kənˈvɪkt/. In the first case, it is stressed initially while in the second instance the stress is on -/vɪkt/. Because of the shift in the stress, the corresponding vowel sound and consequently the pronunciation changes completely. Therefore, it is imperative for us to be aware of the rules that decide which word is to be stressed where. However, since there are only a few rules pertaining to word stress, it is advisable to refer to a standard dictionary to find out where a particular word receives its primary and secondary stress. Therefore, rather than being an exhaustive list covering all possible accentual patterns, the rules that follow can only give you an idea about how certain words in English are stressed.

Let us see the various word stress rules with examples. In English there are a large number of words with two syllables and in these words the stress depends on whether the word is used as a noun or a verb.

(a) When the word is used as a noun or adjective, the stress is on the first syllable. When the word is used as a verb, the stress is on the second syllable. Here are a few examples:

Noun/Adjective	Verb	Noun/Adjective	Verb
'produce	pro'duce	'record	re'cord
'object	ob'ject	'perfect	per'fect
'subject	sub'ject		

(b) Words with weak prefixes are accented on the root. For example:

a'go	be'low	re'duce	ad'mit
a'bove	re'vise	de'velop	be'gin
a'bout	be'neath		

(c) Verbs of two syllables beginning with the prefix dis- are stressed on the last syllable. For example:

dis'arm	dis'pel	dis'close	dis'turb
dis'guise	dis'tress	dis'miss	dis'may

(d) Verbs that have two syllables and end in -ate, -ise/ize, -ct are stressed on the last syllable. For example:

at'tract	nar'rate	de'bate	bap'tize
cre'mate	in'ject	cap'size	

(e) Words ending in -ion are stressed on the last but one syllable. For example:

appli'cation	intro'duction	assimi'lation	exami'nation
simplifi'cation	pro'duction	expla'nation	repe'tition

(f) Words ending in -ic/-ical/-ically, -ial/-ially, and -ian are stressed on the syllable before the suffix.

charis'matic	me'morial	'special	sub'stantial
li'brarian	mu'sician	es'sential	of'ficially

(g) Words ending in -ious and -eous are stressed on the last but one or penultimate syllable. For example:

'curious	mys'terious	la'borious	re'bellious
spon'taneous	cou'rageous	s'purious	'serious

(h) Words ending in -ate, -ise/-ize, -fy are stressed on the third syllable from the end. For example:

'duplicate	'modernize	'justify	'cultivate
'beautify	'educate	'criticize	'satisfy

(i) Words ending in -ity, -cracy, and -crat are stressed on the third syllable from the end. For example:

a'bility	de'mocracy	au'tocracy	curi'osity
crea'tivity	sim'plicity	e'quality	possi'bility
'autocrat	bu'reaucracy	mag'nanimity	ratio'nality

(j) Words ending in -graph, -graphy, -meter, and -logy are stressed on the third syllable from the end. For example:

'autograph	psy'chology	bi'ography	zo'ology
crimi'nology	bi'ology	'paragraph	pa'rameter
'photograph	anthro'pology	ba'rometer	soci'ology

(k) Words ending with the suffixes -aire, -eer, -ental, -ential, -ese, -ese, -esce, -escence, -escent, -esque, -ique, -ee, -ette, -ete, and -ade are stressed on the suffix. For example:

pio'neer	ca'reer	millio'naire	question'naire
pay'ee	barri'cade	de'lete	exi'stential
ga'zette	com'plete	mi'stique	ado'lescent
re'plete	gro'tesque	billion'naire	pictu'resque

(l) In case of compound words, i.e., words that are made up of two words and are written as one word, the stress is on the first element. For example:

'blacksmith	'dining-room	'tea-party	'blackbird

However, in compound words with -ever and -self, the stress is usually on the second element. For example:

how'ever	when'ever	him'self	her'self

Sometimes both the elements are stressed, but the primary stress remains on the second element. For example:

after'noon	old-'fashioned	absent-'minded

> **NOTE** The inflectional suffixes such as -es, -ing, and -ed and the derivational suffixes -age, -ance, -en, -er, -ess, -ful, -hood, -ice, -ish, -ive, -less, -ly, -ment, -ness, -or, -ship, -ter, -ure, and -zen do not affect the stress. For example:

'match	'matches	'want	'wanted
'box	'boxes	'fine	'finely
'write	'writer	'god	'goddess
'play	'player	'waiter	'waitress
'aim	'aimless	'bad	'badly
'good	'goodness	'child	'childish
'bright	'brighten	'care	'careful
'bitter	'bitterness	'blood	'bloody
'create	'creator	'home	'homeless
'city	'citizen	'laugh	'laughter

7.10 HOW TO TRANSCRIBE

Now that you are familiar with the stress rules for words, let us start learning how to transcribe words.

Step 1 Always begin by dividing a word into parts. For example:

daughter	>	two parts of the word-'daugh' and 'ter'
explain	>	two parts of the word-ex and plain
dominate	>	three parts of the word-do, mi, and nate

Step 2 This division is based on the two distinct vowel sounds in the words *daughter* and *explain*, whereas *dominate* has three vowel sounds.

The number of vowel sounds determine the number of syllables in a word. So, while transcribing a word, you need to divide the word into syllables.

Step 3 Then, fix the stress and finally go on to transcribe the word.

Now look at the following examples carefully. The exercise given below should be a handy help in getting you acquainted with how to transcribe the words properly:

daughter	/ˈdɔːtə/	dictionary	/ˈdɪkʃənrɪ/
explain	/ɪkˈspleɪn/	advance	/ədˈvaːns/
dominate	/ˈdomɪneɪt/	arrange	/əˈreɪndʒ/
guide	/ˈgaɪd/	picture	/ˈpɪktʃə/

PRACTICE TEST 7.7

Transcribe the following words (division of syllable also shown) using IPA symbols:

Word	Division	Transcription	Word	Division	Transcription
image	i-mage		serious	se-rious	
society	so-ci-e-ty		human	hu-man	
observe	ob-serve		because	be-cause	
development	de-ve-lop-ment		surface	sur-face	
breakfast	break-fast		hunger	hun-ger	
famous	fa-mous		measure	mea-sure	
creature	crea-ture		island	is-land	
sure	sure		silence	si-lence	
nuisance	nui-sance		wear (V)	wear	
heard	heard		bottle	bo-ttle	
morning	mor-ning		character	cha-rac-ter	
master	mas-ter		musician	mu-si-cian	
enough	e-nough		occasion	o-cca-sion	
force	force				

PRACTICE TEST 7.8

Now keeping in mind the rules of word stress and recalling the forty four sounds you have learnt, transcribe the following words, marking the stress on the right place:

Word	Transcription	Word	Transcription
computer		dictionary	
justify		statement	
suicide		freedom	
army		firm	
yellow		envelop (V)	
win		terminate	
beautiful		regularize	
dictation		women	
reality		skirt	
war		approach	
obtain		eyes	
bird		master	
possibility		laughter	
frustration		registration	
tour		father	
flight		remember	
pencil		dumb	
item		isolation	
barrage		quality	
uncle			

7.11 WEAK FORMS

The functional words are generally used in their weak form in connected speech.

Example

In the following sentence, the first 'do' is a weak form and the second is stressed.

What do you want to do this evening?

/'wɒt dəju 'wɔːntt ə 'duːðɪs iːvnɪŋ/

 Functional words, such as prepositions, conjunctions, auxiliaries, and articles are often pronounced in their weak forms, since they do not carry the main content, and are therefore not normally stressed.

Determiners/Quantifiers

Orthography	Strong Form(s)	Weak Form(s)
the	ðiː	ðɪ, ðə
a/an	an	ə, ən
some	sʌm	səm, sm

Pronouns

Orthography	Strong Form(s)	Weak Form(s)
his	hɪz	ɪz
him	hɪm	ɪm
her	h ɜ:	ə
you	ju:	Jʊ, jə
your	jo:	jə
she	ʃi:	ʃɪ
he	hi:	ɪ
we	wi:	wɪ
them	ðem	ðəm, əm
us	ʌs	əs, s

Prepositions

Orthography	Strong Form(s)	Weak Form(s)
than	ðan	ðən
at	æt	ət
for	fo:	fə
from	frɒm	frəm, fəm, fm
of	ɒv	əv, v
to	tu:	tə, tʊ
as	æz	əz, z
there	ðeə	ðə

Conjunctions

Orthography	Strong Form(s)	Weak Form(s)
and	ænd	ənd, ən, nd, n
but	bʌt	bət
that	ðat	ðət

Auxiliaries

Orthography	Strong Form(s)	Weak Form(s)
can	Kan	Kən, kn
could	kʊd	kəd
have	hæv	əv, v
has	hæz	əz, z
had	hæd	əd, d
will	wɪl	L
shall	ʃal	ʃəl, ʃl, l

Orthography	Strong Form(s)	Weak Form(s)
should	ʃʊd	ʃəd
must	mʌst	məs, məst
do	duː	də, d
does	dʌz	dəz, z
am	æm	əm, m
are	ɑː	ə
was	wɒz	wəz
were	wɜː	wə

Source: http://ell.phil.tu-chemnitz.de/phon/connect/weakForms.html, accessed on 22 May 2010.

7.12 STRESS, INTONATION, AND RHYTHM

Stress, rhythm, and intonation are inextricably linked. It is almost impossible to speak of any one of these aspects of spoken English without referring to the others. However, it is necessary for the sake of clarity to deal with each one individually. Look at the following sentence.

There was an elephant at the corner of the street.

So in the sentence we have three words 'e-le-phant', '**cor**-ner', and '**street**' (highlighted syllables are stressed) stressed and rest are unstressed syllables.

Let us now come to *rhythm*. In English, there is a clear propensity to pronounce stressed syllables according to a relatively regular rhythm which is organised around the stressed syllables. As known, English is a *stress-timed language* which tends toward a regular rhythm of broadly equal-length beats on stressed syllables, the unstressed syllables being 'compressed/squeezed in' to fit the available time, and frequently reduced to a weak form. In other words, it means that the stressed syllables follow each other at intervals of about the same length, which sounds like a pulsating rhythm. This is how rhythm is created.

Then, of course, you have to decide on the appropriate *intonation*. What sort of tone do you use? If the sentence is a simple statement, 'There was an **elephant** at the **cor** ner of the **street**.' the appropriate intonation is a falling tone on the last stressed syllable.

In day to day conversation, when we hear someone speaking, we observe that the person does not speak on the same note throughout. We find frequent rises and falls in the person's voice. This variation in the pitch patterns of voice is called *intonation*. While speaking, we glide over the less important words such as pronouns, articles, auxiliary verbs, prepositions, and conjunctions which are called *functional words* in English, whereas nouns, principal verbs, adjectives, and adverbs which are called *content words* are stressed more. This *quality of quickly gliding over less important words is the characteristic feature of connected speech*.

The syllable on which there is change in pitch is marked in the following ways in English.

7.12.1 Rules for Intonation

[`] **Falling tone**

(a) **Statements have a falling tone at the end**. It signals a sense of finality, completion, or belief in the content of the utterance, and so on. For example:

(i) I went to 'Delhi.

(ii) They have saved enough money to buy a 'new car.

(b) **WH-question** (who? what? why? when? where? how?) **have a falling tone at the end**. For example:

(i)   How did you spend your va'cation?

(ii) When did he 'come?

(c) **Tag questions take the falling tone**.

(i)   He was operated on 'yesterday, 'wasn't he?

(ii) Radha comes here 'every day, 'doesn't she?

(d) **Imperative statements have a falling tone**.

(i)   Go and see a 'doctor.

(ii) Come and wash your 'face.

[´] **Rising tone**

(a) **Yes–No questions** (question you can answer with 'yes' or 'no') **usually have a rising tone**. For example:

(i)   Could you pass me the curd, please?

(ii) Was it expensive?

(b) **Rising intonation is used at the end of the questions which do not have an interrogative word**. For example:

(i)   You are coming tomorrow?

(ii) He has enough money to buy a new house?

(c) **Requests have a rising tone too**.

(i)   Please post this letter.

(ii) Please calm down.

[ˇ] **Falling-rising tone**

Fall-rise signals dependency, continuity, and non-finality. For example:

(i)   Private ˇenterprises are mostly successful.

(ii) Preˇsumably he thinks he can.

PRACTICE TEST 7.9

Read the following sentences and mark the tones that you will use while speaking:

1. No. The woman with the plastic bag.
2. That's the person who robbed the bank!
3. Do you mean the man with the black pants?
4. He drove to work after he had finished working in the garden.
5. Suresh bought new shoes today.
6. Hey, have you seen the new film with Bruce Willis?
7. Do you want some coffee?
8. Would you like some ice cream and cake?
9. Is he going to the dentist?
10. Yes. He has a toothache.
11. Usually, he comes on Sundays.
12. She's totally confused, isn't she?
13. I would be really happy if I could get more input on this.
14. I want a purse on my birthday.
15. Occasionally, I plan my budget.

7.12.2 Contrastive Stress in Sentences

Contrastive stress is used to contrast a word or syllable with an alternative word or syllable which is normally unstressed otherwise. It is used with determiners such as 'this', 'that', 'these' and 'those'.

Example

*I prefer **this** purse not the other one.*
*Do you want these or **those** curtains?*

Contrastive stress is also used to bring out a given word in a sentence which also slightly changes the meaning.

- **She** came to the gym yesterday. (It was **she**, not someone else.)
- She **walked** to the gym yesterday. (He **walked** rather than drove.)
- She came to the **gym** yesterday. (It was **gym** not a meeting or something else.)
- She came to the gym **yesterday**. (It was **yesterday** not two weeks ago or some other time.)

7.13 DIFFERENCE BETWEEN BRITISH, AMERICAN, AND INDIAN SPOKEN ENGLISH

Nowadays, English is probably the most frequently spoken language in the world, either as mother tongue, official language, or foreign language. Speaking English has become more than a trend or a fashion. But the question is what kind of English do we speak? Since it is the language of the professional world, it is essential for all of us to know the differences between the three major varieties of English. When a person does not understand the differences, you may hear him or her using phrases such as 'pardon', 'come again,' or 'I didn't get you'. These ruin the effectiveness of communication. Therefore, it is essential for us to understand these three varieties of English viz. British, American and Indian, briefly.

British English (BrE) is the form of English used in the United Kingdom, and includes all English dialects used there. The pronunciation of standard English is called *Received Pronunciation (RP)*. It is referred to colloquially as 'the Queen's English,' 'Oxford, English' and 'BBC English'. *American English (AmE)* is the form of English used in the United States. Though Indians speak British English, yet because of the regional language or vernacular which an Indian speaks brings a significant change in the pronunciation and usage of English language. Thus, *Indian English (IE)* is considered a group of English dialects, or regional language varieties, spoken primarily on the Indian subcontinent. Indian English has generally absorbed the idiomatic forms derived from Indian literary languages and vernaculars.

Now let us take up a few examples which help us understand how, if you are not familiar with correct words, may lead to some confusion while communicating across cultures. Sometimes, the spelling of the words may be the same but they are pronounced differently. For example, 'schedule' is pronounced /skedʒuːl/ and 'vitamin' is pronounced /vaɪtamɪn/ in American English whereas in British English they are pronounced /ʃeduːl/ and /vɪtamɪn/ respectively. Given below are a few more such examples:

- vase: **/vaːz/** as in cars (BrE); **/vaːs/** as in face (AmE)
- route: /ruːt/ as in shoot (BrE); **/raut/** as in shout (AmE)
- ate: /et/ as in let (BrE), /eɪt/ as in late (AmE)
- leisure: as in /leʒə/ (BrE); /lɪʒər/ with /lɪ/ as in she (AmE)

7.13.1 Other Differences in Pronunciation

- In BrE /a:/ before -f, -s, - m, -n is pronounced /æ/ in AmE eg. Ask, task, after, pass, calf
- In BrE /ɒ / in words such as Not, block, cross, stop, college, is pronounced /a:/ in AmE.
- AmE does not drop the /r/sound in words like better, perceive, bird, here, poor, chair, dare whereas BrE does.
- BrE /ju:/ after consonants /d/, /t/, /n/ is pronounced /u:/ in AmE, eg. duty, tune, new
- The past tense forms of the two following verbs are pronounced differently.

BrE	AmE
shine – shone /ʃɒn/	shine – shone /ʃoʊn/
eat – ate /et/	eat – ate /eit/

- Here are a few examples of words which are pronounced differently in the AmE than in the BrE.

Word	BrE	AmE
Resource	/rɪsɔ:s/	/ri:sɔ:rs/
figure	/fɪgə/	/fɪgjər/
either	/aɪðə/	/i:ðər/
research	/rɪsɜ:tʃ/	/ri:sɜ:rtʃ/
glacier	/gⱵæsɪə/	/gⱵeɪʃər/
Asia	/eɪʒə/	/eɪʃə/
can't	/ka:nt/	/kænt/

- The **BrE** /əʊ/ is pronounced as /ɓ/ in AmE in words such as go, no, crow, romantic.
- Words pronounced with /æ/ in AmE with /a: / in **BrE**: bath, lath, path, aunt, plant, can't, advantage Exceptions: hath, maths, athlete, ant, banter, scant, mantle.

7.13.2 Characteristics of Indian English (IE)

- Many Indian English speakers do not make a clear distinction between /ɒ/ and /ɔ:/. Eg. cot, caught.
- Unlike British speakers, some Indian speakers, especially in the South, often do not pronounce the rounded /ɒ/ or /ɔ:/, and substitute /a/ instead. eg. in South India *coffee* will be pronounced *kaafi*, *copy* will be *kaapi* etc.
- Words such as *class*, *staff*, and *last* would be pronounced /kla:s/, /sta:f/, and /la:st/ in British English, whereas it is pronounced /klæ:s/, /stæ:f/, and /læ:st/ in American English.
- Standard Hindi and most other vernaculars (except Punjabi, Marathi & Bengali) do not differentiate between /v/ and /w/ sounds.
- The voiceless plosives /p/, /t/, /k/ are always unaspirated in Indian English which are aspirated in RP. For example "pin" is pronounced /pɪn/ in Indian English but /pʰɪn/ in British and American English.
- Unlike native speakers Indian English speakers do not make use of the consonant sound /ʒ/. Instead, the sounds /dʒ/, /z/ are used by Indian speakers of English.

RECAPITULATION

✓ English is spoken the world over. As there is such a wide range of variation in pronunciation and accent, it is essential for us to follow a standard, and the one that we follow is the RP of English.

✓ There are *forty four* sounds in English, out of which twenty four are consonant sounds and twenty are vowel sounds. Out of these twenty vowel sounds, twelve are pure vowels and eight are diphthongal glides.

✓ A vowel is a sound spoken with an open vocal tract, so that there is no build-up of air pressure at any point above the glottis.

✓ A consonant is a speech sound that is articulated with complete or partial closure of the vocal tract.

✓ A syllable always has one vowel sound. So a word has as many syllables as there are vowel sounds.

✓ English is an accent-based language, and in a word not all the syllables, that is, the number of vowel sounds in the word, are pronounced with equal emphasis.

✓ Weak forms are monosyllabic words that become unstressed in connected speech.

✓ The rise and fall in a person's voice, that is, variation in the pitch patterns of voice, is called intonation.

 EXERCISES _____

I. Identify the consonant cluster in the following words. Also, name the type of consonant cluster:

price	green
practice	private
appropriate	cry
break	crazy
journalism	great
bring	students
advance	understand
brother	grow
transcript	scratch
create	scream

II. Transcribe the following words into phonetic script using IPA symbols:

1. joy
2. conquest
3. doctor
4. therefore
5. office
6. excellent
7. ghosts
8. attached
9. decision
10. examination

III. Transcribe the following words:

1. sorrow
2. machine
3. sorry
4. design
5. child
6. motivate
7. rest
8. reason
9. service
10. awareness
11. battle
12. fancy
13. market
14. huge
15. fear
16. cultivate
17. taste
18. relatives
19. hopeless
20. about
21. thought
22. stupidity
23. sensational
24. purchase
25. notion
26. quickly
27. musician
28. writing
29. engineer
30. manager
31. object (Verb)
32. opponent
33. magician
34. report
35. doctor
36. employee

IV. Mark stress in the following words as shown in the example below:

Example: Object(n) = ´Object

1. atmosphere
2. comment
3. contribute
4. demonstration
5. support
6. complicate
7. electricity
8. photography
9. industrial
10. departmental

V. What is onset? Provide five words with onset.

VI. What is a consonant cluster and how they contribute to study for effective speaking? Support your answer with appropriate examples.

VII. Differentiate between onset and coda. Provide three examples for each.

VIII. What is intonation? What is the difference between intonation and word stress?

IX. What is a syllable and how are words divided into different syllables?

X. What is the difference between a vowel and a consonant sound?

XI. What is a diphthong? How are diphthongs different from vowel sounds?

XII. How are sounds different from alphabet in English?

XIII. What is the difference in the pronunciation of the following words in British English and American English?

Calf, Graph, Giraffe, Half, Laugh, Staff, After, Craft, Draft, Laughter, Raft, Shaft. Sample, Example,

ANSWER KEY

Practice Tests

7.1
1. /ɪə/
2. /əʊ/
3. /ɒ/
4. /ə/
5. /z/
6. /ŋ/
7. /uː/
8. /eə(r)/
9. /ɔɪ/
10. /θ/

7.2
1. ago   below
2. judge   nudge
3. big   fig
4. then   thus

7.3
1. /ð/
2. /bʊ/
3. /zɪz/
4. /vz/
5. /ə/
6. /zd/
7. /aɪ/
8. /əns/

7.4
1. of-fer
2. sud-den
3. dif-fer-ent
4. Sep-tem-ber
5. Jan-u-ary
6. chil-dren
7. col-lege
8. dis-ap-pear
9. ac-ci-dent
10. pro-per
11. neigh-bour
12. con-fi-dent
13. in-tro-duc-tion
14. faith-ful-ly
15. den-tist

7.5
1. win-dow
2. sym-pa-thy
3. per-haps
4. in-for-ma-tion
5. ac-ti-vi-ty
6. te-le-phone
7. ma-nage-ment
8. e-lec-tri-cian
9. dis-ap-pear-ance
10. con-cen-trate

7.6   3, 5, 6, 8, 9, 10, 14, and 15 are the words with one syllable and the rest have more than one syllable.

7.7.

Transcription	Transcription	Transcription	Transcription
/ˈɪmɪdʒ/	/ʃʊə/	/ˈsɪərɪəs/	/ˈsaɪləns/
/səˈsaɪətɪ/	/ˈnjuːsns/	/ˈhjuːmən/	/ˈweə/
/əbzˈɜːˈv/	/ˈhɜːd/	/bɪˈkɒz/	/ˈbɒtl/
/dɪˈveləpmənt/	/ˈmɔːnɪŋ/	/ˈsɜːfɪs/	/ˈkærəktə/
/ˈbrekfəst/	/ˈmɑːstə/	/ˈhʌŋgə/	/mjuːˈzɪʃən/
/ˈfeɪməs/	/ɪˈnʌf/	/ˈmeʒə/	/əˈkeɪʒn/
/ˈkriːtʃə/	/ˈfɔːs/	/ˈaɪlənd/	

7.8

Transcription	Transcription	Transcription	Transcription
/kəmˈpjuːtə/	/ˈbɜːd/	/ˈɑːmɪ/	/pɒsɪˈbɪlɪtɪ/
/ˈdʒʌstɪfaɪ/	/frʌˈstreɪʃn/	/ˈjeləʊ/	/ˈlɑːftə/
/ˈfriːdəm/	/ˈtʊə/	/ˈwɪn/	/redʒɪˈstreɪʃn/
/ˈfɜːm/	/ˈflaɪt/	/ˈbjuːtəfʊl/	/ˈfɑːðə/
/ɪnˈveləp/	/ˈpensl/	/dɪkˈteɪʃn/	/rɪˈmembə/
/ˈtɜːmɪneɪt/	/aɪsəˈleɪʃn/	/ˈrɪəltɪ/	/ˈdʌm/
/ˈregjʊləraɪz/	/ˈkwɒlɪtɪ/	/ˈwɔː/	/ˈaɪtəm/
/ˈwɪmɪn/	/dˈɪkʃənrɪ/	/əˈprəʊtʃ/	/ˈbærɑːʒ/
/ˈskɜːt/	/ˈsteɪtmənt/	/ˈaɪz/	/ˈʌŋkl/
/əbˈteɪn/	/ˈsʊːɪsaɪd/	/ˈmɑːstə/	

7.9 **1.** `No. The woman with the plastic `bag. (Both the highlighted words have falling tone.)
  2. That's the person who robbed the `bank! (falling tone)
  3. Do you mean the man with the **black pants**? (rising tone)
  4. He drove to `work after he had finished working in the `garden. (Both the highlighted words have falling tone.)
  5. Suresh bought new `shoes today. (falling tone)
  6. Hey, have you seen the **new film** with Bruce Willis? (rising tone)
  7. Do you want some **coffee**? (rising tone)
  8. Would you like some **ice cream and cake**? (rising tone)
  9. Is he going to the **dentist**? (rising tone)
  10. `Yes. He has a `toothache. (Both the highlighted words have falling tone.)
  11. **Usually**, he comes on `Sundays. (falling rising tone)
  12. She's totally con`fused, `isn't **she**? (falling tone)
  13. I would be **really** happy if I could get more input on this. (falling rising tone)
  14. I want a purse on my `**birthday**. (falling tone)
  15. **Occcasionally**, I plan my `budget. (falling rising tone)

Exercises

I.

Price	onset	**gr**een	onset
practice	onset	**pr**ivate	onset
Appro**pri**ate	medial	**cr**y	onset
break	onset	**cr**azy	onset
journali**sm**	**coda**	**gr**eat	onset
bring	onset	**st**ude**nts**	onset, coda
adva**nce**	**coda**	under**st**and	medial
brother	onset	**gr**ow	onset
tran**script**	**coda**	**scr**atch	onset
create	onset	**scr**eam	onset
		screen	onset

Building Advanced Vocabulary

Learning Objectives   After reading this chapter, you will be able to

- understand how important it is to have a good vocabulary
- know the various ways to learn new words
- learn a large number of words through their roots, prefixes, suffixes, synonyms, and antonyms
- know how words can be learnt within a context by working out some situations
- develop technical words by using them in appropriate sentences
- use phrasal verbs and idiomatic expressions in an apt manner
- learn one-word substitution, homonyms, homophones, and eponyms
- differentiate between the words that are generally confused

8.1 INTRODUCTION

Our world is a world of words. For every single idea, belief, emotion, sentiment, and thought we need words. Words colour our lives; they empower us; they distinguish us from one another. Indeed, without words, humans are just like any other animal—dumb and inexpressive. In all walks of life, we need to have words to keep us meaningfully engaged in our human affairs.

When it comes to the professional front, the importance of words grows manifold. In an age of technological advancements, what establishes our credentials is our ability to use words—the powerful words. It is in this sense that a professional requires powerful words much more than other common people do. However, it is not just the power of words but also their appropriate usage that is required for us to be good communicators.

The section that follows gives you ample ideas and practice, going through which should help you attain an effective and powerful vocabulary. Different strategies that help us achieve better vocabulary are suggested, and a lot of exercises are added to the discussion for you to capture new words with their distinct usage.

Read the conversation between two friends on a college campus:

Ashish: Preeti, where were you at 3.00 yesterday?

Preeti: Well, I was where I was supposed to be; in class.

Ashish:	In class? Which class?
Preeti:	Don't you remember? Yesterday, there was supposed to be a special class on personality development.
Ashish:	Oh that! How was it?
Preeti:	**Terrific**!
Ashish:	What about the speaker?
Preeti:	I would say, he was truly **accomplished**.
Ashish:	You mean to say he was really good? What about his speech making?
Preeti:	Matchless! He had a **baritone** voice; his **inflections** were **immaculate**; his pronunciation really **flawless** and articulation quite **distinct**.
Ashish:	Of course, then he has to be really good.
Preeti:	Absolutely! You know Ashish, he spoke to us for more than three hours but his talk was so **engrossing** that it never became **drab**.
Ashish:	Three hours and not boring! He must have been quite good. Did he use jokes?
Preeti:	Jokes? Why? Well, he didn't have to! The way he **unfolded** the **subtleties** of the topic itself was quite **engaging**. Moreover, he had a **puckish** wit and his sense of humour was simply **disarming**. Throughout his talk, he tickled us with his **mild**, almost **caressing** sense of humour.
Ashish:	What did he look like? Was he really good-looking?
Preeti:	Not really! In fact, he appeared so **lanky** and **enervated** that he even seemed **gauche** and **emaciated**.
Ashish:	And even then you appreciate him so much?
Preeti:	Appreciate? I feel I have started **adoring** that person.
Ashish:	So, if he comes again, you are not going to miss his talk?
Preeti:	No way, I would just pounce upon the opportunity to listen to him.
Ashish:	He was so good? Was it only his manner that impressed you?
Preeti:	Oh no! Even the matter was quite **profound**. It was a perfect blend of matter and manner, a **mesmerizing** mix of what he said and how he said it.
Ashish:	Then it must have been really good. Preeti, you know, he is going to speak to the MBA guys today. In fact, his lecture is going to begin in a short while from now.
Preeti:	Really! Let's go and **grab** the front seats…

What do you understand from this conversation about the two friends? Who between Ashish and Preeti seemed more expressive and also impressive in his/her speech? We feel, the contest is rather one-sided, for it is clearly Preeti who seems to possess a superior vocabulary and expression. By the way, have you come across the impressive words Preeti uses in her expressions?

Let us see whether you can make out the meaning of these words by finding them in other contexts:

1. It's a **terrific** shot!
2. Gandhi was an **accomplished** leader.
3. Amitabh Bachchan has a **baritone** voice.
4. In order to sound interesting to your audience, you need to have **inflections** in your voice.
5. Every morning you can see scores of school going kids **immaculately** dressed in school uniform.
6. Artists are **perfectionists**; the work that seems **flawless** to us appears seriously flawed to them.
7. There is a **distinct** change in his behaviour these days.
8. Horror stories, though scary, are usually **engrossing**.
9. I slept in the class on grammar; it was so **drab**.
10. As the story **unfolds**, we find the heroine moving to London while the hero returns to India.
11. **Subtleties** of art are not for everyone to understand and appreciate.
12. An air hostess greets you with an **engaging** smile.

13. The actor knew how to put off the media queries in a **puckish** manner.
14. The **innocent** child looked at the murderer with a **disarming** smile.
15. Surprisingly, when the **burly** person spoke to me, he sounded quite soft and mild.
16. Keats's poetry is known for its **caressing**, classic touch of wisdom.
17. In his earlier movies, the star appeared to be quite **lanky**.
18. His **enervated** face told me that he had not been keeping well.
19. Wilber never knew how to be graceful and always remained **gauche** and **clumsy**.
20. The patient whom they brought on the wheelchair appeared quite **emaciated** and sick.
21. In times of corruption, we not only love but also **adore** those who do their duties properly.
22. The hungry dog **pounced** upon the porridge.
23. With **profound** grief, we have to announce that he is no more.
24. The dancer gave a **mesmerizing** performance.
25. Opportunity knocks at every door, only a few however **grab** it.

Don't you feel that coming across the words highlighted in the above sentences helps you figure out the meaning in a better way? Of course, it does, for it is always the repeated exposure to the words that makes us familiar to them.

PRACTICE TEST 8.1

Based on the understanding formed thus far, state whether the following statements are true or false:

1. A **terrific** act is one that terrifies you.
2. An **accomplished** speaker is quite impressive.
3. A **baritone** voice is deep and powerful.
4. Rise and fall in a speaker's voice stands for the **inflections** in his voice.
5. If people are dressed **immaculately**, they are dressed shabbily.
6. A **flawless** performance is almost perfect.
7. Things which are **distinct** are hardly visible.
8. An **engrossing** play is quite boring.
9. A **drab** life is quite monotonous.
10. When you **unfold** a secret, you uncover it.
11. A **subtle** joke is not direct.
12. An **engaging** smile often keeps you hooked.
13. Something **puckish** is mischievous.
14. A **disarming** reply makes you angry.
15. **Mild** humour is quite harsh and pungent.
16. When you **caress** someone, you stroke them gently.
17. A **lanky** person is lean and thin.
18. An **enervated** body is quite supple and healthy.
19. If someone considers you **gauche**, they consider you clumsy and awkward.
20. An **emaciated** person appears sturdy and well built.
21. When you **adore** someone, you like them to the extent of worshipping them.
22. When you **pounce** on food, you are usually quite hungry.
23. Something **profound** is superficial and non-serious.
24. A **mesmerizing** performance is quite captivating.
25. When you **grab** something, you intend to hold it as you desire it very much.

Let's look at some other ways which can help us learn more words.

8.2 LEARNING THROUGH SPEECHES, DESCRIPTIONS, ETC.

We come across words in almost all walks of life. Whenever we read something or listen to some speaker, we can capture words. For this however, we need to be inquisitive about them. For example, take a look at the following humorous speech and try to figure out meaning of the words highlighted in the text:

Hello, my dear friends

Today, I wish to share with you some thoughts on the **legendary sluggishness** of my friend. Well, my friend's day starts mostly in the afternoon but never before noon. And it begins with a **slight flicker** of his eyelid. Without letting his body trapped in **inertness**, his **reluctant** eyes **grope** for a sense of time. If the **perpetually** ticking wall clock reveals it is around 11 or 12 O'clock, he prefers to remain **shrouded** in his cosy bed for some more time. After more of such **languid** efforts, he finally **resurrects** himself out of his bed.

Though now awake, he is far from being **agile**. **Indolently**, he drags his feet around to the bathroom. Even a visit to the bathroom fails to **rejuvenate** his spirit. The only activity that he ventures to take up at all with some interest is browsing through the net, munching the crunchy wafers, and **sprawling** like a **couch potato** on his cushioned bed enjoying the **maddening screams** and **hoary** laughter of the television set.

Moving further, let's go through these entire target words once more. This time, involve a bit more and try to guess their meaning by suggesting whether the statement given below is true or false:

PRACTICE TEST 8.2

Read the following sentences carefully, and based on your understanding, state whether they are true or false:

1. Something **perpetual** is short lived.
2. An **agile** fielder is sharp and athletic on the field.
3. When you **sprawl** on a sofa seat, you sit very alert and attentive.
4. When a child goes to school **reluctantly**, it suggests his/her lack of interest.
5. A **sluggish** approach towards work is half hearted and full of laze.
6. If something **rejuvenates** us, it infuses in us a fresh wave of energy.
7. When you **scream,** it suggests you are quietly meditating.
8. A **slight** pain in the head is the one that is intensely felt.
9. An **indolent** person is lazy and unwilling to move.
10. A **hoary** laughter is quite original and exciting.
11. A **maddening** place is far too crowded and noisy.
12. A **couch potato** is a very active and energetic individual.
13. A **hoary** tale is trite, repetitive and clichéd.
14. **Flicker** is a type of flower.
15. A **legendary** tale is very famous and well known.
16. A **languid** walk is brisk and energetic.
17. If someone is lying **inert**, it suggests their lack of life, activity and interest in the task ahead.
18. When you **grope** for something, you search for it.
19. To **resurrect** something is to bring it back to life and activity.
20. When a dead body is **shrouded**, it is wrapped up in a cloth.

Actually, when you see a word in different contexts and associations, your affinity with it is likely to grow. In fact, there are various ways to understand and master good and powerful words. Let us look at these strategies in greater detail.

8.3 WORD FORMATION

A simple way to define, understand, and utilize a word is by coming to know its formation—the root that it comes from and the various prefixes and suffixes that change not only the shape of a word but also its shade and meaning.

8.3.1 Roots

Understanding the meanings of the common word roots can help us deduce the meanings of new words that we encounter. But be careful; though different words may emanate from the same root, they will each have a distinct meaning of their own. Further, each of these words may have different meanings in different contexts. In addition, words that look similar may also derive from different roots. So, when you come across a new word, be sure to refer to a dictionary to establish its meaning and usage correctly.

Read the following sentences and figure out the meaning of the highlighted words by focusing on their beginnings:

1. The journey did not seem taxing at all as the old woman sitting beside me chatted **amiably** all the way.
2. They knew that they were not meant for each other and hence decided to part ways **amicably**.
3. She was hurt as her boss spoke to her in an **acerbic** tone.
4. Ameena felt an **acute** pain in her stomach and collapsed in the class.
5. PETA would certainly object to receiving a **carnivorous** Guest of Honour for the function.
6. Man's **carnal** desires are not easily satisfied.
7. The **corpus** of the author's work is quite impressive.
8. Your **malevolence** finally ruins you.
9. The idiotic villain smiles at the heroine **maliciously** not knowing that the hero is standing right behind him and is getting ready to knock him over.

Let us see how we can get closer to the words and the roots that define their nature:

1. An **amiable** person is friendly.
2. An **amicable** solution is peacefully arrived at.
3. An **acute** pain is sharp and intense.
4. An **acerbic** tone is harsh and unpleasant.
5. A **carnivorous** person eats flesh.
6. **Carnal** desires are physical.
7. A **corpus** of work is the body of the work.
8. **Malevolent** people are wicked.
9. A **malicious** smile is villainous and characterized by ill-will.

This is how we can associate the words with their roots:

- am = love, friendliness: amiable, amicable
- ac = sharpness: acute, acerbic
- carn = flesh: carnivorous, carnal
- mal = bad: malevolent, malicious

PRACTICE TEST 8.3

Go ahead and think of more words that have similar beginnings and can fit into these blanks. The intended sense is conveyed within parentheses:

1. Speak _____ (in a friendly way) and win friends.
2. Speak _____ (in a harsh way) and lose friends.
3. The deer is not a _____ (flesh eating) animal.
4. The cheap woman smiled at him _____ (expressing sexual love).
5. She is humility _____ (embodiment).
6. The _____ (friendly emotion) between the brothers is permanently soured.
7. A _____ (cancerous) growth of the tumour is life-threatening.
8. An _____ (unpleasantly sharp) smell hit my nostrils the moment I stepped into the room.
9. Despite all the education and claims to modernity, a girl child's birth is seen as _____ (curse) in many families.
10. An _____ (peaceful) solution to the problem is badly required.

Read the following sentences and guess the meaning of the highlighted words:

1. When people feel vulnerable, they become more **circumspect**.
2. If you want to become a good speaker, try to avoid **circumlocutions** in your speech.
3. No sooner had the bell gone than all the children came out running in a **euphoric** mood.
4. The organizers kept uttering **eulogies** in praise of the chief-guest while the latter slept in his armchair.
5. 'She can't be let off! She has committed a **culpable** crime by letting her Mercedes run over the sleeping labourers', the inspector told the lawyer.
6. The minister was first **inculpated** in the scam but was later **exculpated** as there wasn't enough evidence available.
7. 'Please don't leave home!', the mother spoke to her son in a **placating** tone.
8. The Aussies seem to have become **complacent** about their champion status.
9. God is supposed to be **omnipotent**, **omniscient**, and **omnipresent**.
10. A couple of days back I bought a Hardy **omnibus**.

Euphoric Children!

Let us try and get a closer look at all these words again:

1. A circumspect approach is cautious and careful.
2. If you write with circumlocutions, you beat about the bush a lot.
3. When people are euphoric, they are happy and excited about something.
4. When you eulogize someone, you praise him/her.
5. A culpable crime is one that calls for punishment.
6. When people are accused of having committed a crime, they are inculpated and when they are set free, they are exculpated.
7. When someone speaks in a placating tone, they try to appease you.
8. When you are complacent about something, you are self-satisfied and don't want to improve further.

9. God can do all, see all and is present everywhere. Hence God is seen as omnipotent, omniscient and omnipresent.

10. And omnibus contains the collected works of an artist.

• circ	= around:	circumspect, circumlocutions
• eu	= good:	euphoric, eulogize
• culp	= punish:	culpable, inculpate, exculpate
• plac	= please, peace:	placating, complacent
• tomnis	= all:	omnipotent, omniscient, omnipresent, omnibus

This is how we can associate these words with their roots:

PRACTICE TEST 8.4

Can you think of more such words which have similar root structures and can fit into the blanks in the following sentences? The intended sense is conveyed within parentheses:

1. When annoyed, James becomes almost _____ (not ready to be pleased).
2. The place was too _____ (calm) to be liked.
3. When you say *someone is no more*, you are only _____ (in a polite and good manner) saying that the person is dead.
4. First he _____ (sailed around) the entire world and then settled down to run his business empire.
5. Most of us fail to grow as we live in crippling, _____ (limiting from all sides) situations all the time.
6. He always projected himself to be an _____ (all powerful) leader of the party.
7. How can you _____ (set free) someone who is clearly an offender of the law?
8. It will still probably take years for us to accept the concept of _____ (mercy killing; good, painless death).
9. Bertrand Russell is known to be a _____ (peacemaker).
10. The chief guest dozed off in his chair as the anchor kept showering _____ (song of praise).

Going further, read the following sentences and try to guess the meaning of the highlighted words:

1. She **scribbled** something hastily in her notebook and walked out.
2. When you have diabetes, the doctor **proscribes** sugar and **prescribes** medicines.
3. In order to improve your vocabulary, you need to be a **voracious** reader.
4. The giant **devoured** the poor peasant and demanded more food.
5. Don't be **verbose** in life.
6. A **verbatim** reproduction of the event is required to convince the jury.
7. Mughal-e-Azam is truly a **magnum opus** by K. Asif.
8. The **magnanimous** poet donated the shawl, conferred on him as an honour, to the beggar.
9. Hamlet has as many as eight **soliloquies**.
10. The saint would go to the graveyard to seek **solitude**.

Let us try and get a closer look at all these words:

1. When you **scribble** something, you write hastily and mostly in illegible form.
2. To **proscribe** something is to ban that and to **prescribe** something is to advise the use of it.

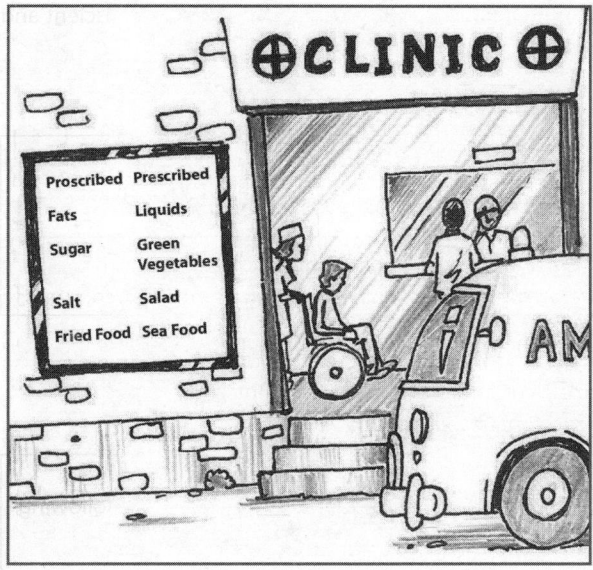

For Healthy Living

3. When you are a **voracious** reader, you read a lot. A **voracious** reader is hungry for more and more reading.
4. If you **devour** something, you eat that very hungrily.
5. If you are considered **verbose**, you seem to be using more words than are required.
6. A **verbatim** account of something is the word for word account of that.
7. **Magnum opus** is a term used to refer to some outstanding and the great work of an artist.
8. A **magnanimous** person is one with a large, generous heart.
9. **Soliloquy** is a speech made by a character when left alone on stage.
10. **Solitude** is the state of being alone.

This is how we can associate these words with their roots:

- scribo, scryptus = to write: scribble, proscribe, prescribe
- vor = eat: voracious, devour
- verbum = word; verbose, verbatim
- magnus = big: magnum opus, magnanimous
- solo = alone: soliloquy, solitude

Read the following sentences and try to guess the meaning of the highlighted words:

1. King Khan's statement became **controversial** soon afterwards.
2. My Lord! There is **incontrovertible** evidence available against the criminal and hence he deserves the punishment.
3. No thoroughfare! **Transgressors** will be severely punished.
4. If we go by his view, we can't progress. He has a **retrogressive** attitude.
5. The **loquacious** little girl kept prattling all the while.
6. His **eloquence** and sense of humour is admired all across the institute.
7. Immediately after being subjected to **corporal** punishment, the child fainted in the class.
8. An impressive car pulled in and a **corpulent** figure emerged out of it.
9. Marlowe was one of the most notable of Shakespeare's **precursors**.
10. You don't have to spend much time reading it; just give it a **cursory** look.

Let us try and get a closer look at these words:

1. Something **controversial** sparks off debate of some kind.
2. An **incontrovertible** evidence cannot be challenged and questioned.
3. A **transgressor** of the law is the one who violates it.
4. A **retrogressive** attitude is suggestive of a mental backwardness.
5. A **loquacious** person is talkative and **garrulous**.

6. Someone's **eloquence** is suggestive of their spontaneity.
7. **Corporal** punishment is physical.
8. A **corpulent** person is fat.
9. A **precursor** is someone or something that exists before.
10. A **cursory** look is casual, speedy, and perfunctory.

This is how we can associate some of these words with their roots and other root words:

- contra = against: contravene, controversial, incontrovertible, contradict
- gress, grad = step: transgress, retrogressive, retrograde, gradual, digress, ingress, regress, congress, degrade
- loqu, loc, log = speech: loquacious, eloquent, colloquial, interlocutor, monologue
- corp = body: corporate, corpus, corpse, corporal, corporeal, corpulent
- curs = run: precursor, cursory, current, concur, incursion

PRACTICE TEST 8.5

Fill in the blanks with the correct form of the verb(s) given below, as per the tense:
(magnanimous, proscribe, scribble, retrograde, verbose, loquacious, voracious, verbatim, eloquent, solitude, corpulent)

1. Don't _____ in an illegible way; write properly and distinctly.
2. Saints relish their _____ as they are not scared of being alone.
3. Once he starts speaking, he doesn't stop; he is such a _____ man.
4. Mother Teresa was truly _____.
5. A _____ man was seen running to catch the bus and all stopped to enjoy the spectacle.
6. An _____ man need not be _____ in his speech.
7. He was such a _____ that he could not reconcile to any change around him.
8. If you reproduce _____ what your books suggest, you are a mugger.
9. When a doctor _____ something, there is a reason behind it.
10. If you want to master new words, be a _____ reader.

8.3.2 Prefixes and Suffixes

Just as it is important to understand the root of a word to understand its meaning, it is also worthwhile to learn how certain beginnings can add to the meaning or change the existing meaning of certain words. It is so because we form words by adding prefixes and suffixes to different words. Therefore, it is not only the roots but also the prefixes and suffixes which help us arrive at the meaning of a word.

In fact, if you study the shape of a word, you can divide it into three parts—the prefix, the suffix, and the root word. Since we have studied some roots and the related words, let us now study some prefixes and suffixes and see the words that can be formed by using them:

in- (A negative prefix)

Examples:

Insufficient	(in + sufficient)
Insane	(in + sane)
Invalid	(in + valid)
Inappropriate	(in + appropriate)

un- (A negative prefix)

Examples:

Unpleasant	(un + pleasant)
Unproductive	(un + productive)
Unadvisable	(un + advisable)
Unhappiness	(un + happiness)

im- (A negative prefix)

Examples:

Impolite	(im + polite)
Improper	(im + proper)
Imperfect	(im + perfect)
Impious	(im + pious)

-fy (A verb suffix)

Examples:

Class	Classify
Pure	Purify
Note	Notify
Terror	Terrify

-ation (A noun suffix)

Examples:

Educate	Education
Indicate	Indication
Authenticate	Authentication
Purify	Purification

-er (A noun suffix)

Examples:

Teach	Teacher
Inform	Informer
Lead	Leader
Manage	Manager

-ment (A noun suffix)

Examples:

Involve	Involvement
State	Statement
Enjoy	Enjoyment
Entertain	Entertainment

-ion (A noun suffix)

Examples:

Provide	Provision
Decide	Decision
Illustrate	Illustration
Appreciate	Appreciation

-ness (A noun suffix)

Examples:

Kind	Kindness
Small	Smallness
Useful	Usefulness
New	Newness

-less (An adjective suffix)

Examples:

Value	Valueless
Mercy	Merciless
Tire	Tireless
Effort	Effortless

-ful (An adjective suffix)

Examples:

Mind	Mindful
Disdain	Disdainful
Fear	Fearful
Regret	Regretful

-able (An adjective suffix)

Examples:

Objection	Objectionable
Achieve	Achievable
Identify	Identifiable
Favour	Favourable

-ental (An adjective suffix)

Examples:

Judgement	Judgemental
Sentiment	Sentimental
Instrument	Instrumental

-ly (An adverbial suffix)

Examples:

Loud	Loudly
Cheap	Cheaply
Suggestive	Suggestively
Imaginative	Imaginatively

Now, look at the following sentences and try to figure out the meaning of the words highlighted in them:

1. He can never speak the truth; he is an **incorrigible** liar.
2. We consider God to be **infallible**.
3. **Ignoble** acts chase you, not just in one life but in many lives.
4. There was an **impregnable** silence all around; it was so uneasy to sit there.

5. An **intractable** child is a nuisance for parents.
6. The teacher was annoyed as he felt that I was being too **impertinent**.
7. Many celebrities lose their standing when their hidden **illicit** relations come to the fore.
8. Though frail in shape, the union leader confronted the management in an **intrepid** manner.
9. Terrorism is gradually becoming an **interminable** problem in many parts of the world.
10. Lata Mangeshkar has an **impeccable** voice.

Look at the following expressions which further establish the sense of these words:

1. An **incorrigible** liar can never be truthful; he is a habitual liar.
2. Someone **infallible** is beyond reproach and fault.
3. **Ignoble** persons are neither virtuous nor noble.
4. An **impregnable** situation is one that cannot be penetrated into.
5. An **intractable** person is difficult to control and manage.
6. An **impertinent** enquiry is considered to be impolite and disrespectful.
7. An **illicit** relationship is not lawful.
8. An **intrepid** person is fearless and courageous.
9. Something **interminable** is considered never ending.
10. Something **impeccable** is flawless and perfect.

Now try to discover further the meaning of the highlighted words by attempting the following exercise.

PRACTICE TEST 8.6

Suggest whether the words highlighted below make sense in the given contexts. Write *Yes* if you feel that the word is appropriate and *No* if it does not fit well:

1. None of us is **infallible**, all of us err.
2. The song that he sang was horrible; his voice so harsh and **impeccable**.
3. You don't have to carry an umbrella. The clouds are patchy and distant; so it would just rain **interminably**.
4. The person who is accused of maintaining **illicit** relations is generally condemned.
5. Don't worry, he is so **intractable**; I can manage him easily.
6. If you maintain such an **impregnable** silence, no one can speak to you.
7. Poets, thinkers, and philosophers are punished for asking questions considered **impertinent**.
8. He can never give up smoking; he is an **incorrigible** smoker.
9. The Bollywood villain is essentially **ignoble** and wicked.
10. He was so cowardly that everybody called him **intrepid**.

8.4 SYNONYMS

A synonym is a word or expression that has almost the same meaning as another word or expression. The word 'synonym' is a composite of two Greek words, namely the prefix 'syn' meaning 'together' and 'onym' meaning 'name'. In English, almost all words have more than one synonym. For example:

Enormous	— big, huge, massive, giant, immense
Drill	— accustom, exercise, habituate, hone, practice, rehearse, tune up, work out
See	— watch, observe, notice, envisage, spot

By learning various synonyms of words, you will be able to increase your word power which will surely help you improve your writing and speaking skills.

PRACTICE TEST 8.7

Look at the following words—the ones that we saw in Practice Test 8.5—and suggest their synonyms:

1. Infallible
2. Impeccable
3. Interminable
4. Illicit
5. Intractable
6. Impregnable
7. Impertinent
8. Incorrigible
9. Ignoble
10. Intrepid

8.5 ANTONYMS

Just as it is possible to learn new words by getting to know their synonyms, it is also worthwhile to understand their antonyms—the words opposite in meaning—to understand their usage. An antonym is a word which has the opposite meaning to a particular word, although not necessarily in all its senses.

Antonym is a word or phrase that is opposite in meaning to a particular word or a phrase in the same language. For example:

Like   — hate, dislike, detest, loathe, despise, abominate

Sharp — blunt, dull, even, moderate, blurred, dim, slow, stupid

Knowing antonyms will help you not only learn the difference between words but also equip you to improve your expression while writing and speaking. Let's learn some antonym with the help of practice tests and exercises given in the chapter.

PRACTICE TEST 8.8

Look at the following words and suggest antonyms for them by adding or removing (or both) a prefix or a suffix:

1. Introvert
2. Exculpate
3. Consistent
4. Enviable
5. Disposed
6. Inter
7. Satiable
8. Legible
9. Illegitimate
10. Intemperate

8.6 LEARNING WORDS THROUGH SITUATIONS

Though it is important to understand how words are formed, it is all the more important to understand how to use them. In this section, we will learn how to use the right word at the right place in different situations. For this purpose, different situations have been contrived. Just as you come across new words while reading newspapers, magazines, journals, novels, etc., given below are a few situations which will introduce to you some words which might be new to you.

Try to guess the meaning of each of the words highlighted in these situations in the first instance; and going further, with the help of the discussion and the exercises that follow, make an effort to capture their meaning and usage in specific contexts.

Situation I

When you grow from childhood to **adolescence**, there are **numerous** complexities that **confront** you. You are neither a child nor a man. Parents **chide** you for being **petulant** and **admonish** you for pretending to be a **savant**. At times your suggestions are brushed aside as **naïve** ideas and on other occasions your hopes are seen as **chimerical** and **fanciful**.

Let us get closer to our understanding of these words:

1. **Adolescence** refers to the age when you are neither a child nor a grown up person. Clinically it refers to someone being between the age group 13–19.
2. **Numerous** things are in large number.
3. When you **confront** a situation, you face it.
4. When someone **chides** you, they scold you.
5. A **petulant** person is the one who behaves in a childish manner.
6. When you **admonish** someone, you rebuke and warn them.
7. A **savant** person is a learned person.
8. A **naïve** person is generally immature as he/she lacks experience and expertise.
9. A **chimerical** idea is fanciful and unrealistic.
10. **Fanciful** hopes are not realistic.

PRACTICE TEST 8.9

Find out whether the following sets of words are same or opposite in meaning:

1.	Chide	= Appreciate	6.	Naïve	= Novice
2.	Confront	= Escape	7.	Petulant	= Childish
3.	Chimerical	= Unrealistic	8.	Adolescence	= Teenage
4.	Numerous	= Plentiful	9.	Savant	= Knowledgeable
5.	Admonish	= Adore	10.	Fanciful	= Realistic

Situation II

In India, cricket is **deemed** to be a religion. Cricketers in our country enjoy a **cult** image and some of them are simply **deified** as gods and goddesses. Though **revered** widely, cricketers are not **infallible** in the public eye. The mob's ways, much like the **destinies** of their **protagonists**, are really **capricious**. When our cricketers beat a champion team like Australia, we sing **paeans** in their praise, shower on them heaps of **adulations** and **felicitate** their every single **whim** and **caprice**. The situation reverses the moment they lose to a side which is seen as **vulnerable** as Bangladesh or as **inimical** to our national interest as Pakistan. Losing to such teams creates a **furore** all over the country and the **erstwhile** heroes are seen as **contemptible** creatures. It is in such **fury**, that we choose to **castigate**, **deride**, and **disparage** our cricketers. All this does not last for long; as soon as one more series is won we are back to our **eulogizing** best, casting **panegyrics** in the honour of our cricketers.

Why do you have to be petulant all the time? You think you are quite a savant? No, you are just naïve! All your ideas are too chimerical and all your dreams too fanciful to be real!

Work on Your Vocabulary!

Let us try to get closer to the target words:

1. If something is **deemed** to be some way, it is believed to be that way.
2. A **cult** status is quite influential and powerful.
3. If you **deify** someone, you worship them and treat them as deities.
4. Those who are **revered** by others, are respected by them immensely.
5. If you find someone **infallible**, you see them as perfect, complete and consummate.
6. Our fate is seen as our **destiny**.
7. The **protagonist** of a play or story is the central, main, and the most important character.
8. A **capricious** person changes his/her ideas very quickly.
9. **Paeans** are songs of praise.
10. Your **adulation** for someone is your appreciation for them.
11. When you **felicitate** someone, you honour them.
12. People's **whims** and **caprices** are the unpredictable patterns of their behaviour.
13. **Vulnerable** people are quite weak and unable to stand pressure.
14. If something is **inimical** to your interest or desires, it is against that.
15. A **furore** is a loud commotion and loud voice of protest.
16. An **erstwhile** position is the former one.
17. Those who are considered **contemptible** are seen as worthy of condemnation and contempt.
18. Somebody's **fury** refers to their anger.
19. When you **castigate** someone, you criticize them severely.
20. When people are **derided**, they are mocked at.
21. If someone is **disparaged**, they are insulted.
22. When you **eulogize** someone, you praise them lavishly.
23. **Panegyrics** are songs of praise and appreciation.

PRACTICE TEST 8.10

Find out whether the following sets of words are same or opposite in meaning:

1. Contemptible = Beautiful
2. Fury = Anger
3. Capricious = Unstable
4. Protagonist = A marginal figure
5. Infallible = Beyond reproach
6. Paeans = Songs of praise
7. Felicitate = Dishonour
8. Revered = Respected
9. Vulnerable = Powerful
10. Whims/Caprices = Quirks
11. Furore = Unrest
12. Castigate = Appreciate
13. Derided = Mocked
14. Disparage = Deprecate
15. Panegyrics = Songs of praise
16. Deify = To be seen as a deity

Situation III

People kept blowing horns but not even a single vehicle moved an inch. The traffic **snarled up** and could move at an **excruciatingly** slow pace. Men, women and children **caged in** tiny vehicles looked **exasperated**. The man in the driving seat of an **azure** car wore a **pensive** look while the woman sitting on the other seat, seemed **oblivious** of his **anguish** and kept **fiddling** with her gold locket that **dangled** across her neck. The question whether it was a genuine or **meretricious** piece of jewellery, was irrelevant to her.

Let us try getting closer to the target words by looking at them in other sentences:

1. The journey of the sky from turning vermillion to **azure** is fascinating.
2. A truly religious person remains totally **oblivious** to the challenges of the world.
3. "Why do you keep such a **pensive** look all the time?" the concerned wife asked anxiously.
4. How can you say that the jewellery is **meretricious**; it is 22-carat gold.
5. The modern man keeps **dangling** between several choices.
6. "Why can't you keep your fingers rested? You keep **fiddling with** something or the other" the teacher said angrily.
7. It is so unfortunate that we keep the animals, born in wild, **in cages** for our entertainment.
8. Being overambitious leads to **anxiousness**.
9. He doubled up as an **excruciating pain** shot through his stomach.
10. Suddenly the dog **snarled at** the intruder and moved towards him menacingly.

Apart from the expression 'snarled at', which in the later context stands for an animal's growl, the rest of the words have been used in the same way.

Situation IV

One of the most important debates of contemporary times could be the question whether television can make us any better, leave us as we are, or just make us worse than what we are. If television programmes, at least most of them, are to go by, the first **proposition** is certainly ruled out. As regards the second question, it seems that even a **status quo** is difficult to maintain when you **confront** a **belligerently** blaring idiot box all through the day. In such a **din** that only a television can create, we are sure to lose our **poise** and **equanimity**. After all, how can you refuse to **degenerate** when you are **plied** with **melodramatic** soap-operas, **excessively maudlin** reality shows, absolute **buffoonery** in the name of laughter and **vulgar** dances and **deceitfully alluring** skin shows in the name of art and **subtlety**?

Let us try and get closer to the meaning of these words:

1. A **contemporary** situation relates to your present times.
2. A **proposition** is a suggestion, task, an idea, or an activity.
3. If something maintains its **status quo**, it stays as it is.
4. When you **confront** a situation, you face it.
5. A **belligerent** walk is aggressive.
6. **Din** is suggestive of the loudness and unpleasantness of the noises around you.
7. If you maintain your **poise**, you generally retain your composure.
8. Someone's **equanimity** refers to their peace, poise, and harmony.

9. When something **degenerates**, it deteriorates.
10. When you are **plied** with something, you are served with that in abundance.
11. A **melodramatic** serial, movie, or song is extremely and artlessly sentimental.
12. If something is **excessive**, it is too much.
13. Something **maudlin** is sentimental and mawkish.
14. **Buffoonery** refers to clowning and cheap laughter.
15. A **vulgar** statement or show is indecent.
16. Something **deceitful** is cunningly misleading.
17. An **alluring** show attracts the viewers to itself in some cheap and baser way.
18. **Subtlety** in art refers to its delicate nuances and finesse.

PRACTICE TEST 8.11

Find out whether the following sets of words are same or opposite in meaning:

1. Vulgar   – Lewd
2. Melodramatic – Tear jerker
3. Deceitful   – Crafty
4. Belligerent   – Peaceful
5. Contemporary – Obsolete
6. Confront   – Evade
7. Degeneration – Deterioration
8. Poise   – Calmness

9. Alluring   – Tempting
10. Buffoonery – Cheap laughter
11. Din   – Noise
12. Equanimity – Disturbance
13. Subtle   – Deft
14. Maudlin   – Emotional
15. Ply   – Offer

PRACTICE TEST 8.12

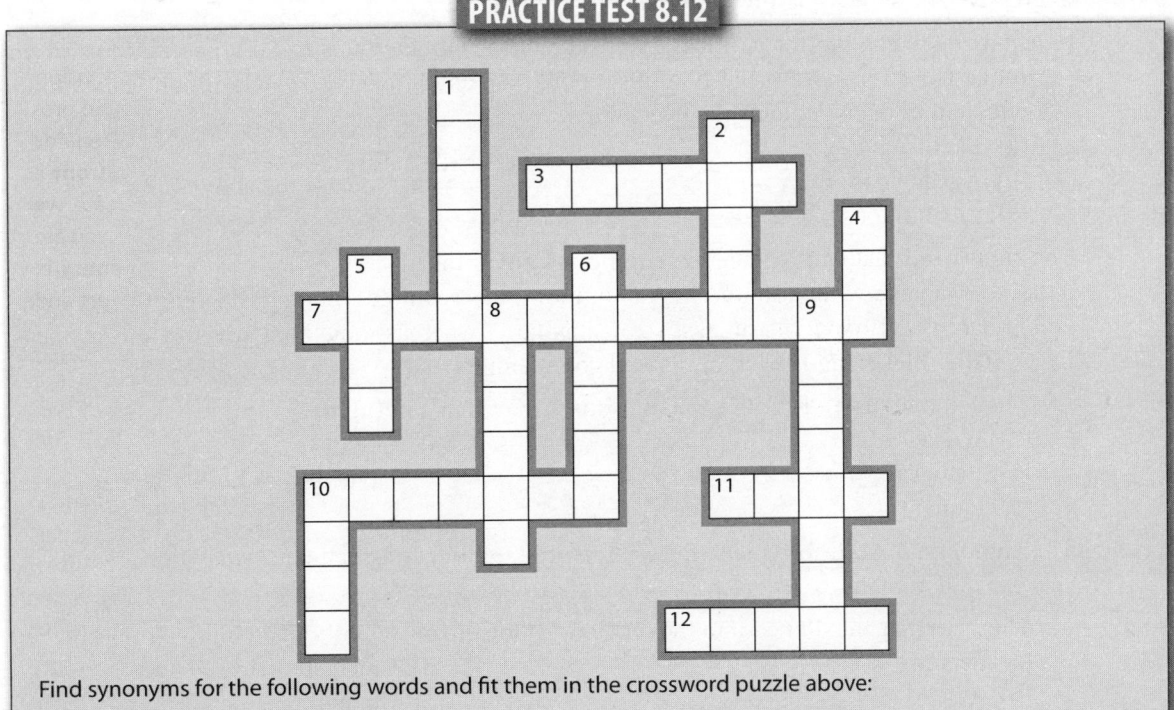

Find synonyms for the following words and fit them in the crossword puzzle above:

Word bank:

calm, deft, deterioration, din, languid, lewd, obsolete, offer, peeved, rescue, subtle tempt, torrid

Across:

3. angry	7. descent	10. lazy	11. peaceful	12. give

Down:

1. slight	4. noise	6. hot	9. redundant
2. allure	5. skilful	8. save	10. vulgar

Going further, look at the following sentences and note the highlighted word in each of these sentences. Four options follow and you are required to pick the word that can replace the highlighted word. Read the explanatory answers that follow and understand how words can be chosen out of the given options.

1. Despite his **boisterous** laughter, he nurtures a deeply unsettling and darkly melancholic nature.

   (a) Vicarious   (d) Bovine

   (b) Raucous   (e) None of the above

   (c) Bellicose

The only word that can replace the head word **boisterous** is **raucous** as both these words stand for the noisy nature of the laughter of the person referred to in the context. Other words are not appropriate in the context as a **vicarious** pleasure or experience is indirect; a **bellicose** walk is aggressive and something **bovine** is harmless and docile; hence (b).

2. In a mad scramble for political gains, alliances and patch-ups of all types can be seen in India. However most of these tie-ups, more than anything else, smack of political **expediency** rather than ideological partnerships.

   (a) Exclusivity   (d) Bickering

   (b) Advantage   (e) None of the above

   (c) Adventitious

The right word that can replace **expediency** is **advantage** as *political expediency* is an effort to seek advantage without bothering about moralities required to be considered. Other words don't relate as **exclusivity** is suggestive of singularity whereas something **adventitious** happens incidentally. **Bickering** is back biting and complaining.

3. Though into his mid-thirties, Salman still behaves in a **petulant** way at times.

   (a) Wicked   (d) Childish

   (b) Innocent   (e) None of the above

   (c) Piquant

The appropriate word to replace the head word is childish as petulant behaviour is childish, peevish, and immature.

4. The spying agent slipped into the territory of the enemy country in a **clandestine** way.

   (a) Secret   (d) Calculated

   (b) Absurd   (e) None of the above

   (c) Crafty

A clandestine move is secretly carried out; hence option (a).

5. Since human life is **ephemeral** nothing should cause us anxieties and frustrations beyond our capacity to bear.
   - (a) Permanent
   - (b) Everlasting
   - (c) Transient
   - (d) Turbulent
   - (e) None of the above

Ephemeral means something that is short lived and transient; hence option (c).

6. Steve Waugh was a great cricketing character. As a **pugnacious** player and inspirational leader, he took the Australian team to dizzy heights of success.
   - (a) Charismatic
   - (b) Combative
   - (c) Clever
   - (d) Composed
   - (e) None of the above

The most appropriate word to replace the word is (b), i.e., **combative** as a **pugnacious** person is competitive and combative in spirit.

7. Though always keen on scenic beauty, it was difficult for me to hold my nerves when the shaky, over crowded bus moved gingerly up the **meandering** road in the hills.
   - (a) Torturous
   - (b) Troublesome
   - (c) Tortuous
   - (d) Tremulous
   - (e) None of the above

The most appropriate word to replace **meandering** is **tortuous** both of which mean serpentine and not straight.

8. Instead of mixing with others, the bride sat in a corner and sulking silently, she looked unusually **irascible**.
   - (a) Reticent
   - (b) Irritated
   - (c) Amused
   - (d) Bemused
   - (e) None of the above

If someone is **irascible**, they are irritated and annoyed; hence the second choice.

9. If you don't know how to keep quiet, people will start calling you **garrulous**.
   - (a) Taciturn
   - (b) Eloquent
   - (c) Talkative
   - (d) Somnolent
   - (e) None of the above

The context suggests that the one of who speaks continuously is **garrulous**. Hence **talkative**, which also means someone who speaks too much, is the answer. The other words don't match. **Taciturn** is the opposite of garrulous. An **eloquent** speaker is spontaneous and effortless, and hence not suited to the context. Similarly, **somnolent** is something sleep inducing, hence inappropriate in the context.

8.7 HOMONYMS AND HOMOPHONES

Homonyms are distinct words that have the same form. For example:

Bank   (where money is deposited)

Bank   (of a river)

Homophones, on the other hand, are distinct words that are spelled differently but pronounced alike or sound alike. For example, **one** and **won**.

Now consider the following expressions:

1. Why don't you **write** to us?
2. You have no **right** to speak to me like that?

Though both 'write' and 'right' are pronounced in the same way, they differ in spelling and meanings. Thus 'write' and 'right' are homophones—the words that are pronounced in the same way but are spelt differently and have different meanings. Interestingly, the word 'right' is a homonym too as it gives different meanings in different contexts even with the same spellings. For instance, look at the following sentences:

1. Will you move to your **right** a bit? ('right' stands for direction)
2. That's a **right** suggestion. ('right' stands for correct).

Look at the following set of homophones:

peace, piece	sale, sail
one, won	die, dye
weak, week	cell, sell
hair, hare	knew, new

Now read the following set of examples which are homonyms:

duck: a water bird that swims	duck: bend quickly
bear: a large mammal	bear: carry
face: part of a body	face: encounter difficulties

8.8 WORDS OFTEN CONFUSED

Besides homonyms and homophones, there are a large number of words that are confused with one another.

Following are some such pairs of words which sound or appear similar to each other and hence are commonly confused. Each of these words is used in a sentence so that the confusion pertaining to these may be avoided.

1. Envious (someone who feels envy for others): Avoid being **envious**, be competitive.
   Enviable (worthy of admiration): Robert's achievement has made him **enviable** in his family.
2. Urban (not rural): The **urban** area of the city is developing quite fast.
   Urbane (sophisticated): His **urbane** manners could not win the intelligent audience.
3. Imaginary (unreal): Don't tell me **imaginary** stories; I trust none of them.
   Imaginative (creative): The **imaginative** director shot the scene with great skill and depth.
4. Honorary (without pay): Though the post is **honorary**, go for it. It will fetch you a lot of recognition.
   Honourable (admired and respected): **Honourable** Sir, may I request your attention for a while?
5. Ghastly (causing fear): After the bomb blast, the site looked **ghastly**.
   Ghostly (relating to ghosts): Don't narrate **ghostly** stories, they scare me.
6. Industrious (laborious): Only the **industrious** students succeed in life.
   Industrial (relating to industry): The modern Indian society is fast becoming an **industrial** one.
7. Ordinance (law, bill): A new **ordinance** is likely to be passed in the parliamentary session.
   Ordnance (military supplies): The **ordnance** factory was blown away by the rebels.
8. Corporal (physical): According to new norms, **corporal** punishment is banned in schools.
   Corporeal (not spiritual): All drives of modern man have by and large become **corporeal**.

9. Comprehensive (exhaustive): Just a **comprehensive** examination and the student passes; we need a continuous evaluation system instead.

Comprehensible (that which can be understood): When we speak Indian English, it is not usually **comprehensible** to a native speaker of the language.

10. Graceful (dignified and polished): Though rich, he knows not how to be **graceful** in speech and manners.

Gracious (large hearted): The **gracious** king was liked by one and all.

11. Childish (silly): No one likes **childish** behaviour.

Childlike (innocent and simple): Nature inspires you to be **childlike**.

12. Lovable (worthy of love): Children have **lovable** innocence.

Lovely (nice): It indeed was a **lovely** present from her.

13. Momentous (very important): The **momentous** occasion has come and we must be there to celebrate it.

Memorial (built in someone's memory): The Bradman **Memorial** Trophy is likely to start at some point of time.

14. Refuge (shelter): After Partition, millions of people became homeless and had to find **refuge** in alien lands.

Refuse (rubbish): Don't scatter the **refuse** carelessly; it is uncivilized behaviour.

15. Exceed (surpass): His behaviour has **exceeded** all limits.

Accede (agree): He finally **acceded** to the idea of discussing the sensitive issue.

16. Allusion (reference): If you can interpret some of the **allusions** in his poetry, you can actually understand Eliot.

Illusion (deceptive appearance): Don't have any **illusions**; be realistic and practical.

17. Dual (double): These days many students pursue a **dual** degree course.

Duel (fight): The movie relates to the **duel** between the famous brothers.

18. Eminent (prominent): The **eminent** scholar delivered the keynote address at the conference.

Imminent (about to happen): The **imminent** disaster is likely to sweep out the city.

19. Collision (clash): Thirty passengers died in the **collision** of the buses.

Collusion (secret agreement): The leader is said to have been in **collusion** with the smugglers.

20. Fare (travel charges): The train **fare** will be reimbursed.

Fair (just): Be **fair** to others.

21. Elicit (to draw an answer): When students are not listening to you, you can't **elicit** an answer for your questions.

Illicit (unlawful): The actor's **illicit** relationship with the junior artiste was much publicized by the media.

22. Hoard (accumulate): Man always intends to **hoard** more and more.

Horde (group): The **horde** rushed into the hall the moment the gates were opened.

23. Ingenious (imaginative): Give **ingenious** answers in the interview and get the job.

Ingenuous (lacking imagination): Answer **ingenuously** and people start laughing at you.

24. Persecute (punish): He was **persecuted** right in front of the crowd.

Prosecute (to indict and charge): The court has given the permission to **prosecute** the chairman of the selection board.

25. Pray (a form of worship): It is considered good to **pray** in the morning.

Prey (victim): Though he struggled well for a while, he finally fell **prey** to circumstances.

26. Prescribe (to direct): The medicines that the doctor has **prescribed** are not available.

Proscribe (to prohibit): The doctor **proscribed** sugar so that he could control his diabetes.

27. Draught (small quantity): A refreshing **draught** of breeze soothed my tense nerves.

Drought (want of rain): Some part of India or the other faces **drought** almost every year.

28. Diseased (ill and sick): The **diseased** dog was kicked out of the house.

Deceased (dead): The parents of the **deceased** had to fight hard to claim the body.

29. Compliment (regards): Wherever she goes, she is **complimented** for her sincerity and alertness. Complement (something that completes the other): Husband and wife have a **complementary** existence in social life.

30. Zealous (enthusiastic): The **zealous** crowd welcomed the singer with thunderous applause. Zealot (fanatic): **Zealots** are hard to appease.

PRACTICE TEST 8.13

Read the sentences given below and choose the word that fits the context from among the given homophones/homonyms:

1. The little boy surprised everyone with his ingenious/ingenuous replies.
2. Mr Sheldon was persecuted/prosecuted on the charges of conspiracy and fraud.
3. Let's prescribe/proscribe peace and prescribe/proscribe violence.
4. Don't run in pursuit of hoarding/hording more and more.
5. Pope's poetic achievements made him envious/enviable in his close circle.
6. Let's work hard to broaden the child's imaginative/imaginary faculty.
7. In the modern industrious/industrial society there is little scope for spiritualism.
8. He presented an exhaustive/exhausting analysis of the situation.
9. The report gives detailed and comprehensive/comprehensible suggestions.
10. The little girl had a loveable/lovely face.
11. Haunted by media, the poet took refuge/refuse in a small village.
12. Milton's *Paradise Lost* is full of literary illusions/allusions.
13. He is a cantankerous man; he is always ready for a duel/dual.
14. The Kareena–Shahid deo/duo features in *Jab We Met*.
15. The imminent/eminent scholar died of cancer.
16. We don't see it as a fair/fare deal.
17. It is difficult to elicit/illicit an honest reply from a crafty man.
18. Those days, slaves were denied even a drought/draught of water.
19. A floral tribute was offered to the diseased/deceased leader.
20. James received rich complements/compliments for his recitation.

8.9 ONE WORD SUBSTITUTION

Another way to increase your vocabulary would be to know how a one-word substitution can replace an entire idea. Once we are able to master quite a few of these one-word substitutions, we can express the idea in a more concise way. Besides that, it provides to us the precision that lends brevity and exactness to our expression.

Let us see how different notions, ideas, and definitions can be substituted by one-word expressions:

1.	A person who believes in God	Theist
2.	A student who stays away from school without telling his/her parents	Truant
3.	The short remaining part of a cigarette	Stub
4.	Someone's life history written by another person	Biography
5.	A tendency to favour one's relatives	Nepotism
6.	A series of three novels/works of art produced by a writer/artist	Trilogy

7.	One who takes pleasure in torturing others	Sadist
8.	A story in which ideas are symbolized as characters	Allegory
9.	Killing one's father	Patricide
10.	A medicine that cures all ailments	Panacea
11.	A person too proud of his/her race, nation, gender, etc.	Chauvinist
12.	A place where wild animals/birds are kept in a protected area	Sanctuary
13.	One who totally abstains from drinking	Teetotaller
14.	One who understands the finer aspects of art, music, etc.	Connoisseur
15.	Someone whose philosophy in life is to eat, drink, and be merry	Epicurean
16.	One who eats indiscriminately and in large quantities	Glutton
17.	Someone with an unprejudiced and accommodating nature	Cosmopolitan
18.	One who deserts one's religion	Apostate
19.	A screen or pen name adopted by an artist or writer	Pseudonym
20.	Deliberate suffering for one's sins	Expiation
21.	Something that is considered redundant and unnecessary	Superfluous
22.	Saying things in a roundabout way	Circumlocution
23.	Someone who travels with devotion to a sacred place	Pilgrim
24.	A person who has beautiful and elegant handwriting	Calligrapher
25.	A woman who never gets married	Spinster
26.	Impulsive stealing	Kleptomania
27.	Easily convinced, cheated and gulled	Credulous
28.	A person with strange and peculiar habits	Eccentric
29.	Someone who is interested in the welfare of women	Feminist
30.	Someone who compiles a dictionary	Lexicographer

PRACTICE TEST 8.14

Given below are certain ideas and definitions. Replace them by providing one-word substitutions for each of them:

1. One who dedicates his life to a selfless pursuit of helping others
2. A general official pardon given to many in one go
3. Someone who is made to suffer because of others
4. A remark or work considered disrespectful to God or religion
5. A person who pays too much attention to personal appearance
6. A woman who displays tantrums to attract men's attention
7. An artist's most outstanding and memorable artistic creation
8. An animal story with a moral
9. A long narrative poem written in a grand style
10. Someone who loves collecting stamps

PRACTICE TEST 8.15

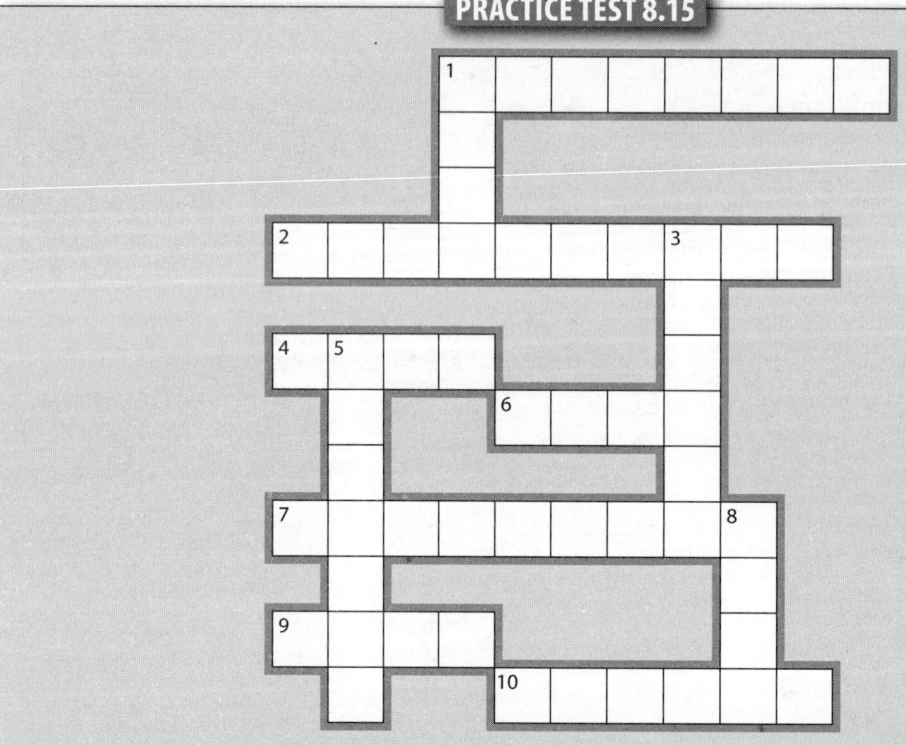

Find one word from the word bank for the following expressions and fit in the crossword puzzle above:

Word bank:

altruistic, amnesty, epic, mast, pact, penitence, solace, spinster, spur, stab, tactic

Across:

1. An unmarried lady
2. One who dedicates his/her life to a selfless pursuit
4. A tall pole on a boat or ship that supports the sails
6. An agreement between two parties
7. A feeling of being sorry because one has done something wrong
9. To injure someone with a sharp pointed object
10. A feeling of emotional comfort when somebody is sad or disappointed

Down:

1. Suddenly without planning in advance
3. A particular method used to achieve something
5. A general official pardon given to many in one go
8. A long poem about the actions of great men or women or a nation

8.10 PHRASAL VERBS

Phrasal verbs spice up a person's communication skills, written as well as spoken. The use of phrasal verbs lends a communicative and interactive touch to a person's expression. Moreover, since phrasal verbs essentially express the notion in action, they assign to the tone the force and vitality required for good and impressive communication. Learning phrasal verbs and using them appropriately, however, is a challenge. It is so because a phrasal verb can be used in a variety of ways. Further, since many phrasal verbs are used in many similar and closely connected ideas, a lot of confusion occurs over the choice of the best phrasal verb in a particular situation.

The examples and the discussion that follow will help you understand the meaning of some phrasal verbs, different in meaning but similarly sounding.

1. 'Though we have sorted out our differences, she still keeps _____ me,' Marx said dolefully.
   (a) Going with
   (b) Going at
   (c) Going for
   (d) Going on
   (e) None of the above

Key: The context here suggests that despite the reconciliation, someone keeps attacking the speaker. Hence, the option *going for* fits the context. *To go* at something is to work on that with great enthusiasm. When you *go with* something, you support that. Something that *goes on* continues to happen over a period of time.

2. I just don't trust a word you say; everything looks so _____.
   (a) Made off
   (b) Made in
   (c) Made up
   (d) Made up with
   (e) None of the above

The context suggests that the speaker does not believe anything that the other one is telling him/her. It seems the speaker finds the other person's narration to be imaginary, concocted, and hence *made up*. Therefore, (c) *made up* is the right option. *Made off* suggests running away with stolen valuables. Something made in a particular image or metal is cast in that. When you *make up* with someone, you reconcile with them.

3. She _____ the report indignantly.
   (a) Cast away
   (b) Cast off
   (c) Cast down
   (d) Cast aside
   (e) None of the above

Key: The sentence suggests that the report annoyed and shocked the reader. She, therefore, is likely to *cast aside* the newspaper, which means to reject someone or to discard something. Hence, the right option is (d). Regarding other options, you may note that *cast away* is to be left somewhere following a ship wreckage. *Cast off* is to discard things, especially clothes, because they are no longer in use. *Cast down* is to look down or be sad.

4. Knowing his immaculate driving skills, it is hard to imagine how the accident _____.
   (a) Came across
   (b) Came along
   (c) Came apart
   (d) Came about
   (e) None of the above

Key: The right choice is *came about* (d) which fulfils the sense of happened or took place required by the sentence. Regarding other options, *come across* somebody is to meet them by chance. *Come along* is to arrive or appear somewhere, whereas to *come apart* is to break or fall to pieces.

5. Thanks to intensely aggravated terrorism, many countries that love playing cricket anywhere in the world have decided to _____ of the competitions organized in Pakistan.

(a) Pull out
(b) Pull up
(c) Pull through
(d) Pull in
(e) None of the above

Key: The answer is option (a) *pull out* as to pull out of something is to stop being involved in that thing. *To pull* up someone is to scold them. To *pull through* is to succeed in doing something very difficult and if a train or a bus *pulls in*, it arrives at a station or a bus stand somewhere and stops.

6. I felt a shiver down my spine as the bulldog suddenly _____ me.
    (a) Snapped back
    (b) Snapped at
    (c) Snapped off
    (d) Snapped up
    (e) None of the above

Key: The right option is (b) *snapped* at as if an animal *snaps at* you, it opens and shuts its jaw quickly near you as if it were going to bite you. To *snap back* is to bounce back from something shocking or unpleasant. If something *snaps off* it breaks with a sharp noise, for example, some branch of a tree. To *snap something up* is to buy, take, or claim something quickly and with enthusiasm, for example, *Kumble snapped up both the remaining wickets in the Aussies' innings without much fuss.*

7. Courage, determination, and charity are the virtues _____ to the Sikhs from generation to generation.
    (a) Handed out
    (b) Handed over
    (c) Handed down
    (d) Handed back
    (e) None of the above

Key: The right option is *handed down*. It fits in with the context as when certain skills, traits, characteristics, and features are handed down to the succeeding generations, they are passed down, i.e., given to them as a legacy. To *hand over* a power or responsibility is the act of moving it from one person or a group to another. To *hand out* something to somebody is to give that to each person in a group. To *hand back* is to return to someone to whom it belongs.

8. Ashamed of her unsavoury past, she tried to cast it _____.
    (a) Aside
    (b) Off
    (c) Up
    (d) Out
    (e) None of the above

Key: The right proposition to go with the verb *cast* would be off (b) as to *cast off* something is also to get rid of something that you consider bad. While *cast aside* is to discard or reject, *cast out* means to drive somebody, for example, demons or something away by using force and if the sea *casts something* or somebody on the land, it carries it or them and leaves it or them there.

9. The terrorist might now try to _____ historical monuments.
    (a) Blow up
    (b) Blow over
    (c) Blow off
    (d) Blow out
    (e) None of the above

Key: The correct option in this context would be (a) *blow up* as to *blow up* something is to destroy that with bombs and guns. To *blow something out* is to put out (a flame, etc.) by blowing. If a storm *blows over* it becomes less intense and forceful. To *blow off* a relation is to bring it to an end.

10. Let's not waste any time now and _____ business straightaway.

    (a) Get off   (d) Get over

    (b) Get into   (e) None of the above

    (c) Get down to

Key: The right option is (c) *get down to* as if you get down to something or doing something you start giving it some serious attention: It's high time we got down to business. The other options don't fit in as to *get into* something is to reach a particular state or condition which is usually unpleasant, for example, Not essentially a clod, he still has an uncanny knack of *getting himself* into trouble every now and then. To *get off* a horse, train, or bus is to leave it, for example, He broke his ankle as he could not get off the horse in time. And if a boring movie *gets over*, everyone is relieved.

11. After a series of failures, the scientist finally _____ an idea as to how to keep a room cool without using air conditioners.

    (a) Hit back   (d) Hit upon

    (b) Hit out   (e) None of the above

    (c) Hit off

Key: The correct option in this context would be (d) *hit upon* as to hit upon an idea is to think of a solution to a problem often by chance or coincidentally. The phrasal verb *hit back* in a sentence such as 'Dr Manmohan Singh has finally started hitting back at his opponents' suggests that he has started criticizing those who had been criticizing him for a long while. If you *hit it off with your new neighbors*, you strike an affinity with them. To hit out at something or somebody is to attack violently: It seems the girl hit out at her assailant with all her might before eventually capitulating to him.

12. The sick child _____ while waiting at the doctor's.

    (a) Threw off   (d) Threw up

    (b) Threw down   (e) None of the above

    (c) Threw out

Key: The correct option is (d) *threw up* as to throw up is to vomit which goes with the rest of the sentence. You usually *throw down* a challenge at your adversary. Modern women are trying to *throw off* their old image. To *throw out* is to ask people to leave a place or job forcibly or get rid of something, for example, 'During recession, scores of professionals were getting thrown out of their jobs.'

13. Oh my God! He's getting breathless; _____ the doctor immediately.

    (a) Send for   (d) Send back

    (b) Send in   (e) None of the above

    (c) Send off

Key: The right choice would be (a) *send for*. When you *send for* somebody you ask them to come and help you, exactly the sense required by the sentence. However, when some forces are asked to help and combat a difficult situation, they are *sent in*, for example, 'Trained commandoes were *sent in* to combat the terrorists holed out at Taj Hotel, Mumbai.' To give a *send off* (noun) is to throw a party in the honour of somebody who is leaving, e.g., he could never forget that memorable *send off*. To send somebody off is to ask them to go somewhere or to make them leave for they have breached some code, e.g., 'Despite all the stardom they enjoy, the soccer players are at times sent off.' When you send something back you return it to where it came from, for example, 'You can send back the exerciser if it does not help you shape up your body better.'

14. After retirement, Kamal Kant plans to open a showroom that would _____ exclusively _____ the Rajasthani art.
    (a) Deal in
    (b) Deal with
    (c) Deal out
    (d) Deal without
    (e) None of the above

Key: The correct option is (a) *deal in* which means doing business by selling a particular product or kind of goods. Regarding other options, see the usage below:

Don't worry; I can deal with such issues. (To solve/tackle a problem)

Just deal out the cards! What are you waiting for? (distribute)

Deal without is not a phrasal verb.

15. Jimmy Porter, the disillusioned protagonist of Osborne's *Look Back in Anger*, seems to strike the keynote of his anguish as he blurts out, 'We have lost all the good causes to _____'.
    (a) Live on
    (b) Live through
    (c) Live by
    (d) Live for
    (e) None of the above

Key: The right phrasal verb to be used here should be (d) *live for* as when someone lives for something/someone, they consider that to be the most important thing or person in life. For instance, 'He lived and died for his country.' Regarding other options, note the following uses:

• Caught up in a pandemic patriarchal social structure all around them, women have to more or less *live by* men's rules. (To follow a particular set of rules)
• Fat people continue to be fat, for *living on just* raw vegetables sounds impossible to them. (To eat a particular food/survive on something meagre)
• During the recession, James had to *live through* horrible financial crunch. (To experience something difficult and survive)

16. It was Tyson's monumental punch that _____ his opponent _____.
    (a) Knocked off
    (b) Knocked out
    (c) Knocked over
    (d) Knocked down
    (e) None of the above

Key: The right option is (b) *knocked out* as in boxing to *knock out* your opponent means to hit him so hard that he falls to the ground and cannot lift himself/herself up in time to continue the fight. Regarding other options, note the uses below:

• As a manager, you are expected to *knock off* a report of several thousand words in just an hour or so. (To complete something quickly)
• On his way back from office, David was *knocked over* by a speeding wagon. (To be hit and injured/killed by a vehicle).
• As Preeti did not open the door, the anxious hotel staff had to finally *knock down* the door of her room. (To hit something so hard that it falls to the ground)

17. What _____ Rajiv Gandhi _____ from other politicians was the loveable candidness with which he admitted that the bureaucratic set up of the nation did not allow the government funds to reach the common man.
    (a) Marked out
    (b) Marked down
    (c) Marked off
    (d) Marked by
    (e) None of the above

Key: The correct option is (a) *marked out* as if something marks a person or thing from others, makes him or that thing seem different from others. As regards other options, see the usage below:

- Though he writes originally, Jim often gets *marked down* because of his poor handwriting. (to reduce something in value, price or mark)
- Before we start playing, let's *mark the boundary line off* with a chalk. (To separate something by drawing a line)
- The Assembly polls were *marked by* rigging and booth capturing. (Characterized by)

18. Recession has forced a lot of companies to think of *resorting to* unprecedented _____.
    - (a) Cut downs
    - (b) Cut backs
    - (c) Cut offs
    - (d) Cut ups
    - (e) None of the above

Key: The correct expression required to be used here would be (b) *cut backs* which means the measures adopted to reduce the cost. The expression in this context is used as a noun, as the phrasal verb is *cut back* which means to reduce something, like costs and expenditures.

19. I suppose a fourteen year old boy like you is not a toddler and should be able to _____ himself.
    - (a) Look into
    - (b) Look upto
    - (c) Look after
    - (d) Look through
    - (e) None of the above

Key: The correct phrasal verb to be used here is *look after* (c) which means to take care of somebody/something. Regarding other phrasal verbs, note the uses given below:

- After much media protest, the police had to finally agree to *look into* the complaint. (To examine or investigate)
- Veermati *looks upto* her cousin, Shakuntala Phenji, for inspiration. (To admire someone and seek inspiration from them)
- He was caught *looking through* confidential files. (To go through/look at the pages of a file or magazine)

20. If you can just _____ me for a little while longer, I would explain to you the nuances of modern poetry.
    - (a) Bear up
    - (b) Bear out
    - (c) Bear on
    - (d) Bear with
    - (e) None of the above

Key: The correct option is (d) *bear with* which means to be patient with someone. Note the other phrasal verbs starting with 'bear' through the uses given below:

- How is Kate doing during recession?
    Oh, nothing to worry! She is *bearing up* quite well.
- The shoddy evidence presented in the court could not *bear out* the plaintiff's claims.
- Capitalism is not just a philosophical notion. It's a way of living. It directly *bears on* our daily lives.

8.11 IDIOMATIC EXPRESSIONS

Just like phrasal verbs, idiomatic expressions too add to one's style the warmth, intensity, and a personal tinge which is an essential feature of an emphatic and effective expression. In this section, we will list some idiomatic expressions. The context in which they are used is also conveyed in the sentences that follow the idiomatic expressions.

1. Jump bail: Run away while being tried in court
   He was on parole but he **jumped bail** and was never traced.

2. Get going: Start doing something, start working
   When Rohit **gets going**, not many bowlers know where to bowl to him.

3. Keep the ball rolling: Continue something
   Though there is no love left between them Jimmy **keeps the ball rolling** by calling up Alice sometimes.

4. Spill the beans: Tell everything
   When the police exerted a little more force, the thief **spilled the beans**.

5. Walk all over someone: Treat someone very badly
   How can you allow someone to **walk all over you** like that?

6. Come in handy: Be useful
   It is good to have a torch while travelling; it often **comes in handy** at times.

7. Backhanded compliment: Ambiguous compliment
   Coming from a haughty man like him, I took his words to be just **backhanded compliments**.

8. Axe to grind: Selfish purpose
   If you really are searching for a true friend, try to find someone who doesn't have an **axe to grind**.

9. Bad blood: Unfriendly feelings
   The movie shows how **bad blood** between families can cause havoc.

10. Mum's the word: Keep quiet
    The moment you betray even a single emotion, they will get after you; so remember that **mum's the word** while you attend the engagement.

11. Stand a chance: Have a possibility
    Though he is not all that confident, he still **stands a chance** of doing well in group discussion.

12. Hold one's ground: Stay where one is and not give up inspite of adversity
    Though we were attacked from all the sides, we **held our ground** and kept fighting.

13. A calculated risk: Attempting something knowing the danger involved
    Knowing fully well that his views on Marxism will not be appreciated, he still took **a calculated risk** and kept talking about it.

14. Call it a day: To quit
    Immediately after the defeat, the famous tennis star **called it a day**.

15. Out of one's element: To be in an unfamiliar and unpleasant situation
    An old man generally finds himself **out of his element** among youngsters.

16. Spitting image: Exact likeness
    When he walked in, we were simply astounded; he was a **spitting image** of Boris Becker.

17. Hand in glove: Close cooperation in something wrong or bad
    Criminals succeed because most of the time the police are **hand in glove** with them.

18. Halcyon days: Happy and peaceful times
    Groping for his roots, he just craves for the **halcyon days** when he was a child.

19. On one's guard: Ready to defend oneself
    You simply can't survive in a ruthless system unless you are constantly **on your guard**.

20. On one's knees: Begging desperately
    The weak have to be **on their knees** even while being in the right.

21. On one's last legs: Extremely exhausted, tired
    After a week of grueling schedule and routine, one is generally **on one's last legs**.

22. Follow in someone's footsteps: Doing something someone else has done

    Though always at loggerheads with his father, Kevin followed his **father's footsteps** after his death.

23. Asking for trouble: Inviting punishment

    By confessing to his boss that he had been pretending to be sick, Aston **asked for trouble**.

24. Maiden speech: First speech

    In his **maiden speech**, he appeared nervous and completely out of sorts.

25. Speak one's mind: Say what one really thinks

    The democratic leader that he is, Mr Lawrence allows us all to **speak our minds** during meetings.

26. Kick against the pricks: Complain about things that cannot be improved

    She will have to learn to accept things as they are; how long is she going to **kick against the pricks**?

27. Keep one's head above water: Managing to stay with difficulty

    Though the market crashed in a big way, our organization exceeded in **keeping its head above water**.

28. Make heavy weather of: Have difficulty

    Even the normal human troubles caused him big agony and he often **made heavy weather of** the most ordinary looking everyday problems.

29. Bring into question: Create doubt

    Though the government claims that the law and order has improved in the state, what **brings into question** all such claims is the steep rise in the rape and murder cases in the recent past.

30. Fly at someone's throat: Attack someone

    Once the thief was discovered to be hiding in the attic, the owner straightway **flew at his throat**.

8.12 DEVELOPING TECHNICAL VOCABULARY

Whatever be the field of study you choose to pursue, you are bound to come across a large number of words which are particular to that subject or field of knowledge. For achieving professional success, it is crucial that you possess the technical vocabulary to be able to communicate the nuances of your own field of study and interest. In order to broaden your technical knowledge, you are advised to read the books that relate to it. However, in order to give you a glimpse of how different types of technical words belonging to variegated fields of study can be used in sentences, a brief note to that effect follows:

Laser beams—Most laser beams are just beams of light but they have properties that distinguish them from ordinary light.

Blade—After you push down on the arm of the hand-held stapler, the top-leaf spring raises the blade from the magazine, and the magazine and base move apart.

Oxymoron—An oxymoron combines two terms that are normally contradictory to each other such as pleasant nightmare, living death.

Galvanized—Steel often gets galvanized after individual parts have been formed, such as braces, nails, and screws.

Gravity—Essentially, gravity is an attractive force between objects.

Photosynthesis—Photosynthesis is a technique for converting sunlight into energy, which is utilized by certain organisms.

Electroplating—If you have ever purchased something inexpensive with a fine coating of precious metal, then you have witnessed the end result of electroplating.

Juvenile crime—Juvenile crime occurs when an individual under the age of majority acts against the law.

Shoplifting—Shoplifting is considered a petty or minor type of crime, and may carry less serious penalties than other types of thefts.

Cookie—A temporary cookie solved this problem in the short term by setting aside a little bit of browser memory to save information.

Tsunami—Tsunami is the Japanese word for 'harbour wave', but is actually a series of waves usually generated in the deep ocean, causing massive amounts of damage upon land.

Amputate—To save his life, the doctors amputated his legs.

Biotechnology—Early advances in the growing of crops for food or sale, as well as the breeding of animals, can both be said to be developments in biotechnology.

Anthropologist—Anthropologists seek to understand all the cultures, customs, artifacts, knowledge, habits, history, etc. of the world.

Heat exchanger—Gas furnaces use a heat exchanger to warm the room air.

Telepathy—Many of us find it very difficult to state our needs. We expect people to know what we are feeling through telepathy.

Teleconference—Managers at their factory in Bengaluru hold a two-hour teleconference with head office in Santiago every day.

Combustion—The two principal combustion products are water vapor and carbon dioxide.

Hazard—Oil leaking from a barge in the river Ganges poses a hazard to the drinking water of Haridwar.

Excrete—Calcium is excreted in the urine and stools.

8.13 EPONYMS

Eponyms are the terms or names given to a particular place, tribe, era, discovery, or situation. Usually such terms and names emanate from some historical characters, mythological figures, or legendary or fictional characters. In using an eponym, the author intends to convey the mystery, implicity, controversy, or any other peculiarity about a person, place, or situation. Picking up eponyms is certainly useful as they enrich our vocabulary. Following are some of the eponyms with their meaning and usage (Table 8.1).

Table 8.1 Eponyms and Their Uses

Word/Term	Meaning and Background	Usage
Achilles heel	A term used to describe the vulnerable point on the body of an otherwise invulnerable Achilles, the strongest Greek warrior in the Trojan War. The legend suggests that by dipping the infant Achilles into the river Styx, his mother had made him invulnerable. However, the infant Achilles' heel, by which he was held by his mother, remained a vulnerable spot on his body where he was hit by Prince Paris with a poisoned arrow.	Despite all the professed preparation for the overseas tours, the erratic form attack may turn out to be the Indian team's Achilles heel.

(Contd)

Table 8.1 *(Contd)*

Word/Term	Meaning and Background	Usage
Narcissistic attitude	Excessive admiration for oneself, self-love, egocentrism, and self-centredness. The legend is derived from the story of Narcissus, a handsome Greek youth, who, while seeing his reflection in a water pond, fell in love with his own image.	We can certainly do a lot for the world around us. For this, however, we need to rise above our narcissistic tendencies and think beyond our petty, selfish interests.
Sisyphean task/ Herculean task/ Mammoth task	A never ending labour, a repetitive task full of tedium, drudgery, and boredom. The legend is derived from the story of Sisyphus, who was punished for his misdeeds and was asked to carry a huge boulder uphill. Every time Sisyphus somehow carried the boulder and reached the top of the hill, the boulder would slip down, forcing Sisyphus to go in its chase once again.	Caught in the Sisyphean task of meeting out everyday tedium, the creative soul in him pushed, nudged, and forced him for a change.
Malapropism	Wrong use of words, often creating a humorous effect. Derived from Mrs Malaprop, a humorous character in Sheridan's *The Rivals*.	The book is full of errors and malapropism of various types.
Bowdlerize	To edit a text for expurgating or weeding out the words and expressions considered indecent. The expression came into being since Dr Thomas Bowdler, an English editor, brought out in 1818 *The Family Shakespeare*, an edition of Shakespeare's plays in which all those expressions were expurgated which were considered improper or offensive by the editor.	Though the original text itself had enough promise, it is the bowdlerized version of the novel which makes it a real work of art.
Machiavellian	Cunning, crafty, and deceitful. It has been named after Niccolo Machiavelli, who promulgated the theory that any means including those of craftiness, deceit, and duplicity could be used to serve one's selfish ends. The term is used to establish the nature of one such person in life or fiction.	No change in the country's politics is foreseeable unless we are able to screen out the Machiavellian characters from our political system.

(Contd)

Table 8.1 (*Contd*)

Word/Term	Meaning and Background	Usage
Chimerical	Unreal, fanciful, and fantastic. Taken from Chimera, a mythological monster who had the head of a lion, the body of a goat, and the tail of a serpent, the word is used to refer to anything that appears to be unrealistic, imaginary, or fanciful.	'The suggestions given in the report submitted by you are far too chimerical to be followed!' the boss said sternly and the subordinate seemed dismayed at the revelation.
Cassandra	A prophet who predicts a disaster or misfortune but is generally not trusted. The legend is derived from the story of Cassandra, the daughter of Priam and Hecuba, who was loved by Apollo. In love, Apollo gave Cassandra the power to prophesize calamity and woe. Later on, when Cassandra failed to carry out her promise of love for Apollo, the god decreed that though she would be able to prophesize the impending doom, nobody would believe her.	Over the years, we have developed a tendency to keep our eye shut to the truth. The Cassandras like Gandhi are always remembered but are hardly ever followed.

EXERCISES

I. Each of the words listed below is followed by five choices. Pick the word that is closest in meaning to the listed word. In case you don't find a synonym for the head word among the words given as options, mark the last option.

1. Punctilious
   (a) Meticulous
   (b) Casual
   (c) Perfunctory
   (d) Final
   (e) None of the above

2. Opulence
   (a) Poverty
   (b) Penury
   (c) Affluence
   (d) Indigence
   (e) None of the above

3. Stolid
   (a) Stupid
   (b) Stylish
   (c) Impressive
   (d) Impassive
   (e) None of the above

4. Buoyant
   (a) Energetic
   (b) Blissful
   (c) Tedious
   (d) Enticing
   (e) None of the above

5. Momentous
   (a) Mesmerizing
   (b) Stormy
   (c) Memorable
   (d) Magnificent
   (e) None of the above

6. Smother
   (a) Stupefy
   (b) Simplify
   (c) Choke
   (d) Chaff
   (e) None of the above

7. Profligate
   (a) Prolific
   (b) Proliferate
   (c) Extravagant
   (d) Exaggerate
   (e) None of the above

8. Babble
   (a) Talk
   (b) Scribble
   (c) Believe
   (d) Sight
   (e) None of the above

9. Retaliate
   (a) Reveal
   (b) Repeal
   (c) Respect
   (d) Retort
   (e) None of the above

10. Superfluous
    (a) Redundant
    (b) Superficial
    (c) Essential
    (d) Superb
    (e) None of the above

11. Kickback
    - (a) Slapstick
    - (b) Attack
    - (c) Bribe
    - (d) Payment
    - (e) None of the above

12. Specious
    - (a) Spacious
    - (b) Supportive
    - (c) Misleading
    - (d) Mistimed
    - (e) None of the above

13. Insouciant
    - (a) Intrepid
    - (b) Instructive
    - (c) Undiluted
    - (d) Unconcerned
    - (e) None of the above

14. Florid
    - (a) Elaborate
    - (b) Elusive
    - (c) Floral
    - (d) Flexible
    - (e) None of the above

15. Consummate
    - (a) Supreme
    - (b) Consecrate
    - (c) Captivate
    - (d) Flawless
    - (e) None of the above

16. Boorish
    - (a) Rude
    - (b) Polished
    - (c) Beseeching
    - (d) Bedazzled
    - (e) None of the above

17. Dazed
    - (a) Fazed
    - (b) Confused
    - (c) Drowned
    - (d) Doomed
    - (e) None of the above

18. Articulate
    - (a) Express
    - (b) Coherent
    - (c) Agitate
    - (d) Adulate
    - (e) None of the above

19. Exhort
    - (a) Extol
    - (b) Urge
    - (c) Repel
    - (d) Expel
    - (e) None of the above

20. Comply
    - (a) Follow
    - (b) Confound
    - (c) Fulfil
    - (d) Complicate
    - (e) None of the above

II. Fill in the blanks with appropriate phrasal verbs:

1. You will have to work hard to _____ with other players.
2. Since the journey was long, we _____ at five in the morning.
3. By losing the match he _____ his supporters _____.
4. Didn't you notice, he was _____ you in the party?
5. The Babri Masjid was _____ on 6 December.
6. _____ something for adverse circumstances.
7. Driving blindly, he _____ a tree.
8. We didn't come here to _____ old issues.
9. She was so nervous that she totally _____ at the interview.
10. I alone know how I _____ those days of separation and alienation!
11. Her enthusiasm _____ very quickly.
12. You can't _____ like that.
13. Always alert, he easily _____ the entire plot.
14. I would like to _____ this report and then take a final decision.
15. It is useful to _____ the important ideas during a meeting.

III. Fill in the blanks with appropriate idiomatic expressions:

1. We need reliable supporters and not the _____ flatterers.
2. Many parties crop up during election times but just like _____ are not seen afterwards.
3. The management has _____ that some of us can be retrenched.
4. Most of the political alliances are expedient; there is no _____ for their togetherness.
5. We are grateful to you _____ for all your support and encouragement.
6. It is _____ that most of our employees come late and go early.
7. Once you move from a small to a big city, you always find yourself to be _____.
8. Ever since he married that woman, he has been _____.
9. Yesterday, I _____ a strange fellow.
10. As there was nothing new in the newspaper, he _____ impatiently.

IV. Choose the right option in the following sentences:

1. I am not sure how to decide this; I am quite ambivalent/ambiguous about it.
2. One can just wonder how he is pulling on with such a meagre income; he has a large number of dependents/dependants to feed and support.
3. The girl did not take the compliment/complement kindly.
4. He was prosecuted/persecuted brutally for a minor crime.
5. The movie was really insipid/incipient; I simply dozed off during the first half an hour itself.
6. Cancer is an invidious/insidious disease; it kills you silently.
7. What hurt me most was the tone of the person; it was so boorish and degenerating/deprecating.
8. You really stand a chance to clear the interview; our boss is a/an uninterested/disinterested man; all he needs is a good, hard working person.
9. The Prime Minister has been apprised/appraised of the latest developments in the state by the Chief Minister.

10. He considers others to be dirt; whenever he speaks he sounds quite contemptible/contemptuous towards others.
11. Keats's poetry—a great work of art—is known for its sensuality/sensuousness.
12. Don't sound so moribund/morbid; it is not good to be so pessimistic and negative in life.
13. Only a waver/waiver in the fees delighted him.
14. Mr Kapoor died in a rail collision/collusion.
15. The chief guest's speech was not apposite/opposite to the occasion.

V. Fill in the blanks in the following sentences with appropriate words from among those listed below. Choose the correct form of verb(s), if any, as per the tense and the subject of the verb:

(allegations, absorb, aspire, bovine, amplify, betrayed, acquaint, bestow, assemble, befriend, accelerate, bemoan, associate, adjourn, bewildered, annoyed, amend, admonished, besiege, affirmed)

1. All the employees are going to be _____ in the new company.
2. Let's try to _____ the pace of growth in our organization.
3. Our engineers are required to get _____ with the latest developments in science and technology.
4. The committee meeting was _____ without reaching any decision.
5. The branch manager resigned after being _____ by the CEO.
6. The Chairman _____ that the company policy in this regard is fixed and final.
7. We need to _____ the laws to help the daily wagers.
8. Smith resigned amidst _____ of misappropriation of funds.
9. Unless the union leader _____ our concerns, the management won't do anything.
10. Karl Reckman, the new boss, seemed particularly _____ with the non-performing staff.
11. Though quite young, he _____ to be on the Board of Governors.
12. Unless we _____ ourselves with others in business, we can't progress.
13. All the members _____ in the conference room and waited for the Director join them.
14. The young manager has _____ many of his subordinates.
15. There is no sense in _____ the loss; the entire market has crashed.
16. It seems that the multi-nationals will _____ all the local smaller players.
17. I tried to please my boss, but he refused to _____ any consideration on me.

18. We all felt _____ when the company secretary joined the rival group.
19. He can never be the manager; he always looks _____.
20. Throughout the meeting, he displayed a _____ face.

VI. Fill in the blanks in the following sentences with appropriate words from among those listed below. Choose the appropriate from of the verb(s) if any, according to the tense used:

(blinked, buzz, culminates, befuddled, candidly, capsizing, compensate, boom, concludes, consoled, brooded, committing, collision, condone, contrives, confined, conduct, conspire, conserve)

1. When the police raided our office, everyone seemed _____.
2. When the production officer was questioned about his production projections, he simply _____.
3. There is a _____ in the market, let's invest without inhibitions and fear.
4. The CEO _____ deeply over the issue before announcing his decision.
5. Ever since she has become our CEO, there is a _____ all around.
6. When Nora's rage _____, she leaves her husband.
7. In the movies a hero can do anything; he can even save a ship from _____.
8. The actor _____ admitted that the one who jumped the rope was actually his duplicate.
9. The story begins by showing a head-on _____ between the two.
10. Idiotic movies would never show a star _____ even a small mistake.
11. The match was so boring that the spectators needed to be _____ for the loss of the money they spent in buying tickets.
12. When one idea _____, another begins.
13. People have a very big heart in India, they can _____ a bad leader a number of times.
14. The report criticized the _____ and character of the people in the forum.
15. It was such a poor meeting that we felt we were being _____ and contained inside a dark room.
16. The child _____ a story and the simpleton mother believes that.
17. Unless we want to kill our offspring, we need to _____ water.
18. Do anything in life but never _____ against your friends.
19. Everyone _____ the parents of the deceased soldier.
20. The committee _____ seven members.

ANSWER KEY

Practice Tests

8.1
1. False	6. True	11. True	16. True	21. True
2. True	7. False	12. True	17. True	22. True
3. True	8. False	13. True	18. False	23. False
4. True	9. True	14. False	19. True	24. True
5. False	10. True	15. False	20. False	25. True

8.2
1. F	5. T	9. T	13. T	17. T
2. T	6. T	10. F	14. F	18. T
3. F	7. F	11. T	15. T	19. T
4. T	8. F	12. F	16. F	20. T

8.3
1. amiably	3. carnivorous	6. amity	9. anathema
2. acerbically, acrimoniously	4. amorously	7. malignant	10. amicable
	5. incarnate	8. acrid	

8.4
1. implacable	4. circumnavigated	7. exculpate/acquit	10. eulogies/paeans
2. placid	5. circumscribing	8. euthanasia	
3. euphemistically	6. omnipotent	9. pacifist	

8.5
1. scribble	4. magnanimous	7. retrograde	10. voracious
2. solitude	5. corpulent	8. verbatim	
3. loquacious	6. eloquent, verbose	9. proscribes	

8.6
1. Yes	3. No	5. No	7. Yes	9. Yes
2. No	4. Yes	6. Yes	8. Yes	10. No

8.7
1. infallible: unerring, trustworthy, unfailing
2. impeccable: perfect, consummate, irreproachable, faultless
3. interminable: endless, continuous, ceaseless, perennial
4. illicit: illegal, unlawful, secret, clandestine
5. intractable: unmanageable, uncontrollable, ungovernable, restive
6. impregnable: invincible, unconquerable, indomitable
7. impertinent: insolent, impudent, disrespectful, irreverent
8. incorrigible: beyond redemption, irreclaimable, intractable
9. ignoble: vile, base, knavish, wicked
10. intrepid: fearless, undaunted, courageous, audacious

8.8
1. extrovert	3. inconsistent	5. indisposed	7. insatiable	9. legitimate
2. inculpate	4. unenviable	6. disinter	8. illegible	10. temperate

8.9
1. Opposite	3. Same	5. Opposite	7. Same	9. Same
2. Opposite	4. Same	6. Same	8. Same	10. Opposite

8.10
1. Opposite	5. Same	9. Opposite	13. Same
2. Same	6. Same	10. Same	14. Same
3. Same	7. Opposite	11. Same	15. Same
4. Opposite	8. Same	12. Opposite	16. Same

8.11

1. Same	**4.** Opposite	**7.** Same	**10.** Same	**13.** Same
2. Same	**5.** Opposite	**8.** Same	**11.** Same	**14.** Same
3. Same	**6.** Opposite	**9.** Same	**12.** Opposite	**15.** Same

8.12

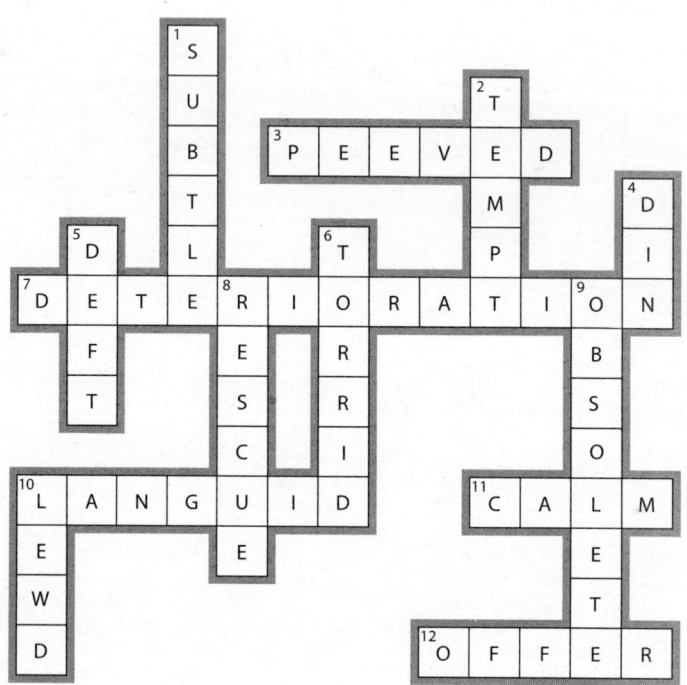

Across

3. PEEVED   **7.** DETERIORATION   **10.** LANGUID   **11.** CALM   **12.** OFFER

Down

1. SUBTLE	**4.** DIN	**6.** TORRID	**9.** OBSOLETE
2. TEMPT	**5.** DEFT	**8.** RESCUE	**10.** LEWD

8.13

1. Ingenious	**6.** Imaginative	**11.** Refuge	**16.** Fair
2. Prosecuted	**7.** Industrial	**12.** Allusions	**17.** Elicit
3. Prescribe, Proscribe	**8.** Exhaustive	**13.** Duel	**18.** Draught
4. Hoarding	**9.** Comprehensive	**14.** Duo	**19.** Deceased
5. Enviable	**10.** Lovely	**15.** Eminent	**20.** Compliments

8.14

1. Altruistic	**4.** Blasphemy	**7.** Magnum opus	**10.** Philatelist
2. Amnesty	**5.** Dandy	**8.** Fable	
3. Scapegoat	**6.** Coquette	**9.** Epic	

8.15

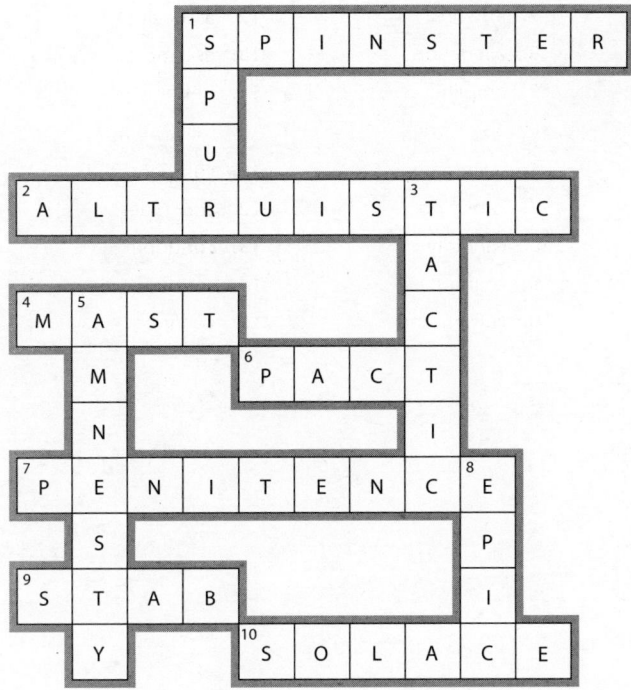

Across

1. SPINSTER	**4.** MAST	**7.** PENITENCE	**10.** SOLACE
2. ALTRUISTIC	**6.** PACT	**9.** STAB	

Down

1. SPUR	**3.** TACTIC	**5.** AMNESTY	**8.** EPIC

Exercises

I.

1. (a)	**6.** (c)	**11.** (c)	**16.** (a)
2. (c)	**7.** (c)	**12.** (c)	**17.** (b)
3. (d)	**8.** (a)	**13.** (d)	**18.** (b)
4. (a)	**9.** (d)	**14.** (a)	**19.** (b)
5. (c)	**10.** (a)	**15.** (d)	**20.** (a)

II.

1. catch up	**4.** drooling over	**8.** rake up	**12.** go on
2. set off	**5.** pulled down	**9.** messed up	**13.** saw through
3. let…down	**6.** keep aside	**10.** got through	**14.** go through
	7. rammed into	**11.** faded away	**15.** take down

III.

1. fly-by-night	**4.** common ground	**7.** a little fish in a big pond	**9.** came across
2. a flash in the pan	**5.** beyond measure	**8.** leading a dog's life	**10.** cast it aside
3. dropped a hint	**6.** no laughing matter		

IV.

1. ambivalent	**5.** insipid	**9.** apprised	**13.** waiver
2. dependents	**6.** insidious	**10.** contemptuous	**14.** collision
3. compliment	**7.** deprecating	**11.** sensuousness	**15.** apposite
4. persecuted	**8.** a disinterested	**12.** morbid	

V.
1. absorbed
2. accelerate
3. acquainted
4. adjourned
5. admonished

6. affirmed
7. amend
8. allegations
9. amplifies
10. annoyed

11. aspires
12. associate
13. assembled
14. befriended
15. bemoaning

16. besiege
17. bestow
18. betrayed
19. bewildered
20. bovine

VI.
1. befuddled
2. blinked
3. boom
4. brooded
5. buzz

6. culminates
7. capsizing
8. candidly
9. collision
10. committing

11. compensated
12. concludes
13. condone
14. conduct
15. confined

16. contrives
17. conserve
18. conspire
19. consoled
20. comprises

Developing Effective Listening Skills

Learning Objectives   After reading this chapter, you will be able to

- understand how listening is different from hearing
- learn about the different types of listening so that you can use them effectively, as and when required
- identify the major causes of poor listening
- know the various techniques to improve your listening skills
- hone your listening ability with the help of exercises given at the end of the chapter

If speaking is silver, listening is gold.

9.1 INTRODUCTION

Rose, Siddharth, Ishita, and Dushyant attended an expert lecture on 'Soft Skills for Entrepreneurs' delivered by Paul Kimura, a management guru. While talking about the role of IQ and EQ he said, 'IQ represents abstract intelligence which gets the entrepreneurs started, whereas EQ helps them become successful.' After the lecture, Ishita shared the idea that the statement is applicable in every field of life. Siddharth endorsed the idea and told them that besides IQ and EQ, HQ is also becoming very essential, and he referred to an article that he had read in the *open page* of *The Hindu* sometime back. Rose remarked that she had found the lecture quite boring, and that she had in fact 'dozed off for a while.' Dushyant said, 'I do not even know what you are talking about.'

You can see that though all the four attended the same talk, only two of them paid attention to the speaker and listened carefully, whereas the other two did not listen to the speaker at all. In other words, they just faked listening. This is a very common problem. Have you ever pondered why it happens?

Listening is an everyday affair. Despite that, or probably owing to that, many of us shirk listening. In fact, most listeners see listening as a challenging task, so much so that the greatest challenge that a speaker faces is to make his/her speech worth listening to. In other words, when a speaker envisages his/her speech making endeavour, he/she grows increasingly obsessed with the idea of making his/

her speech interesting. This is simply because if he/she does not make his/her speech engrossing enough, he/she will not be listened to. After all, isn't it very common to react the way Rose did?

Many a time, we find listeners yawning, much to the anguish of the speakers who go out of their way to keep their audiences engaged. To keep the listeners attentive to their speech, most speakers use a variety of tactics. Yet, some of the listeners emerge from the lecture theatre declaring, 'It was quite boring. I dozed off!' Some others keep themselves completely away from the task of listening by mocking their fellow listeners saying, 'I don't know what you are talking about!' Now, why does it happen? Let us delve further.

9.2 LISTENING IS AN ART

Research corroborates the view that human beings spend more time listening than speaking. It is a skill most used by all of us, but we get little training in this. On the other hand, right from our school days, we are formally trained in all the other three language skills, namely speaking, reading, and writing. It is primarily so because listening is usually taken for granted. It is presumed that by making a child sit while being spoken to, we can also make him/her listen. So, the child sits, but does not listen to the teacher or the parent. While attending a lecture or talk, some of the listeners just tune out, or get caught up in an internal dialogue trying to translate a specific word, or think of something more interesting happening somewhere else outside the room.

However, it is not just psychology that makes way for such lack of interest in the listening activity. Research has established that human beings can speak at the rate of about 125–150 words per minute, whereas human brains can process 500–700 words per minute. This means that we are able to use just 25 per cent of our mind power and 75 per cent is left unused while we listen to others. Consequently, most of us are more interested in speaking than listening to others. So in the name of *listening* what people do most of the time is *hearing*. Let us see how these two terms can be distinguished.

9.3 LISTENING VS HEARING

Hearing takes place when something disturbs the atmosphere, and that disturbance takes the form of pressure waves that strike our eardrums as sound. For example, a truck rolling by on the road in front of our house would be just *heard* and not *listened to*.

Listening is different. It expands on hearing when we pay attention to the meaning of what we hear. Therefore, listening is all about consciously, actively, and systematically processing information. Listening demands perfect coordination between the ears and the brain, which results in decoding the speaker's message aptly. Regular practise and consistent efforts are required if we want to improve our listening skills.

Effective listening is a dynamic activity that seeks out the meaning intended in messages, considers their motivation, evaluates the soundness of their reasoning and the reliability of their supporting material, calculates the value and risk of accepting their recommendations, and integrates them creatively into the world of the listener.

Thus, we quite often merely hear the words someone else speaks. They are just vibrations in the atmosphere. We nod, smile, perhaps even respond, but do we listen to the speakers around us? Hardly. Listening requires us to be open to the meaning of the other person's words. It is no longer just about sound but about the thoughts, feelings, point of view, expectations, memories, beliefs—in fact, the whole of the other person and his or her ideas.

Now let us identify the major differences between a poor listener and a good listener.

9.4 POOR LISTENING VS EFFECTIVE LISTENING

The major differences between a poor listener and an effective listener are as follows.

Poor Listener	Effective Listener
Either tries to blame the speaker or considers the subject to be dry.	Thinks and mentally summarizes, weighs the evidence, listens between the lines to the tone of voice and evidence.
Gets distracted easily.	Fights against distractions and knows how to concentrate.
Finds it difficult to listen to complex material; has the tendency to read light and recreational materials.	Keeps listening on a regular basis; not averse to listening to matter which requires critical listening.
Tends to enter into unnecessary arguments.	Takes notes and organizes important information.
Resists new ideas.	Listens for ideas.
Pays too much attention to appearance and delivery.	Pays attention to the body language, tone, and style, along with the message being conveyed by the speaker.
Waits for his/her turn to speak.	Patiently listens to the speaker and responds as and when required.

9.5 ADVANTAGES OF GOOD LISTENING

Following are the advantages of good listening:

1. We generally find that good listeners are better performers. Thus, listening starkly differentiates between a poor and a good performer.
2. Listening is a vital skill which helps our learning. Good listening ability increases knowledge, develops critical thinking, and broadens opportunities.
3. Listening skills help us build effective relationships in our personal as well as professional life.
4. It prevents miscommunication.
5. It also facilitates solving problems in our personal life and at workplace.
6. Effective listening helps in sharing emotions, ideas, and experiences.
7. Good listening also improves decision-making and critical thinking.

9.6 PROCESS OF LISTENING

Becoming a good listener requires us to understand how listening happens. Essentially a cognitive process, listening involves the following stages.

Sensing　At this stage, the listener has physical hearing of the message because the sound waves fall on the eardrum, as a result of which he/she perceives the sounds.

Recognizing　After hearing the physical sounds, the listener identifies and recognizes the pattern of sounds. After this, sounds are recognized in a specific context. Here, the listener makes a conscious effort to recognize the word symbols that he/she hears.

Interpreting　Now the listener starts decoding the message. As he/she listens, he/she employs his/her own values, beliefs, needs, ideas, etc. to interpret the speaker's message. Since he/she also pays attention to non-verbal messages, the accuracy of his/her interpretation of the message also increases.

Evaluating   After he/she understands what the message actually means, he/she critically evaluates it. He/she assesses its strengths and weaknesses, its accuracy, reliability, and feasibility.

Responding   At this stage, the listener is ready to respond and react. He/she shows his/her rejection or acceptance, or understanding or confusion, or even indifference through his/her non-verbal cues.

Remembering or memorizing   This is the final stage of listening. Good listening enables the listener to retain the information for future reference. In order to increase the retention, you need to make conscious efforts by taking down proper notes, organizing the matter sequentially, or retaining the information by using analogies or other associated visual symbols.

9.7 TYPES OF LISTENING

In order to hone our listening ability, it is advisable to know the different types of listening that we need to employ on different occasions.

Content listening   In this type of listening, the primary focus is on understanding the message sent by the speaker, and to gather and understand the information. Therefore, it is also known as *informative listening*. We listen to reports, briefings, instructions, speeches, and conversations to obtain the desired information.

Empathetic listening   This is also known as *therapeutic/relationship listening* for the obvious reason that it is used in times of crisis. This is mostly done by us when we allow a troubled friend to express his/her feelings. It does not mean that empathetic listening is not employed by professionals. In fact, all good leaders and managers always try to empathize with their subordinates when they approach them with some problem. Counsellors, doctors, and psychiatrists also use this type of listening in their profession. In this type of listening, the listener is required to empathize with the speaker and help him/her get things off his/her chest. Since you try to understand the speaker's situation as an empathetic listener, it helps in strengthening the relationship between you and the speaker.

Appreciative listening   You do not employ content listening or empathetic listening when you listen to music or watch a movie. In these situations, you use appreciative listening since you appreciate the lyrics, direction, melody, style, or dialogue delivery. Thus, it is listening for pleasure or enjoyment.

Analytical listening   The purpose of this listening is two-fold. On one hand, you try to absorb the message and on the other, you attempt to analyse the ideas or facts and make critical judgement. In fact, this type of listening helps you evaluate the strength of argument, accuracy of evidence or facts, validity of inferences and reveals gaps in thinking.

Although all the skills referred to so far will help you evolve as a good listener, for the purpose of acquiring language skills, you need to pay special attention to intensive and extensive listening.

9.8 INTENSIVE LISTENING VS EXTENSIVE LISTENING

Intensive listening is listening to a small amount of material a number of times. For example, when you listen to a recorded lesson again and again, you may be able to acquire the correct pronunciation or intonation, or understand the hidden meaning of the text. When you listen, you concentrate and keep your mind focused on the text. This is the perfect technique to improve your listening comprehension. This helps you develop an intuitive feel of the grammar, structure, and words of the language which you listen to in this type of repeated listening. By regularly listening to news in English, audio articles, audio books, movie scenes, speeches, etc., you can easily develop your intensive listening.

However, you will take a lot of time to learn the language by using this technique. So, you also require *extensive listening* in order to learn a language. It gives you the opportunity to hear different voices and different styles. Thus, it helps you get used to the natural flow of the language. This also helps in developing effortless listening, since listeners listen to the ideas and opinions on a wide range of topics. However, when listeners employ only extensive listening, they passively pick up the erroneous expressions with the correct ones and make the same errors when they themselves speak and write. That is why, we are required to employ both intensive and extensive listening in our day-to-day interactions with others.

9.9 BARRIERS TO EFFECTIVE LISTENING

Listening thus is a very creative, interactive, and interpretive process. At times, however, it fails to click as some barrier may block the process. Here, we will discuss various barriers that impede us from being good listeners.

9.9.1 Forged Attention

This is one of the most common barriers to effective listening. We usually find audiences staring at the speaker but their minds are preoccupied with something else. They have a very attentive listening posture with their hands below their chins and eyes wide open, but they are not listening at that point of time, only faking it.

9.9.2 Premature Evaluation of the Subject Matter and Speaker

We often find that poor listeners convince themselves that the topic is uninteresting even before listening to the speaker and the complete message, and thereby a chain of negative thoughts start mushrooming. The next preconceived notion is that the speaker is dull and boring, and finally they turn to the many other thoughts and concerns stored in their minds for such an occasion. Such listeners tend to mentally criticize the speaker for not speaking distinctly, for talking too softly, or for not looking at the audience. They often do the same with the speaker's appearance. If speakers are not dressed as they should be, they probably tend not to listen. On the other hand, good listeners try to get something good even out of a dull discussion or talk.

9.9.3 Hard Listening

Poor listeners try very hard to listen to and absorb every word the speaker utters. Such type of listening is called *hard listening*. By employing this, listeners lose sight of the main idea by concentrating too hard on details. In such cases, the listeners pay more heed to the individual words and expressions rather than concentrating on the actual essence of the message. Thus, most of us who listen for facts may recall some isolated facts, but may miss the primary thrust or idea the speaker was trying to convey. Remember that in order to overcome this problem, we must listen to the speaker's primary points.

9.9.4 Poor Interpersonal Relations

Human beings base their reactions on the type of relationship they have with the other person. They interpret the message according to their past or present relation with the speaker. A lack of confidence, or a sense of superiority or inferiority, prevents them from having proper involvement in the listening task. These prejudices affect the sense, interpretation, and evaluation of the message.

9.9.5 Over Excitement

Whenever we hear something with which we happen to disagree, we get swayed by a chain of thoughts related to that. We spend too much time on thinking about the counter arguments and we tend to lose track of the speaker's ideas. In such cases, our listening efficiency drops to nearly zero because of over-excitement.

9.9.6 Different Language Variety and Accent

When the speaker uses a different accent which the audience is unable to understand, it becomes a great barrier to listening. For instance, the Indian audience can follow the British accent easily, whereas when they listen to a speaker from California who has an American accent, they find it difficult to follow him/her.

9.9.7 Distractions

Some listeners have very poor concentration while listening. They actually get distracted even with the slightest sounds of opening and closing of doors, people whispering to each other, or vehicles outside.

9.9.8 Evading the Difficult Types

Poor Listening Leads to Poor Learning

We have a tendency to listen to whatever is easy and familiar, and avoid whatever seems to be difficult and unfamiliar. Poor listeners become easy victims of this in classrooms, meetings, interviews, or group discussions. This in turn leads to poor or inadequate performance.

So, to combat this problem, we should train ourselves and develop enough patience to listen.

9.9.9 Non-attentive State of Mind

The listener often fails to listen to the speaker's message because he/she is preoccupied with certain thoughts, or is tense or exhausted. Various other distractions, such as hunger or drowsiness, and some other discomforts of a similar sort, can stop him/her from being attentive while listening. Thus, the closed mind becomes a major barrier to listening.

9.9.10 Different Levels of Perception

The speaker at times presumes that all his/her listeners have the same level of understanding that he/she enjoys, which might lead to information redundancy, or at times complex information, which may not be easily comprehended by the listeners. It is likely that the listener with a lower level of understanding could then face a problem in decoding the message aptly, and the intelligent listener may find it redundant and not pay proper attention to what the speaker says.

Now, let us see how we can develop better listening skills.

9.10 FIVE STEPS TO ACTIVE LISTENING

Following are the five steps that will help us develop active listening skills:

1. Look the speaker in the eye as far as possible, or follow his/her movements.
2. Summarize what the speaker is saying.
3. Take down notes.
4. Link what you are listening to what you already know.
5. Ask and answer questions in your mind for clarity in your understanding, regarding the problem under discussion.

9.11 TECHNIQUES FOR EFFECTIVE LISTENING

Following are the important techniques for effective listening:

1. To improve your listening skills, you should have an open mind.
   You should sit alert and look at the speaker with a view to establish your interest in him/her.
2. The effectiveness of listening generally depends on the intensity of the interest taken. So, take interest in the discussion or talk.
3. Do not prejudge the speaker, or his/her message, until you have listened to it completely.
4. Employ your critical thinking while you are listening.
5. Stop talking and do not interrupt the speaker unnecessarily.
6. Observe the non-verbal clues of the speaker, as this will enable you to grasp the message completely.
7. Take advantage of the *lag time* that we get in terms of small pauses between two different ideas.
8. Ask relevant questions to yourself, so that you keep on track with the ideas presented by the speaker.
9. Take down notes or paraphrase the message in simple words. This will certainly enable you to grasp it quicker.

9.12 LISTENING AND NOTE TAKING

In this section, we will learn in detail how listening helps in taking down notes effectively and the important tips to prepare notes efficiently. So while listening, observe the following points:

1. Listen for ways to relate ideas to previous lectures, or to a specific chapter of a book, or to previous experiences.
2. Listen to what is being said, and not how it is being said.
3. Do not try to write down everything.
4. Look for clues from the teacher/professor who indicates what he/she considers important. (vocal, postural, and visual clues)

5. As you listen, categorize the lecture according to its different parts, that is, introduction, body, and summary.
6. Try to restate what is being said in your own words.
7. Use abbreviations as much as possible to increase your note-taking speed.
8. Take down notes in one word or phrases or 'one-liners' as much as possible.
9. Mark your notes with checks (✓), asterisks or stars (*), question marks (?), by circling dates and names, etc. to bring out the important facts.
10. Learn to write quickly.
11. Write down definitions. If your instructor defines a term, make sure you write it down and understand what it means.
12. If you write down every word, you will stop listening to the lecture and might also miss out on information provided in the lecture.
13. If your instructor is going too fast or is unclear, say so. There is nothing wrong in asking your teacher to clarify a point.
14. After the class is over, go through your notes as early as possible. This does not take very long, and in fact helps you contextualize the new information.

 ## RECAPITULATION

✓ Effective listening is a dynamic activity that seeks out the meaning intended in the messages sent by the speaker.

✓ There are a number of barriers to listening, for example, sometimes we pay too much attention to the speaker's body language or delivery rather than the message being communicated by the speaker.

✓ In order to sharpen our listening skills, we need to overcome these barriers.

✓ If we improve our listening, it will increase our knowledge, develop critical thinking, and broaden opportunities.

✓ Improved listening will also help us build effective relationships in our personal as well as professional lives.

EXERCISES

Concept Review Questions

1. What are the major differences between hearing and listening? How is listening important for a professional?

2. What are the different types of listening that we use in our day-to-day life? Provide two examples for each type.

3. Explain the process of listening and also describe how listening helps in increasing the understanding and knowledge of the listener.

4. Discuss the advantages of effective listening briefly and also state the various tips which may help you improve your listening skills.

5. Lack of effective listening skills results in loss of time, lowering of productivity, and missed opportunities. Do you agree with the statement? If yes, substantiate your answer with appropriate examples.

6. How do good listening skills add to the ease of working in a team-based environment? Explain with the help of two examples.

7. One of your friends tells you, 'I feel that you must not waste your time in listening to the personal problems of your classmates.' Do you agree with your friend? If not, state reasons and elucidate with appropriate examples.

8. Effective listening is a rather daunting task. Do you think so? If yes, why? Substantiate your answer with appropriate examples.

LISTENING ACTIVITIES

The listening activities provided here offer a wide range of listening practices. They cover a variety of formal and informal styles of language, from interviews and presentations to conversational dialogues. The passages are of varying lengths, so as to train you for listening to different types of conversations, talks, and lectures. Teachers may read each of these activities for students and ask the given questions to train them for effective listening.

Each listening activity includes some questions, the answers for which are suggested immediately.

Listening for Making Inferences

I. Listen carefully to the story and then answer the questions:

A lazy grasshopper laughed at a little ant, as she was always busy gathering goods. 'Why are you working so hard?' he asked, 'come into the sunshine and listen to the merry notes, play, and have fun'. But the ant went on with her work. She softly replied, 'I am collecting food and laying aside a store for winter. These sunny days won't last for ever.'

'Winter is so far away and you are so worried right now', laughed the grasshopper loudly. And, when the winter came, the ant settled down in her snug house, as she had plenty of food to last the entire season. On the other hand, the grasshopper had nothing to eat. In fact, he remained hungry for a couple of days, and then went to the ant and begged her for a little corn. 'No', replied the ant, 'you laughed at me when I worked. You sang through the summer. So you'd better dance the winter away.'

1. What is the moral of the story?

Answer: Idleness can be a curse.

2. State two differences between the grasshopper and the ant.

Answer:

(a) The ant is hardworking and the grasshopper is lazy.

(b) The ant had a vision. She could think of the future and plan in advance, whereas the grasshopper lacked this kind of vision.

II. Listen carefully to the conversations and then answer the questions:

1. Sahil: My roommate and I have decided to do our own cooking next semester.

Pratiksha: Then, I hope you will have a lighter schedule next semester.

Question: What do you infer from Pratiksha's response?

(a) He may not have enough time to cook.

(b) He may not really enjoy cooking.

(c) He may spend too much of money to cook food on his own.

(d) He may go hungry quite often if he decides so.

Answer: (a)

2. Gaurav: Do you have to play that music so loud? Don't you know I have got a test tomorrow?

Garima: Sorry, I did not realize that you were studying.

Question: What will Garima probably do?

(a) Play some different music

(b) Close the door

(c) Turn down the volume

(d) Help Gaurav in his studies

Answer: (c)

3. Mr Verma: Should I add some spices to it?

Mrs Verma: First add some cumin seeds and then wait till I wash all these ingredients. They will then be ready to be put into the pot.

Question: What are the people discussing?

(a) Chopping of the onions

(b) Ingredients of a dish

(c) Pouring of spices

(d) Cooking a meal

Answer: (d)

Listening for Specific Details

I. Listen carefully to the following news broadcast and then answer the questions:

This is All India Radio

Today's News Headlines

Australian PM Kevin Rudd to Meet Manmohan Singh Today

The Australian Prime Minister, Kevin Rudd, will meet Prime Minister Manmohan Singh on Thursday evening and will discuss issues of bilateral and international importance.

Bombs Kill 13 in Pakistan, Intel Agency Targeted

Suicide car bombs tore through security offices in Pakistan on Friday, killing at least thirteen people and heavily damaging the Peshawar headquarters of the country's top intelligence agency.

The deadly assaults on Pakistan's police and intelligence agents, came with 30,000 troops pressing their most ambitious offensive to date, against home-grown Taliban networks in their mountain strongholds on the Afghan border.

Swine Flu Deaths in India Jump to 514; over 15,000 Affected

Six swine flu deaths, including two each from Rajasthan and Gujarat, were reported on Thursday, pushing the total toll to 514 in India, health authorities said here in New Delhi.

Cyclone Weakens into Storm, Keeps Mumbai on Its Toes

Cyclone Phyan, on Wednesday afternoon, crossed the west coast north of Mumbai, and weakened into a storm. However, it managed to lash the city with both rain and winds.

Rajnath, RSS Chief Discuss BJP's Future

Rajnath Singh, whose term as BJP president ends this year, discussed a host of issues with the RSS chief. BJP chief Rajnath Singh, on Wednesday met RSS chief Mohan Bhagwat to discuss 'a host of issues' related to the party and its future.

Now, Election News

The Firozabad Lok Sabha seat—where the voter turnout jumped to 53 per cent from 48 per cent in May, and where Dimple Yadav suffered a humiliating defeat—reflects the deep challenge SP chief Mulayam Singh Yadav faces in Uttar Pradesh.

1. Which of the following is not among the Top News Stories today?

(a) Election for Firozabad Lok Sabha seat

(b) Bombs kill 13 in Punjab

(c) Total toll 514 in India due to swine flu

(d) Australian PM Rudd meets Manmohan Singh today

2. State whether the following statements are true or false.

(i) Suicide car bombs have heavily damaged the Peshawar headquarters of the country's top intelligence agency.

(ii) Cyclone Phyan on Wednesday afternoon crossed the south-west coast of Mumbai.

(iii) BJP chief Rajnath Singh met MNS chief Mohan Bhagwat to discuss 'a host of issues' related to the party and its future.

(iv) Australian Prime Minister (PM) Kevin Rudd meets Prime Minister Manmohan Singh on Thursday evening to discuss bilateral issues.

Answer:

1. b
2. (i) True   (ii) False
   (iii) False   (iv) True

Listening for Gaining Knowledge, Content, and Information

Now listen carefully to the brief talk on the Exploration of Mars and then answer the questions:

Exploration of Mars

Spacecraft exploration of Mars began in the early 1960s. Numerous missions were launched, but most ended in failure—some spacecrafts did not even manage to leave, or even reach, earth's orbit. A few spectacular successes, such as Mariner 9 and the two Viking missions, produced tens of thousands of pictures. After the Vikings, no launches were sent to Mars for over a decade. Humans started new efforts to reach Mars in the late 1980s, but the first successful missions did not arrive at the planet until some two decades after the Vikings. One of those missions was Mars Pathfinder, which landed on the planet's surface on July 4, 1997. It released a roving robot vehicle, called Sojourner, that carried out some geological studies. The mission also produced remarkable pictures. Another success was the probe Mars Global Surveyor, assigned to do mapping work. It was actually launched a month before Pathfinder, but did not go into orbit around Mars until September 1997. One of the findings of the Mars Global Surveyor included evidence that a massive flood had swept through an area of Mars as recently as perhaps 10 million years ago.

1. When did spacecraft exploration of Mars begin?
2. Name two successful missions sent to Mars.
3. What was Sojourner?
4. What was the major·finding of the Mars Global Surveyor?

Answer:

1. In the early 1960s
2. (i) Mariner 9   (ii) Viking
3. A roving robot vehicle
4. One of the findings of the Mars Global Surveyor included evidence that a massive flood had swept through an area of Mars as recently as perhaps 10 million years ago.

Listening for Main Ideas

I. First listen carefully to the talk on the Origin of Life and then answer the questions:

Origin of Life

Since time immemorial, one of the most complicated and fascinating problems before biologists is, 'How did life originate on earth?' In fact, philosophers, scientists, and other persons have thought about this problem for many centuries. Have you ever thought of this problem? Did you ever ask yourself—when and where did these living organisms originate? Did life originate only once or several times at different places?

In the book of Hindu mythology, it is written that Brahma is the God of creation and it is He who has made all the plants and animals of today. Gods, demons, and men are said to have arisen from His head; birds from His breast; goats from His mouth; and herbs, roots, and fruits from the hair of His body.

The chapter on Genesis in the Bible tells that life was created in six days. On the first day, heaven and earth were made; on the second day, the firmament was separated from the water; on the third day, plants and land were formed; on the fourth day, the sun, moon, and stars were made; on the fifth day, birds and fishes; and finally on the sixth day, animals and men were created. The first man who came to this earth was Adam.

Then came the concept of *Spontaneous Generation*. Many people thought that living things such as insects, worms, and mice could not be born by themselves but from the mud and dung of the earth. It was believed that the mud of the Nile in Egypt could give rise to living creatures when warmed in the sun. Frogs, toads, snakes, and crocodiles were believed to arise in this way.

The idea of spontaneous generation came to an end with the experiments of the Italian physician Francesco Redi (1621–1697). He decided to test the idea of spontaneous generation, and in 1668 he published the results of his experiments. He cooked the flesh of various animals so that nothing could remain alive in them. He kept them in three separate jars. One he left uncovered, the second was covered with parchment paper, and the third with muslin cloth or gauze. After a few days, maggots (housefly larvae) appeared in the first jar but not in the second. In the third jar, larvae appeared only when some eggs fell through the muslin cloth pores. So, it was proved that maggots arose only from the eggs that fell into the flesh.

If any doubts still existed, they were set aside by the painstaking experiments of the French scientist Louis Pasteur. He proved that microorganisms developed only when something from the air is in the medium. It proved that life comes from pre-existing life only.

Modern Concept of Origin of Life

This concept was given by Oparin in 1930. According to him, the earth originated about 6 billion years ago. At that time, it was very hot and had various gases and vapours of several elements. The earth then gradually cooled down and the gases condensed. Thus, a solid crust of earth was formed. The three molecules—methane, water, and ammonia—constituted the atmosphere of the primitive Earth. These molecules combined to form complex organic compounds under the influence of energy from the sun. These compounds then interacted to produce the first living cell. So, according to him, the first living cell arose from simple inorganic and organic non-living elements.

1. According to Hindu mythology, how did life originate on earth?
2. What does the Bible say regarding the genesis of life on earth?
3. Explain the modern concept of the origin of life on earth?
4. How were Louis Pasteur's views different from Francesco Redi's regarding the development of microorganisms?

Answer:

1. According to Hindu mythology, Brahma is the God of creation and it is He who has made all the plants and animals of today. Gods, demons, and men are said to have arisen from His head; birds from His breast; goats from His mouth; and herbs, roots, and fruits from the hairs of His body.

2. The Bible tells that life was created in six days. On the first day, heaven and earth were made; on the second day, the firmament was separated from the water; on the third day, plants and land were formed; on the fourth day, the sun, moon, and stars were made; on the fifth day, birds and fishes; and finally on the sixth day, animals and men were created. The first man who came to this earth was Adam.

3. Oparin gave us the modern concept of origin of life in 1930. According to him, the earth originated about 6 billion years ago. At that time, it was very hot and had various gases and vapours of several elements. The earth then gradually cooled down and the gases condensed. Thus a solid crust of earth was formed. The three molecules—methane, water, and ammonia—constituted the atmosphere of the primitive earth. These molecules combined to form complex organic compounds under the influence of energy from the sun. These compounds then interacted to produce the first living cell. So, according to him, the first living cell arose from simple inorganic and organic non-living elements.

4. Louis Pasteur was of the view that microorganisms developed only when something from the air is in the medium. It proved that life only comes from pre-existing life, whereas Francesco Redi tried to prove that new life originated only through organic living elements.

10

Non-verbal Communication

Learning Objectives   After reading this chapter, you will be able to
- understand what is meant by non-verbal communication
- learn in detail about the various aspects of non-verbal communication, namely body language, paralinguistic features, proxemics, etc.
- understand the different nuances of body language conveyed through one's personal appearance, body posture, walk, hand movements, eyes, facial expressions, etc.
- understand how paralinguistic features such as voice, volume, articulation, pronunciation, inflexions, pauses, etc. create an added impact in interpersonal, mass, and media communication
- learn to appreciate the subtleties of non-verbal communication and use them in your personal and professional communication

10.1 INTRODUCTION

Write the text of a speech on global warming. Ask the best speaker in your class to deliver it. Let his/her speech be followed by the most mediocre speaker in your class. Although the text is the same, do you think that the impact that each speaker creates would also be the same? Not really. Even if both the speakers speak the same thing, they will not speak it in the same way. The best speaker would use his/her mental agility, voice, and body to communicate to you the essence of the message. The mediocre speaker, however, would not really make such an effort.

Does it mean that the real difference between a good speaker and a poor one is the difference in the manner of speaking? Yes, it does mean that. In fact, in most of the situations, the speakers do not stand out because of an outstanding text, but because of the manner in which the text is put before the audience. It is so because when people communicate, they do not merely share some words, but also bring into play the subtle aspects of their voice and body to play an important part in communicating their ideas. Those who do it well realize their objective of being successful communicators, and those who are clumsy at it fail to register an impact. Therefore, as we prepare ourselves for a professional career, let us be perceptive about how the human body—through gestures, posture, expressions, hand movements, appearance, and eye contact—communicates the entire range of emotions, expressions, moods, and attitudes.

For example, when you see passengers sitting at the airport waiting for their flight, candidates sitting in the foyer waiting to be called in for the interview, players getting ready to replace or join their teammates playing on the field, or a friend of yours waiting for his/her turn to express his/her view during a group discussion, you can figure out that their obvious actions underline similar emotions such as restlessness, boredom, fatigue, or agitation. For displaying these emotions, no words are required. It is their sitting or standing posture, their hand or leg movements, their eyes or facial expressions which tell us in no unclear terms that they are going through the ordeal of waiting. The language that shares the unspoken emotion in a silent way is called body language, and together with the other paralinguistic features and proxemics, forms an integral part of non-verbal communication.

10.2 BODY LANGUAGE

Body language refers to all the expressions that we share by means of our body movements and not through words. Interestingly, the vocabulary of body language is universally common as it is interpreted and understood with the help of universally acknowledged body signs, cues, and symbols. The study of body language is known as *Kinesics*.

Since body language acquires a universal appeal and impact, it becomes imperative on our part to attain a certain level of proficiency in this. In fact, just like any other language, body language too needs to be improved with conscious efforts, especially because it is more trustworthy than any number of spoken or written words.

Let us take a look at some of the important elements of body language.

10.2.1 Personal Appearance

A person's external appearance is as important as the anterior of a house. Do you think you would be impressed with a house or its people if you found the anterior of a house shabby?

Dress Well to Make a Good First Impression

Certainly not. If the anterior is not well kept, most of us are likely to conclude that the house is not well-maintained.

Similarly, a speaker who does not seem to be maintaining himself/herself well is not likely to win the appreciation of the audience. Let us have a close look at the nuances of the term and understand how to score well on this front.

Dress, make-up, shoes, and hair style Nothing in the professional world happens by chance. Professionals make their own choices. For example, if you choose to wear a rumpled dress on the day you are being interviewed, it is sure to spoil your chances of landing into a job. Therefore, it is important that you select a dress that is neatly washed, properly ironed, and fits you well. Being properly dressed does not mean being fashionably or glamorously dressed. Dressing well and keeping a good posture is not something you should do only on special days to make an impression on someone. Make it a habit to remain well groomed and keyed up for a professional assignment.

10.2.2 Posture

Posture refers to the way we sit, stand, and carry ourselves. Our posture communicates the way we visualize the world around us. For instance, what do you think of a person who keeps his/her head down while walking? You must have seen people looking down while walking. Such people don't exude confidence and ease. On the other hand, the person who stands, sits, and walks upright commands respect and attention. Therefore, a professional has to cultivate and maintain elegance in his/her sitting, standing, and walking posture.

Given below are some important tips, following which you can maintain an impressive posture during professional meetings, interviews, group discussions, presentations, and other formal occasions:

1. Look straight while walking; avoid looking down at the floor, outside the window or door, or up at the ceiling.
2. Don't let your shoulders droop.
3. Lift your feet clearly off the floor while walking; avoid dragging them.
4. Avoid being too slow or aggressively fast while walking up to the podium or dais.
5. Don't slouch while walking, or sprawl while sitting.
6. Don't sit on the edge of the chair; it communicates unease and discomfiture.
7. Avoid crossing your legs while sitting or standing before your audience.
8. Avoid leaning on the lectern or reclining against the back of the chair.
9. Keep shifting your body weight as you stand before your audience.
10. Feel and communicate ease through your sitting and standing posture, and also the way you carry yourself at professional gatherings.
11. Avoid keeping your feet at attention or parallel.
12. Keep one foot ahead of the other; this helps you feel and appear at ease.

  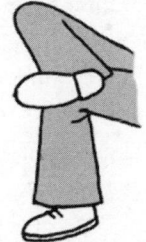  

| Crossed at the Ankle | Crossed at the Knees | Open Crossed with One Ankle on the Other Thigh | Uncrossed and Straight Closed Together | Uncrossed and Straight Apart |

10.2.3 Gestures and Hand Movements

Just as a picture can silently speak a thousand words; a gesture can communicate all that the speaker feels, consciously or unconsciously.

Crossed Arms Indicate
Defensive Attitude

Common gestures and their commonly understood meanings

- Waving indicates saying hello or goodbye.
- Making a fist indicates anger.
- Thumbs up shows appreciation or agreement.
- Pointing means showing something.
- Crossed arms indicate submissiveness, defence, and negativity.
- Hands on knees indicates readiness.
- Locking hands behind one's back indicates one's arrogance.
- Rubbing the eye indicates doubt and disbelief.

A speaker or listener's gestures and hand movements can support and emphasize their state of mind. Imagine a speaker who keeps rubbing his/her palms while delivering a speech; imagine a listener who sits cross-legged and also clamps his/her arms against the chest; think of a communicator who keeps rubbing his/her face. What do you make of them? Do these people create a favourable impression on others? Clearly, they don't.

It is so because, though the gestures do not overtly convey anything, the impact created by them is telling enough. Therefore, it is advisable to use gestures and hand movements appropriately, so that the impact created by them is graceful and suits the occasion.

Given below are a few tips which should be borne in mind while using gestures and other hand movements:

1. Keep your hands in control; don't let them have a life of their own.
2. Don't let your arms wave below your waist or allow them to loosely move about.
3. Use graceful and socially acceptable gestures.
4. Avoid aggressive and provoking gestures.
5. Don't rub your palms or your face while speaking or listening to others; it suggests lack of confidence and uncertainty.
6. Don't keep your arms folded against your chest; it suggests evasion and fear.
7. Don't keep your hands locked behind you; it suggests concealment of your true personality.
8. Avoid twitching or rubbing your nose.
9. Don't scratch your forehead, or eyebrows, or head; it suggests that you are unsure of yourself.
10. Don't lean on to a lectern; it reveals lack of confidence.
11. Avoid keeping your hands in your pocket; it suggests that you are hiding something from others.
12. Avoid playing with key rings, etc.; it distracts your listeners.
13. Don't wring your hands or play with rings on your fingers.

Aggressive Gestures

14. Don't tug on your shirt-sleeves or shirt collars; it reveals your discomfiture.
15. Don't scratch or crane your neck; it reveals uncertainty and doubt.

10.2.4 Eye Contact

Eyes are the windows to the soul. They truthfully convey the emotions and feelings one goes through. Therefore, looking into a person's eye is the best way to understand his/her attitude or reaction to all that you speak. Hence, maintaining an eye contact with your speaker and listener is the most important part of your non-verbal communication skills.

As a professional speaker, try to look into the eyes of the people in front of you. At times, the crowd that we face is huge and we feel nervous and hence start avoiding eye contact. Remember however, that it is bound to spoil all the impact of your otherwise well-written and well-articulated speech. Similarly, if you avoid eye contact during a job interview, you are more likely to lose rather than gain. It is so because someone who is not able to look into the eyes of their interviewers is considered edgy, nervous, and lacking in self-confidence. Even during group discussions and other meetings, the speakers become increasingly conscious of those who do not look at them as they speak. Again, speakers who do not look into the eyes of their listeners during meetings and other discussions are also likely to lose their credibility.

Here are a few suggestions, following which you will be able to use your eyes to support your effort in communicating your ideas effectively:

1. Maintain good eye contact with your listeners.
2. While addressing a large gathering, ensure that you keep looking in all directions.
3. While others speak, observe them carefully and try to understand the non-verbal cues they emit.
4. Exude confidence through your eyes.
5. Feel warmth for your fellow listeners/speakers; it is likely to improve your eye contact with them.
6. Feel and express a willingness to connect and communicate through your eyes.

10.2.5 Facial Expressions

Just as eyes are regarded as the windows to the soul, the face is considered an index of our mind. If there are unpleasant, sad, and gloomy expressions on your face, you are likely to create a very negative impact on your listeners or speakers. In the entire communication process, it is the person's face that we get to see most of the time. Therefore, if a face reflects negativity of any type and expresses dejection, irritation, indifference, fear, confusion, inhibition, vulnerability, or doubt, it is likely to severely affect the effectiveness of communication.

Since your face gives an indication of what you experience while communicating with others, it is of great significance to all the people involved in the process of communication. Therefore, use your face for expressing your confidence and ease. Start with a smile; a smile can light up your face. A smile is more often than not likely to help you establish a rapport with your co-communicators. You may have noticed that speakers who spoke to you with a smile on their faces were well accepted and better listened to. If a smile can work wonders, there are in contrast many negative expressions also which may completely wreck your communication with others. Think of a face that has a three day stubble or a perpetual frown or smirk. Would

 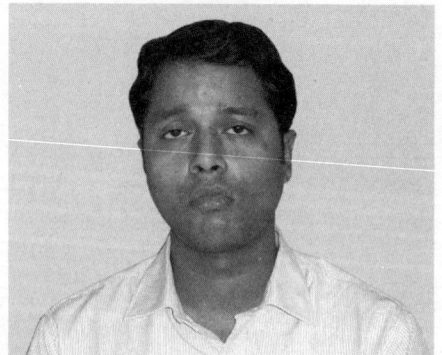

A Smiling Face and a Grumpy Face

you, as an audience, associate with such a person who runs you down or doubts you through his/her facial expressions? No way. Hence it is important to present a pleasant disposition through your facial expressions. Let there be emotions of confidence, zeal, and enthusiasm. Let your face reveal a heart that is willing to associate and communicate; express this attitude both while being a speaker and a listener.

Given below are some tips which may help you maintain proper facial expressions while speaking or listening to others:

1. Start with a smile but don't keep smiling throughout.
2. Don't have a frown on your face; it suggests arrogance.
3. Avoid raising your eyebrows while speaking or listening to others.
4. Don't purse your lips while speaking; it reveals your lack of confidence.
5. Don't narrow your eyebrows; this too suggests your lack of trust in others.
6. Avoid being dull in the face; express confidence and ease.
7. Avoid expressing dejection, sadness, or indifference.
8. Avoid reflecting strong emotions on your face.
9. Let your face suggest your honesty, integrity, and conviction in what you say.
10. Don't smirk; it suggests arrogance.
11. Don't express any kind of disrespect or contempt for your listeners.
12. Let your face suggest a willingness to associate yourself with others.

10.3 PARALINGUISTIC FEATURES

Just as we can communicate various attitudes through our gestures, posture, expressions, and body movements, eyes, and hands, we can express emotions and feelings with the help of different aspects of our voice. Though we cannot radically change our voice, there are different aspects of voice which can be carefully worked on to create the right type of impact on our listeners while we deliver a speech, make a presentation, participate in a group discussion, or appear for a job interview.

10.3.1 Rate

Rate refers to the number of words we utter per minute. When you speak in professional situations, try to assess whether you speak too fast or too slow. Speaking too fast is related to lack of comfort. A speaker who does not feel sure of himself/herself generally feels intimidated by the challenge of speaking in professional situations. This leads to a feeling of nervousness, and the best solution seems to speak as fast as one can and be finished with the frightening prospect of standing in front of the audience. Such a speaker, however, fails to win the audience as the breakneck speed of delivery not only reveals the speaker's lack of confidence but also makes it difficult for the audience to comprehend, assimilate, and digest what is being said by the speaker.

Just as too fast a pace causes inconvenience to the audience, so does a pace far too slow. In fact, too slow a pace of your speech is most likely to cause monotony and boredom to such an extent that the audience start feeling sleepy and lose interest in the speech. Moreover, too slow a rate suggests lack of preparedness on the part of the speaker.

Now the question arises: How does one understand what a slow or fast rate is? Studies in this regard suggest that a rate between 125 and 150 words per minute is ideal in professional situations. However, if the matter needs deep thought and meditative attention, the rate is generally a little slower. Similarly, when we have to share something in an exciting or casual way, the rate of delivery can accelerate.

10.3.2 Pauses

Pauses are an essential part of all human interactions. We pause between different thought units in our day-to-day interactions with others. Therefore, if we do not pause while we speak in professional situations, it only makes our speech appear unnatural and hasty. Pauses lend credibility to the text of the speech. The speakers who pause suggest that they are quite accomplished, poised, and composed, and are not *really worried about not being able to locate an idea* once they have paused. Thus, if we pause, we display a sense of security and feeling of assurance that we know how to go further in our speech after a pause. On the contrary, those who do not pause seem to be in a hurry. Moreover, those who rush through their speeches and presentations are nervous about using pauses, as once they stop, they feel they would not know how to resume or reconnect.

By all means, we must use pauses while speaking in professional situations. They make our speech sound natural. Moreover, pauses are also required for the audiences to comprehend what you say, relate it to your earlier statement, and critically participate in the act of communication.

The most crucial thing about pauses is their timing. A rightly timed pause is as important as a rightly placed word. Since a pause has to indicate either the emphasis or the conclusion of a thought unit, it is important not to put them at wrong places. Therefore, whenever you pause, pause at the conclusion of a certain thought unit and not in between. Remember, a rightly timed pause adds to the value of what you say and makes it adequately natural and emphatic. A wrongly placed pause, however, distracts the audience. Also remember that though a pause is always a natural breather, both to the speaker and the listeners, silence—a longer pause—makes the audience feel impatient. To understand the difference between a pause and silence, let us look at a speech situation. When a speaker comes to speak, he/she first takes his/her position, walks up to the lectern, waits for things to be in order, and then starts. All this while, nothing is spoken and heard; this is what silence is. It is a long pause which indicates the beginning of a new momentum, whereas a pause is a short silence which indicates a natural gap between different thought units, and is meant to secure emphasis at certain places.

Just as silences or wrongly placed pauses distract the audiences, so do the *vocalized pauses* which truly spoil the impact of an otherwise effective speech. Vocalized pauses are sounds such as 'umm...', 'err...', 'aa...', etc. In professional situations, they act as a nuisance since they do not add to the meaning of what you say. They only suggest that we struggle with ideas and are not in control of our matter and manner. So, if we use vocalized pauses frequently in professional situations, we are likely to be mocked at.

Similar to vocalized pauses, is the overuse of repetitive expressions. Many a time, we come across a speaker who adds a phrase such as 'you know...', 'I mean...', 'actually...', 'basically...', 'in fact...', 'okay...', 'well...', or 'right...', to almost all the ideas he/she communicates. Using an expression once in a while is not distracting, but when we start putting up a string of such expressions to begin or end all that we have to say, it surely distracts the audience. In such instances, rather than focusing on your speech, some of them start looking forward to hearing that typical refrain of yours.

10.3.3 Volume

A speaker's volume often decides how he/she is likely to be received by the audience. The speaker who speaks at a low volume is likely to be seen as someone who lacks confidence, whereas a speaker whose volume is too high suggests his/her boorishness. Low volume is essentially associated with diffidence, and once you reveal that you lack confidence, you cannot gain control of your audience or command their respect as a speaker.

At the other extreme is the speaker who speaks so loudly that the people in the front rows start dreading him/her. A speaker of this type is also likely to be rejected by the audience, simply because it suggests his/her arrogance.

Now the question arises: How do we understand whether the volume we maintain is adequate or not? In order to understand this, carefully observe the reactions of the audience while you speak. If you see some smirks or mocking expressions on the faces of the people sitting in the first couple of rows, the chances are that you are speaking far too loud. On the other hand, if you observe people in the last rows craning their necks and their faces registering confused expressions, it means that you are not audible enough. Remember, maintaining an adequate volume is extremely crucial for creating the right kind of impact on your audience, and if you are found wanting in this, you are likely to be rejected by them.

10.3.4 Pitch/Intonation/Cadence/Voice Modulation

Pitch refers to the rise and fall in human voice. Just like the other aspects of voice, pitch too plays a crucial role in communicating your ideas to others. In fact, it is the pitch—the rise and fall—in your voice which can express all the emotions that are to be conveyed. So that you do not confuse volume and pitch, let us understand the difference between the two.

All of us listen to songs on our stereos, CD players, iPods, etc. When someone asks you to make less noise, what do we do? We simply lower something. What is it that you lower? Is it the pitch or the volume? Obviously, it is the volume of a song that you can decrease or increase, but you cannot increase or decrease the pitch. However, what characterizes a song is not its volume, but the pitch which the gifted singers so meticulously vary. And not just singers, but all speakers need to effectively employ variation in pitch patterns in their speeches. Pitch is something that adds colour and lustre to your voice, and hence plays an important role in the overall communication process.

Since pitch can express and convey all our moods, emotions, and sentiments, it becomes really important for us to carefully employ the desired pitch patterns. Quite often, we find the speech of a person quite boring because he/she does not use the variety of pitch patterns as per the requirement of the situation. In order to understand this, listen carefully to those songs in which the singer keeps singing in a solemn way until he/she reaches the crescendo. The change in the initial note and the later part is the change in the pitch patterns. The changes in pitch patterns can be observed even while we continue to listen to the song at the same volume. It is, therefore, suggested that you employ the variegated pitch patterns quite judiciously in order to keep your listeners engaged and interested in your expression.

It is, in fact, the variations in the pitch which provide colour and lustre to a singer or a speaker's voice.

10.3.5 Pronunciation and Articulation

Pronunciation plays an important role in expressing our ideas. As discussed earlier, English is not our native language and hence the pronunciation of Indian speakers of English is different from that of the native speaker. An effort should constantly be made to make our speech as close to standard English as possible. In terms of pronunciation, we should stick to RP English as it is recognized as the standard pronunciation of English worldwide. The chapter on *phonetics* discusses the different RP sounds of English and also gives you sufficient information regarding word stress, weak forms, and intonation patterns, following which you can make your spoken English intelligible to a native listener.

Articulation is also as important as pronunciation. To highlight the difference in these two commonly confused terms, let us consider the following words and expressions: 'psychology', 'mythology', 'rendezvous', 'clerk', 'sample', etc.: all such words and many more in English are pronounced in different ways, but there is only one standard way to pronounce these. Therefore, the difference is in the different ways of pronunciation—some acceptable and others not acceptable. However, when we listen to expressions such as 'lemme...', 'yeah...', 'dint...', etc., it is not the pronunciation but the person's articulation that is at fault.

Articulation refers to our ability to speak different sounds distinctly. If we are able to speak and enunciate different sounds in a distinct and crisp manner, our articulation is considered appropriate and impressive. On the other hand, if we mix or mumble words, it is regarded as sloppy and inelegant. Among youngsters, the problem of sloppy articulation is quite common, probably partly because of the influence of American movies and mannerisms, but mainly owing to a cyber-savvy mobile culture that believes in chopping, truncating, and abbreviating all that is elaborate and requires more effort in reading, writing, speaking, and listening.

Consequently, quite often you find a friend of yours slurring and mumbling his/her expressions and reducing 'let me' to 'lemme...', 'have to' to 'hafta...', 'I didn't...' to 'I dint...', 'yes' to 'yeah...', 'you ought to' to 'you otta...', and so on and so forth. Some of our young friends really find their own speech becoming 'stylish', and hence 'impressive', with the help of such chopping, slurring, and mumbling expressions. A speaker who tries to sound 'trendy' and 'stylish' is likely to be ignored as someone trivial, immature, and 'funky' by a knowledgeable audience.

10.4 PROXEMICS/SPACE DISTANCE

Have you ever observed lions and tigers in a zoo? Do they appear to be comfortable with their caged existence? Don't you often find them moving restlessly inside their cage? Tigers and lions—and none of the other animals or birds—seem to be happy inside a cage. Therefore, they appear to be restless and disquieted most of the time. The situation gets worse if many of them are put inside the same cage. You often see them attacking and mauling each other. Why does it happen? And it is not just animals or birds but also humans that detest being inside cages. We all love our freedom and want to protect it at any cost.

Observed closely however, it seems that it is not just freedom but also space that matters to us. Physically, all of us are free. But when we see a crowded place, we do not feel comfortable. We do not want to board a crowded bus or train; sit on a waiting bench where others are sitting; stand in a long queue; sleep in a room that seems crammed and crowded with things. In fact, these are only a few instances which suggest how we all want our own territory and space to feel relaxed and enjoy a comfort that is lost if we are surrounded by things or people.

While communicating in formal situations, therefore, it becomes quite important for us to understand and respect the territories of other professionals, and see to it that they never feel intruded. In fact, if you stand too close to people while speaking to them in formal situations, they are likely to resist and resent your presence. Standing or sitting too far away from your listeners or speakers, on the other hand, is also not all that advisable. Just as standing or sitting too close to others may make them feel intruded upon, and violated or choked, standing or sitting too far away may communicate a sense of alienation and lack of warmth. Therefore, it becomes important for us to understand the different zones into which the psychological territories of human beings can be divided. How we can appreciate the various psychological zones maintained by most of us is given below.

10.4.1 Intimate Zone

No stranger is welcome into the intimate zone which is shared only by spouses, lovers, children, parents, and very close relatives and friends. Anyone who tries to enter someone's intimate zone in professional situations is more likely to seem like an intruder.

10.4.2 Personal Zone

Watch carefully the distance maintained by people while they interact with one another during business gatherings, social functions, parties, and other friendly get-togethers. The distance maintained by people in a zone varies from a couple of inches to a couple of feet and is indicative of the warmth or the necessity to maintain formality in relations. When the personal and the professional relations seamlessly fuse, it becomes possible for professionals to enter each other's personal zone without appearing to be intruders.

10.4.3 Social Zone

The distance maintained between a couple of feet to several feet is suggestive of the social zone that we maintain while interacting with strangers or occasional visitors such as laundry persons, gardeners, plumbers, electricians, etc. In professional gatherings, people sometimes are seen maintaining this distance. Social distance is effectively maintained in situations where professional needs overweigh the personal.

10.4.4 Public Zone

In most professional communication situations, public zone is most commonly maintained by the speakers and their audience. Consequently, we find a defined area from where the speaker has to address his/her listeners. Though a distance of some feet is usually maintained between the speaker and the listeners while they share a public zone, the actual distance maintained differs from culture to culture. For instance, it is quite possible for a teacher in India to walk up to his/her students and reduce the distance of several foot to barely half a foot or so. It may not be possible for them to do so while addressing students in some other countries.

10.5 HAPTICS

Just as it is important to learn and maintain proper distance in professional situations, it is also worthwhile to understand the subtleties of haptics, which in such situations is limited to hand-shakes or occasional hugging and patting. Of course, a handshake is a very common etiquette that people in almost all situations seem to observe. Very often we see people exchanging handshakes while communicating a welcome greeting as well as signalling a departing greeting. In some very formal situations, we see the heads of states, such as the prime ministers of two countries, shaking hands with their counterparts. In such situations, a handshake generally signals the sealing of a contract or agreement.

Respect Other People's Personal Space

Besides that, a handshake among professionals is regarded as a sign of the warmth that is required to be communicated in almost all human interactions. Hence, as professionals, we must not hesitate while being offered a hand-shake by a superior or senior. At times, some nervousness is seen on the face of youngsters who are offered a handshake by their senior officers or boss. Many of us tend to put our hands together to form a *namaste*. However, not responding to a handshake with a handshake may appear to be curt and it is advisable not to shy away from such opportunities. Similarly, many men are seen simpering and displaying a nervous smile when offered a handshake by a woman colleague or professional. In professional situations, it augurs well to suggest that you have a cosmopolitan upbringing, and that you do not obsess yourself much with the distinctions of caste, colour, creed, and gender. Therefore, you are advised not to spurn a hand that is extended in your direction.

There is hardly any sense in running to the other side of the shore as well, and shaking hands with everyone who comes your way in professional situations. In fact, before extending our hand for a handshake it is required that we analyse the situation and assess whether extending a hand for a handshake is appropriate or not. However, whenever you shake hands with others, let it be a firm handshake and not a clammy and limp one. Extending a lifeless hand for a handshake is similar to suggesting your lack of interest or rudeness towards others. Therefore, let your haptics too suggest your zeal and enthusiasm while dealing with others in professional situations.

Though handshakes are common and easily acceptable in professional situations, other forms of haptics such as hugging and patting on shoulders are rarely practised. Such signals are important in some situations where a professional has to emit warmth and personal interest, particularly to his/her subordinates. Therefore, a team leader is often seen patting the shoulders of his/her subordinates while sharing a sense of celebration or achievement with them. Since both such situations and leaders are few and far between, hugging and patting is not very commonly displayed in professional situations. The most important thing to bear in mind with regard to haptics is to observe and adopt the pattern followed by other senior professionals around you.

 ## RECAPITULATION

✓ Non-verbal communication plays a very important role in all professional situations.

✓ Non-verbal communication includes aspects such as body language and paralinguistic features.

✓ In professional situations, we have to take care of our body language in terms of personal appearance, gestures, posture, body movements, walk, facial expressions, hand movements, eye contact, etc.

✓ Effective communicators enter professional situations in a well-groomed manner; they take care of their personal appearance; walk confidently in a business like manner; use graceful gestures; maintain elegant sitting and standing postures; use hand movements to emphasize their ideas; display conviction and friendliness through their facial expressions; maintain eye contact with their listeners throughout their speech and presentation.

✓ We have to bear in mind the importance of the paralinguistic features of communication such as volume, voice modulation, pronunciation, articulation, pauses, etc. to create the right kind of impact on our listeners.

✓ In order to become a practised communicator, we have to maintain an audible but not too loud a volume during our presentations and speeches. Besides volume, we also have to employ proper voice modulation, correct pronunciation, well-timed pauses, and crisp articulation in order to communicate the desired mood, tenor, and tone of our speech.

✓ Space distancing and haptics too form an essential part of our non-verbal communication, and hence we must practise these as per the common code of professional behaviour.

WISEWELL QUIPS

EXERCISES

Objective Questions

State whether the following statements are true (T) or false (F).

1. The way you look—your clothing, grooming, and posture—telegraph a delayed non-verbal message about you.

2. All the gestures and postures have a fixed and permanent meaning in communication situations.

3. Study of the non-verbal cues helps in identifying whether people actually mean what they say while delivering their messages.

4. The intimate space zone for social interaction is reserved for colleagues.

5. Voice modulation is essential for effective speech making.

6. Rate refers to the number of variations per second of your voice.

7. Vocalized pauses enhance the impact of an oral presentation.

8. Non-verbal communication effectively reinforces verbal communication.

9. A person well prepared for a presentation is able to create a profound impact even if he does not pay attention to the effective presentation strategies.

10. Actions speak louder than words and thus, Kinesics provides a deeper insight into the sender's message.

11. The most effective way to hold the audience's attention is to make proper use of paralinguistic features.

12. Slurry and choppy articulation leaves a bad impact on the audience.

Concept Review Questions

1. 'Body language can make or mar a speech.' Discuss and elucidate.

2. 'It is not just the body language but other paralinguistic features also which determine the effectiveness of a professional speech or presentation.' Comment on this statement and support your views with proper examples.

3. Write a brief note on each of the following:

   (i) Haptics

   (ii) Proxemics

4. 'Like body language, your voice carries both intentional and unintentional messages.' In the light of the statement, discuss the various aspects of voice that determine the success of any speech or presentation.

5. 'The most effective way to hold an audience's attention is to make proper use of paralinguistic features.' In the light of the statement, discuss the importance of various paralinguistic features.

6. 'Actions speak louder than words and thus, Kinesics provides a deeper insight into the sender's message.' Elucidate the statement in about 350 words, giving suitable examples from your own experiences.

7. A person well prepared for a presentation fails to leave a profound impact if he does not pay attention to effective presentation strategies. State how body language and voice help the speaker create a long lasting impact on the audience.

8. Write short notes on each of the following in about 200 words:

   (i) Posture
   (ii) Gestures
   (iii) Facial expressions

ANSWER KEY

Exercises

1. F	2. F	3. T	4. F	5. T	6. F
7. F	8. T	9. F	10. T	11. T	12. T

CHAPTER 11

Dynamics of Professional Presentations

Learning Objectives After reading this chapter, you will be able to

- combat stage fright while making professional presentations
- prepare appropriate slides for your presentations
- describe objects/situations/people in formal settings
- participate in individual/group role play situations
- deliver effective just-a-minute (JAM) presentations

11.1 INTRODUCTION

Some two decades back, a group of Australian students came to our institute. They were working on a project and were supposed to make presentations regarding the project from time to time. The day Julia, one of the students, was supposed to make her presentation, she was very nervous. She could speak fluent English, yet she was so scared, that the night before her presentation, she could not sleep and kept tossing in bed. She walked in for her presentation bleary eyed, seemed tired and unsure of herself, spoke briefly and ended hurriedly. That evening, she was so upset with her performance that she wept uncontrollably.

This can happen to any of us. Quite often, students feel so fidgety about the prospect of making a presentation that they wish to either evade the whole process or just end up giving a dismal performance. None of this can help us in the long run because as we prepare ourselves for a professional career, it becomes essential for us to understand that we cannot wish away the responsibility of making a professional presentation.

This is so because in today's professional world it is essential to be a communicator and, regardless of the job or position one holds in an organization, one has to frequently deal with various communication-related assignments. For example, professionals are quite frequently asked to speak formally to a group to discuss the progress of their project, highlight the innovations they propose the company to consider, give suggestions to improve the working of a particular plant or unit, and so on and so forth. In fact, as professionals, we have to always be prepared for a communication assignment, and therefore it is important that we learn and master its nuances so as to develop ourselves as effective communicators.

Before anything else, however, we must learn how to overcome nervousness and stage fright, which does not allow us to realize our potential as speakers and often severely affect our presentation tasks.

11.2 COMBATING STAGE FRIGHT

Most of us are scared of a situation in which we have to stand up in front of our audience and say something. This fear is so widespread that as many as 70 per cent of the university students regard it as a very big challenge and seek to avoid the task of making a presentation as long as they can. However, just as most of our fears are baseless and can be overcome with some guidance and training, it is possible to overcome stage fright too. Here are a few suggestions which can help you overcome this fear:

1. Recognize your nervousness. Don't run away from the task of accepting the fact that you feel nervous when you speak in formal situations.
2. Understand what happens to you physically when you feel nervous. Actually, nervousness is a type of chemical movement which is caused by a sudden rush of adrenaline in your body.
3. Recognize the fact that the flow of adrenaline in the body only gives you more glucose, which actually provides your body with greater energy.
4. Regard your nervousness, therefore, as a positive phenomenon and a form of energy which you can turn to your advantage.
5. Since nervousness makes you feel more energetic, you are not likely to feel timid when you experience it.
6. Nervousness is a normal feeling with most people who have to perform in professional situations. Not just the speakers, but even other professionals experience it from time to time. For instance, actors are nervous when they face the camera; singers are nervous when they are required to sing in front of their audience; politicians are nervous when they have to address a crowd.
7. Regard nervousness as a positive, healthy sign, which rather than spoiling your performance can actually improve it.
8. Visualize yourself giving a good and strong presentation. Positive imagination infuses freshness and confidence.

9. Work hard on your content. If you are well prepared, you will feel excited about sharing what you know with others.
10. Work hard on your opening lines and the rest of the introduction. Use humour and wit if possible. Nervousness is most disquieting during the initial part of a presentation. Once we are able to put a couple of opening sentences in the right place, we start feeling better almost instantly.
11. Rather than bothering about your nervousness, focus on what you have to speak to your listeners. Once you are caught up in the task of telling the audience what you know, nervousness starts leaving you.
12. Rehearse your presentation in front of your friends, parents, or siblings. Rehearsal and

Combat Stage Fright

practice not only polishes your performance, but also makes you feel more confident and assured of yourself.

13. Look at your audience; maintaining eye contact with them helps you feel rooted and related to them.

14. Start your presentation with a smile and a warmth you feel when you do things you like doing.

11.3 PREPARING POWERPOINT SLIDES FOR PRESENTATIONS

Besides taking care of stage fright, it is also important for a professional to plan his/her presentation appropriately. One of the key areas of preparation relates to the planning of PowerPoint slides that the speaker intends his/her audience to view. Many a time, an otherwise good presentation is spoiled because of poor presentation slides. Over the years, PowerPoint slides have almost replaced all other types of presentation aids. Very often, we see speakers walking in with their laptops, connecting wires to the LCD projector and displaying the slides one by one.

Nothing seems to be wrong with this strategy, but like many other misuses of technology, a PowerPoint presentation does have its disadvantages, and if not planned properly the slides can actually derail a presentation. Remember the tips given below as you plan your presentation on PowerPoint slides:

1. Reach the presentation area much before the audience and adjust material on the laptop or pen drive well before you speak. When speakers fiddle with technological gadgets in the presence of the audience, they start getting panicky, especially if some technological glitch confronts them.

2. Time your slides to perfection. It is important to know when to show a slide. At times, the discussion and the slide do not match, which looks odd and distracts the audience.

3. Keep the lens of the LCD covered with its lid or a handkerchief, so that the audience do not necessarily have to look at a blank screen before you choose to show them your first slide.

4. Even while displaying a slide, avoid standing or walking in front of the LCD lens. A shadow looming large on a projected slide is a sight weird enough for the audience to feel distracted.

5. Don't clutter your slides with too much detail. Avoid writing long paragraphs or lengthy sentences on slides. A cluttered slide is ultimately going to irritate and distract your audience.

6. Give all your material in bulleted form, with a single slide not exhibiting more than eight to ten points.

7. Don't read your own slides by looking at the screen. It certainly makes you lose eye contact with your audience. Moreover, it reveals lack of preparation on your part. To aid your memory, you can keep some flash cards with you, or look at the computer screen if required. At the most, you can look at the slides perfunctorily; however, avoid reading them. Remember that you display slides for your audience and not for yourself.

8. Keep your slides to the minimum number. Many a time, speakers pour slide after slide on their audience. This not only makes them feel weary of it but also causes monotony. Moreover, it also reveals that you, as a speaker, depend far too much on slides for all your ideas and discussion.

9. Make your slides as captivating and innovative as possible for your audience. Avoid, however, unnecessary frills and ostentation.

Both in your academic and professional career, you are required to give presentations of various types. Given below are some such presentations, with suggestions that will help in performing these tasks with confidence, clarity, and conviction.

11.4 DESCRIBING OBJECTS/SITUATIONS/PEOPLE

Many a time, you are required to describe a picture, object, situation, or a person to others. This is a task that all of us do. In fact, it could be one of the routine class assignments that your teacher may assign you every month. Just because it is a common task does not mean that all of us do perfectly well. On the contrary, not many of us perform such tasks in an accomplished way. Therefore, we need to cultivate the skills required to make such descriptions lively and interesting.

In order to properly understand the complexity of the challenge involved in such tasks, read the stanza given below and decide whether it can be called an effective description:

> Friends, this is a picture in front of you. This picture is beautiful, isn't it? This was made by the great American artist Richard Ashwood. The picture shows how a simple, poor family is enjoying an evening. It shows kids playing around. The mother of these kids is a beautiful woman who is shown knitting. The grandparents of the children are looking at the kids and smiling at them. The colours on this picture are quite suitable. They depict reality. Though the picture is quite old, it is still very good. Do you know when it was painted for the first time?

You certainly cannot listen to such descriptions for a very long time. It is bland, unfocused, and superfluous. The description lacks the vitality, intensity, and the details which can make the audience sit up and not just look at the picture but also enjoy its description. Here is an alternative way to discuss the same picture:

> Ladies and Gentlemen, the picture before you is one of the most stunning works of art. Painted in 1899 by the great American artist Richard Ashwood, the picture exhibits a simple, poor family lost in a languorous evening. These kids, who are shown to be playing around with gleeful ease, are remarkably created with all their innocent faces and the small pranks that give them extreme delight and pleasure. The grandparents' indulgence in those small acts by the little kids speaks for the innocent joys of both the young and the aged. In a way, the picture reconciles the past and the future, with the mother busy with the needlework. Friends, I find it quite symbolic, the young mother's focus on her knitting stands for the duties of those who have to shoulder the responsibilities of the household. In a way, her continuance of needlework stands for the necessities of the practical life which requires focus, work, and a profession to keep the world around them going. Despite that message however, the faint smile with which she goes about doing her work suggests harmony and assurance around her. The modest setting of the household and the absence of a masculine adult around, suggests a deprived existence the entire family is reconciled to living. Although a picture of a humble, simple life, the painting does not, in any way, present a cynical view of life. On the contrary, it makes us realize the pleasure of living with the minimum, as in the want of things, human love flows spontaneously.

What do you think is the difference between these two descriptions? Don't you find the second description to be more effective than the first one? You certainly do! Now, think about what makes the second description more effective. Can we understand some features of an effective description? Of course, we can. This is what has made the second description—and would also make any other description—effective:

1. A description should start with some captivating statement. By introducing the picture as *one of the most stunning works of art*, the speaker easily secures the attention and focus of the audience.

2. Connect the objects, situations, and all other descriptions to the audience. Help the audience see the vital points of importance with lively, intense descriptions. The speaker in the second description helps the audience appreciate the picture by bringing into focus its subtle nuances and symbolic worth.

3. Maintain a warm and intimate style while describing an object, picture, or person. Carefully observe the audience-oriented tone that the speaker maintains while describing the picture the second time. Moreover, he/she seems to offer views on the picture with enthusiasm and excitement.

4. Your descriptions should be not only lively but also incisive and thought provoking. The second speaker brings to the fore the hidden message of the painting and thus leaves a positive impact on the audience.

5. A good description, by all means, needs an effective use of words. The second description does that and hence creates a better impact on the audience.

Go further and read the description of a movie:

> Ladies and Gentlemen, I am here to discuss with you one of my favourite movies, *A Wednesday*. It's a great movie and I believe you must have watched it already. If you haven't, go ahead and watch it, because the movie is wonderful. The movie has a message. It tells us about the feelings of a common man. Because the system doesn't help, the common man has to take the law into his hands. Throughout the movie, the suspense is maintained. We keep guessing about what is happening. At the end of the movie, the suspense is clarified. The movie is well directed and the roles performed by Naseeruddin Shah and Anupam Kher are wonderful. In this movie there is violence, but we like it. There are no songs. But that's good again. As I said, the movie has a message and in such movies, you don't have songs. But even then, it is really wonderful. You must go and watch it.

Do you find the description of *A Wednesday* to be appropriate? Is it the way to describe a movie? Look at another description of the same movie and then decide.

> Ladies and Gentlemen, for an incorrigible movie fan like me, discussing a movie is by all means a matter of great delight. The delight becomes manifold when you have the freedom to choose a favourite movie of yours, and discuss it. The movie that I have chosen for discussion is *A Wednesday* by a young, talented director, Neeraj Pandey. *A Wednesday* has all that makes a movie meaningful and memorable. It has a very tight script, penetrative camera shots, crisp dialogues, controlled performances by good actors, and above all, a flawless direction. A seminal movie by all standards, *A Wednesday* expresses the deep-seated anger and distress of a common man who finds himself helpless and suffocated in an immune socio-political system, and sets out to decide a course of action on his own. Though the movie seemingly supports violent ways to be the justified means for a peaceful end, it does not sound gross and overt. On the contrary, it subtly symbolizes the vulnerable, frustrated, scared, and anguished lives of millions of Indians. On the whole, the movie *A Wednesday* has a relevance and appeal much beyond its celluloid charm, and serves to redefine not just Indian cinema but also our ethos, laws, and practices.

Sometimes, we are required to discuss a person. When required to do so, bear in mind the instructions given for drafting descriptions of objects, pictures, and movies. While discussing a person however, try to sound as objective as possible. Even when you immensely admire the person you choose to talk about, don't sound lopsided and emotional in your adulation. Consider the description of a public figure and political leader. Though the speaker admires and supports the leader, he/she never loses sight of reason.

Good morning Ladies and Gentlemen, it's a matter of great privilege and delight for me to share my thoughts on a person whose life and actions have always inspired my humble existence. Friends, it is Mrs Indira Gandhi, the first and by far the only woman Prime Minister of India, about whose determination and vision I would like you to be briefed about.

Known as the *Iron Lady* of Indian politics, Mrs Gandhi was groomed into a political career by her father Pandit Nehru, the first Prime Minister of independent India. Right from her childhood, Indira Gandhi displayed a zeal and passion to serve the country, and participated in the freedom struggle of our nation even at a very young age. Our country became independent, but troubles continued to haunt us. Indira had the clarity, focus, and a plan to serve the needy and the poor. As the Prime Minister, she worked hard for the removal of poverty; she started the family planning mission, the impact and relevance of which we can appreciate even now; she nationalized banks and bravely fought war against Pakistan. Indira was known for her no-nonsense approach and once she took a decision, she stuck to it. Many of her decisions were criticized; particularly, the emergency declared by her in 1975, which made her very unpopular for a while. Strong in her notions, she forced her way back to fame and popularity and continued to serve the nation with tenacity and guts. As a leader, she had a vision and the calibre to not only articulate but also execute it. Steeped in nationalistic feelings, she resisted with ferocity all attempts that were targeted to divide our country. The operation Blue Star is a shining example of her bravery and commitment to serve the right cause in trying circumstances. It was this audacity of hers that cost her her life. On 31st October 1984, she was gunned down by her own security personnel. Today, Mrs Gandhi is no more, but her zeal, dedication, and commitment to her cause will continue to inspire us forever.

11.5 INDIVIDUAL AND GROUP PRESENTATIONS

In present day professional situations, it is quite common for a group of people to be involved in combined projects. Many such projects require a presentation to be prepared and delivered by a group to another group of professionals. At times, all the members of the team are required to present some segment of their study, experiment, or research involved in such a group project. Such presentations made by different members of a group that jointly worked on a project are referred to as group presentations.

Apparently, a group presentation seems to be an attractive alternative to an individual presentation where the entire task is to be carried out by a single person. In a group situation, the workload is shared; and the members have to speak about only a part of the whole project. With each member having to prepare a smaller part of the entire exercise, the amount of work, the labour involved, the time spent in preparation, and above all the amount of time to be spent in front of the listening group seem less daunting and challenging. This, however, is only one aspect of the reality. For, on the other side, making a group presentation becomes all the more challenging, considering the various factors which come into play.

Some such factors are enumerated below, with suggestions as to how to avoid the pitfalls which may hinder a group presentation from becoming effective and memorable:

1. It is seen that a group presentation is quite often loosely structured. The group members presenting their part of the presentation are often unable to maintain continuity in the discussion. It is, therefore, important to pick up the discussion from the point where your group member has left and connect well to his/her ideas before taking the audience to newer realms.

2. It is also seen that each of the group members starts with a greeting such as 'Good Morning, Ladies and Gentlemen', 'Welcome to the presentation on...', etc. As a group member speaking second or third, we must realize that the discussion has already been initiated and a series of greetings would only leave a ludicrous impact on the listeners.

3. On the other extreme, quite a few speakers in a group presentation situation tend to start abruptly without making any effort to connect with the preceding speaker. Such presentations fail to leave a unified impression on the audience.

4. In order to maintain continuation and connectivity to the preceding speaker, all the members of the team must thoroughly prepare the entire presentation. Remember, though each member has to present only a part of the entire presentation, all of them must be acquainted with the entire presentation.

5. Once some member of your group has finished his/her part, you need to start with a brief recapitulation of what he/she has said. For instance, in a presentation on 'consumerism', one can always connect to the preceding idea by saying something like *We have just seen the advantages of a consumerist culture but not everything about consumerism is worth adulation and celebration. There are some very grave concerns associated with this phenomenon, a few of which I would be discussing with you...*

6. At times, group members consciously or unconsciously end up contradicting each other. While making a group presentation, we have to keep in mind that *though there are different speakers in the presentation, they are all making one single presentation*. Therefore, it is important that their ideas cohere and integrate. In no way a contradictory statement such as 'I really don't agree with my friend who has presented his views prior to me...' has a place in a team presentation. Of course, the different aspects of the same issue are required to be brought into perspective, but contradicting your own team member would only serve to distract and confuse the audience.

7. Improper distribution of text and slides also makes a group presentation go haywire. In a group situation, it is common that somebody starts a presentation, some others develop it, and finally some member of the team brings it to a conclusion. An unequal distribution of the task may however leave the audience bemused and distracted. For instance, a speaker getting up to say: 'Finally, I would like to end by saying that democracy is the ultimate form of governance...,' would hardly make a presentation sound convincing and emphatic. Even when a member is supposed to give a conclusion, it has to be well crafted and meticulously planned. Standing up as the last member in the crew and just throwing about a perfunctorily uttered concluding remark does not help. Such hasty and casual participation would only serve to harm the presentation rather than help it, as it sounds artificial and forced.

8. Plan well in advance regarding who is going to start, who is going to cover which points, who is going to focus on which aspect of the problem, and who is going to give a proper, authentic conclusion to the presentation.

9. Rehearse within the group so as to ensure that no member loses sight of the common thread that runs through a well constructed presentation.

10. Remember to use connectives; avoid abrupt transitions and sudden changes in thought.
11. Let the entire presentation develop coherently through different speakers in your group. Don't allow inconsistencies, contradictions, and dissonance of any sort to develop alongside the discussion.
12. Stick to meticulous time management. Identify the strengths and weaknesses of the group members, and share time within the group with intelligence and objectivity.

11.6 DELIVERING JUST-A-MINUTE (JAM) SESSIONS

Besides group discussions and personal interviews, just-a-minute (JAM) session, which is a form of presentation lasting only for a minute or so, has emerged as an important recruitment tool in today's professional world. Just like group discussions and job interviews, JAM presentations require professional skills to be acquired and demonstrated.

Since JAM sessions are short and crisp, they require a lot of imagination and presence of mind. The challenge in such presentations is to make the point immediately, without appearing shallow and perfunctory. Therefore, the speaker in such situations is required to be creative, prompt, relevant, and witty enough to capture the idea in an arresting and emphatic manner. Before we proceed to read examples of effective JAM sessions, following are the important points that are required to be borne in mind by the speaker when entrusted with the task of making a JAM presentation:

1. Start imaginatively; say something interesting and arresting to capture the attention of the audience right from the start.
2. Come to the point immediately; avoid long introductory ideas.
3. Stay focused on one point; avoid digressions.
4. Avoid examples and instances to substantiate.
5. Use vivid language; don't leave the audience guessing.
6. Avoid elaborate data and statistics as there is no time in JAM sessions for such substantiations.
7. Be witty, creative, and humorous.
8. Avoid great expectations from yourself as a speaker.
9. Avoid being panicky or nervous.
10. Manage your time well.

Let us study the following examples and learn how to make your JAM sessions interactive, interesting, and impressive:

Example I
Religion: An Obstacle to God?

Good Morning Ladies and Gentlemen, the topic of my JAM session is really intriguing and sounds paradoxical. It is so because many of us, while imagining God, imagine religion and its many forms. Suggesting therefore that religion rather than being a gateway to God is actually an obstacle to it certainly sounds quite strange. Viewed closely however, the statement appears to be absolutely correct. This is so because we have lost touch with the real sense of religion. In modern times, religion is reduced to either a divisive or a ritualistic formula. Most religions today, rather than emphasizing the eternal human values, try to stress their fixation for a particular doctrine, and therefore alienate the believers of other sects. Moreover, in the name of faith and belief, superstitions are blindly followed, and hatred and separation is induced in various segments of the society. Certainly, if religion has to be the gateway to God, it has to emphasize the universal values of love, compassion, tolerance, and non-violence as its central idea; otherwise religion will be reduced to dogmatic, superstitious, and parochial beliefs and hence would only impede us from discovering the ultimate reality, truth, or God. Thank you for your kind attention and patient listening.

Example II

My Purpose on This Mother Earth

Friends, I am Siddharth Arora and I would like to discuss with you *My Purpose on This Mother Earth*. Well friends, we come to this earth without seeking life and go from here without desiring death. In between, we have a time period called life, and to me the greatest challenge in life is to find a purposeful living on this mother earth. Most of our attention is focused on searching for a meaningful career and making materialistic and worldly progress in the world. I, for one, would like to concentrate more on a life of purpose rather than just achieving worldly milestones. Looking at the greatest crisis that has confronted us in the form of global warming, I would like to dedicate my life to saving human life on this earth. For this, I would first like to work in the direction of saving water and trees. After graduating from here, I am going to form an NGO or work for an organization that focuses on reinventing the ancient techniques used by our forefathers for water conservation. Further, I am quite aware of the fact that trees are as vital as water on this earth, and my endeavour would be to plant at least one thousand trees in my life time, and to be able to rear them too. Since such tasks cannot be performed by individuals, I would like to form or join a group of like-minded people, and work in the direction of saving our mother earth. I know I may not sound very practical, but I believe that life on earth is more important than my small life and a smaller professional growth. Of course, you can save a career, only when you are able to save life.
  Thanks for your patience!

Example III

Science: Man's Best Friend or His Worst Enemy?

Friends, asking a science student whether science is man's best friend or worst enemy is probably like asking someone whether they want to succeed or fail in life. Of course, all of them would like to succeed. Similarly, all of us here, the students of science and technology, know for sure that science is man's best friend. After all, science has worked wonders for human life. It has brought us unbelievable comforts, and has helped us achieve incredible milestones, not only on this earth but also in the firmament above us. The growth and development, the speed and connectivity, the luxury and enjoyment, the medical, transport, communication, and educational facilities that we enjoy today; all this has been possible to come our way only because of the newest inventions and discoveries made in the field of science and technology in the past half a century or so. In fact, you remove the word 'science' from man's dictionary and all his history would remain primitive and shorn of all achievements. Therefore, science certainly is man's best friend, but it is imperative for us to use it properly. Science gives us luxury but an unmindful bondage to it can make us lazy and slothful. Similarly, technology brings us growth and development, but misusing the same would certainly result in widespread destruction and calamity. So, it all depends on how we use it. And it is not only science; anything that you misuse, destroys you. In this context, I would like you to recall the old saying which suggests that science is a good servant but a bad master. So, enslave science if you want it to serve you well. Otherwise, if allowed to go out of hand, the same scientific inventions which were meant to comfort you, will start harming you.
  Thank you!

@   Refer to the Online Resource Centre for more examples on JAM presentations.

    In all presentations, hence, good research about the topic, vivid language, effective slides, substantial data, good use of wit and humour, a captivating beginning, and an emphatic ending create a lasting impact on the listeners. Eight important ways of making a presentation work is @ given in the Online Resource Centre in the section on professional presentations. Go through them to visualize how these eight best practices help in adding life to your presentations.

    For more discussion and examples related to this topic, refer to Chapter 14 on *public speaking*, which also highlights several other strategies which can be utilized for making presentations more effective and memorable.

RECAPITULATION

✓ Making presentations is a common, everyday affair in a professional's life. Hence, we need to be prepared for such a task.

✓ There are various types of presentations that we have to make in our academic and professional life.

✓ In all presentation situations, most of us experience nervousness, which with some conscious efforts we can not only overcome but also use to our advantage.

✓ Tasks such as professional presentations, public speeches, descriptions of an object, person, book, picture, etc. are routinely required both in your academic and professional career, besides just-a-minute (JAM) sessions, which have become an important means of evaluation in the recruitment process nowadays.

✓ Effective description requires penetrative perception into the finer aspects of the object being described. Besides, a controlled, and interesting way chosen to enunciate the points helps you achieve success and appreciation from the audience.

✓ In JAM sessions, alertness of mind serves you well. For this, you must have good reading habits and analytical skills. You can spice up such presentations by starting on an arresting note, by using wit and humour, and by employing forceful and vivid language. Moreover, while dealing with JAM sessions, try to stay focused on the issue being discussed; avoid digressions; and try to analyse the topic imaginatively.

✓ For making a presentation, we need to prepare well and practise well in advance.

WISEWELL QUIPS

EXERCISES

Critical Thinking Questions

1. 'Stage fright or nervousness helps us achieve better performance in professional presentation situations.' Do you subscribe to this view? Discuss and elucidate.

2. Discuss the points you would bear in mind while making a group presentation. Provide examples to substantiate your views.

3. As a part of your class assignment, you are required to discuss a book you have recently read. Write the complete text of your presentation. Invent the necessary details.

4. Assume that you have associated yourself with Nokia as its sales executive. The company has asked you to give a presentation on the latest mobile set launched by it, to a group of potential customers. Write the text of your speech, discussing the essential features of this mobile set and describing its functions, utility, and added features.

5. 'PowerPoint slides are used not just for decorative purposes; they must be functional.' Discuss and substantiate.

6. Imagine you have to deliver a team presentation on the topic 'Global Warming.' Prepare twelve slides on the topic related to its different aspects. The full text of the presentation should not exceed 300 words.

7. What tips would help you the most in preparing and delivering persuasive, powerful presentations?

8. Choose a handful of topics from the list given below. Prepare just-a-minute (JAM) presentations on them and perform all these presentations within a group of your classmates, friends, or family members:

   (a) My objective in life
   (b) To be or not to be
   (c) Attitude determines the altitude
   (d) Where the mind is without fear
   (e) My worst phobia
   (f) Life without water
   (g) Love is blind
   (h) War and peace
   (i) Freedom vs responsibility
   (j) Borderless world
   (k) Dynasty in democracy
   (l) Mission without vision

Group Discussions

Learning Objectives   After reading this chapter, you will be able to

- understand what group discussion is and how important it is in the selection process
- learn in detail about the various personality traits, namely awareness, initiation, body language, paralinguistic features, confidence, etc. that are assessed during group discussions
- familiarize yourself with the different types of group discussions
- understand how to have an effective opening and a proper conclusion for a group discussion
- learn how to perform as a team player and also emerge as a leader in a group

12.1 INTRODUCTION

Group discussions (GDs) are now being used as an important step in the selection of candidates both in private and government organizations. Regarded as an effective tool in the recruitment process besides job interviews, a GD plays a pivotal role in selecting the most suitable candidates from many who apply for the same post. It is also used as a tool to study the behavioural and attitudinal responses of the participants.

12.2 DEFINITION

A GD is a formal discussion which involves six to fifteen participants who sit in a group to discuss a topic or a case given for this purpose. It is a methodology used by an organization to gauge whether a candidate possesses certain personality traits and/or skills that are desired of him/her. It is like a football or hockey game where all the players pass the ball to their team players and aim for a common goal. In these games, the team which has better coordination and skills wins the game and so is the case with GDs. In GDs, the group members have to interpret, analyse, and argue, so as to discuss the topic or case threadbare as a team.

12.3 DIFFERENCE BETWEEN GD AND DEBATE

A GD is not the same as a formal debate. In a debate, you are supposed to speak either for or against a motion. In GDs, on the other hand, all the members of the

Group Discussion

group are expected to deliberate upon the issue extensively, and it is possible for any of them to change their stand if they find themselves convinced about the other side of the perspective. This kind of alteration in the stance does not find a place in debates where those who speak for the motion or against it prepare their argument well in advance, and the contestant is not supposed to argue for both the sides. Such is not the case in GDs, where the discussion just evolves naturally without anything to be proved, from the onset. The very nature of GD, therefore, demands flexibility on the part of the participants, and a lack of it, or a consequent stubbornness or rigidity is seen as a serious flaw in their personality.

12.4 NUMBER AND DURATION

In a formal GD, there are six to fifteen members in a group and they are asked to sit in a circular, semi-circular, or U-shaped seating style (senate room sitting). They may be familiar or unfamiliar to each other. They are given fifteen to forty-five minutes to discuss a topic or a case study depending on its nature.

12.5 PERSONALITY TRAITS TO BE EVALUATED

Following are the most important personality traits that a candidate should possess to do well in GDs:

- Reasoning ability
- Leadership
- Openness

- Assertiveness
- Initiative
- Motivation

- Attentive listening
- Awareness

12.5.1 Reasoning Ability

If you possess good reasoning skills, it helps you express your ideas and opinions in a convincing and rational manner. The golden rule is that when you present your ideas with proper reasoning and argument, you have a better score.

The following example serves to suggest how to rationally express your views on the topic *Democracy is the Best Form of Government*.

Friends, in my view, compared to any other form of governance, there is a greater possibility of peace, justice, and harmony in a democratic set-up. In democratic governments, people are connected with the system directly as well as indirectly at every level, and it provides maximum opportunity to people for development. Above all, the people in authority are answerable to the common people. I think that is why, a big portion of the world today happens to be under a democratic system of governance.

While participating in a discussion, use facts and figures to lend credence and conviction to your arguments, as is shown in the following example:

> Friends, I think it is wrong to assume that terrorism and unrest in Jammu and Kashmir is the problem of the last one decade alone. If you take a look at the statistics from 1985 to 2001, you can easily observe that even during those years, tens of thousands of innocent people were killed in such incidents. In fact, Pakistan once claimed that over 75,000 Kashmiris had been killed between 1985 and 2001. India holds that the total number of casualties was around 30,000. In any case, such a large number of casualties conclusively proves that the problem is of a chronic nature.

As you can see in this situation, citing statistics lends credence to what the participant has said. It also shows the speaker's general awareness. So, supporting your arguments with facts and figures always helps.

12.5.2 Leadership

An effective leader discusses the topic assertively by touching on all its nuances and tries to help the group reach the objective of the GD. Leadership in GDs is never pre-defined. It is through the person's performance that he/she emerges as a leader in a GD. A person aspiring for that, however, must display leadership qualities such as clarity, objectivity, perception, poise, and communication skills. So, a leader would be someone who facilitates discussion on a GD topic in a constructive manner. A leader shows direction to the group whenever the group drifts away from the topic. He/she coordinates the efforts made by different team members in the GD. He/she also contributes to the GD at regular intervals with valuable insights. He/she continuously inspires and motivates his/her team members to express their views. Being a mere coordinator in a GD does not help, because it is a secondary role. So, a leader should not only contribute to the GD with his/her ideas and opinions, but also try and steer the conversation towards a goal.

12.5.3 Openness

In GDs, you must be open to the ideas of others as well as to the evaluation of your own ideas—that is what flexibility is all about. For example, presume that the topic of a GD is *Military Services are Not for Women*. While discussing a controversial topic such as this, some participants tend to get emotional about the topic and take a stand either in favour or against the topic, that is, announcing 'Yes, women are not meant for military services', or declaring 'No, women too can contribute significantly even in military services.' Whatever stand you might have taken, if you encounter an opposition with a very strong point, you end up in a typical *catch-22* situation. If you change your stand, you are seen as a fickle-minded or a whimsical person. If you do not change your stand, you are seen as an inflexible, stubborn, and obstinate person. However, if you maintain a stand that is open and not averse to either side of the view, it will reflect your openness. In fact, denying the possibility of a change is always suggestive of dogmatism.

12.5.4 Assertiveness

You must put forth your point to the group in a very emphatic, positive, and confident manner. Participants often confuse assertiveness with *aggressiveness*. Aggressiveness is all about forcing your point on the other person, which can be a threat to the group. An aggressive person uses negative body language while putting forth his/her point, whereas an assertive person displays positive body language, both while speaking and listening to others.

12.5.5 Initiative

Participants have a tendency to start a GD to get initial benefit of the points. But that is a high risk–high return strategy. You should initiate a GD only if you are well-versed with the topic. If you start and fail to contribute at regular intervals, it gives the impression that you started the GD just for the sake of the initial points. Also, if you fumble or stammer, it may work against you. Moreover, a choppy, cluttered, and prejudiced beginning is a definite spoilsport.

12.5.6 Motivation

In order to exhibit good leadership skills, always try to encourage the inert participants. This will reflect your positive trait. It is seen that in GDs, participants are more keen to speak rather than listen to others. Similarly, many of them try to speak for most of the time. This, however, is a negative trait of one's personality. Remember, a leader and good team member not only participates in the discussion but also encourages others to do so.

12.5.7 Attentive Listening

You should listen carefully to others when they present their views. This will help you in two ways. First, it will help you understand the ideas presented and second, you can get your ideas analysed by others, which in turn enables you to critically ascertain their validation. Remember, it is only a good listener who can present himself/herself as a good speaker in a GD.

12.5.8 Awareness

You must be aware of the things that are happening around you, be it a political, religious, financial, or social development. As an educated person, you are required to be aware of that. Your awareness helps you provide proper examples, adequate facts, and proper analysis in GDs. Moreover, it is not about just being aware but also about having a view. So, it is your perspective of the issue and not just your being vaguely aware of something which actually matters in GDs.

12.6 DYNAMICS OF GROUP BEHAVIOUR/GROUP ETIQUETTE AND MANNERISMS

In any group task—be it a project or presentation or discussion—the behaviour of the group really matters. At times, the group members tend to have difference of opinions and go off the track now and then. It is important to keep the purpose, goal, or task in mind, and bring the discussion back to the stated focus. This is the shared responsibility of all, because as a participant in a GD, every member has to ensure that the discussion takes place in a smooth and proper manner. Logical ideas, poised demeanour, supportive attitude, balanced view, and team spirit are some of the most important ingredients of a successful GD. Besides these, however well behaved, a group without a leader is a rudderless ship that moves about perpetually without ever arriving at the destination. Therefore, the following section highlights the importance of proper group etiquette and leadership, both of which are crucial to the success of a GD. Following are some of the points you should keep in mind to exhibit your positive group etiquette:

Being friendly and approachable   It is important that your co-participants find you approachable and easy to talk to. People who are grumpy, haughty, or impassive are often left out in @ group activities. Take a look at the GD section in the Online Resource Centre of the book, and learn how a friendly approach helps you in group situations.

Remember, even with your non-verbal language you can send out the message that you care for what others have to say; that you are friendly and approachable.

Encouraging participation from co-participants   Only a cooperative and conducive environment will encourage healthy participation from the group members in a GD. What do you do when a participant does not contribute to the discussion at all or is dominated by other participants when he/she tries to say something? Would you ignore him/her or would you say something to encourage such a participant? Surely, one should try and encourage the silent participant to present his/her views.

Not hurting anyone   Humility is a virtue that is always appreciated. Even if you disagree with what a co-participant is saying, you should never rebuff or belittle the other person. Since such behaviour finds takers neither in GDs nor in real life, it is best to be humble, though assertive. Remember that you may have a difference of opinion, but that does not give you the right to raise your voice or ridicule others. GDs are held not just to test your language skills and general knowledge, but also to see how you tackle the varied situations. Your presence of mind, humility, tranquility, tolerance, and ability to adapt and respond to impromptu situations are also under a scanner in such group exercises.

Not being dominative or dismissive   Since a GD is meant to test your team skills and leadership qualities, apart from other qualities, it is good to support your ideas with facts, figures, data, and experiences. But do not try to dominate others or emphasize an idea beyond the point of tolerance. Also, don't dismiss another person's point of view to score points. Picture this situation where one person in the group tries to dominate another.

> *Participant*: I don't think global warming is going to kill us all. Those who create rumours, always try to exaggerate the facts. Even in the past, predictions have been made about the end of life on this very earth. And what can you do to tackle global warming? Can you stop moving in cars, flying in planes, running industries, using mobiles, air conditioners, and all that we cannot survive without? I don't think what you are saying is practical. I don't think we need to create such a fuss about global warming.

Here, you can feel that the speaker was really dismissive in his/her attitude. Now look how the same argument can be said in a better way:

> *Participant*: Recently, I read an editorial in *The Times of India* that said that global warming may not be as disastrous as it is projected to be. The author had also quoted instances when such predictions were made in the past about bigger disasters but nothing of that sort happened. That, of course, is one point of view. There are others who have a completely different opinion. Therefore, we simply cannot brush aside all this concern in a whiff of over-confidence. I think we need to carefully analyse how drastically we have to change our lifestyle, in order to be able to live up to the demanding adjustments and sacrifices, if at all we have to stem the tide.

As you can see, this participant speaks in a calm and collected manner and does not harp on his/her view. Neither does he/she ridicule or override any other participant's view, though it is clear that the position he/she has taken is different from others. Always remember, in a GD it is not about right or wrong arguments, it is about how convincing you sound and how well you are able to deliver your argument.

Avoiding emotions   Emotions are common and natural to human nature. However, it is not desirable to overplay your emotions, particularly in a GD. A surfeit of emotions makes a person irrational. It is best, therefore, to stay calm even when provoked. Do not make any personal comments during GDs.

Read the following example to observe how the participant in this situation loses her cool and reacts to her co-participant's attitude:

> *Participant 1*: You know, I agree with all this women's liberation thing, but you see, sometimes, women take this liberation thing a little too seriously. I mean, a baby needs her mother more than her father. So, a woman will have to compromise and may be even have to give up her job after she has a baby. Men can't be expected to give up their jobs and sit at home and take care of babies ... it doesn't help.

Here, another participant responds by saying the following in an intensely emotional and agitated tone. She responds thus:

> *Participant 2*: You are such a chauvinist! Do you think children are only the responsibilities of the mother and not the father? Tell me, what is wrong if men sit at home and look after babies? You think a woman's professional career is not important to her? What if she too wants to become rich, famous, and important in life? Why should she alone put a break in her career because of the baby?

This lady here is completely overcome by emotions, and because of this, she is not able to convey her views clearly. Let us see how similar views can be expressed in a more graceful and composed manner:

> *Participant 3*: I think just like men, women too crave for equality, economic independence, and recognition at the workplace. Their contribution in the workplace is as important as that of their male colleagues. And as far as babies are concerned, they are a huge responsibility of both the parents. Times have changed and there are a lot of cases where men have willingly agreed to be house-husbands and are taking care of the kids while the woman earns for the family. It is a decision which has to be taken by the couple as to who can afford to take a break from work or whether they could alternatively take a break from work, or whether help can be hired and both can continue to work. So, my friend, I would say that there is nothing like men can't take care of kids; and that it is only women who know how to do it. Times have changed and a lot of men around you are doing it already.

Avoiding peer discussion   A GD is meant to test your team skills. Hence, whatever information you intend to share, it should be addressed to the entire group and not to one or two members. Don't start talking or arguing with one or two members ignoring others, something that the two participants are doing in the following GD:

Participant 1: No, I don't think India is spending too much on defence. After all, it is the question of our sovereignty. We need to protect our boundaries.

Participant 2: And that should be done at the cost of keeping our people hungry! Do you know that the billions of rupees we spend in protecting our border can give food to millions?

Participant 1: So, what should we do? Allow neighbouring countries to enter our territories and capture our land? Make us their slaves?

Participant 2: So, we should protect our borders but let our people die!

Participant 1: Is it only because of expenditure on defence that our people go hungry?

Participant 2: Yes, of course, what else?

A heated debate like this must be avoided. Otherwise, it can create a bad impression about you and your co-participant. But sometimes it so happens that two people are so much in appreciation of each other's views that they forget that they are in a GD and there are other participants as well. Such lopsided expressions of admiration and depreciation should both be avoided.

It is seen at times that members lose sight of the original purpose and may get sidetracked. Arriving at a decision can thus be a difficult process for a group. Sometimes, groups resort to voting to decide issues. When possible, it is useful to work towards achieving a consensus or conclusion. Group members should respect each other, and each person should recognize the potential contributions made by the others. The two most common types of people affecting a group's effectiveness are the persons who dominate and the persons who remain silent. Each person should realize how he/she can contribute to the solution. It can be as difficult to get the quiet person to speak as it is to get the talkative person to talk less. It is the group's responsibility to help manage the group's dynamics; for example, to help the shy person contribute, and to help the dominant person make time for others to speak. It is because of such peculiarities of group etiquette that a leader's space is naturally created in a group.

Leadership   As suggested earlier, a group without a leader can be as wayward as a ship without a captain. Since in a GD no one walks in with a designated tag to be the leader of the group, anyone can emerge as the leader, provided he/she showcases some such qualities during the discussion. Now, what makes a leader? Of course, there can't be a defined formula to suggest what makes this magical concoction. However, there are certain attributes. Some of these are as follows:

- Clarity
- Objectivity
- Discernment
- Expression
- Composure
- Erudition
- Maturity
- Amiability
- Patience
- Motivation

12.7 TYPES

GDs are of two types:

1. Topic-based
2. Case-based

Topic-based GDs can be divided into three types:

(i) Factual topics   (ii) Abstract topics   (iii) Controversial topics

Factual topics   Mostly groups are given topics which are factual in nature. These are related to day-to-day socio-economic facts or environmental issues. For example, *Growth of Tourism in India* and *Higher Education in India* are factual topics.

Abstract topics   Abstract topics are given at the higher level. These are usually intangible in nature. You need to approach these topics with innovative and lateral thinking. For example, topics such as *Blue is Better than Green, All are Equal but Some are More Equal than Others, Money Makes You Poor*, etc. are some abstract topics.

Controversial topics   These topics are controversial in nature. Participants are bound to have divided opinions. These topics are given so as to observe the maturity level of participants on such issues. You should not lose your temper or give a narrow interpretation of issues being discussed. For example, you may be asked to discuss debatable issues such as *Reservation should be Abolished in India* or *Women are Unfit for Defence Services*.

Case studies-based discussion   These are real-life simulated situations. Usually, these involve some kind of problems which are to be resolved. The key to such topics is that there is no right or wrong answer, but your approach to the solution is highly important.

12.8 OPENING OF A GD

It is seen at times that some participants try to open the GD in a hurry. They think that the one who opens the discussion surely gets selected or gets better marks. This is a wrong notion. In fact, despite your zeal to start the discussion on a flyer, you should open the GD only when you have enough points to set it in motion. Any of the participants can initiate the discussion, but always try to speak and contribute as early as possible. Many GDs fail to develop because of poor beginnings or hasty endings. Many a time, a speaker initiates the discussion without realizing that he/she does not know enough about the topic. Take a look at how an opening sounds when the speaker begins without any preparation:

> *Participant 1*: Good morning, we are here to discuss the topic *Cloning should be Banned*. In this regard, I feel that cloning as such is not bad. I don't think it needs to be banned.

Don't you think the discussion has ended before it began? Consider this as an alternative beginning for the same discussion.

> *Participant 2*: Friends, good morning. We are here to discuss the topic *Cloning should be Banned*. As we all know, cloning is a technique of producing identical copies of cells or organisms from body cells. The first animal to be successfully cloned was Dolly, the sheep. In light of this we need to discuss points such as—Does cloning hold new promises for the future? If yes, what are they? If no, what are the major threats it may pose to humanity? Are the fears so great that cloning should be banned?

Here, the participant has not only introduced the topic by providing the background but also raised pertinent questions to steer the discussion. Given below is another such beginning that sets a GD moving appropriately:

> *Participant 1*: Good morning friends! We have been asked to discuss the topic *Quality is a Myth in India*. In other words, it means that we are unable to sustain our products and services against global

competition. Is the change in quality really happening in India? Are we not quality conscious? Is the growth in our foreign trade not a reflection of our quality goods? What is our mindset towards quality? We also need to see how far our Indian attitude of '*chalta hai*' lets people get away with substandard products and services. So friends! Today competition is forcing our industry to adhere to global standards, so we are getting there in terms of quality....

12.9 SUMMARIZING A DISCUSSION

Just as the beginning of a GD is crucial, so is its ending. If you plan to conclude the GD, keep it brief and concise. You should avoid raising new points. You should not state only your viewpoint. In fact, you should try to include the major points discussed by the whole group. Moreover, an abrupt ending in a GD is perplexing and annoying.

See how badly this discussion comes to an end:

Participant 1: So, we have covered all the points. (Looks around) Is there anything else? No? So, we can conclude that all of us here feel that women's lib is certainly a fib.

Don't you find the ending in this discussion to be choppy, unconvincing, and abrupt? Consider a better one:

Participant 2: Friends, I guess we have explored enough facts and instances which make us believe that though it is not easy to harmonize business and ethics together in modern times, the task is both possible and desirable with some efforts, sacrifice, and patience. On this note of optimism, I think we can bring the discussion to an end. Thank you!

While concluding, also remember not to reflect only on one aspect of the GD. It must incorporate all the important points that have come out during the GD. If the examiner asks you to summarize a GD, it means the GD has come to an end. Do not add anything once the GD has been summarized.

Take a look at this example on the topic *Formation of Telangana: A New State*.

Participant 1: Now it's time to sum up. At the outset of the group discussion we discussed the background of the problem. As we have discussed, way back in December 1953, the States Reorganization Commission was appointed to prepare for the creation of states on linguistic lines. The States Reorganization Commission (SRC) was not in favour of an immediate merger of the Telangana region with the Andhra state, despite the common language between the two. The Telangana people had a number of complaints. However, the Bharatiya Janata Party (BJP), promised a separate Telangana state if they came to power in1990. The Congress party MLAs from the Telangana region, supported a separate Telangana state and formed the Telangana Congress Legislators Forum. In another development, a new party called Telangana Rashtra Samithi (or TRS) was formed with the single-point agenda of creating a separate Telangana state. While tracing the history, we have also found that there was pressure

(Contd)

on the Congress party to create a Telangana state in 2008. On 9 December 2009, Mr P. Chidambaram, Union Minister of Home Affairs, announced that the Indian government has started the process of forming a separate Telangana state, and that a resolution would be introduced in the Andhra Pradesh assembly for this soon. However, some of us were also of the view that rather than creating a separate state, the Telangana people can be given adequate opportunities to grow. Finally, a majority of us are of the opinion that earlier the states of Haryana, Jharkhand, Chhattisgarh, and Uttarkhand were created, and when the formation of these states have helped and contained the unnecessary tensions, it is better that the government should do as promised.

Thanks!

You can observe that the above conclusion is quite apt, since it includes all the points that were discussed during the discussion. Moreover, the speaker has tried to talk about the point over which participants had a difference of opinion. The language used is also suitable for a formal occasion such as this.

Generally speaking, there are a few points which you should bear in mind while participating in a GD. We will look at these in the next section.

12.10 SOME TIPS FOR GROUP DISCUSSIONS

Dos	Don'ts
• Sit comfortably	• Be in a hurry
• Keep track of time	• Be silent
• Share time fairly	• Dominate vocally/physically
• Encourage participation from others	• Assume the role of the chairperson
• Rope in the reticent/diffident ones	• Be belligerent
• Listen to the topic	• Take extreme stance
• Organize ideas	• Look at evaluators
• Speak at the earliest	• Put up a lacklustre performance
• Exude ebullience and spark	• Be curt and dismissive
• Allow supporters to back your ideas	• Appear to be impatient/restless
• Sound cogent and convincing	• Indulge in peer discussion
• Avoid reproach	• Look stubborn/snobbish
• If derailed, bring it back to the track	• Move/shift excessively
• Look relaxed and comfortable	• Speak fast
• Be friendly and approachable	• Digress and deviate
• identify supporters/opponents	• Indulge in debate and altercation
• Maintain eye contact	• Get emotional
• Connect to the ideas of others	• Use slang
• Avoid skirmishes and heated debates	• Thrust greatness upon yourself
• Aim for a conclusion in the absence of consensus	• Be overawed by bulldozers
• Reveal and induce camaraderie	• Ever start your GD with a decisive, firm stand or a conclusion
• Feel and reveal keenness to share	• Throw all ideas at one shot
• Transcend personal choices	• Feel trapped or scared
• Take mental notes	• Appear immune or nonchalant
• Provide vital points	
• Steer the discussion smoothly	

RECAPITULATION

✓ Many companies and institutes are making GD the first criterion for screening the candidates for face-to-face interviews.

✓ Essentially different from debates, GDs are more flexible, natural, and spontaneous in nature.

✓ A GD is used for mass elimination. GDs are the best tools for assessing a candidate's communication skills and other personality traits such as interpersonal skills, time management, and decision-making skills.

✓ While participating in a GD, you should also use effective body language to communicate your ideas. You should try to speak at the earliest and contribute significantly.

✓ To score well, always try to substantiate your stand with proper arguments and facts.

WISEWELL QUIPS

EXERCISES

Concept Review Questions

1. What do you mean by a group discussion (GD)? Why are these so important for university students and professionals?

2. What are the major differences between a debate and a GD? Also discuss the different types of GDs that are used in the selection procedure.

3. Discuss the personality traits of participants that are evaluated in a GD.

4. What do you mean by group dynamics? While participating in a GD, what aspects of group dynamics will you keep in mind? Discuss in detail.

5. What are the different leadership styles that may emerge during a GD? Discuss in detail.

6. Read and prepare arguments for the following topics:

(a) To lead a successful life one should live in a hostel for at least a year.
(b) There is no right or wrong in life.
(c) Peace cannot be attained through violent means.
(d) Children of today are not the same as those of yesteryears.
(e) Corruption is a necessary evil in a democratic system.
(f) Television and computers are stealing the warmth and creativity of people.

7. Think and write the opening and closing for the following topics:
(a) In the present scenario, Gandhi and his principles have become irrelevant.
(b) What India requires is microchips and not potato chips.
(c) Art is better than science.

(d) Working mothers are better than mothers who are just housewives.

(e) For leading a good quality life, one needs to have spiritual awakening too.

(f) Quality is a myth in India.

(g) Smoking advertisements should be banned first before banning smoking itself.

8. Participate in a GD based on a topic. You will be given a twenty-minute duration for this discussion. The topic will be provided to you on the spot. You will be given five minutes to organize your thoughts before the discussion commences. A pool of topics is given below; try to prepare your arguments, opening, and closing for these topics:

(a) Sky is the limit.

(b) Laptops are better than computers.

(c) It is easier to speak than listen.

(d) Green is better than white.

(e) The importance of zero.

(f) World War III for water.

(g) Encroachment of media in personal lives of celebrities.

(h) Aggression has become a way of life.

(i) History has no relevance in school studies.

(j) Classical music and dance are on the verge of extinction.

(k) A borderless world is the need of the hour.

(l) Teachers can easily be replaced by computers in higher education.

(m) Vegetarian food is better for both physical and mental health.

(n) The Twenty-Twenty matches have killed the spirit of cricket.

(o) Life is like a blank page.

(p) Mobile phones are necessary evils today.

(q) The Internet has increased crimes all over the globe.

(r) It is better to have the depth and grandeur of the sea than the heights of mountains in life.

(s) It is better to be happy than successful in life.

Job Interviews

Learning Objectives After reading this chapter, you will be able to

- understand what a job interview is and its importance in the selection procedure
- familiarize yourself with the job interview process
- learn in detail about the various personality traits that are assessed during job interviews
- understand the different stages of job interviews that are held before recruiting a candidate
- get to know the different types of job interviews held
- learn the dos and don'ts for achieving success in job interviews

13.1 INTRODUCTION

MT College of Engineering and Management is quite a busy place these days like all other colleges. The students in their last semester are undergoing the placements. They are preparing their level best. Some of the students of the college have been picked by the companies, but some others have not been so lucky. Amit Dubey is also one of them. He has become depressed for he has not been selected by any of the companies. His academic record is very good. But why is it so that he is unable to make it in the interviews? He has already been interviewed by three companies. This failure is bothering him so much that he has become a diffident person and is also suffering from insomnia. However, the solution to the problem does not lie in getting disturbed. After knowing this, Siddhant, his senior, met him and tried to know about the questions that were asked and how Amit had responded. To Siddhant's surprise, the answers given by Amit were not all that bad. He analysed the whole situation and tried to make him understand the proper use of body language, how to improve his enthusiasm level, and develop the art of becoming assertive while answering. After a rigorous training of two weeks, he got through an interview in one of the multinational companies, with a handsome salary package.

Amit Dubey's case establishes the fact that in order to emerge successful in job interviews, some special preparation is required. There certainly are certain traits that are desirable in job interviews and some others that are not so. Getting to know what stands us in good stead in job interviews is quite important to us. However, before going further, let us understand what an interview is and how it is conducted.

13.2 DEFINITION

The word 'interview' comes from 'inter' and 'view'. 'Inter' means *in between*, and 'view' means *to see*. In fact, an interview is a process in which the employer gets an opportunity to see whether the candidate is suitable for the position vacant, and the candidate tries to prove that he/she possesses the desired skills and knowledge.

The crux of the matter is that your prospective employer is interested in you only if you have the desired skills, qualification, and knowledge. Therefore, before you appear for an interview, you have to make sure that you possess these. Further, it is crucial that you are well prepared, so that you can confidently stake your claims for the slated position. In fact, it would be quite worthwhile to understand the whole process of the interview so that you can prepare yourself for all its stages.

13.3 PROCESS

Even if many of us are not sure of how to perform during an interview, most of us are aware of the process. During the interview process, the employer attempts to determine whether the applicant is suitable for the job or not. In a typical job interview, there is a panel which consists of three to four people who sit on one side of the table and a candidate who sits on the other side. During a job interview, the interviewers ask the candidate questions about his/her job history, personality, work style, and other factors relevant to the job. For instance, a few common interview questions are 'Tell me about yourself', 'What are your strengths and weaknesses?', 'Why should we hire you?', 'How will you contribute to our company?', etc. All such questions aim at finding out your strengths, your motivation to work, and your suitability for the job. You will usually be given a chance to ask any questions at the end of the interview. These questions are strongly encouraged since they allow the interviewee to acquire more information about the job and the company, but they can also exhibit the candidate's confidence and strong interest in the company.

Though the interview process appears to be simple, it involves a lot of money on the part of the company. In fact, most of the companies want to select the candidates they are interviewing, simply because they do not want to repeat the costly affair. So, that gives the candidates an advantage, as the panel interviewing them is keen to see them succeeding and not failing. So, if you do reasonably well and showcase your potential appropriately, the chances for success are much higher than those of failure. At the same time, the companies do not want to select a wrong candidate, as righting a wrong would result in further wastage of time, money, and energy on their part. In many companies, therefore, you can observe that the duration of an interview is increasing, particularly for high ranked positions, and the interview process may involve analytical tasks, group activities, presentation exercises, and psychometric tests, besides the usual interaction between the candidate and the selection panel. Therefore, it is advisable to understand and be prepared, because interviews are becoming grilling and challenging with the passage of time.

Keeping the above mentioned facts in mind, the first thing you should do is to keep your résumé updated and practise the frequently asked interview questions. These simple exercises can be very effective and can go a long way in getting the most out of your interview. Broadly speaking, we can say that this is a three-step process which includes the following steps:

- Gathering information
- Establishing a rapport
- Closing and follow up

The interview process requires you to first gather information regarding the company where you are going to appear at an interview. It is essential for you to know the total turnover, the products and services, number of employees, its branches, work culture, and future plans of your prospective company. This will enable you to establish a quick rapport with the panel members. And finally, you need to be well prepared for the closing. It is possible that you might be asked to ask a question. Be ready with one, and think about your closing remarks when the interviewers tell you that your interview is over.

You should be able to achieve four objectives at the close of the interview:

1. Make it clear that you want the job.
2. Set the stage for the next step.
3. Create a final good impression.
4. Get an actual offer.

13.4 STAGES IN JOB INTERVIEWS

Before you appear for your first job interview, it is advisable to know the various stages through which your suitability for the job is assessed. These stages may include the following steps/ stages of screening:

- Screening of application
- Group discussion (GD)
- Appraisal of curriculum vitae (CV)
- Negotiations
- Competency tests/technical know-how
- Medical test
- Psychometric tests/aptitude test

13.5 TYPES

There are various types of interviews that are held for different fields and positions. Here are a few types which are held for middle level managers and fresh engineers.

13.5.1 Telephonic/Phone interview

A common initial form of interview is the telephonic interview. This is an interview conducted over the telephone or mobile phone. This has the advantage of keeping costs low for both the employer and the candidate. In case of a large number of candidates, this method is used as a tool for the first round of screening. Though a telephonic interview sounds easy when compared to a face-to-face interview, the task requires thorough preparation on the part of the candidate. In this type of interview, the candidate's voice plays a key role.

At times, telephonic interviews may finally decide a candidate's suitability for a position. Mostly, however, it is followed by other rounds of the interview which aim at finding out a candidate's suitability for the job from various perspectives. At times, these rounds may not be preceded by a telephonic interview at all.

13.5.2 Interview through Video Conferencing and Teleconferencing

Companies hiring at all levels, from entry-level to experienced professionals, are more comfortable using technology, thus making video interviewing and teleconferencing the preferred ways for the online recruiting process these days. For the job seekers though, it is often a little intimidating. It may appear to be a daunting proposition, even to those who are familiar with technology and use it on a regular basis.

Though teleconferencing and videoconferencing are used synonymously, yet there is a difference between the two.

Video conferencing	Teleconferencing
People who you are talking to are essentially present there on the screen and the interaction becomes more authenticated.	People who you are talking to are not necessarily visible to you and hence the communication is not similarly authenticated.
Videos and webcams are essential for a video conferencing to take place. In video conferencing interview, you either have to physically go to a place where company has established its videoconferencing facility or you may be asked to use your own facility such as webcam on your computer, laptop or smart phone.	Videos and webcams are not essential for teleconferencing.
During video conferencing, you may see people nodding their heads and feeling excited about what you are saying. Therefore, a continuous feedback is possible all through the interaction.	During teleconferencing, when persons on the other end stop saying anything, it can be nerve wrecking for you at times.
Video conferencing is more revealing. Even if we put it on mute, people in conference can observe each other's expressions, gestures and responses and the process of communication continues.	Once a teleconference is put on mute, the whole process of communication comes to a standstill.

Keeping in mind the fact that video conferencing is edging out teleconferencing, let us learn a few tips for preparing well which will help you to succeed in such interviews.

Tips for Video Interviewing

Advance planning

- Make sure that you send your résumé and any other supporting material required by the recruiter in advance.
- If the interview is at a company office, arrive early so that you have time to get settled.
- Ask for assistance if you're not sure how to use the equipment. Even if you think you can figure it out, it's good to ask for a quick overview.

What to wear

- Dress professionally. Wear the same formal attire as you would do for an in-person interview.
- Don't clad top half in formals thinking only your torso would be visible. You will feel embarrassed when you would require standing up for something and a full view of your attire is visible including the jeans you might be wearing.

During the video interview

- Make sure the table is neat and clean.
- Be aware that the microphone picks up all the noise in the room. Don't tap your pen or shuffle papers.

- Make an eye contact, otherwise you will not sound convincing
- Use the Picture-in-Picture feature so that you too can see how you appear on the screen.

The interview process

- The interview process will be the same as an in-person interview. The interviewer's objective (to screen candidates for employment) is the same.
- You will be asked the same type of interview questions. Also, be prepared to ask questions as well.
- If you're not sure about how the interview is proceeding, it's fine to ask the interviewer about how you are doing.

What's most important is to consider this type of interview just as important as if you were meeting the interviewer in his or her office. The value, for you as well as for the hiring manager, is equivalent, and a good, convincing performance is crucial to getting hired.

13.5.3 Technical Interview

This is an essential round of screening. In this part of the interview, the experts on the panel try to assess your knowledge in the subject domain. They ask you questions related to various fundamental concepts involved, their application, and your ability to relate your knowledge in other related fields. Look at a few technical questions:

Question: What is the name of the first clone? Do you think cloning is dangerous for mankind?
Answer: It was Dorset sheep which was named Dolly. This was created in Roslin Institute, Edinburgh, 1992. This new genetic technology was a major breakthrough, but surely unrestricted cloning is dangerous for mankind for obvious reasons. The probabilities of its misuse are more serious. The responsibilities of children and parents, the concept of identity, etc. will become the points of bitterness. It will collapse the social order and will give rise to many legal problems. So I feel cloning can do more harm than good to society.

Question: How do aeroplanes fly?
Answer: Sir, in the flying of an aeroplane, the theorem of Bernoulli is applicable, and by virtue of the drag and lift force it goes in the air.

Question: What do you mean by the term 'lead bank'? Is it anyway applicable in the industrial area?
Answer: Sir, this is a concept related to area banking. Each bank is specified a district in which it has to survey the district and prepare a development programme, including the credit scheme and other requirements. Coming to the second part of the question—yes, it exists in the industrial area as well. When two banks jointly fund an industrial project, the bank which has the major share is considered to be the lead bank.

13.5.4 Behavioural Interview

A common type of job interview in the modern workplace is *behavioural interview*. This type of interview is based on the notion that a candidate's past behaviour is the best indicator of his/her future performance. In behavioural interviews, the interviewer asks the candidates to recall specific instances where they were faced with a set of circumstances, and how they reacted. Typical behavioural interview questions are usually worded like this:

1. Tell me about a project you worked on where the requirements changed midstream. What did you do?
2. Tell me about a time when you took the lead on a project. What did you do?
3. Describe the worst project you worked on.

> **Question**: Describe a time you had to work with someone you didn't like.
>
> **Answer**: Sir, everyone has his/her own likes and dislikes. However, one has to work even when you do not like a person because of certain things which you do not appreciate in him/her. This situation calls for tremendous patience and understanding. This happened with me when I was to give a team presentation with a classmate who had a laid-back attitude. This was a real challenge for me because I am a hard worker and I am quite committed to my work. This gave me an opportunity to understand him well. That teammate was actually sharp and intelligent but lazy and complacent. So I started to get his inputs by having discussions with him and then by sitting with him. Then he used to feel like working. In fact, we delivered an excellent presentation and I found a friend in the person who I did not like sometime back.

13.5.5 Stress or Skeet Shoot Interview

The candidate is asked a series of questions by panelists in rapid succession to test his/her ability to handle stress filled situations. You need to be mentally alert since you are asked more than one question at a time. You should stay calm during such sessions. Stress interviews might involve testing an applicant's behaviour in a busy environment. Questions about handling work overload, dealing with multiple projects, and handling conflict are typical.

Another type of stress interview may involve only a single interviewer who behaves in an uninterested or hostile manner. For example, the interviewer may not make eye contact, may roll his eyes or sigh at the candidate's answers, interrupt, turn his back, take phone calls during the interview, or ask questions in a demeaning or challenging style. The goal is to assess how the interviewee handles pressure or to purposely evoke emotional responses. Given below are a few sample questions of a stress interview:

> **Interviewer 1**: I have gone through your dossier and found that you have done two courses, one on technical communication and the other on mass communication. Now tell me what technical communication is and how it is different from mass communication.
>
> **Interviewer 2**: Also tell us whether you find the application of technical communication in some way in the area of mass communication.

13.5.6 Psychometric/Aptitude Test

In order to judge a candidate personally, sometimes, psychometric tests are administered. Gradually, the test is becoming a part of the whole selection procedure in interviews. In this test, almost fifty to sixty questions are asked to assess whether the candidate has the desired aptitude and knowledge. If the candidate clears this round, he/she is asked to appear for a GD round, and finally an HR (human resource) round.

13.6 DESIRABLE QUALITIES

While appearing at job interviews, the prospective candidates must aim at reflecting the following traits:

- Clarity of thought
- Presence of mind
- Balanced point of view
- Cool composure
- Logical thinking

- Maturity
- Sincerity
- Openness
- Capacity to conceptualize
- Good understanding of fundamentals

13.7 PREPARATION

Any fact facing us is not as important as our attitude towards it, for that determines our success or failure.

–Norman Vincent Peale

Preparation for a successful job interview requires the candidate to do the following:

13.7.1 Know the Company

Researching a company about its products and services is essential before you go for an interview. This can easily be done by browsing the site of the company, by going through its brochures and report, and by getting to know the value of its shares and debentures.

History Gather information about the company you want to work for. Visit the company's website and talk to anyone you might know who works there or had worked there. Find out details such as: What products or services does the company offer or sell? When did it start? Who are its promoters? What is the total strength of employees in the company? What is its position in the market? Who are its major competitors in the market?

Projects undertaken What are its significant projects? What kind of benefits does it reap from those projects?—These are certain questions that the candidate should seek answers for.

Growth The candidate should try to find out details such as: What are its growth prospects in future? Does it have any plans to expand in the near future? What is its growth rate?

It is important to know about the company because if the candidate does not know anything about the company, it reflects his/her lack of preparation for the interview as well as lack of enthusiasm to associate himself/herself with the company.

13.7.2 Know Yourself

Before you set foot on your job-hunting expedition, take some time to know yourself. The more self-aware you are, the more confident and comfortable you will be in job interviews.

Strengths and uniqueness While preparing for an interview, you should always try to know your own strengths and weaknesses. In the following situation, you can see how impressively the interviewee responds.

> **Interviewer**: What is your greatest strength?
>
> **Interviewee**: Sir, I think my abilities to motivate people and to adapt with changing circumstances are my key strengths. These qualities actually help me lead a team and accomplish tasks within the stipulated time.

As you can see from this example, you need to know about yourself. For this, you need to think of the possible answers for the most frequently asked questions. Read the following example carefully.

> **Interviewer**: What is your major weakness?
>
> **Interviewee**: I sometimes go into greater details of the task assigned to me which makes the task a little unmanageable at times. But, lately I have realized that I need to keep track of both my ideas and the time allotted for the task, and I am working towards being better at time management.

Blurting out Your Answers Unmindfully can be Devastating to Your Prospects

Here, you can see a well thought out answer. The candidate is able to tell the interviewers about one of his/her qualities even when he/she talks about his/her weakness.

Competitive advantage It is essential that you make a good lasting impression on the people who meet you at any job interview. Showing yourself as qualified for the job is one important element, but you need to win the person over and ensure that he/she is also impressed, both by your knowledge and your personality.

To be a great champion, you must believe that you are the best. If you're not, pretend you are.
—Muhammad Ali

13.7.3 Review Common Interview Questions

There are a few questions which are invariably asked in most of the interviews. A good technique is to write out your answers to the questions you anticipate and then read your polished answers out loud, over and over. Go through mock interviews with the help of your friends. Most questions will relate either to your ability to do the job or to the type of employee you will be.

Some such questions are given below. For these questions, try to find out and craft your own answers. Do not try to imitate others. A few questions have been answered, and the purpose behind these questions have also been discussed briefly to guide you, so as to help you get to know how to prepare answers for the frequently asked questions on your own.

1. Tell me about yourself.

Approach: This is the most often asked question in interviews. You need to have a short statement prepared in your mind. Be careful that it does not sound rehearsed. Limit it to work-related items unless instructed otherwise. Talk about things you have done and jobs you have held that relate to the position you are interviewing for. Start with the item farthest back and work up to the present.

2. What is your greatest strength?

Approach: This is a great chance to highlight your best skills. Don't pick just one; focus on your top three or four strengths. Some of the qualities you can mention are persistence, dedication, punctiliousness, commitment, leadership skills, team-building skills, and organizational skills. Determine which strengths would fit best with the position for which you are applying.

Answer: Sir, as far as my strengths are concerned, I'm good at organizational skills, prioritization, and time management. But my greatest strength is my ability to effectively handle multiple projects within deadlines.

3. What is your greatest weakness?

Approach: Be careful with this one. To stand out, be more original and state a weakness, and then emphasize what you have done to overcome it. Be sure the weakness you talk about is not a key element of the position.

Answer: Since I am a hard worker, I've had trouble delegating duties to others. The reason being, I felt I could do things better myself. This has sometimes backfired because I ended up with more than what I could handle and consequently, the quality of my work would suffer. But I've realized this lately and attended courses in time management and learned effective delegation techniques, and I feel I am able to overcome this weakness.

4. Are you a team player?

Approach: You are, of course, a team player. Be sure to have examples ready. Provide the example that shows you often perform for the good of the team rather than for yourself to establish your team attitude. Do not brag.

Answer: Sir, I am a good team player. I was part of a team of six members for the Automated Traffic Control System project which was to be demonstrated at the IIT Mumbai Tech Fest. This was ready well in time, but the system did not work an evening before when we were supposed to leave for Mumbai. As a first reaction, all of us felt dejected and low because we checked everything but could not trace the problem. Our project guide was out of town. We were helpless. Then I called up my uncle who is a mechanical engineer in LNT and explained the problem. He guided us in that crucial hour. We worked throughout the night and finally we could demonstrate it. I mean, I believe that the task that the team has been assigned or taken up should not suffer even if I have to walk an extra mile.

5. Explain how you would be an asset to this organization.

Approach: You should be keenly looking forward to this question. It gives you a chance to highlight your best points as they relate to the position being discussed. Give a little advance thought to this relationship.

Answer: Sir, I will definitely be an asset to your organization, because I am a person who possesses skills both in networking and banking, which are the major requirements of this position. Moreover, the experience that I gained during my industrial internship in JP Morgan will help me do my work efficiently. So, in this way I will contribute significantly for the growth of the organization.

6. Why should we hire you?

Approach: Point out how your qualities and skills meet what the organization needs. Do not mention any other candidate's name to make a comparison.

7. Why do you want this position?

Approach: Here is where your research about the company will help you stand out among the other candidates. Explain how you have always wanted the opportunity to work with a company that provides a vital public service and leads the industry in innovative products. Explain how your qualifications and goals complement the company's mission, vision, and values (use specific examples).

Answer: Sir, I have gone through the job profile that was mentioned in the advertisement. I understand that you are looking for a person who is an expert in both networking and banking. Since I have done a couple of courses in banking, and computer science is my discipline, I know networking very well. I shall be able to apply and expand on the knowledge and experience I've gained during my internship, and will be able to increase my contributions that will add value to the company.

8. **Where do you see yourself five years down the line?**

Approach: Don't tell them that you want to be where they are sitting right now—the interviewer may feel threatened. Also, don't tell them that by that time you would be renouncing all your worldly pursuits and would set out on a search for truth. This is an opportunity for you to demonstrate your long-term planning capacities. They are asking about your career aspirations. Tell them that you see yourself in a role in which you will be handling more responsibilities effectively and capably, because the current job will provide you with a lot of learning and experience to do so.

List of questions asked frequently during interviews Following is a list of questions often asked by interviewers during interviews:

1. Tell me about yourself.
2. Why have you applied for this job?
3. What do you know about this job or company?
4. What are your major strengths?
5. What is your greatest weakness?
6. What type of work do you like to do best?
7. What motivates you to do your best on the job?
8. What do you feel has been your greatest work-related accomplishment?
9. What are your interests apart from your professional interests?
10. What accomplishment gave you the greatest satisfaction?
11. How does your education or experience relate to this job?
12. Do your skills match this job or another job more closely?
13. Tell me about your ability to work under pressure.
14. Where do you see yourself five years from now?
15. How do you handle stressful situations?
16. Describe a time you had to work with someone you did not like.
17. Describe the worst project you worked on.
18. Are you willing to put the interests of the organization ahead of your own?
19. Tell me about a time when you had to deal with a co-worker who was not doing his/her fair share of the work. What did you do and what was the outcome?
20. What is the toughest problem you've had to face, and how did you overcome it?
21. Describe your management style.
22. Do you have plans to go for higher studies? If yes, when do you want to do it?
23. Have you appeared for any of the competitive exams such as GRE or CAT?
24. How long do you plan to stay with us?
25. In case you get a job in a company which offers you a higher package, what will you do?
26. You have witnessed economic depression recently. Suppose it occurs again, how will you help the organization in that crucial hour?
27. How do you define success?

28. Who is an effective leader according to you?
29. What can we learn from China's economic reforms that have taken place recently?
30. Are you willing to work overtime? Nights? Weekends?
31. What are your hobbies?
32. Where do you see yourself ten years down the line?
33. Why are you here?
34. What do you know about corporate social responsibility (CSR)? How can our company go for it?
35. According to you, which colour is better—red or white, and why?
36. What is the major difference between a manager and a leader?
37. Are you a manager or a leader?
38. What can you do for us?
39. What kind of person are you?
40. What salary do you expect?
41. What distinguishes you from nineteen other people who have the same skills as you have?
42. Can I afford you?
43. Do you read newspaper every day? If yes, what is the major news item today?
44. Who is your role model other than your family member(s)?
45. What motivates you to work?
46. How do you handle an ethical dilemma?
47. Tell us a time when you had to finish too many things at a time and how you went about doing that?
48. We would like to know your views regarding the naxalite problem in India?
49. Why did you choose this discipline particularly?
50. Imagine that you are not lucky enough to get this job and fail in the subsequent attempts, how will you take it and what will you do then?

13.7.4 Prepare Questions You Want to Ask the Interviewer

Employers are as interested in your questions as they are in your answers. Ask intelligent questions. This is your opportunity to separate yourself from the other interviewees. Here are a few examples of some smart questions you may ask the interviewers.

1. If you hire me, what would be my first assignment?
2. Would you please tell me about the people I will be working with?
3. Other than yourself, who else is involved in the hiring process? Is it possible for me to meet them today?
4. What are the prospects of my growth in the company?

Most importantly, you should learn from your mistakes. If you don't get an offer from this company, you will succeed another time. Do not allow rejection to defeat you.

13.8 USING PROPER VERBAL AND NON-VERBAL CUES

Using effective non-verbal communication techniques is essential for you to get your dream job. It is believed that over 90 per cent of the message you send during your job interview is through non-verbal cues. However, you need to effectively communicate your professionalism, both verbally and non-verbally. If you come to an interview reeking of cigarette smoke, chewing gum, too much perfume, or not enough deodorant, it will not help you succeed. Not being dressed appropriately, not having proper leather shoes, talking on your cell phone, or listening to an iPod while waiting to be called for the interview may also prove fatal.

What is important when being interviewed is to appear professional and attentive throughout the interview process. Here are a few verbal and non-verbal tips required before, during, and after an interview:

1. Before you leave for the interview, make sure you are dressed professionally, neatly groomed, and your shoes are well polished.
2. When you are entering an interview room, it is always better to walk with your head up to show your confidence.
3. When you enter the interview room, shake hands with your interviewer(s). Your hand-shake should be strong and firm. A weak, limp handshake signifies nervousness and lack of enthusiasm.
4. Sit up straight with your hands relaxing completely and lean slightly forward in your chair to exhibit your confidence as well as interest.
5. Don't sit on the edge of your chair. It shows that you are tense.
6. Maintain an eye contact with the interviewer or interviewers while answering questions but don't stare at them constantly.
7. It is also essential to have proper eye contact while your interviewer is speaking to you, it will ensure that you are listening to and understanding him/her.
8. Don't forget to smile occasionally since it will help you show your enthusiasm and interest in the job.
9. Don't move your legs a lot. It is distracting and shows how uncomfortable you are.
10. Do not rest one leg or ankle on top of your other knee. It makes you look too casual or even arrogant.
11. Avoid speaking in a monotone; express yourself all the time by bringing variation to your tone and pitch.
12. Ensure that your voice does not sound apologetic or nervous.
13. At the end of an interview, stand up and shake hands while you thank the interviewer for the opportunity.

13.9 EXHIBITING CONFIDENCE

It is very common to be nervous before an interview. It is fine to be nervous. You are nervous because you want to do well. Therefore, being anxious can raise your energy level and that is a good thing.

Lack of confidence will mess up your chances of getting the job. Make sure that you exhibit confidence when you are being interviewed for a job. The interviewer will take your lack of confidence as a sign that you may not be able to handle the job.

When your name is called, walk in confidently and briskly; this shows confidence and enthusiasm. Extend your hand first; this shows a genuine interest in the person you are greeting. Shake hands firmly with interviewers and introduce yourself in a courteous and confident manner. Tell them that you are excited about this opportunity to share your job skills with them. You should appear assertive during your job interview so that you can ace it. Keep your eyes and mind focused and also try to keep a charismatic facial expression with a gentle smile. If you believe in yourself, it will be easier for you to be yourself in the interview. If you rehearse your answers, you will feel more confident. Record your mock interviews, watch yourself in front of a mirror, or get a friend to practise with you.

And finally, practise, practise, practise—it will make a difference.

13.10 TIPS FOR SUCCESS

Following are the dos and don'ts for the success of an interviewee.

Dos

1. Find out about the company.
2. Practise.
3. Greet interviewers enthusiastically and sit comfortably.
4. Dress smartly to make a good first impression.
5. Be mentally alert.
6. Stay positive.
7. Focus on what you have to offer, not what you want.
8. Appear confident.
9. Be prepared to ask the interviewer questions.
10. Thank the interviewers before leaving.

Don'ts

1. Don't tell lies.
2. Don't blame your circumstances.
3. Don't find faults with your earlier employer or company.
4. Don't make tall claims about your skills.
5. Don't fidget about in your chair.
6. Don't use vocalized pauses while answering.
7. Don't look down or make furtive eye contact with interviewers.
8. Don't bluff about issues you are not aware of.
9. Don't keep on simpering sheepishly or answer emotionally.
10. Don't exhibit your nervousness.

@ Watch the video on job interviews in the Online Resource Centre of the book for a better understanding of the above stated points.

RECAPITULATION

✓ A job interview is considered one of the most useful tools for evaluating potential employees.

✓ It gives a chance to the potential employee and employer to learn more about each other.

✓ If you are interested in getting a good job, they are also looking forward to getting a suitable employee for their firm. Therefore, how you present yourself becomes very important. So, be well prepared and brush up your subject and general knowledge. Sending the right message during your job interview is extremely important for your success.

✓ Using effective non-verbal communication techniques is essential for you to get the desired job. You need to dress appropriately and limit your make up, perfume/aftershave, and trinkets so that you look smart, clean, and well groomed.

✓ You should maintain eye contact, as it is essential to use appropriate body language to establish a rapport with your interviewer.

✓ Moreover, speaking in a clear and controlled voice is necessary to convey confidence.

✓ Above all, practise your answers for frequently asked questions so that you face the interview with full confidence.

WISEWELL QUIPS

EXERCISES

Concept Review Questions

1. What is a job interview? Discuss the process of a job interview in detail.

2. Imagine that you are going to sit for an interview for JP Morgan for the position of middle level manager. Along with other frequently asked questions, prepare answers for the following set of questions.
   (a) Tell me about a time when you took the lead on a project. What did you do?
   (b) Describe the worst project you worked on.
   (c) Describe a time you had to work with someone you did not like.
   (d) Tell me about a time when you had to stick to a decision you had made, even though it made you very unpopular.
   (e) Give us an example of something particularly innovative that you have done, which made a difference in the workplace.
   (f) What happened the last time you missed the deadline of a project?
   (g) Have you ever witnessed a person doing something that you felt was against company policy? What did you do then and why?

3. How far is body language important to succeed in an interview? Discuss in detail.

4. Your placements are going to commence next month. What preparations will you make to get through the job interview?

5. Write short notes in about 150 words each on the following:
   (a) Technical interview
   (b) Eye contact during an interview
   (c) Psychometric test
   (d) Skeet shoot interview
   (e) Phone interview

6. Discuss the ways, adopting which, you will exhibit confidence during an interview?

7. Write short notes on the following points to get ready for your interview to be held next week:
   (a) Your strengths and weaknesses
   (b) Your academic achievements
   (c) Your extracurricular activities
   (d) Your hobbies
   (e) Your unique selling proposition (USP)

8. Discuss the various qualities of a candidate that are evaluated during an interview.

9. How will you combat your nervousness before and during an interview?

10. Discuss the interview process and the various stages that a candidate has to face for getting a job.

Public Speaking

Learning Objectives   After reading this chapter, you will be able to

- figure out the nuances of public speaking tasks as well as learn about other modes of addressing the audience
- understand and use the different patterns for structuring speeches
- recognize and utilize the methods for delivering your speeches according to the requirement
- explore ways to make your speeches interesting and captivating
- learn how to deliver different types of speeches

14.1 INTRODUCTION

You are a final year student and you have been asked to address your juniors. What do you do? Do you straightaway go to the podium and start speaking without any prior preparation? No. For all such occasions, you need to prepare well, before you start addressing others. Preparation is one of the features of public speaking, which relates to addressing a group of people on such formal occasions. Public speaking, like any other professional task, requires planning, preparation, and perfect execution of ideas. All this requires a lot of hard work and patience on the part of the speaker.

Before discussing the essential features of public speaking, and learn the ways to make it interesting, purposeful, and rewarding, let's understand other modes of addressing a large audience.

Storytelling   Passed on to humans through generations, storytelling has always been regarded as a wonderful tool to communicate with others. Regardless of our academic qualifications, each one of us has some storytelling skills but in order to be an interesting storyteller, this skill needs to be sharpened to create an impact on the listener.

Though apparently a routine, everyday activity, storytelling helps us develop our imagination, creativity and expression. It also develops our powers of description and narration, and teaches us how to be captivating in our expression so as to engage the audience in an interesting and imaginative manner. Since making the audience listen is a challenging task, storytelling can build in us the essential capability to hold the audience while we speak to them. Besides these, it also equips us with the skill of appreciating others. Through effective story telling technique, we gain more confidence and self-esteem, as well as learn to empathize with people and enjoy the world of simplicity. Stories can be very long or very short but regardless of its length, it should be relevant and interesting to hear.

Now, let us learn a few tips for developing the art of effective storytelling:

- Before beginning to narrate a story, look at the audience with a welcoming smile and bright eyes.
- Try to create an atmosphere, like casting a good spell. Set the scene for your audience. Start with the time, place and weather of the story.
- Maintain proper volume and bring adequate modulation.
- Involve yourself with the task wholeheartedly. Begin to live the characters and situations to intend to describe or narrate.
- Use facial expressions, reflecting the feelings of the imagined characters, their nature or personality.
- Use some of the techniques employed in the role-play exercise with proper dialogue delivery.
- Vary the speed, pace and volume of your voice where appropriate. Make your voice melodic and interesting as per the situation, helping the audience feel involved with the characters.
- Speak more slowly and loudly than normal, so that everyone can hear. This will help you keep them attuned to your story.
- Use your hands, shoulders and body as much as you can to show the shapes of objects, scenery, actions and feelings. Use mime and gesture to "paint the story," like a picture.
- While narrating the story to kids, use other sounds, for example, weather sounds like wind or rain sounds; explosions or rustling sounds; animal sounds; emotional sounds, such as sighs, sobs, yawns, etc.
- Leave a space between your words or sentences sometimes in order to create an atmosphere of suspense and anticipation.
- Collect stories from magazines, books, films, videos, TV, and people, besides your own experience and imagination.
- Since storytelling helps us improve our comprehension; adds words to our vocabulary, develops language and expression; and increases confidence; read humorous, witty, suspenseful and well written stories, plays, novels and descriptions by authors such as George Orwell, Oscar Wilde, Katherine Mansfield, R.K. Narayan, Leo Tolstoy, O' Henry, Rabindranath Tagore, Somerset Maugham besides the classics by Charles Dickens, Jane Austen, William Shakespeare, G.B. Shaw, Prem Chand, etc.

Elocution Elocution stands for the art of clear and concise manner of speaking, with clarity of meaning and thought. Elocution originates from the word 'eloquence' which stands for fluent, elegant, or persuasive speaking. It refers to one's power of expressing strong emotions in striking and appropriate language, with a view to influencing them to one's ideas.

Effective speech has deep roots in elocution as it includes pronunciation, accent, grammar, tone and gestures which play a key role in forming a meaningful and emphatic message. Elocution has been considered a key aspect of learning the art of communication. In fact, poor and unintelligent speech makes the situation unpleasant for both the speakers and their listeners whereas elocution empowers the speakers with apt expression and enunciation which help them engage their audience keenly and create a positive impact.

Difference between Public Speaking and Elocution

Though employed in public speaking situations quite frequently by speakers, elocution in itself stands for the art of enunciation and power of effective delivery. The basic difference between the two is in terms of the content of the speech. This means that a speech is mostly written by the speaker himself or herself, whereas in elocution, the content may not be original. Moreover, public speaking includes only speech making whereas elocution includes poetry recitation, prose reading, dramatic presentation of a speech, and rendering of a monologue or soliloquy in a play.

Following are some tips for developing the art of elocution:

- Enunciate your part crisply and precisely.
- Speak naturally with well-controlled and modulated voice.
- Put yourself in the situation and emotions of the character you intend to live.
- Combat your fear of public speaking and be enthusiastic about the task at hand.
- Acquire right pronunciation, stress, and inflections in the speech by listening carefully to the trainers and teachers and by referring to the websites devoted to such material.
- Develop good reading habits as it broadens our perspectives.
- Show a keenness to compere programmes organized by your institute/organization from time to time.
- Learn proper use of gestures, stance, and dress in public speaking performances.
- Test the microphone before beginning to speak.
- Read aloud your part repeatedly. Eventually, it is the practice that takes us to perfection.

Suggested topics for Elocution

- Why I am glad to be/ not to be …
- Recite 'Where the Head is Held High' – a poem by Rabindranath Tagore
- Life of a blind/deaf/disabled, etc.
- Life of an insect/animal/fish/plant, etc.
- Life of an actor/ actress/ political leader/ sports person, etc.
- Life on another planet
- Antony's Speech in *Julius Caesar* by William Shakespeare
- What I would change if I could and why …
- Recite the poem 'Road Not Taken', and 'Stopping by Woods on a Snowy Evening,' by Robert Frost
- Speeches by famous speakers such as Pt. Jawaharlal Nehru, Winston Churchill, Martin Luther King, Bill Clinton, Barack Obama, etc.
- Unequal Educational Opportunities
- Cultural Misunderstanding
- Cultural Decadence
- My mother/grandmother/older sister/aunt always says …
- I'm Enthralled by …
- A one-eyed man is the king amongst blind men

- A cat in a cage becomes a lion
- A book is like garden carried in a pocket
- Internet: Modern day god of all information

Extempore   Extempore refers to the task of delivering a talk or address without much preparation. It tests the speaker's presence of mind, clarity of thoughts and command of language. At times, a minute or two are given to the speakers to arrange their thoughts before speaking. In other situations, one is required to speak without any preparation.

Though seemingly a challenging task, participation in extempore can be a rewarding and entertaining experience. Saying words and phrases in a short span of time is where we can showcase our confidence and the capacity to articulate and think. Extempore enables speakers to think out of the box and provides an opportunity to develop both communication skills and time management. It pushes one to envision and produce ideas without any preparation, as well as improves one's logical thinking and reasoning ability. It is an excellent training exercise for future leaders.

Let us look at some easy and workable tips to develop an effective extempore speech making ability:

- Be a good reader; this will help you develop your thoughts and feelings on various issues.
- Feel excited by the task and not get encumbered with it.
- Think about the first couple of sentences which can immediately capture the attention of the audience.
- Since expressions do matter, express your understanding about the topic as best as you can. As this activity is stage-oriented, remember to render best impression on your audience.
- Use your presence of mind, wit and humour, and stay confident while you speak and deliver your speech.
- Since body language plays an important part in all public speaking situations, use your non-verbal ability effectively by looking into eyes of the audience, using appropriate hand movements, displaying an occasional smile, keeping the head straight and body upright, and maintaining required distance from the microphone.
- Voice depicts a person's character, mood, feelings, emotions, and tone. Bring variation in the pitch pattern according to the message you are delivering. The message delivered and manner of utterance should match well to create the desired impact.
- Along with competence, enthusiasm, and delivery, you must keep 'time' in mind. Delivering quality and valuable speech in the specified duration is the best rendition that the audience generally expects and appreciates.
- Above all, practising extempore before you participate in the actual competition would be very helpful.
- These days, many companies use extempore as a recruitment tool to judge a candidate's presence of mind, communication skills, expression, and confidence level. Therefore, participating in such events in your college and university is likely to bring rich rewards in times to come.

14.2 CHOOSING AN APPROPRIATE PATTERN

As a speaker, the first thing that one has to do is to structure one's speech. Before speeches are delivered formally, the speaker has to decide the pattern in which the idea will be put across to the audience. There are some widely recognized patterns, and while planning to make a speech, you may choose any one of these or use some of these. Let us look at the finer aspects of all such patterns in some detail.

14.2.1 Chronological Pattern

The chronological pattern is one of the most commonly employed patterns for organizing a speech. In this pattern, we arrange ideas, keeping in mind the chronology of events. For example, if we are required to give a speech on 'The History of Indian Cricket', 'The History of Mughal Emperors', 'The Development of Psychology', 'The Progress of Democracy in the World', etc., we are likely to choose a series of events and speak about their development over a period of time. Whenever you choose to utilize the chronological pattern, you are required to match the sequence of events and time. The speeches organized through this pattern require a clear link to be established between the events and their time sequence. Often, the style of presentation structured in this pattern is detached with more focus on sharing with audience the information on how a particular system, organization, or situation has evolved over a period of time.

14.2.2 Causal Pattern

In this pattern, the ideas are divided into two major components—causes and their effects. Normally, this pattern is chosen to highlight the relationship between a problem and the reasons for its existence. While employing this pattern, some speakers choose to reverse the pattern and highlight first the effects of a problem, and then explain the causes behind it. Some of the topics for which you may require to structure the speech on this format are 'Corruption in the Indian Bureaucracy', 'Global Warming: Challenges and Perspectives', 'Impact of Advertisements on Young Minds', 'From the Joint Family System to the Nuclear Family Structure', 'Alcoholism: Its Causes and Effects', etc.

Unlike the speeches arranged in the chronological pattern, the speeches arranged in a causal pattern are more involved and emotive in approach. While choosing this pattern, the speaker intends to highlight the gravity of a situation by focusing either on its causes or their effects. The causal pattern is preferred in situations when the speaker intends to create a lasting impact on his/her listeners.

14.2.3 Spatial Pattern

Spatial pattern is best suited to speeches which have a geographical or structural orientation. For instance, topics such as 'Amber Fort: Its Structure and Splendour', 'The Birla Temple, Delhi: A Structural Description', 'Design of an Automatic Washing Machine', etc. would fall under this category. Therefore, the speeches that require us to discuss the components and structure of a particular building, machine, organization, etc. are arranged in this pattern. While choosing this pattern, we can move the discussion from top to bottom or right to left or front to back. As the spatial pattern chooses to arrange the discussion in different directions, it is also referred to as *directional pattern*.

14.2.4 Topical Pattern

Another commonly employed pattern, the topical pattern chooses to divide and arrange the different parts of a speech into various headings and sub-headings. When the speaker has to inform the audience about the various kinds or types of something, he/she chooses this pattern. Of all the patterns, the topical pattern is the most widely utilized. Some of the topics for which you may choose the topical pattern for arranging the ideas of your speech are 'Types of Cancers', 'Importance of Sex Education in Schools', 'Differences in Marxist and Capitalistic Economies', 'Kinds of Cyclonic Storms', etc.

14.2.5 Psychological Pattern

At times, a situation requires the speaker to structure the speech according to the needs of his/her listeners. In such situations, the speaker arranges ideas in a manner most likely to create an immediate impact on the listeners. Essentially persuasive in appeal, the speeches structured in the psychological pattern are quite emotive in style and sense. Some of the topics speaking on which the speakers may use this pattern are 'Loneliness: A By-product of Modern Living?', 'Are We All Getting Americanized?', 'Pakistan Nuclear Programme: A Threat to Our Survival?', etc.

14.3 SELECTING AN APPROPRIATE METHOD

In public speaking, not only the pattern but the method of delivery also matters a lot. All such methods have their own relative advantages and disadvantages. A brief discussion on these methods would help you figure out which of these methods suits your purpose on a particular occasion. Broadly categorized, there are four major ways of delivering a speech. These are as follows:

- Speaking from Memory
- Speaking Impromptu
- Speaking from the Manuscript
- Speaking from Notes

14.3.1 Speaking from Memory

Some public speakers rely heavily on their memory for delivering their speeches. Speaking from memory suggests that the speaker has to memorize all the text of his/her speech and deliver it verbatim. While choosing this method, one has to not only prepare really well, but also rehearse it a couple of times to deliver the speech effortlessly. In this method of speech delivery, the speaker's concentration and memory play a very crucial part. This mode of delivery has some advantages and disadvantages, a few of which are listed below.

Advantages Following are the advantages of speaking from memory:

1. Memorizing an entire speech helps the speaker put across his/her ideas with requisite flair, tone, and tenor.
2. The method requires a lot of practice and rehearsal, which in turn, helps the speaker to be well prepared for the speech.
3. Since the speaker is usually well prepared while speaking from memory, he/she can maintain better eye contact with the audience while delivering the speech.
4. With this method, the speaker has the advantage of casting and recasting the entire text of the speech, and making it as impressive and emphatic as possible.

Disadvantages Following are the disadvantages of speaking from memory:

1. It is generally seen that speaking from memory makes a speaker rely too much on it.
2. Since the entire speech is memorized by the speaker, forgetting some part of it is tantamount to losing an entire thread and can derail a speaker's momentum.

3. Normally, we do not memorize things and speak on the basis of it. Therefore, when a speaker chooses this method of delivery, he/she is bound to appear unnatural.
4. As the entire text is already memorized by the speaker, it does not give him/her much room for creativity and originality.
5. While choosing this pattern, the speaker binds himself/herself to expressing certain views which he/she cannot change even if the situation so warrants.
6. Adopted generally by novices, the method of speaking from memory often smacks of a person's lack of experience when he/she endeavours to choose this method.

14.3.2 Speaking from the Manuscript

In sharp contrast to speaking from memory is the method of speaking from the manuscript. When speaking from memory, the speaker has no paper for support, whereas in speaking from the manuscript, the speaker walks in with the entire manuscript of the speech written, which is read out word for word. Just like the memory method, speaking from the manuscript also has its advantages and disadvantages.

Advantages Following are the advantages of speaking by looking at the manuscript:

1. Since the entire speech is written to be read out from the manuscript, it adds to the confidence of an inexperienced speaker.
2. As the entire text is already written, the margin of error is minimal.
3. In situations where accuracy is extremely important, this method of delivery is quite useful.
4. As the entire speech is to be first written and then delivered, it helps the speaker prepare thoroughly for the occasion.
5. Since the text is already written, it can be edited and rewritten many times to create the right impact on the audience.

Disadvantages Following are the disadvantages of speaking by reading out from the manuscript:

1. As the speaker reads from the manuscript, the entire speech-making process looks too formal and monotonous at times.

Reading from Manuscript without Eye Contact can be Boring

2. Since the speech is already written, the speaker does not have the chance to make changes at the time of delivery, if required.
3. As the entire speech has to be uttered verbatim from the script, it lacks originality and spontaneity.

14.3.3 Speaking Impromptu

Unlike speaking from a manuscript and speaking from memory, impromptu speeches are those that are delivered at the spur of the moment. Though in one's professional life most of the things are planned in advance, an impromptu speech does not emanate from any such planning. You are at times invited to 'say a few words' without any intimation or prior notice. A speech thus delivered without preparation is considered an impromptu speech. Like all other methods, this method too has some advantages and disadvantages which are listed here.

Advantages   Here are the advantages of speaking impromptu:

1. Since in impromptu speeches you are called to speak a few words, nothing much is expected; your listeners know that you were not given any chance to prepare, so they do not judge you strictly.
2. While listening to an impromptu speech, audiences are always keen to see their speaker succeed.
3. Since the speaker delivers his/her thoughts without much preparation, he/she enjoys tacit public consideration and sympathy.
4. Delivering well in impromptu situations is like investing minimum and accruing maximum, as the expectations are low and the level of emotional approbation quite high.

Disadvantages   Following are the disadvantages of speaking at the spur of the moment:

1. When someone is asked all of a sudden to walk upto the dais and say a few words, they are bound to feel a little nervous and jittery. Because of this suddenness, sometimes the speakers are not able to speak with conviction or clarity.
2. Impromptu speeches often lack organization of ideas, simply because the speaker has no chance for arranging his/her thoughts in proper order.
3. One crippling disadvantage of the impromptu method of speaking is that even if the speaker says something meaningful and profound, it is not taken very seriously.
4. In impromptu speeches, the audiences' patience runs short and they are in a hurry to judge the speaker and his/her ideas. Therefore, one can't speak at length in an impromptu speech.

14.3.4 Speaking from Notes

One of the best known ways of delivering a speech—speaking from notes—has been regarded as the most favoured method chosen by most of the speakers. In this method of delivery, the speaker prepares notes/points/slides and enumerates the main ideas with the help of these main points. Regarded as probably the best method of delivering a speech, this method has few disadvantages and numerous advantages, some of which are given below:

1. While speaking from notes, the speaker can afford to look natural and spontaneous.
2. Speaking from notes makes the speaker also look prepared and yet flexible in approach.
3. While speaking from notes, the speaker can also maintain eye contact with the audience most of the time.
4. Since the speaker does not either read out from the manuscript or speak from memory, he/she has an added advantage of making necessary changes at the time of delivery, if required.
5. As the speaker also does not go into a speech situation as abruptly as he/she ventures into it while asked to make an impromptu speech, he/she is both prepared and yet flexible to changes wherever required.
6. While speaking from notes, the speaker gives the impression of being professionally prepared and in control.

7. Choosing this method also gives him/her an opportunity to appear more spontaneous and natural.
8. Unlike impromptu or memorized speeches, the speaker in this method has the facility to adapt to the situation better and manoeuvre his/her ideas accordingly.
9. In this method of delivery, the speaker has the advantage of figuring out the response of the audience as he/she speaks, which enables him/her to change his/her ideas as per the expectations of the audience.

Regardless of the method or the pattern chosen for a speech, you need to concentrate on making it work. Since public speaking is essentially a creative process, all speeches are judged on the scale of persuasion and interest. It is believed that every time a speaker walks onto the dais, he/she faces the twin challenge of persuading and influencing his/her audience and keeping his/her speech interesting and captivating. Let us briefly look at these two vital aspects of successful public speaking and learn how to develop these two features in our speeches.

14.4 ART OF PERSUASION

As already suggested, speech making is a creative process by which the speaker intends to influence the audience with his/her ideas. To be able to achieve this, the speakers have to keep their speeches well planned, properly substantiated, adequately convincing, and unquestionably relevant to the needs of the audience. It is through these that the speaker tries to persuade his/her audience to understand, appreciate, and possibly follow his/her line of thought. If a speech falters on any of these finer aspects of its appraisal, it fails to live up to the task of achieving its objective. In order to make your ideas persuasive, bear in mind the following points:

1. Research your topic thoroughly.
2. Use proper data and figures to sound convincing.
3. Keep an audience-oriented approach.
4. Employ personal examples to establish rapport with the audience.
5. Create a verbal imagery to leave the desired impact on the audience.
6. Use language suited to the occasion.
7. Keep your audience entertained during your speech.

Apart from persuasion, speeches are also required to be interesting enough to be listened to and followed. In fact, to be able to convince the audience about your ideas, you should be able to first of all make them listen to you. Keeping your audience engaged is a greater challenge than convincing them about your ideas. Let us learn how to keep our speeches interesting.

14.5 MAKING SPEECHES INTERESTING

In the professional world, a number of speeches are delivered everyday. However, most of these speeches are forgotten as soon as they are completed. What do you think they lack? Information? Ideas? Data? No. What they actually lack is the quality of arousing and sustaining the interest of the audience. In the entire public speaking situation, the toughest challenge that a speaker actually faces is to be able to make his/her audience listen to his/her speech. Therefore, keeping your audience interested in you is an art that needs to be carefully developed and gradually inculcated in your speech-making efforts. There are some tried and tested ways of achieving this, a few of which are discussed in detail in the remaining part of the text.

Broadly speaking, you can render your speeches interesting and captivating by working on the following strategies:

1. Make your beginnings catchy.
2. Use humour and wit.
3. Use body language appropriately.
4. Employ proper voice modulation.
5. Use examples and instances.
6. End emphatically.

Let us look at all such crucial aspects of a speech in greater detail.

14.5.1 Make Your Beginnings Catchy

Almost all good speeches begin on an impressive note. Regardless of the occasion, a good, catchy start immediately establishes a good rapport with the audience. Moreover, it adds to the confidence of the speaker, as a few initial words put in the right place in the beginning of the speech make the speaker feel related, authentic, and at home. Most of the impressive beginnings are regarded so because the speakers begin with an anecdote, an interesting statement, a joke, an arresting idea, or something else that immediately captures the attention of the audience. Look at the following example and see how the speaker starts innovatively to cast a captivating spell on his/her audience.

Prepare Well for Your Introduction: A Good Beginning is a Battle Half Won!

Situation: A Speech on Intellectual Property Rights

Friends, a couple of days back, I went to see my physician. As I waited for my turn, my eyes fell on a few lines of a poem that was put up in a gilded frame against the wall of the doctor's cabin. Truly a nice poem on positive thinking—of course of great worth to an ailing person—I silently read the lines and admired both the doctor who had put it up to help his patients and also the 'invisible' poet who had penned it. But wait; was the poet all that invisible? No! I knew the poet! I recalled clearly … after all it wasn't all that long ago … it was Soniya, my classmate, who had written it! How did I know? Of course, I knew. Hadn't she showed me this very poem just last week saying, 'Look Manish, this is the poem I wrote last night!' Suddenly, I felt so good for my friend. She had become such a popular poet almost overnight! But then, my heart sank as I went close to the framed piece hung on the wall. The name of the poet read differently; it wasn't Soniya Batra, it was—Katherine Porter. And the poem had not been written just 'last night'. It was written long ago; in fact, as early as 1934. Suddenly, Soniya appeared to be a cheat, a plagiarizer, who unlawfully had tried to own someone else's property. Friends, the question that I want to raise today is related to intellectual property rights.

In order to make the speeches interesting thus, you can make an interesting statement in the beginning, start with a striking statement, put a startling question to the audience, or narrate an anecdote. Remember that starting on an innovative note helps you arrest the attention of the audience and gives you an immediate edge to build up the rest of your speech equally convincingly.

14.5.2 Use Wit and Humour

Humour and wit are essential components of a good speech. To some extent, they are so crucial that without them it is not possible to conceive of an effective and memorable speech. Regardless of the situation, if properly employed, humour and wit can enliven any speech. A couple of examples would illustrate this point. Let us look at the following example:

> **Situation:** A Speech on Psychological Health Fears
>
> I told my doctor all that I could think of was wrong with me—I told him that my heart often skipped a beat; that my pulse reported stoppage at times; that my stomach always ached whenever I chose to overeat; so much so that even my legs trembled when I spoke in front of a crowd! I told him all that I had. I also told him what I didn't have in the name of disease was the housemaid knee. Thanks probably to the fact that for, being a man, it was not possible for me to be a housemaid and have a housemaid knee!

14.5.3 Use Appropriate Body Language

In all professional situations, body language plays an important part. While delivering a speech too, body language plays a very crucial role. It has already been discussed in Chapter 10 in the book as to how to maintain positive and impressive body language in professional situations. The tenets of effective body language for other professional situations also hold good for effective public speaking. To understand how proper body language can help a speaker perform better and how the lack of it can lead to poor and unconvincing performances, please refer to @ the professional presentations section in the Online Resource Centre.

14.5.4 Use Proper Voice Modulation

Besides effective body language, it is important for a speaker to introduce inflections in his/her voice and make proper voice modulation. As the importance of voice modulation has also been highlighted in Chapter 10 of the text, some instances of proper voice modulation have been highlighted so as to help you understand its importance and utility in public speaking situations.

Let us look at the following examples:

> **Situation:** A Presentation on the Advantages of Vegetarianism
>
> Not only this (The speaker pauses slightly.), non-vegetarian food also affects our overall system and results in other health problems such as constipation (emphatic pause), drowsiness, and even high blood pressure (With a little raised voice, the speaker speaks to register the climactic effect.). Not just that, but it also affects our mental health since we tend to have dull thinking, we become aggressive, and increasingly insensitive to other creatures including our own brethren.
>
> In short, being carnivorous dehumanizes us! (The speaker ends the sentence with a rising tone to secure emphasis and attention.)

Situation: A Presentation on Child Labour

(The speaker speaks the following with a deep voice, maintains a slow pace, and uses pauses to emphasize successively):

He greets you with a faint smile on his dry lips; (pause) though aged twelve or thirteen, his sunken belly, and in fact the whole frail structure, can easily slip into a vest meant for a well-fed child of six or seven; he nods feebly at you; (pause) remains unsure of the response his desperately prompt service evokes in you; even then, he *sprints away* (emphasizes these two words) to the canteen counter to fetch yet another burger for you. Having known hunger at close quarters, he knows fully well that in a well-fed stomach, (pause) hunger never waits with patience and grace! (The speaker ends with a rising tone to underline irony and arouse compassion in the listeners.)

14.5.5 Use Examples and Instances

Using examples from one's personal life or professional experiences in a speech always adds to the interest of the audience. Instances, anecdotes, sayings, quotations, and personal examples, besides statistics, facts, data, and testimony, really substantiate the ideas of a speaker. Moreover, such strategic allusions relate the audience to the speaker and evoke strong responses from them. Read through the example below to see how a team manager convinces and inspires his/her team to work hard and keep abreast of the latest developments in their field, to be successful in their professional lives, by citing his/her own example.

Friends, we are in an era of cut-throat competition. People who cannot keep up with the pace will perish and only the fittest will survive. The rule of the game is to learn fast, adapt to changes quickly, and be in the know-how of the latest developments around you. Let me give you an example. When I came into this field, I was a complete fresher; I had no knowledge of this profession. I interacted with people who had superior knowledge in their areas and learnt a lot from them about the different processes in this organization. I worked hard on acquiring skills that were required and I kept myself updated with the changes around me. When I learnt that there was a vacancy for a specialist in quality assurance, I applied for that as I could sense that in a new role the prospects for growth are abundant. I worked on my skills and developed new quality assurance techniques by following the best practices in other established organizations. Soon we became a small team and as you can see, this team has now grown in size to a full-fledged department. Other branches of our company are replicating this concept and new quality assurance teams are functional in some branches, and in other branches this idea is in the pipeline.

See further how the speaker, by quoting relevant sources in the following example, creates the emotional impact he intends to leave on his listeners:

Friends, if you refer to *The Week* of 30th May, you would realize that I am not narrating the stories of the dark ages when women were branded as witches and were raped, tortured, and even killed, but am citing before you facts from the present times. The title chosen by the magazine really highlights India's hidden shame and brings to the fore the fact that despite all our claims to modernity, awareness, and development, millions in our country continue to live even today in perpetual darkness…

It is thus important that the speeches we make have the required substantiation and authentication from sources, facts, instances, and examples.

14.5.6 End on an Emphatic Note

Just as starting innovatively is important, so is ending emphatically. In fact, many a speech becomes memorable because the speaker ends the speech on an emphatic note which, in turn, leaves the audience enthralled. Ending your speech emphatically is important also because of the fact that it is your one last chance to strike a chord with your audience. When you end your speech with a statement such as 'In the end…', 'To conclude…', 'Finally…', 'One last word…', etc., the audience expects you to speak about the most important part of your speech. As a speaker, you cannot take lightly what to your audience appears to be a significant part of your speech. Therefore, it is quite important that you end your speech on an emphatic note so that you can create on your listeners a lasting and impressive impact.

Given below are some such strategic endings which help the speakers achieve required emphasis and reiteration at the end of their speech:

> **Situation:** A Speech on Leadership
>
> Friends, we have seen how leadership is more than what meets the eye; how leadership is not just assigning tasks to subordinates; how leadership studiously avoids being seen as dictatorship; how leadership has a vision; how it inspires others; and how it differs from simple management. Finally, we can say that a leader is the one who has not only a vision and pursues it with passion, but also has the imagination to make others see and follow it.

The ending here acquires emphatic connotations as first, the speaker summarizes the main ideas of his speech and then winds up the discussion on a note of purpose and intensity.

Here is another example:

> **Situation:** A Speech on Drug Addiction
>
> My friend is back from the rehabilitation centre. He intends to start his life all over again. He spends a lot of time with his Mom, Dad, and Sis. He told me that he is planning to resume his studies. Of course, with the awesome percentage that he had upto the 12th standard, getting into a degree programme is not going to be all that difficult. What, however, is going to be difficult would be to overlook the sum of losses accumulated over the four years of frenzy, madness, and the subsequent darkness and damnation in the name of drug addiction. Mercifully for my friend, his darkness and damnation were not permanent. The love and support he got from his family members was amazing. He was also fortunate enough to get an opportunity in his life to stage a comeback. I know my friend is a lucky person. Not every other drug addict around us is so lucky as he. Scores of them, in fact, are lost to the darkness and are damned forever.

By alluding to his/her friend's example, the speaker reminds us about the tragic, dark pit that a drug addict can fall into and be perpetually lost in. Look at the profound imagery of darkness the speaker creates through a vivid language that maintains the intensity of expression throughout the ending.

14.6 DELIVERING DIFFERENT TYPES OF SPEECHES

In this section, we will see different types of speeches.

14.6.1 Welcome/Introductory Speech

Of all the speeches that a professional has to deliver in his/her career, *welcome speeches*, also known as *introductory speeches*, are the most common ones. It is so because many a time, your organizations receive guest speakers who are to be introduced to the audience. Whenever required to introduce a speaker to the gathering, keep in mind the following points:

1. Introduce the speaker by referring to his/her full, formal name, including all the titles that go with her/his name, namely Ms/Mr/Prof./Dr/Jr…, etc.
2. Highlight the achievements of the speaker by mentioning the expertise, distinctions, degrees, awards, recognitions, etc. achieved by him/her.
3. Don't just introduce the speaker, but also the area that he/she is going to touch upon, highlighting the speaker's expertise relevant to his/her message.
4. Don't sound fulsome in your praise for the speaker; it may embarrass the speaker and he/she may find it difficult to raise themselves to the expectations thus raised. Moreover, praising the speaker beyond a limit may make you seem insincere and sycophantic.
5. Don't go on speaking for a long time. While introducing the speaker, you should be crisp and brief. Remember, a long-winded introductory or welcome speech is always a nuisance for the audience. Moreover, if you speak too much, you may eat into the time allotted for the invited speaker's address. So, be focused and to the point.

Take a look at this example:

> Ladies and Gentlemen, I feel honoured to have been given this opportunity to introduce to you Prof. M.K. Bhatnagar, the keynote Speaker for the Inaugural Session of the Seminar. However, when I venture into doing that, I know that neither does Prof. Bhatnagar require any such introduction, nor can my brief introduction sufficiently highlight his multifaceted genius. Nevertheless, I can make a modest attempt by saying that Prof. Bhatnagar is a scholar and intellectual of enormous distinction and repute, who in his academic career spanning over two decades, has served several prestigious academic institutes, both in India and abroad. He has also been the Chairman, Common Syllabi Committee of the universities in the state. Prof. Bhatnagar has to his credit over two dozen books, more than sixty research articles, and scores of featured articles in anthologies, journals, and newspapers. He has guided nearly fifty scholars in their doctoral research and has delivered keynote addresses in hundreds of academic gatherings such as this. Prof. Bhatnagar has also been a research fellow and principal investigator for various projects sponsored by prestigious academic bodies such as AICTE, UGC, etc. He himself has been the convenor of several national and international workshops, seminars, and conferences. Today, Prof. Bhatnagar is going to speak on 'Multiculturlism: Possibilities and Perspectives,' a theme that the eminent scholar has always explored with vision, authenticity, and intensity. Friends, they say a thing of beauty is a joy forever and I don't want to stand between you and the beauty of his ideas and the sheer joy of listening to Prof. Bhatnagar's sublime expression of those ideas.

14.6.2 Vote of Thanks Speeches

Like welcome/introductory speeches, *vote of thanks speeches* are also very commonly required to be delivered by professionals. However, unlike *introductory speeches* which are delivered at the beginning of a speaker's address, the *vote of thanks* is delivered at the end of a particular event, programme, seminar, workshop, or conference. A *vote of thanks*, by its very nature, is quite brief and succinct. When asked to deliver a *vote of thanks*, keep in mind the following points:

1. Make a list of the people to be thanked. Review the list to ensure that no one has been missed out.
2. Use a proper order to express your gratitude. Normally, people in the higher echelons of the profession are thanked first, followed by those who offered real help, then the people who made the event successful, and finally those who provided mechanical help.
3. Whenever you thank someone, also cite the reason for your gratitude and the kind of help received.
4. Be brief; avoid verbosity and exaggerations.
5. Maintain warmth; avoid being mechanical.
6. Vary your expressions to express your gratitude for the various types of help received from a variety of people.
7. Be witty and humorous, if possible.

Let us look at this example:

> Ladies and Gentlemen, my duty is pleasant, primarily because it is brief! Nevertheless, it is always a pleasure to express one's gratitude when you witness a successful spectacle, such as the one we are about to wind up. Well friends, to say that the Workshop has been immensely successful would be an understatement. And, there are quite a few people who have generously contributed to its success. First and foremost, we are grateful to Sri R.K. Patnaik, our Managing Director, for his inspiration, guidance, and administrative approval for the workshop. We are deeply indebted to Mr M. Shankar and Dr S. Vasanthi, our erudite resource persons for their enlightening talks. We appreciate and thank all the participants who came all the way from different parts of the country and participated in the two-day workshop with enthusiasm, commitment, and zeal. We are grateful also to all our technical staff, Mr Venkat and Mr Buddhiraja, who saw that there were no technical glitches during the programme. Thanks are due to M/s Padma Caterers who ensured that we kept getting the delicious food for our body, besides the food for thought, on both the days. And above all, we are really indebted to our august audience for their participative and patient listening. Thank you one and all. Thank you!

14.6.3 Farewell Speeches

In their professional career people often change jobs, leaving one organization and joining another. At times, they are transferred or have to move away from an organization after the completion of a project, a term, or an assignment. In all such eventualities, a *farewell speech* comes into play. When required to deliver such a speech, keep in mind the following points:

1. Thank the organization for the opportunity you received for serving it.
2. Avoid criticizing the company's policies or people, even if you don't appreciate them much.
3. Don't exaggerate your own achievements.
4. Sound polite and humble in your expressions.
5. Be brief in your farewell speech; if you drone on for a long time, you may first lose the interest of the audience and subsequently their sympathy.

Take a look at this example.

My dear friends, it is said that the more intense your emotions, the least articulate they become. That probably is the case with me as I rise to say goodbye to all my seniors, colleagues, and friends present here. No doubt, it has really been an amazing experience to be with Oasis Conglomerations for eight long years. All these years, I thoroughly enjoyed doing the tasks assigned to me, simply because the ambience here has always been quite conducive to work, innovation, and creativity. I am grateful to the company for giving me an opportunity to work on the projects of my interest. I am also obliged to the management for their guidance and motivation throughout my stay here. Moreover, I can never forget the sense of bonding and belonging we as colleagues developed with one another. Really all these years, we worked like a family, and it's painful to bid adieu to you all who have always been wonderfully supportive, understanding, and accommodating. For this and for everything else, Oasis Conglomerations would continue to harbour a very special place in my heart. Thank you one and all for making my stay here such a delightful experience!

In brief, remember to keep the following points in mind in order to become an accomplished public speaker:

1. Prepare well for your speech.
2. Research the topic of the speech thoroughly.
3. Strategically overcome your nervousness. Remember to follow the guidelines given in Chapter 11 on the *dynamics of professional presentations*, to overcome your stage fright.
4. Start your speech innovatively. A good beginning helps you build a rapport with your audience immediately.
5. Use effective body language: maintain eye contact with your audience; use gestures and expressions to support your ideas.
6. Employ proper voice modulation: work on your pronunciation, articulation, volume, pauses, and voice modulation. Follow the points discussed and illustrated in the text and the Online Resource Centre (in the Professional Presentation video).
7. Use humour and wit wherever possible.
8. Express enthusiasm during your speech.
9. Arrange your PowerPoint slides properly. Do not clutter or crowd them.
10. End your speech on an emphatic note. If possible, leave your audience laughing.

@

 ## RECAPITULATION

✓ Public speaking tasks are part and parcel of a professional career.

✓ When required to deliver a speech, plan to structure your ideas by choosing a suitable pattern.

✓ There are five patterns meant for organizing your ideas—chronological pattern, causal pattern, spatial pattern, psychological pattern, and topical pattern. You can choose any of these patterns or blend some of these as per the requirement of your task.

✓ While structuring the speech, you also need to identify the method of delivery to be utilized. A speech can be delivered in an impromptu manner; it can

be delivered with the help of notes; it can also be read from a manuscript or delivered from memory. Writing main points on cards and making the speech with their help is regarded as a mature and most interactive way of delivering a speech.

✓ In professional situations, speeches are not just merely to be delivered, they are also required to be made interesting and worth listening to. There are certain ways, adopting which we can make our speeches more entertaining and captivating, such as working on the introduction and ending of the speech; incorporating instances and examples to carry conviction; employing emphatic body language; using proper voice modulation at proper places, and being witty and humorous at times.

✓ There are some speeches such as introductory/welcome speeches, vote of thanks, farewell speeches, etc. which are quite frequently required to be delivered by professionals. Such speeches are required to be brief, methodical, and engaging.

✓ In welcome speeches, one has to introduce the guest speaker in a graceful and informative way.

✓ While delivering a vote of thanks, remember to sound courteous and polite, besides highlighting the type of help or contribution received.

✓ In making farewell speeches, express your gratitude for the opportunity given to you, and the motivation, guidance, and support received from others.

WISEWELL QUIPS

EXERCISES

Concept Review Questions

1. Imagine that as the Media Relations Officer of Mega Products Pvt. Ltd, Delhi, you are required to deliver a speech on the *Role of Media in Corporate Sector*. Write the full text of your speech. Invent the necessary details.

2. Imagining yourself to be the Deputy Training Officer at Paramount Trainers, Mumbai, prepare a *vote of thanks* to be delivered at the end of a week long personnel training programme that has just concluded. Invent necessary details.

3. Imagine that you have been working as Junior Sales Manager with Exwell Corporations, Pune, for the past five years. Now, you have been offered the position of Sales Manager in Eastwood Corporations, New Delhi. You are on your way to leave Exwell Corporations to join as a Sales Manager with Eastwood Corporations. Prepare the text of the farewell speech you plan to deliver.

4. Write the text of the introductory/welcome speech you are required to deliver to introduce Mr Roget

Warren, the Chief Administrative Officer, Human Resource Development Unit, at Acme Internationals, New Jersey, who has come to deliver an extension lecture at your company headquarters in Hyderabad. Invent the other necessary details.

5. Suppose that as the Development Officer of New Era Visions, New Delhi, you have been asked to deliver a keynote address on *Personality Development*. Write the introduction and ending of your speech, inventing the necessary details.

6. 'Speeches are not just meant to be spoken, they also are required to be made interesting and entertaining to the audience.' What are the strategies that can make a speech interesting and entertaining to the audience? Discuss and substantiate with appropriate examples.

7. In delivering a speech, patterns of organization and modes of delivery play a very significant role. Highlight the importance and suitability of each of these patterns and methods generally employed in the speech-making process.

8. 'In delivering a speech, imagined anecdotes, examples, and instances too serve a significant purpose.' Elucidate the statement augmenting your views with proper examples.

9. 'There is no substitute for wit and humour in public speaking situations.' Do you agree to this statement? Offer elaborate comments to substantiate your point of view.

10. 'Public speaking is as much about non-verbal communication as it is about its verbal aspects.' Elucidate the statement with proper examples.

ANNEXURE 14.1

Checklist for a Professional Presentation/Speech

Before making a professional presentation or a speech in professional situations, answer the following questions:

Q.1 Is the topic of your presentation well defined?   Yes ( ) No ( )

Q.2 Have you collected enough data for making your presentation?   Yes ( ) No ( )

Q.3 Have you also done an audience analysis regarding their age, sex, nationality, education level, etc.?   Yes ( ) No ( )

Q.4 Have you prepared a catchy, interesting beginning for your presentation?   Yes ( ) No ( )

Q.5 Does your presentation/speech have interesting statements, striking ideas, and humorous and witty expressions?   Yes ( ) No ( )

Q.6 Have you written the text of your speech in a warm and interactive manner?   Yes ( ) No ( )

Q.7 Have you finalized the audio-visual aids to be used in your presentation?   Yes ( ) No ( )

Q.8 Are all your slides numbered and arranged serially?   Yes ( ) No ( )

Q.9 Have you worked on the conclusion of your speech and made it emphatic and memorable?   Yes ( ) No ( )

Q.10 Have you prepared answers for the questions likely to be posed by the audience during or after the presentation?   Yes ( ) No ( )

Q.11 Have you rehearsed and practised your speech before a group of friends or fellow professionals? Yes ( ) No ( )

Q.12 Have you chosen an appropriate mode of delivery for your presentation?   Yes ( ) No ( )

Conversations, Dialogues, and Debates

15.1 INTRODUCTION

Conversation is perhaps one of the most commonly employed methods of self-expression that characterizes our everyday speech-making activity. Although conversations occur normally and naturally to us, most of us tend to take for granted our ability to make our conversations work. However, like any other form of communication, conversations require effort, focus, and practice. Before talking about how to become a good conversationalist, let us know the basic aims of conversations.

15.2 PURPOSE OF GENERAL CONVERSATIONS

Broadly speaking, there are three aims and purposes of conversations.

15.2.1 Self-expression and Interaction

The first aim of a conversation is to have the pleasure of self-expression and interaction with other people. We spend time with people whom we like and whose company we find stimulating. This is the driving force behind all our social activities. Whenever we have an opportunity to interact with people over dinner, a party, or some occasion, we wish to express ourselves, share our ideas, and get acknowledged as a good conversationalist.

15.2.2 Getting to Know the Other Person Better

The second purpose of conversation is to get to understand the other person better. In all kinds of business, you require to know the other person so as to get a feel of how he/she thinks, feels, and reacts.

15.2.3 Building Trust and Credibility

The third aim of conversation is to build trust and credibility with the people whom you meet. It is only possible with the kind of conversations we have with one another. In our professional lives, particularly for better teamwork, we need to converse well with others. People who get along very well almost invariably spend a lot of time talking about various subjects.

15.3 FEATURES OF A GOOD CONVERSATION

Here are the important characteristics of a good conversation:

1. The conversation should shift back and forth, with each person getting an opportunity to talk. Conversation in this sense is like a ball that is tossed from one person to another, with no one holding on to it for very long.
2. It should be clearly and concisely worded.
3. The sender should deliver the complete message, leaving no room for confusion.

15.4 TIPS FOR IMPROVING CONVERSATIONS

In this section, we will discuss the important tips for improving conversations.

15.4.1 Spend Unbroken Time

One of the very best ways to learn about other persons is to spend unbroken time in their company. You must have observed that a four or five hour car trip or train journey with another person helps you know him/her better. This is because of the quality and amount of conversation that you have had with the other person.

15.4.2 Listen More than You Speak

In conversations, you should listen twice as much as you talk if you want to get a reputation for being an enjoyable person with whom to converse.

15.4.3 Ask Questions

The art of good conversation centres very much on your ability to ask questions and to listen attentively to the answers. You can garnish conversations with your insights, ideas, and opinions. However, you perfect the art and skill of conversation by perfecting the art and skill of asking good, well-worded questions that direct the conversation and give other people an opportunity to express themselves. Ask open-ended questions that cannot be answered with a simple 'yes' or 'no'. Open-ended questions encourage the speaker to expand on his/her thoughts and comments.

15.4.4 Resist the Urge to Dominate

In order to be an excellent conversationalist, you must resist the urge to dominate the discussion. The best conversationalists seem to be easy-going, cheerful, and genuinely interested in the other person. They seem to be quite content with listening when other people are talking, and they make their own contributions to the dialogue with remarks that are short and to the point.

Listening is the most important of all skills for a successful conversation. The major reason why most people are poor listeners is that they are busy preparing a reply while the other person is still speaking.

15.4.5 Use Appropriate Body Language

You should also nod and smile when you agree to what the person is saying. Be active rather than passive. Suggest non-verbally also that you are totally engaged in the conversation. Throughout the conversation, maintain an eye contact while the other person is talking. A short pause, of three to five seconds, is a very classy thing to do in a conversation. This helps you avoid running the risk of interrupting if the other person wants to still continue.

15.4.6 Paraphrase the Speaker's Words

By paraphrasing the speaker's words, you exhibit that you are genuinely paying attention and making every effort to understand his/her thoughts or feelings. This way, they will find you interesting and fascinating. They will want to be around you. They will feel relaxed and happy in your presence.

15.4.7 Apply the Three Cs

The final key to becoming a great conversationalist is to practise the friendship factor. The friendship factor is based on the three Cs—*care*, *courtesy*, and *consideration*.

Whenever you show another person that you genuinely care about him/her, you come across better as a conversationalist and as a friend. Moreover, courtesy is a magic quality that makes people want to be around you. All good conversationalists make others feel calm and comfortable in their presence. They never do or say anything that could hurt or offend the other person in any way. Moreover, if we respect others and are considerate towards them, we too are respected and considered highly by others.

15.4.8 Be Fluent while Speaking

Fluency is a much desired attribute of a good conversationalist. Therefore, be fluent while talking to others. Apart from good listening and regular practice at conversations, having a good vocabulary also helps you attain fluency in your expression. So, be a good reader and try to learn more and more words in order to express yourself effectively and successfully.

Read the short conversations that follow and learn how to converse effectively with others.
@ Some of these conversations have also been included in the Online Resource Centre of the book. Listening to these conversations will help you understand how to maintain the right kind of tone and tenor during such conversations.

15.5 SHORT CONVERSATIONS

In this section, we will see three examples of short conversations:

Example I
Discussing Health

Aarti: I'm always sick these days, Deepa.

Deepa: What can you expect? During winter, people often fall sick.

Aarti: You're right. Anywhere I go, someone will just sniff, cough, or sneeze. Good Lord, is there anything I can do?

Deepa: Certainly, there is. You can always do something to boost your immunity and fight those invisible germs gliding your way.

(Contd)

Aarti: I'm aware of that. Our immune system relies hugely on the right food we regularly eat.

Deepa: It's good to see you eating peanuts. Do you know that nuts are a great source of energy?

Aarti: Yeah. I also read somewhere that nuts have vitamins, fibre, protein, nutrients, and antioxidants that can boost the immune system. In fact, recent studies reveal that nuts can lower the effects of bad cholesterol.

Deepa: That's correct! Maybe, you need to add more fruits and vegetables to strengthen your immune system. You need to do yoga as well to keep yourself fit.

Aarti: Yes, thanks for reminding me of that. I will surely have fresh fruits and veggies which are rich in vitamins and fibres. And of course, Yoga; I'll certainly try that.

Deepa: Oh! My turn has come; I shall withdraw some money. Bye-bye!

Aarti: Thanks for your wonderful tips. See you. Bye!

Example II

Discussing One's Ambition in Life

In the conversation given below, two girls discuss their ambitions in life:

Shruti (to Snigdha): What do you want to become?

Snigdha: I want to become a singer.

Shruti: Why?

Snigdha: I have great passion for singing. I derive great aesthetic pleasure out of music and when I sing, I am certainly on a different plane.

Shruti: That's right. It is a very satisfying profession. But Snigdha, do you know that it requires a great deal of hard work and one has to sacrifice a lot?

Snigdha: Sacrifice? What do you mean by that?

Shruti: It depends on how great a singer you want to become. Most of them have to sacrifice the taste of the tongue. They never take spicy, oily food, and they rehearse a lot.

Snigdha: That's even otherwise advisable; one should not take spicy and oily food. That I don't think it is a sacrifice.

Shruti: Yes, that does not sound too demanding. But some really great ones go for greater sacrifice. See how Lata Mangeshkar, the melody queen, did not get married but dedicated her entire life to music and singing.

Snigdha: Oh! Come on. It's not that she did not get married because she was a singer. Maybe she had some other reasons for not choosing to lead a married life. Anyway, what is your ambition?

Shruti: Well, I want to become a model.

Snigdha: A model? How interesting! I think you are beautiful enough to become one. Maybe a Miss World also.

Shruti: Thanks for your compliments. But I am not aiming too high at the moment. Let me go up the ladder step by step.

Snighda: But what about your studies? I suppose modeling is such a profession that causes a lot of distraction and one cannot focus on studies and modelling simultaneously.

Shruti: I don't believe this. I can balance and manage studies and modelling together.

Snigdha: Anyway. Let's do our best and leave the rest to God.

Shruti: Yes, OK Snigdha! Goodbye.

In order to become a good conversationalist, remember to observe the following.

15.5.1 Exchange Greetings/Pleasantries

Greetings and pleasantries form an important part of human interaction. Normally, a human interaction starts with some greetings. Therefore, it sounds quite abrupt if it does not. See how the following conversation takes off:

> A boy (starting quite brusquely): I want to know the status of my application. When are we going to get the gas connection?
>
> A lady at the desk (equally dismissive): I don't know; go to the third table.

Don't you think that the boy needed to be a little more polite in his query? See how the same person asks the same question differently:

> The same boy (politely this time): Good morning, Madam.
>
> The same lady (this time equally polite and positive): Good morning.
>
> The boy (very courteously): Madam, may I know the status of the application I filed for getting a gas connection?
>
> The lady (equally polite): Please enquire from the gentleman sitting behind table number three.
>
> The boy (with same level of humbleness in the tone): Thank you very much, Madam.
>
> The lady (warm): You're welcome.

So, remember to use pleasantries, especially when you have to talk to a stranger. It makes you sound well-mannered and cultured. Without pleasantries and greetings, you may sound rustic and unsophisticated in your approach.

15.5.2 Reciprocate

Reciprocation forms an important feature in human conversations. Sometimes, the conversation reduces to the level of an empty claptrap, particularly when some members do not seem much inclined to talk. Read this professional conversation taking place between two company executives. Observe how the lack of interest and reciprocity by one of the speakers sinks the entire conversation:

> Speaker A (with enthusiasm and positivity): Sir, the idea of opening a retail outlet at Jaipur seems to be quite exciting and with some planning, we can really make it work.
>
> Speaker B (somewhat laid-back and circumspect): Let's see.
>
> Speaker A (with more emphasis and attempting to hold on to the point): You know, Jaipur is not just a big city… it's really a happening place. People visit this place from all over the world, and the kinds of products we offer are certainly going to attract a large number of people.
>
> Speaker B (still lukewarm in his approach): Maybe…
>
> Speaker A (trying to convince the listener by citing an example): You know we tried this in Agra and it worked there. I am very sure if it could work in Agra, it sure can work in Jaipur.
>
> Speaker B (still not convinced and forthcoming): Hope so.

We can easily identify that both these speakers do not seem to share the same level of optimism. One person appears to be quite keen, enthusiastic, and hopeful while the other one sounds circumspect, withdrawn, and uninterested. Probably, he is not as convinced about the idea as the other speaker is. Even then, he could have replied better. Read the conversation again and understand how the other speaker might have expressed himself/herself better:

> Speaker A (with enthusiasm and positivity, as suggested earlier): Sir, the idea of opening a retail outlet at Jaipur seems to be quite exciting, and with some planning we can really make it work.
>
> Speaker B (with interest and keenness in the voice): Is it? How do you feel so confident about our doing well in Jaipur?
>
> Speaker A (with encouragement in his/her voice; buoyed by the response of the boss): Sir, you know, Jaipur is not just a big city... it's really a happening place. People visit this place from all over the world, and the kind of products we offer are certainly going to attract a large number of people.
>
> Speaker B (with accommodation of the other's view but analytical at the same time): Of course, you are right; but simply because the place is a tourist attraction, it should not make us believe that our products would work there.
>
> Speaker A (trying to convince the boss with some substantiation and precedence): See, Sir, we tried this in Agra and it worked there. I am very sure if it could work in Agra, it sure can work in Jaipur. After all, we have tourists visiting both these places for almost similar reasons.
>
> Speaker B (exhibiting interest and restraint; also inducing more, vigorous action in subordinates): It sounds fine; but I believe we have to do more research on this front before we actually can decide about that. Why not call a meeting and discuss the matter with the other members of our unit as well?

Remember, lack of interest can kill not just the spirit of conversation, it can actually smother the idea that sparks enthusiasm and interest in team members. So, remember to reciprocate even if you have to spurn an idea, quash a view, or postpone a decision. Don't forget that if you do not reciprocate in personal and professional conversations, you may sound boorish, bored, uninterested, and rude.

15.5.3 Be Courteous and Polite

Further, all conversations require courtesy and politeness. Take a look at the following conversation and see if it sounds courteous enough:

> Woman (somewhat impertinent and lacking sophistication): Is this the place where we can lodge our complaint?
>
> Man (dismissive and authoritative): Yes, you can put the complaint in the complaint box.
>
> Woman (curtly): Will I be getting any receipt for that?
>
> Man (bluntly): Nothing.
>
> Woman (challenging): How do I know that it will not be lost anywhere?
>
> Man (matter of fact): It won't be lost.
>
> Woman (probing with all the doubt, almost accusation in the tone): But what is the guarantee that my complaint will not go missing?

(Contd)

> Man (providing laid-back, matter-of-fact assurance): It won't; there is a system.
>
> Woman (with unpleasant insistence in the tone): Can you give it to me in writing?
>
> Man (staying officious and aloof): No, we don't do that.

Situation 1

The conversation does not sound all that pleasant, does it? Remember that courtesy demands nothing but can always make you sound pleasant and approachable. Follow the same conversation and observe how the introduction of courtesy can change the entire mood this time:

> Woman (polite and humble): Excuse me, I have been told this is the office which registers people's complaints. Could you guide me how to go about it?
>
> Man (with courtesy): Welcome Madam; the procedure is very simple. You can just drop your complaint in the complaint box kept over there.
>
> Woman (somewhat unsure about dropping her complaint into the complaint box): Will I be given any receipt for having dropped my complaint?
>
> Man (politely clarifying official limitation but at the same time with assurance in the voice): Sorry Madam; as per the procedure it is not possible for us to give you a receipt. But don't worry, it would be quite safe.
>
> Woman (feeling accommodated and pleased with the response, though still somewhat circumspect): Thanks, Sir. But how do I feel assured that it won't go missing?
>
> Man (explaining the procedure painstakingly and thereby convincing the lady): Actually, there is a procedure for it. The moment you drop your complaint in the box, it would flash a number. Now, when we retrieve all the applications at the end of the day, we know exactly how many complaints there are. We count the applications and tally it with the number flashed on its screen. So the system keeps track of all the applications dropped into the box.
>
> Woman (with gratitude and confidence in her voice): That's fine then. Thank you very much for your help.
>
> Man (maintaining courtesy and politeness): The pleasure is ours, Madam.

Situation 2

In all human conversations, particularly the formal ones, one needs to be careful enough to maintain courtesy in one's tone. When courtesy is lost in your speech, nothing much can be gained from the conversations. Take a look at the telephonic conversation between the counsellor and the customer. Do you find it courteous enough? Find out who needs counselling—the customer or the counsellor?

> Woman Customer (with customary politeness): Good morning, I am Amrita Daswani. I bought a food processor from your company last month…
>
> Counsellor (somewhat boorish and rude): Please specify when you bought it…
>
> Customer (somewhat taken aback by the rudeness in voice at the other end): OK… I bought it on 20 April 2010…

(Contd)

Counsellor (still blunt and impertinent): Where from?

Customer (somehow maintaining her poise, though not really pleased with the way she is being treated, tries to match the bluntness in the voice of the respondent of her call): I bought it from your showroom at Darya Ganj.

Counsellor (very curt and almost offensive): OK… Now what is your problem?

Customer (clearly offended but still maintains her cool): Actually, whenever I operate the device, it gives a terrifying sound. It seems some part of it is loose for it also shakes violently…

Counsellor (with indifference and lack of concern in the voice): That is with every device.

Customer (trying to make her listener understand her stance, still maintaining her nerves and sticking to the rationality of her view): No, but earlier I had a mixer-grinder, and that never shook this much or made such a noise.

Counsellor (callously): But it is not a mixer-grinder; it's a food processor.

Customer (now really piqued and irritated, with sarcasm in her voice): So, it should be allowed to shake and create a rumble just because it's a food processor!

Counsellor (almost attacking): Have you ever had a food processor in your life?

Customer (shocked, struggling to recover): No…but…

Counsellor (disrespectfully): That's the reason why you are making such a noise!

Customer (losing her nerves finally, irritated, hurt, and annoyed, almost shouting): I'm making a noise! It's that goddam food processor of yours that is making a noise!

Could you observe how lack of courtesy can displace a conversation? Listen to the revised conversation on the same issue, and feel the change in the impact that courtesy and politeness creates on the customer:

Customer (beginning with customary politeness): Good morning, I am Amrita Daswani.

Counsellor (equally customary, polite and suave): Good morning, Madam. What can I do for you?

Customer (matter of fact): I bought a food processor from your company last month…

Counsellor (showing interest): That's right! Now, how is it doing, Madam?

Customer (not challenged, maintains cordiality even while making a complaint): Not properly. Actually, that's the reason I called you up.

Counsellor (composed, concerned, and quite ingratiating to the customer): Don't worry, Madam. We'll take care of that. Please let me know what's wrong with the food processor.

Customer (encouraged, comes out easily): Actually, whenever I operate the device, it gives a terrifying sound. It seems some part of it is loose for it also shakes violently…

Counsellor (maintaining urbanity, assurance, and composure in the voice): Not to worry; it seems the problem is of a minor nature. Anyway, we'll send our engineer; he'll come and thoroughly check the device.

Customer (quite satisfied but still needing assurance): When can I expect him?

Counsellor (suggesting action, promptness, and willingness to help): Within no time, Madam. May I know where you are calling from?

Customer (with difficulty, somewhat unsure of the response): See, we live at South Extension, Phase III.

Counsellor (still very confident of getting the needful done): No problem, Madam. Please let me have your address. Our engineer will reach you within three hours.

(Contd)

Customer (quite satisfied): That would be so nice of you!

Counsellor (maintaining suaveness and professional sophistication): Pleasure is entirely ours! Thanks for calling. Have a nice day, Madam.

Customer (quite pleased): Thank you!

15.5.4 Be Specific and Use Vivid Language

Sometimes, conversations fail to leave the desired impact simply because the language employed is not clear, precise, or vivid enough. Lack of clarity in your expression reveals confusion and uncertainty.

Take a look at the conversation and observe carefully the lack of clarity in the deputy's voice, something that his manager tries to induce in him all through the conversation:

Manager: We can send our front desk staff for training.

Deputy: Yes, Sir, we should.

Manager: They have to work on their communication skills… they terribly lack that.

Deputy: Sure, Sir, they do lack that.

Manager: When do we send them?

Deputy: Any time, Sir… the day you tell me, I'll send them for training.

Manager: And for how many days?

Deputy: Probably a month, Sir.

Manager: A month! That would too much!

Deputy: Yeah, Sir… Maybe we can send them for a fortnight.

Manager: Do you think they require that long a period at training?

Deputy: Yeah, Sir… but may be a week would do.

Manager: See, I feel they only need some polishing.

Deputy: You are right, Sir.

Manager: Should we send them together?

Deputy: Yes, Sir, together they will learn better.

Manager: But in their absence who will run the front desk business?

Deputy: Yes, Sir, you are right Sir… business will suffer if they all go together.

Manager: Which month do you think will suit them?

Deputy: Any month, Sir.

Manager: But I believe that January would be better. We don't have many people walking in during that period. So, maybe we can retain one or two and the remaining staff can complete the training during that period.

Deputy: And when they return, the remaining two can be relieved for their training.

Manager: Exactly.

As you can see, the deputy hardly knows a thing about his people. The conversation between two better professionals might have gone something like this, particularly if the deputy would have known how observant, clear, and specific he/she has to be in such conversations with his/her boss.

Manager: We can send our front desk staff for training.

Deputy (with promptness in voice): Yes, Sir… I too feel the same… they need to be trained. And more than anything else, they need training in effective communication skills.

Manager: You are right… so when can we send them?

Deputy (sounds prepared, clear, and purposeful in is approach): Sir, I have worked out a plan for that. Actually, I noticed that our staff lacks these skills quite a few months back. I have identified the personnel and the type of training they require… In fact, I have prepared a report on that and would like to submit that to you.

Manager (judging his deputy's preparedness): But I don't want a report that just contains problems… I need solutions too.

Deputy (not nervous, quite clear and well prepared): Sure, Sir. The report analyses the whole situation and also gives some solutions.

Manager: So, you too believe that they need some such training?

Deputy (confidence in voice): Sir, I believe that out of fifteen people, ten would certainly require training.

Manager: And how long should the period be?

Deputy (sounds methodical, purposeful, and focused): Sir, I have categorized these ten people in three brackets. One such lot requires intensive training for a month or so. This bunch has five people. Three of them require a training of a fortnight or so. The other two can come up well even with a week-long training.

Manager: So, where do you propose to send them?

Deputy (confident because of the homework done well in advance): Sir, I have identified three spoken English and personality development institutes located in Patel Nagar; they have good faculty and they charge a reasonable fee.

Manager: But our work will suffer if we send them for as long as a month.

Deputy (expresses clarity and confidence in voice): No, Sir. I have worked out the training schedule in such a way that we will have at least eight people looking after the front desk operations even during the training period.

Manager: How will you do that?

Deputy (speaks with conviction and planning): Sir, not all of them will be sent for training together. I have given the proposed training programme in detail, and since most of them would be sent alternatively, the front desk operation will not suffer even for a single day.

Manager: When can we start?

Deputy: I would suggest that we send the first batch by 1 July.

Manager: Roughly, when will it be over?

Deputy: By August end.

Manager: Have you worked out the cost to company?

Deputy: Yes Sir, the whole training programme will cost us about ₹35,000.

Manager: That sounds affordable… When can I see the report?

Deputy: Right away Sir, I have brought it.

Manager: Good, you have planned smartly.

Deputy: Thank you very much, Sir.

15.5.5 Be Concise

Many conversations are spoiled because of the fact that some speakers use needlessly lengthy and full of roundabout expressions. Therefore, in conversations, particularly professional conversations, try to be concise and to the point. Listen to the lengthy expressions commonly used by many of us, each of which is followed by an alternative expression which is concise and to the point.

Lengthy/Roundabout Expression	Direct and Concise
• In the present moment of time…	• At present…
• In the event of our not being able to convene the meeting in time…	• In case we are unable to conduct the meeting in time…
• May I take this pleasure of informing you that the company has decided to…	• You will be pleased to know that…
• You are requested not only to make a consideration but also make a selection of…	• You are requested to consider and select…
• Last but not the least; I would like to offer a suggestion that…	• Finally, I would also like to suggest that…
• Make a calculation of all the facts and report back to me…	• Calculate the facts and report to me…

15.5.6 Use Appropriate Body Language

@ Since all conversations are human interactions, our body language plays an extremely crucial role in our personal and professional conversations. Refer to the video section of the Online Resource Centre of the book where different aspects of body language have been analysed with proper visual examples.

15.5.7 Don't End Abruptly

Regardless of the issue whether the occasion is personal or professional, all conversations need to start appropriately, develop logically, and end appropriately. An abrupt ending to a

Patient: Good Morning, Doctor.
Doctor: Good Morning, how are you?
Patient: Better Sir, but not really quite fine.
Doctor: Tell me what's happened?
Patient: Sir, actually as long as I take medicines, I don't experience any pain in my knee but even if I miss one dose, the pain returns.
Doctor: See, I told you that you will have to take medicines for two months. But tell me, are you going for a walk in the morning and doing the knee exercise regularly?
Patient: Normally I do go, but sometimes I miss…

(Contd)

Doctor: No, that won't do… you have to do that at least six days a week.

Patient: OK Sir, thank you…

Don't you feel that the conversation ended quite abruptly? It seems that the patient suddenly feels too embarrassed to discuss his/her problem further with the doctor. Listen to the revised ending and observe how appropriate it now sounds.

Patient: Normally I do go, but sometimes I miss…

Doctor: No, that won't do… you have to do that at least six days a week.

Patient: I understand Sir, but sometimes I get up late and cannot afford to go for a walk or take exercises.

Doctor: See, if you are not regular in your routine, your problem may aggravate. Actually, the medicines we give in such cases give only temporary relief; the real cure is regular walk and knee-exercise.

Patient: Is it that important, Sir?

Doctor: Yes, dear. It is that important. Miss neither the medicine nor the exercise and walk.

Patient: Will my knee improve then, doctor?

Doctor: Surely, it will. Just be nice to your knee and it will be nice to you.

Patient: Now I understand, Sir. I'll not miss either of these.

Doctor: That's like a good buddy! If you do that, very soon you will get results.

Patient: Thanks a lot, Sir…

Doctor: Wish you all the best…

15.6 TELEPHONIC SKILLS

In the contemporary world, most of our conversations, interactions, and discussions with others take place over the phone. Telephone or cell phone is, in fact, the most commonly used tool of communication, among professionals staying in different parts of the world. Whether they work in a large or a small organization, professionals need to interact mostly through telephones. Although telephonic conversations play an important role in everyday professional world. Not all of us are equally accomplished in displaying appropriate telephone manners. In fact, owing to this pervasive use of telephones, many of us tend to treat it like a trite, everyday affair and hence sound uncouth, discourteous, impatient, or annoyed to our listeners.

Given below are tips following which one can maintain proper etiquette while making or receiving a telephonic call:

1. **Identify yourself and thank them for contacting you**
   Give your full name and the name of your company. Since the caller has taken time to call you, you may answer the phone this way: "Thank you for calling Phonix Infotech, this is Anita De'silva, how can I help you?"

2. **Try to remain positive**
   A positive opening sets the stage for a pleasing experience. A warm and enthusiastic tone tells the caller that he/she is not being regarded as an interruption. Instead of giving him/her the impression that he/she is wasting your time, your tone can make him/her feel like (s)he is the most important person to you at that moment. Establish control of the call by asking the first question.

3. **Consider the tone and speed of your voice**
   Speaking too fast or too slowly or in a very loud or subdued voice might offend anybody. Also if someone places too much emphasis on some specific tone or maintains a bored, flat tone which unmistakably suggests disregard for or disinterest in the caller, it can prove quite risky for your business prospects.

4. **Listen Carefully**
   As discussed earlier, listening to others is not as easy as it appears to be. However, without careful listening, one cannot answer appropriately and with conviction. Therefore, make it a point to take the

calls seriously and listen to the caller very carefully. An empathetic listening can help you achieve success for you and your organization.

6. **Respond**

   Now, some might want to become defensive which brings a change in the tenor of the voice. You may be realistic and practical while answering to a request or complaint, but your reaction should not be rude or dismissive regardless of the caller's intent. While dealing with a complaint, do not start blaming your colleagues or your organization.

7. **Remain Composed**

   Acquire a posture of deliberate politeness with guarded intent. While dealing with a particularly rude, upset or frustrated caller, try to be composed and considerate. One such expression can be:

   "I'm sorry, who were you trying to reach?" Remember, you need to continue to be polite; having done so, however, you need to be still realistic and focused.

8. **End well**

   End the call on a pleasant note. Normally, a pleasantry such as 'Thanks for calling…', 'You are welcome…', 'Feel free to call me if you need anything else…', etc. puts the persons at the other end at ease and helps them create a positive picture about you in their mind.

9. **Time your Call**

   If your client has in-depth and complex queries that will take a while on the phone to sort out, then adequate time should be given to him/her. At the other end of the spectrum, rushing a call or limiting call time can make the other person feel like they aren't important, whereas making the call much longer than necessary can also become an issue.

10. **Be Precise and Clear**

    Ensure that the people who are on the phone actually know what they're talking about. A professional requires knowing the nitty-gritty of his/her subject. Providing clarity to the caller can lead to conviction and credibility.

Tips for an Effective Telephonic Conversation

- Spoken courtesy on a telephone can reap rewards we may not expect.
- Never raise your voice on the phone.
- Try to speak clearly with the receiver a couple inches away from the mouth—not too far, not too close or on the lips.
- Take notes during the call, and record the date and time. Keep all your notes in one place.
- When you call back and get an answering machine, do not hang up. Leave a complete message with your name and telephone number.
- Receive a little formal training in telephonic communication.

15.7 DEBATE

Debate is a form of an argumentative speech that is more of expressing an opinion on a topic related to life, society, politics, and so on. In debates, the contestant knows that there can be opposition from the other side. It is about taking sides on a particular situation and remaining strong about your views. Notwithstanding the issue you intend to support or oppose, the debater has to be smart enough to win over the opposition with proper logical arguments, force of his/her expression, conviction in his/her argument. As a debater has to live upto these challenges, taking part in debates can work miracles for his/her confidence level.

15.7.1 Purpose

Debates are meant for the following purposes:

1. Debates are a means of sharpening critical thinking and improving personal expression.
2. A debate helps in putting across views in a rational manner.
3. It enhances tolerance about other persons' opinions.
4. It helps in inculcating the art of persuasion in a speech.

15.7.2 Features of a Good Debate

Before we go ahead with how to prepare a winning debate, let us see what characterizes a good debate:

1. A good debate is truly convincing and well researched.
2. You must consider the pros and cons of your debate before you plunge into it.
3. A debater should try to get newer ideas.
4. Since the opposition also has their arguments ready, a debater must play this game with conviction.

15.7.3 Preparing for a Debate

Here are some steps that might help you prepare for a debate:

1. Usually, if you are participating in a debate competition, you may get a specific topic to speak on. The topic mostly demands serious thinking and high argumentative skills. These topics may range from political or humanitarian issues to international affairs.
2. Think carefully whether you want to speak for the motion or against it.
   It is advisable that first of all you get convinced about the argument yourself before you move on to convince others. An argument without conviction can lead to chaos.
3. You can stick to the original arguments on a serious mode, or add humour to make it more interesting.
4. Do some research on the topic you are supposed to speak on. Try to collect ample information available and do not leave loose ends, as these can be turned against you by your opponents. Use your resources well, keeping in mind the various advantages and disadvantages your debate topic might have. Study well and prepare an introduction that would bring to the fore the line of thought you have decided to toe in the debate.
5. While preparing your argument, you should also prepare answers to scuttle your opponent's arguments. At times you will find the argument given by your opponent not falling in line with your expectations; even then, the answers or counter-arguments prepared would stand you in good stead to steer through the discussion convincingly. For all this, you need to know your subject quite well; knowing the matter in depth would help you prepare and perform better.

A Debate is Generally Aggressive

6. While preparing the text of your debate, divide the topic into main and sub-points, and pay equal attention to all the segments. Write your points clearly; you can also number your arguments for better referencing.

In brief, observe the following points which you should take into account while preparing a debate and also while delivering it.

15.7.4 A Few Dos and Don'ts

Dos	Don'ts
• Arrive ten minutes prior to the commencement of the debate.	• Be fearful about standing in front of people and delivering your views.
• Ensure that your first and last name is written on the official score sheet and chairperson's sheet.	• Depend solely on stock arguments.
• Dressing style should be formal as it can add to your confidence.	• Wear casual dress.
• Say 'Good morning/evening Mr/Madam Chairperson' at the beginning of your speech.	• If you are sitting, avoiding slouching and if you are standing, avoid a sluggish posture.
• Show alertness and exhibit proper focus, intelligence, interest, and awareness.	• Fidget or fumble while speaking.
• Wait for the chairperson/anchor to introduce you before getting up to express your ideas.	• Use mannerisms that distract the audience.
• Show your appreciation for other speakers.	• Interrupt other speakers while they speak.
• Use notes/cue cards and handle them with confidence. Make sure they are organized, easy to read, and are numbered.	• Converse with the audience or ask the audience's opinions during your speech.
• Shake hands with your opponents after the debate.	• Shout or use personal attacks towards the opposition in your speech.
• Remain quiet while other debaters are presenting their speeches.	• Include false evidence or facts in your speech.
• Be humble in victory and gracious in defeat.	• Use inappropriate or offensive language.

Normally, in debates you are required to speak for three minutes or so, though at times they can run into five minutes. For a three-minute debate you can follow the planner suggested below.

15.7.5 The Three-minute Debate Planner

In this section, we will discuss in detail how to work out a three-minute debate planner.

Topic   The topic given in debates is quite explicitly stated. Just write it down as it is stated, and note which side you are on by writing 'For' or 'Against' at the end of the line.

Issue and justification   The issue is what the topic is really about. This question is very important if the topic is about a metaphor, like 'Grass is always greener on the other side'.

In this example, the issue is not about growing grass, or living on the other side. You must *always* take time to identify the issue.

In the same section, write down your justification for interpreting the issue, whichever way you happen to perceive it. This shows briefly why you have chosen what you have chosen as the issue. At this stage, you can come up with both the 'stock' arguments and the ingenious ones. 'Stock' arguments are those which everyone at the debate is generally aware of. At times they sound repetitive, clichéd, and trite. Therefore, what gives credence and vitality to your stated position are the imaginative ideas and a vibrant, forceful articulation of these ideas.

Definitions and justifications   Having decided your line of thought, you need to understand how to use definitions. Mostly, these will come from your interpretation of the issue, but you must also touch upon the other dimensions of the issue. Sometimes, the definitions will come from what the 'person in the street' would believe, or from the context of the topic.

Case line   Next comes your case line. You need a quick statement of how the debate will proceed from your side. While working on your case line, imagine a structure for your debate and ensure that a structure emerges from your discussion.

Arguments   Then comes a brief points-form listing of your material, the arguments that you wish to put forward. You give arguments for choosing one or not choosing the other. Fill out your arguments by providing facts or evidence so that when articulated, they must ring true to your audience.

Rebuttal arguments   As a separate section, note down rebuttal arguments. When you listen to other debaters' viewpoints, note them down and remember to bring them to the fore when rebutting the other side's arguments. It is repeatedly seen that the winners often clinch the issue in the rebuttal round. The debaters who perform better in this round capture the attention of the audience and influence them immediately. It is so because one cannot prepare a rebuttal beforehand. Therefore, those who perform well in the rebuttal, effortlessly underscore their superior imagination, ideas, expression, and delivery.

Anticipated opposition arguments   Finally, note down what you expect the other side to argue about. There may be more than one way of arguing either side of a case.

Though all the above stated strategies may work for you, in nutshell, it boils down to one single thing: *Practise, Practise, Practise*

Read your case so that it flows flawlessly and consistently. Give yourself enough practice and rehearsals to be able to make hand gestures and eye contact when you read, without thinking about it. Make sure that your emphasis is correctly placed.

Sample debates   Given below are some of the examples which can help you prepare your argument in favour of or against an issue.

Topic: Competition in School is Really Helpful for the Proper Development of Students

For the motion   I think competition in school is absolutely essential to the students for their overall development. It teaches you how life is really going to be. As you know, not everything is always going to be fair and nice. We are all bound to face problems, hurdles, and difficulties in life. Competition in school life helps students understand that they are always going to be competing with others, as well

---

(Contd)

as the standards set by themselves. So, I think competition teaches students how to deal not only with success but also with failure. Students, by going through the competition in school days, are also made to realize that they are not always going to be great at everything, and although that is a tough lesson to learn, I definitely feel it is important for one to learn this art early in life. By this, students learn how to deal with things without losing control every time they don't succeed in their pursuits. Moreover, they also learn how to prepare for various tests, examinations, projects, and exhibitions, properly and systematically for positive results, and what actions, of their own can enhance such an outcome. So, competition is essential in schools, as it is only during the educational process that unbiased feedback and support is given freely, together with a structured approach for improvement. And competition works as a catalyst in bringing about that improvement.

Having said that, however, I also agree with the view that the results of competitions should never be used to lower the self-esteem of students; instead, taking part in something should be valued more. This approach imparts a message that everyone has an equal opportunity to rise to the top, provided they compete with others. Competition thus gives exposure to students. It introduces both the success and the failure, a situation that can lead to great stress levels and illness if not handled properly.

It is in this way, ladies and gentlemen, that competition brings the focus onto important points of learning, as only within a testing situation you are able to clearly judge what you do or do not know. Competition indeed is a tool to mark the incremental development within a subject and indicates whether forward movement or continued parallel studies are necessary, providing anchor points for learning. I would like to end by saying that 'Competition is not the problem, the problem is the way people deal with it.'

Against the motion Competition in school is harmful to students. It induces in them a false sense of superiority or inferiority. Schools are meant to educate students, and just competing with others for more marks, better grades, and a greater number of certificates doesn't really educate them. Competition at a young age breeds in students intolerance, jealousy, and meanness. Competition makes students so blinded and selfish that they lose the sense of community altogether. These very students then have to live in families, work in teams and for communities later on in their lives. And as we can see, the increasing level of competition in schools is making all this increasingly difficult.

Seen objectively, competition hardly helps anyone—neither the winner nor the loser. Those who emerge successful in a competition are declared winners, rankers, or pointers with a superior grade point average, and develop an exaggerated sense of importance. Such students start becoming boastful, arrogant, and egotistical. Such people fail to pull along well in a team. The unsuccessful competitor on the other hand, the loser, or individual with the lowest grades, will develop low self-esteem. Low self-esteem results in the feeling that it isn't worth putting oneself out for anything which lowers the productivity of the work team and leads to unemployment. Moreover, a poor self-image leads to the belief that the individual is disliked by parents, co-workers, and bosses. Students and adults who lack confidence distrust their own abilities and end up being underachievers, derelicts or anti-social elements at times.

After all what are schools meant for? Ladies and Gentlemen, I believe they are meant for educating us. They are meant for taking us from darkness to illumination, from ignorance to enlightenment. And it is not through mindless competition, but through knowledge and wisdom that the gates of enlightenment are unlocked to an individual. That is why, I firmly believe that only when schools stop being competitive, when they stop being judges, graders, and labellers, then alone can they become centres of learning and help students grow in self-esteem, and develop as creative, broadminded, aware, intellectual beings. It is only then that our students of today will grow and become makers of tomorrow. Otherwise, they would always remain parasites; they would just remain hungry, dissatisfied, dejected, and disillusioned people who can merely seek and can never offer anything to anyone. After all, a system of competition that is based on criticizing, reprimanding, coercing, pushing, and punishing, and just looks for flaws and fails to recognize the true potential of young people, builds losers not winners.

Topic: Smoking should be Banned

For the motion   Smoking or Non-smoking…Should there be a choice? Don't answer this categorically. Just imagine sitting in a restaurant. The day is bright, sunny, and beautiful and you want to enjoy a hot pizza with a steaming cup of coffee. But you can't. Because you are unable to enjoy your dish due to the cloud of smoke emanating from your neighbour's table. The fact that there has never been a designated area for smokers has put the smoker and the non-smoker in an uncomfortable situation. Smoking should be banned in public places because non-smokers have a right to clean air, and because passive smoking is as dangerous as active smoking. My opponents say that it is the smoker's choice, whether to smoke or not. Even then, smoking should at least be banned in public places, because just as it is the smoker's choice to smoke, it is the non-smoker's choice not to smoke. By smoking in public places, smokers do not just smoke themselves; they also force the non-smokers to smoke passively. This concern has been raised by many eminent personalities, including the former US President Clinton who is quoted as saying, 'We've got to do more to protect people in public places and clean up the air that all of us share'. Even otherwise, we must understand that for non-smokers, inhaling someone else's cigarette smoke can be pretty nerve-racking, revolting, and repulsive. Can't the smokers understand that it is bad enough that automobiles, processing plants, and other types of industries are continuously polluting our environment? Do they have to add to this nuisance with their smoking, which helps neither them nor others?

Going into a Debate Situation without Preparation can be Disastrous

And, smoking is no longer a mere nuisance, it even risks the life of a passive smoker. Recent reports show that, 'exposure to environmental tobacco smoke (ETS) raises a non-smoker's risk of developing lung cancer by at least 50 percent'. The only effective way to avoid the negative effects of second hand smoke is to insist on a smoke-free environment wherever possible. Those who contravene on this should be taken to task. After all, we are all citizens of a free nation and our constitution encourages the right to freedom of press, freedom of religion, and freedom of speech. But all this is possible only when there is freedom to live and let live. The earlier the smokers understand and respect this basic right of non-smokers, the better it is for them, for others and also for the entire nation.

Against the motion   The decision to smoke is one that adults should be allowed to make for themselves. That decision, however, should not be influenced in any way by those who produce the product.

Alcoholic beverages are also dangerous and life-threatening, causing problems from habitual drunkenness to liver disease, yet few people complain about beer or vodka. In fact, some of these harmful products are ironically designed to help cure diseases, and are known as prescription drugs. Have you ever taken time to read the side effects of your prescribed pain medication, antibiotics, or other drugs? Some prescription drugs can even cause death, yet tobacco products, far from causing immediate harm, are frowned upon. Of course, my opponents may well argue that the side-effects of

(Contd)

such drugs are rare and that the benefits of prescription drugs outweigh the possible harm. I admit, that may be true, but tobacco's effects are not felt till long after a habit has formed, giving the person plenty of time to keep it at bay if he/she so desires.

Today, there is hardly a country where tobacco, in some form or the other, is not consumed. It is smoked, chewed, and dipped (held in the jaw or lower lip). The tobacco industry is one of the world's largest, with annual revenues exceeding the national budgets of all but the richest countries. Despite the industry's efforts to suppress or ignore the facts, tobacco is harmful. Armed with this knowledge, it remains an individual decision as long as it is a legal substance, whether or not to use tobacco.

Given economic and political realities, an outright ban on tobacco production and use is hardly likely or a possible solution to this raging debate. Tax revenues from the sale of tobacco products is something no government, including the United States, is willing to forego. The influence and power of the tobacco industry and their legions of lobbyists, also pose an almost insurmountable barrier to any such effort.

If people choose to smoke, knowing the health hazards tobacco presents, they must live (or die) with that decision.

@ Refer to the Online Resource Centre for more sample debates.

15.8 SITUATIONAL DIALOGUES AND ROLE PLAYS

After learning about conversations, why should we learn about dialogues? Are they same or different? If they are different, how are they different from each other?

Yes, they are same as long as the nature and style are concerned. However, they serve different purposes. In order to understand this in greater detail, let us first understand what dialogues are and how they are written or crafted.

15.8.1 Definition

A dialogue* is a verbal exchange between two or more people that is reported in a drama, movie, or narrative. So, dialogues are the spoken words between two or more characters which serve a purpose within a story.

It is believed that all dialogue should accomplish at least one of the following three things:

- Moving the story forward
- Contributing to characterization
- Giving information

Thus, it can be understood that in the most basic kinds of writings, dialogue serves the interest of moving the story forward. Better writing involves dialogue that contributes to characterization, where what a character says somehow relates to what he/she does as well. Dialogues thus help to portray events which are to come, and make these events more vivid when they do arrive. They also give life to characters, and establish the kind of relationship that exists between them.

It is useful to learn how to construct dialogues. It helps you perform better whenever you wish to participate in a role playing exercise or perform on stage. Moreover, it also helps you appraise the dialogues spoken by characters on stage, in movies, and fiction. Most importantly, learning to appreciate the art of speaking and constructing dialogues helps you improve your overall communication skills.

*Source: http://hollylisle.com/fm/Workshops/dialogue.html, accessed on 19 September 2010.

15.8.2 Tips for Writing Dialogues

Following are important tips for writing dialogues:

1. Dialogue should have a certain verisimilitude. It should seem real to the reader.
2. When quoting a dialogue, put the words of each speaker within quotation marks, and indicate a change in speaker by starting a new paragraph.
3. Use contractions ('don't', 'shouldn't', 'can't') unless a character is very stuffy or speaks in a very formal context.
4. Internal/inner dialogue (thinking) does not need to be within inverted commas.
5. Let characters break off sentences, or speak in phrases rather than sentences. (You might think of these as *verbless sentences*—they are ideal for dialogue.)
6. Have characters interrupt one another.
7. Use the occasional 'um' or 'er', if a character is being particularly hesitant.
8. During a discussion, include the name of the speaker after every five to six pieces of dialogue; otherwise readers may find themselves flipping pages to find out who actually said what.
9. Remember to show who is speaking, it need not be a 'he said' or a 'she said', an action works just as well, provided we are told who is doing it.

15.8.3 Giving Characters Distinct Speech Patterns

Some factors should be taken into consideration when finding each character's 'voice' as well as their personality. These are as follows:

- Educational background of the character
- Likes and dislikes
- Place they belong to
- Speech habits and other behavioural patterns
- Age
- Occupation

All these will decide whether your character is well defined or long-winded, whether they use technical terms or that of a lay person. The factors will also determine the sort of slang that your characters use.

15.8.4 Learn How to Write Dialogues and Perform Role Plays

While learning to write dialogues, start with a conversation between two characters who represent contrasting world views. What all this means is that one character feels one way about the world/life in general and the other character feels quite the opposite. Try to form small groups comprising your friends or classmates, assume a situation, assign some role to each other, and write dialogues for each person who has to play a specific role in the given situation. Once the dialogues have been written for each such character, perform it in the class or before another group. Some such situations follow next and some more are available in the Online Resource Centre of the book.

Before you start however, keep in mind that dialogue thus framed is definitely not generally a literal representation of the way people really speak. Everyday speech is full of repetition and hesitation and mundane utterances which are extremely tedious and boring when written down. Read the following dialogue and see how dull and meaningless it sounds:

> Speaker 1: Good morning Jassi, how are you?
> Speaker 2: Oh I'm fine thanks, how are you?

(Contd)

> Speaker 1: Not too bad, thanks. Lovely weather today, isn't it?
> Speaker 2: Yes, gorgeous.
> Speaker 1: Yes, I thought it would rain again. Did you watch any movie last night?
> Speaker 2: Yes, I saw *Titanic* on STAR Movies, it had that actress in it, what's her name? Oh goodness what is her name? It's on the tip of my tongue, hold on a sec....

Does it really interest you? Will you keep reading this story or abandon reading it? Dialogues should always be used to convey something important to the plot. They convey the rhythm and syntax of real speech. Dialogues need to convey information to the reader, but in a way which sounds natural. For example, if Jassi tells Malti:

> Have you heard Rahul Bose, the film actor, will be the chief guest for our cultural fest?

This line conveys to us that there is a character around called Rahul Bose, who is known, and it also conveys the added information that this character will be the chief guest of some function. So, this piece of information will carry the story further. It does so in a way which sounds perfectly natural.

Don't overload dialogues with information. If you do so, they become conspicuous and sound unnatural. For example:

> Do you know Rahul who acts in movies, I am talking about Rahul Bose... the hero of the movie *Jhankar Beats*, about whom I told you I met in Delhi at some function... he will be the chief guest of our cultural function.

This is an absolutely cluttered dialogue, since it strives to convey too much in one dialogue. In a scene with only two characters, dialogues such as these can largely be dispensed with, but with three or more characters present, the reader will get caught in the whirlpool of information. Therefore, keep the dialogue crisp, brief, and pointed in order to secure the attention of the audience.

 ## RECAPITULATION

✓ Conversations are an essential form of human interaction and constitute an integral part of our everyday speech.

✓ It requires certain skills to be able to emerge as a good conversationalist, both in personal and professional situations.

✓ Good listening, paraphrasing others' views in your own words, resisting a desire to dominate, asking pertinent questions, and using positive body language helps you emerge as a good conversationalist.

✓ Conversations differ from dialogues in the sense that conversations are more general in nature, while dialogues are shared by characters on stage and in movies and fiction.

✓ Debates differ from conversations and dialogues. Debates are argumentative speeches in which the debater presents his/her views in favour of or against an issue.

✓ Since debate is a formal means of communication, it requires a lot of preparation and practice.

✓ Often delivered in an energetic and vibrant mode, debates help one sharpen one's communication skills, particularly the art of persuasion and negotiation that can be well augmented by participating in debates from time to time.

✓ Debaters require preparing meticulously for debates. They have to firmly put into place their own line of thought. They also have to anticipate their opponents' line of thought and keep their argument prepared for any eventuality.

✓ While participating in a debate, the debater has to observe proper etiquette and learn the art of convincing others without bullying or hurting them.

WISEWELL QUIPS

EXERCISES

I. Write dialogues for the following situations in about 150 words each:

1. Discussion between mother and father regarding their son's carelessness towards studies.
2. Career counselling of a student by a counsellor.
3. A discussion between two friends regarding the existence of God, as one believes in the existence of God whereas the other does not.
4. Imagine that you plan to write a story about an exciting train journey wherein you wish to show a dialogue between two passengers. One of them hails from Bihar and the other from Maharashtra.
5. A dialogue between a father and son. Here the son seeks permission for going on a motorcycle excursion with his friends from Bombay to Goa and the father is not willing to allow him.
6. A discussion between two friends (girls) regarding a movie which is liked by one but the other does not like it.
7. Imagine you are writing a play and you wish to include a funny dialogue between a teen age boy and a stranger who is asking him about the location of a house in his locality.

II. Prepare debates on the following topics in about 500 words each. Choose arguments both in favour of and against the topic:

(a) Television does more harm than good.
(b) Homework should be banned.
(c) School uniforms are good for students.
(d) We can do away with cars, an unnecessary luxury.
(e) All our woes have come from science.
(f) The value of Purpose in Life
(g) Education is the key to success.
(h) Real education starts after school.
(i) There is strength in diversity.
(j) Peace cannot be without fear.
(k) Vegetarian food is only good for human beings.
(l) Women's lib is a fib.

(m) Education without Value/Culture is Futile.

(n) Beauty lies in the eye of the beholder.

(o) A thing of beauty is a joy forever.

(p) It is better to plant a cabbage than a rose.

(q) He who hesitates is lost.

(r) We should look before we leap.

(s) Computer games do more harm than good.

(t) We are slaves to conformity.

(u) The media promotes a selfish society.

(v) Smoking advertisements should be banned.

(w) Democracy is the best form of governance.

(x) Professionalism has destroyed sports.

(y) It is better to be feared than to be loved.

(z) A little knowledge is a dangerous thing.

CHAPTER 16

The Art of Negotiation

Learning Objectives After reading this chapter, you will be able to

- understand what negotiation is and how important it is for you to learn this art
- identify the situations that require successful negotiating skills in your personal and professional life
- know the different stages of the negotiation process
- acquaint yourself with the elements of a successful negotiation
- understand the dos and don'ts of a successful negotiation

16.1 INTRODUCTION

Successful negotiation is an art that comes naturally to some, but must be developed by most. Negotiation is a part of everyday life, but in business it is absolutely vital for success. You must have observed while buying a car, furniture, or a house that all of us try to bargain the price that we have to pay. Even during HR interviews, candidates have to negotiate for a better salary. All this requires good negotiation skills. In fact, in business, poor negotiation affects the total health and wealth of the company as it may cripple its growth and result in losing key customers or great financial losses in business deals. At the workplace, we negotiate with our clients, other organizations, or government agencies to obtain their support and cooperation for our business and vice versa. Thus negotiation takes place every day, however it is not easy to perform this task without skill, efficiency, and persuasion. Although every negotiation is different, the basic elements do not change. The question now arises what negotiation is and how we can develop a knack at it.

16.2 DEFINITION

According to the freedictionary.com, 'negotiation is the act or process of conferring or discussing to reach agreement in matters of business or state.' This means it is a kind of bargain with others for a treaty or contract for expansion of business, maximization of profit, or better growth of the state. Etymologically, the word *negotiation* comes from the early 15-century French word *negociacion* which means 'business, trade'

A Negotiator can be Soft, Hard, or Principled

and from Latin word *negotiationem* that stands for 'business, traffic.'

Since emotion, luck, and magic have no place in a successful negotiation, it becomes essential for us to know what steps the negotiation process involves and how to go about develop better negotiation skills in order to be successful.

The process of negotiation comprises the following six stages:

1. Relationship building
2. Information gathering
3. Strategy formulation
4. A win-win solution with persuasive tone
5. Concessions and agreements/contracts
6. Final offer for closing the deal

16.3 DIFFERENT TYPES OF NEGOTIATION STYLES

Kenneth W. Thomas, a famous researcher on the art of negotiation, has identified five styles of negotiation. Negotiators can be soft, hard, or principled, and can have strong dispositions towards numerous styles but their styles can change over time. The style also depends on how far the negotiators are relationship-oriented and deal-oriented. These negotiation styles are as follows:

Accommodating   Such individuals enjoy solving the other party's problems and preserving personal relationships. It is observed that accommodators are sensitive to the emotional states, body language, and verbal signals of the other party with whom they negotiate for business purposes. In this style, the negotiator feels comfortable when the other party also values and lays emphasis on building relationships. They are soft bargainers. The offers they make are usually not in their best interests, they yield to others' demands, avoid confrontation, and maintain good relations with fellow negotiators. Their perception of others is one of friendship and their goal is to reach an agreement.

Avoiding   These people do not like to negotiate unless necessary. Avoiders tend to defer, dodge, or evade the confrontational aspects of negotiating. They avoid contests of wills and insist on agreement. They are at times perceived as tactful and diplomatic because of their reserved behaviour.

Collaborating   Such individuals enjoy negotiations that involve solving tough problems in creative ways. Collaborators are good at using negotiations to value the concerns and interests of the other parties. They can, however, create problems by transforming simple situations into more complex ones because of their creative and innovative approach. They are principled bargainers and base their choices on objective criteria rather than power or pressure. These criteria may be drawn from moral standards, principles of fairness, professional standards, tradition, and so on.

Competing   These negotiators are the ones who enjoy negotiations because they present an opportunity to win something. Competitors have strong instincts for all aspects of negotiation and are often strategic. They often neglect the importance of relationships because they feel this might lead them to lose in the bargaining process. Such people use contentious strategies to influence and use curt phrases such as 'this is my final offer' and 'take it or leave it'. They make threats, are distrustful of others, insist on their position, and also exert pressure on others to negotiate. They see others as opponents and their ultimate goal is victory.

Compromising   Such individuals are eager to close the deal by doing what is fair and equal for all parties involved in the negotiation. Compromisers can be useful when there is limited time to complete the deal. However, they unnecessarily rush the negotiation process, making concessions too quickly, and as a result, suffering losses often.

16.4 TIPS FOR WIN-WIN NEGOTIATION

For making a successful negotiation, you need to possess a zeal for success. As a negotiator do your homework, be street-smart, and have discipline. These keys will unlock your ability to get the best deal possible under any circumstances. Let us know a little more on how to have a win-win situation during negotiations.

Preparation Is Key

As a negotiator, you need to know about the party you are negotiating with so that you can capitalize on your strengths and the other party's weaknesses. If the other party is very experienced, it means they also have a history that could contain useful information. If possible, talk to your business associates who have already dealt with that party. For example, before negotiation, the Japanese ask too many questions to identify the needs and preferences of all the stakeholders involved in the negotiation. They are known for their long drawn-out preliminary ground work.

If you are a buyer, make sure you know thoroughly about the product or service that is the subject of negotiation. If the other party gets to know that you have incomplete details, you may become a prime target for a bluff or some other technique designed to create anxiety and uncertainty. In fact, in order to sail through a successful negotiation, we need to understand the psychology of the other party, size up their vulnerabilities, and anticipate their next move.

While aiming for a negotiation with regard to a price target or goal in mind, you should have realistic expectations considering all the constraints that will undoubtedly surface. These may include budget limits, direction from management, pressure to make sales goals, and multitude of other external forces. During the course of the negotiation, the goal may change based on alterations in scope and other unforeseen actions by either party. Before you start the negotiation, ensure that the other party is fully empowered to make the required commitments. You should not land up in a position where you believe you have signed a deal, only to discover that your agreement will be approved by someone higher in the chain of command and will be subject to change or unnecessary delay.

Have a Strategy

There are certain fundamental principles that apply to every negotiation. It is a well-established fact that you will never get what you do not ask for, therefore make your first offer bold and aggressive. The asking price normally includes a margin to give away during negotiations. As long as your offer is not ridiculous, the other side will continue the negotiations in hopes of settling at a better number. Therefore, always have something to give away without hurting your negotiating position.

If you are submitting a price proposal to a buyer, consider quoting the last price from your side that may be attractive enough for the other party. Also, while making the proposal, list all the items which may be omitted during final negotiations. For example, if you are bidding for a project, consider including some items that are not critical to the success of the project but will be nice to have. You could also include spare parts that may or may not be needed in the end. If the buyer takes those items out to reduce the overall cost, you will not lose anything but it may help the buyer reach his/her price target.

Read Body Language and Understand the Speech Pattern

While sitting at the negotiation table, be vigilant and watch out for clues such as body movements, speech patterns, and reactions of the other party to what you have to say. For example, US negotiators make an effort to modify the views of their negotiating counterparts using persuasive tactics. Also, be prepared to suspend or cancel negotiations if you feel things are getting nowhere or the other party does not seem to budge from their position. Express your reluctance to continue under that situation. Be patient even if the other party is not.

As far as body language is concerned, positive signs include nodding of the head and direct eye contact whereas negative signs include folded arms across chest, eye evasion, or a subtle head shake as if to say 'no'. Besides, words employed are also very important because they shape not only how other people hear you, but also about how they feel about you.

Find a Leverage

In order to find a leverage, exploit the other party's weaknesses and concentrate on taking maximum advantage of your strengths. If you are the only source available for a particular product, you will enjoy an incredible advantage. If economic conditions have created a market in which the product you are selling is in huge demand and short supply, it will give you more bargaining power to quote your price. If you are a buyer in a depressed economy, you will normally have the advantage of too much supply and low demand. The current housing situation is the most suitable example of what happens when supply outweighs demand and market prices fall considerably.

Making the Final Offer

An offer is more than just money that it involves. It must include a basis for a contract that formalizes the agreement. If you make an offer without jotting down all the specifics, you may discover later on that it was hardly a good deal. The basis of the bargain should include offer price, statement of work with its scope, identification and quantities of goods or services, delivery schedule, performance incentives (if any), express warranties (if any), terms and conditions, and any other document incorporated for further reference.

Trading one element for another, such as a lower price for a more relaxed schedule, is a common tactic. If the other party refuses to alter the onerous terms, consider taking your business elsewhere. To avoid misunderstandings, offers should be presented in writing and must include all elements of the deal. It is advisable to keep notes containing the rationale for each offer. If you work for a company or the government, those notes are usually required to document the negotiated outcome and complete the contract file.

Go for a Win-win Solution

Throughout the negotiation, try to determine what you consider an acceptable outcome for the other party. A negotiation can be termed as a win-win negotiation, when both the parties reach their objectives. Such a negotiation is also called integrative negotiation. Both parties cooperate to achieve maximum benefits by integrating their interests into an agreement. For example, the delivery date or price may be the most important thing to the other party, while product quality may be your primary concern.

Understanding the other side's priorities is just as important as understanding your own, so figure out what you would do if you were in their shoes. When formulating your offers, attempt to satisfy some of their priorities if doing so will not weaken your overall position. While you have the power to influence the negotiation process in your favour, your goal should be to secure a good deal from the other party. This is especially true if you are negotiating with the same party on a recurring basis. The most effective negotiators are professionals who know their business and do not allow personalities and unfounded behaviour obscure their business objectives. If both the parties have to compromise, it will undoubtedly lead to a lose-lose situation.

Know Your BATNA

As a negotiator you should be aware of the best alternative to a negotiated agreement (BATNA). It is the most preferred course of action should negotiations do not come to fruition. You should know your BATNA beforehand, otherwise it becomes hard to know when a negotiation makes sense and when it does not. For example, suppose you plan to buy a three-bedroom house for which you wish to invest ₹70 lakh. You know it beforehand that instead of buying a two-bedroom house in that prime location you would like to buy a three-bedroom house in a location that is within five kilometers of the prime location. This idea of buying in the periphery of the prime location is your BATNA.

Pay Close Attention to Closing the Deal

Successful negotiation is like horse-trading as it requires a sense of timing, creativity, keen awareness, and the ability to anticipate the other party's next move. Look for a profitable negotiation, but if it does not happen, be prepared to withdraw. You should not lose heart at the untimely death of your effort. That is the reason why all the stages are equally important and until the final offers are made and the written documents are signed, nothing is for sure. As a final handshake, both the parties should make the agreement legal.

Keep the Discussion Result-oriented

It is recommended that while negotiating, keep the focus of the negotiation solely on results and what would make the best long-term deal for both parties. By taking the emphasis off the people involved and keeping it on the facts, the negotiation is less likely to become a failure.

Be Wise and Concerned

Framing negotiations around mutual interests rather than your interest alone helps both sides feel good about it. If you are respectful, it is more likely that the other will reciprocate the same for you. Rather than looking for short-term gratification such as assuming a power posture or making a joke at another's expense, always exhibit maturity and keep yourself focused on the end result.

Put your Concerns on the Table

It is better to voice your fears and express your worries during the negotiation itself. The other party may help you out with a solution they might have. Long-term negotiation can happen only if there is an open communication and issues and challenges are shared candidly.

Avoid 'I' Statements

When two parties wish to work together instead of one saying 'I need to reach a solution,' the negotiator should say 'we need to reach a solution'. Such inclusivity will keep the discussion focused on the deal and will not be thwarted by unnecessary ego display. As a result of this, the atmosphere will not be suffocating for both the parties.

Exercise your Right to Negotiate

You have worked very hard for your money or the product you wish to negotiate about, therefore it is your right to ask for a better price, more benefits, a free warranty, or a first-time-customer discount. The best part is that most vendors want you to negotiate. They would be surely keen to give you a 10–20 per cent discount if it results in retaining you as a customer. Show them your value and exercise your right to negotiate.

Other Tips

Language and cultural understanding of the fellow negotiator, socialization, humour, listening skills, business etiquette, and shared experience are certain other communication competencies that influence the success of a negotiation.

Don'ts

- Lean back in your chair as it puts a physical distance between you and the other side.
- Use expressions like 'take it or leave it'.
- Lower down your price or esteem unnecessarily.
- Sound dominating while presenting your deal as it will make you sound incredibly haughty or boorish.
- Quote unrealistic prices.
- Be lazy in gathering information about the other party.
- Feel shy or jerky about the legal formalities.
- Overlook essential pieces of information.
- Fold your arms across your chest, avoid eye contact, or give subtle head shake as if saying 'no'.
- Give extended pauses as it usually means that the opposing party is hesitant or is pondering upon the offer.
- Burn your bridges, if a negotiation fails to work out. Keep the lines open for future deals.

 RECAPITULATION

✓ Negotiation is the act or process of conferring or discussing to reach agreement in matters of business or state.

✓ The negotiators can be soft, hard, or principled, their styles can change over time, and can have strong dispositions towards numerous styles.

✓ As a negotiator, you need to know about the party you are negotiating with so that you can get a good deal.

✓ In order to find a leverage, exploit the other party's weaknesses and concentrate on taking maximum advantage of your strengths.

✓ The basis of a bargain should include offer price, statement of work with its scope, identification and quantities of goods or services, delivery schedule, performance incentives (if any), express warranties (if any), terms and conditions, and any documents incorporated for further reference.

✓ Successful negotiation is like horse-trading as it requires a sense of timing, creativity, keen awareness, and the ability to anticipate the other party's next move.

WISEWELL QUIPS

EXERCISES

1. 'A good negotiator is defined as one having a quick mind but unlimited patience, someone who knows how to dissemble without being a liar, inspire trust without trusting others… while remaining indifferent to all temptations…', Fred Ikle in *How Nations Negotiate*.

   Discuss the statement and analyse the qualities that a negotiator should exhibit while negotiating.

2. Describe in brief the different negotiating styles.

3. What is BATNA? Give one example to explain its importance for a negotiator.

4. Provide ten points each for the dos and don'ts of negotiation.

5. Imagine that you have an old house in a prime locality of a city, which you wish to sell and buy a new one in a colony 15 kms away from the city. Discuss the kind of background check that you will do and explain briefly in the form of a dialogue how you will carry the negotiation further.

6. Assume that you wish to buy a second-hand car, for which you plan to visit a showroom of used cars. Before you pay a visit, list the things that you will keep in mind and what will your BATNA be and why?

The Art of Effective Reading

> **Learning Objectives** After reading this chapter, you will be able to
> - understand the importance of reading in achieving success both in academic and professional life
> - acquire the various types of reading skills that one may employ while reading different kinds of texts
> - learn the four basic steps in the process of reading
> - get to know the basic hurdles in efficient reading and the tips to overcome them
> - know the relation between speed reading and success and also learn the ways to increase your reading speed

Resolve to edge in a little reading every day, if it is but a single sentence. If you gain fifteen minutes a day, it will make itself felt at the end of the year.

–Horace Mann (1796–1859)

17.1 INTRODUCTION

Reading maketh a full man, conference a ready man, and writing an exact man—Francis Bacon, the famous essayist, has rightly observed in his well-known essay 'Of Studies'. It is true that of all the four skills—listening, speaking, reading and writing—reading is probably the most crucial skill. It is so because a good reader can rise above the disadvantage of limited opportunity received in terms of listening, speaking, and writing a language. By all means, effective reading skills are vital to achieving success not only in one's academic but also professional life. Usually, it is observed that only good readers are good communicators. It is a skill that cannot be ignored. It is advisable to acquire efficient reading skills as early as possible.

17.2 NEED FOR DEVELOPING EFFICIENT READING SKILLS

You definitely know how to read. But the question is whether you know how to read skilfully and artfully. Skilful reading is reading for specific information for a better learning experience in a short span of time. It is an art which can be learnt easily by using a systematic approach and by undergoing formal training. This chapter will help you learn this art. You will surely be able to know how to use diverse ways so that you can enhance your effectiveness and make your future reading experience more rewarding and enriching.

Have you ever noticed that we do not approach a novel, or a report, or a personal letter, or an email in a similar manner? In fact, the truth is that while reading, you employ different reading speeds and different approaches to them. However, it is also true that different people have varied speeds of reading and understanding. So, as a student, who has to read much for academic pursuance, and as a future professional, there is a dire need to learn the skills and techniques of artful reading. Before we talk about these techniques and types, let us know the major benefits that you may achieve by developing the art of efficient and artful reading.

17.3 BENEFITS OF EFFECTIVE READING

Here are the benefits of effective reading:

1. Effective reading can provide you with a 'comprehensible input' from the book or document that you read.
2. It can enhance your general ability to use other language skills such as listening, speaking, and writing.
3. It can enhance your vocabulary, that is, you can always learn new words, phrases, and expressions.
4. Effective reading also helps you keep your mind focused on the material and prevents it from unnecessary distractions.
5. Moreover, this enables you to extract useful information much more efficiently within a limited time.
6. It can help you consolidate both previously learned language and knowledge.
7. It helps in building confidence as you start reading longer, and later, voluminous texts.
8. It gives you pleasure and relaxation as books are the best and most reliable friends.
9. Effective reading facilitates the development of various other skills, such as making predictions, comparing and contrasting facts, creating samples, hypothesizing, reorganizing the message as transmitted by the text, improving your critical thinking, and hence developing a sharp acumen with passage of time.

17.4 SPEED OF READING

An average college student reads between 150 and 250 words per minute. A 'good' reading speed is around 250 to 400 words per minute, but some people can read even 1000 words per minute or more. However, research shows that there is difference in the rate of reading for different purposes (Table 17.1).

Table 17.1   Rates of Reading

Purpose of Reading	No. of Words (words per minute/wpm)
• Reading for memorization	• Fewer than 100
• Reading for learning	• 100–200
• Reading for comprehension	• 200–400
• Skimming	• 400–700

Understanding the role of speed in the reading process is essential. Research shows that a close relation can be observed between reading speed and understanding. Proper reading training can help you increase both your rate of reading and comprehension. It is a vital fact about reading that plodding, which means reading word by word, reduces understanding rather than increasing it.

Actually, most adults are able to increase their reading rate significantly and rather rapidly without lowering their comprehension. Merely reading more rapidly, without actual improvement in the basic reading habits, however, usually results in lowered comprehension. Broadly categorized, there are three main factors involved in improving reading speed:

- The desire to improve
- Conscious efforts in using newly acquired techniques
- A motivation to practise

With these, almost anyone can double his/her speed of reading while maintaining equal or even better comprehension.

Table 17.2 gives the differences between an efficient and inefficient reader.

Table 17.2 Differences Between Efficient and Inefficient Readers

An Efficient Reader	An Inefficient Reader
• Always reads for ideas or information	• Tries to read words
• Reads group of words/multi-phrases	• Reads word by word
• Quickly adjusts his/her speed of reading to the nature of the text	• Reads the text from the beginning till the end
• Sets the purpose of reading right in the beginning	• Reads everything and deliberately goes slow while reading
• Reads smoothly	• Reads the information again and again to figure out a clear understanding of the text
• Visualizes ideas	• Vocalizes or sub-vocalizes words while reading
• Has a good vocabulary in that subject	• Has limited vocabulary which hampers his/her speed and understanding
• Continuously keeps improving his/her pace of reading	• Rarely attempts speed reading
• Properly tries to sort out the material as critical, interesting, analytical, etc.	• Reads everything indiscriminately

17.5 FOUR BASIC STEPS TO EFFECTIVE READING

While reading a text to learn something, you need to follow the following four basic steps:

1. *Figure out the purpose of reading a particular text.* You can identify suitable reading strategies and use your background knowledge of the topic in order to anticipate the contents.

2. *Spot the parts of the text relevant to the identified purpose and ignore the rest.* This selectivity enables you to focus on particular elements of information from the text. So, you are able to sift through the information, which in turn reduces the amount of information you have to hold in short-term memory.

3. *Choose the appropriate reading strategy that suits your purpose.* Select the strategy that is suited to the reading task in that particular context and use that strategy in an interactive manner. This will develop your understanding as well as confidence.

4. *Test or assess your comprehension during reading and also when the reading task is completed.* Monitoring comprehension helps you make out the inconsistencies and discrepancies in total comprehension of the text. At this step, you can also learn to use alternative strategies.

17.6 OVERCOMING COMMON OBSTACLES

Let us now learn the major hurdles in developing efficient reading. These problems are as discussed below.

17.6.1 Lack of Concentration

Poor concentration directly emanates from poor reading habits. If we are not used to reading, it often appears to be an unpleasant task when we are forced to do it. In such circumstances the mind begins to drift. However, since we know that there is no escape from reading, we need to cultivate an active interest in our reading assignments. Gradually, our concentration begins to improve if we consciously start taking interest in the reading activity.

17.6.2 Eye Fixation

While reading, when one is unable to progress well beyond a thought, expression, or word, it is regarded as *eye fixation*. This is often caused because of lack of training, and cripples our reading efficiency. To overcome this, try to divide a sentence into certain thought units, and don't fixate on a particular word or expression for long.

Don't Let Your Mind Wander while Reading

17.6.3 Regression

Just like eye fixation, regression too creeps into our reading habits because of lack of training. When we are unable to understand an idea, we habitually go back to the part where it occurs in a passage. It clearly suggests incompetence in reading and comprehension skills. Enhanced reading habits with enthusiasm and interest can help you overcome this problem. For efficient and artful reading, you should make your eye move over at least 2–3 words in one go and let it glide over the text in a rhythm. Since reading is an intensive process, if trained well, the eye quickly moves to assimilate text.

17.6.4 Reading Aloud or Turning the Head from Side to Side

Often we see mothers telling their little tots *to read aloud* from their books to them, so that they can also finish their kitchen work while they help their kids improve their reading skills! This, however, sometimes leads to poor concentration while reading, as their primary objective in such reading becomes proper articulation of certain words and expressions. Similarly, moving or gliding a pen, pencil, or scale through the text are also not worth imbibing, as any added activity besides reading and comprehension tends to affect your understanding of the concept. Ideally, our eyes should see and our mind should read the text before us.

17.6.5 Indiscriminate Use of the Dictionary

Some damage to reading habits is also done because of lack of guidance about the proper utilization of the dictionary. Some of our students are even misled into cramming the dictionary they possess. Remember, a dictionary is not meant for this purpose, and you don't learn words by cramming them. For picking up more words, work conscientiously at least on some part of your reading material everyday. Observe carefully how the words and expressions are used in a given context. Look up the meaning of all such words in the dictionary to understand their specific usage, and the words similar and dissimilar to them. It has to be followed by using all such words and expressions frequently in your speaking and writing tasks.

Another misuse of the dictionary is the tendency to refer to it the moment one comes across a word one finds difficult to understand. However, when we rush to look for the meaning of a word in the dictionary the moment we confront it, we lose our spontaneity in reading. Moreover, most of us start with a modest vocabulary; hence, the number of words which appear unfamiliar to us are many in number. Repeatedly looking into the dictionary for every such word also makes us weary of the whole process of reading and we tend to give up very soon. Therefore, go through your reading assignments without taking such breaks; encircle the words and expressions you are unable to understand. Once you have finished the entire assignment, or a least some sizable chunk of it, turn to the dictionary to understand the words. Even as you do so, try to scan the target word from different perspectives. Try to learn its pronunciation, identify its class, its singular and plural forms, its different meanings and usages, and the words which are similar to it and those that are dissimilar.

Given below are a few more obstacles that you need to overcome while reading:

1. Reading becomes a cumbersome activity when you skip the major part of the text and are in a hurry to finish it.
2. When you do not read a longer text at stretch for at least an hour to allow yourself to develop interest in it, you cannot enjoy reading.
3. When your mind is preoccupied with other thoughts rather than the message that you are reading, the time spent is just wasted.
4. Sometimes students have the tendency to read aloud, which is called *vocalization*. They do so because they feel they should not get distracted while reading. However, this kind of reading not only mars their speed of reading but also their comprehension of the text.
5. Reading becomes an uphill task when you have an unfriendly approach towards it.
6. Reading becomes boring when you do not select a book or text for reading according to your linguistic and cognitive level.
7. One cannot read faster in an effort to remember everything rather than to remember selectively.

17.7 TYPES

Now, for learning the art of reading, you should know the four basic types of reading skills that a reader may use in any language. These types will be discussed in this section.

17.7.1 Skimming

Skimming is used to quickly gather the most important information or 'gist' from the document or novel that you read. For this, you should make your eyes run over the text, noting important information. You may also go through the contents of the chapters rapidly, picking out and understanding the key words and concepts. For example, while reading the newspaper in the morning, you quickly try to get the general news of the day and discover which articles you would like to read in more detail.

17.7.2 Scanning

Scanning is used to find a particular piece of information. In this type of reading, you allow your eyes to run over the text, looking for the specific piece of information that you need. You use scanning when you go through your schedule or meeting plans, look for specific advertisements, or some specific information in a book, etc. This is the most effective way of getting information from magazines—scan the table of contents or indexes and turn directly to interesting articles to arrive at the specific information.

17.7.3 Extensive Reading

Extensive reading is used to obtain a general understanding of a subject. It includes reading of the longer texts for pleasure, and business reports to know about the general trends. In this type of reading, you should not worry if you do not understand each and every word. You can use this type of reading to improve your general knowledge of business procedures, or to know the latest marketing strategy. Also, while you read a novel before going to bed, or magazine articles that interest you, you use your extensive reading skills.

17.7.4 Intensive Reading

Intensive reading is used on shorter texts in order to extract detailed information. It includes very close and accurate reading for details. You use this skill to grasp the details of a specific situation. In this case, it is important for you to understand each word, information, or fact. When you read your text book for understanding the concepts or various theories, you have to read intensively. You also require to read every single detail when you want to go for an insurance claim or want to sign a contract. Intensive reading skills help you understand and interpret the text, infer its central idea, observe the common thread that runs through it, and intuit its overall purpose and significance.

17.8 METHODS OF READING

There are several methods of reading, with differing rates, for different kinds of material and purposes.

Sub-vocalized reading   It combines sight reading with internal sounding of the words as if spoken. It is a bad habit that slows down the pace of both reading and comprehension. However, some studies show that it helps the readers better, particularly while reading complex texts.

Speed reading   It is a method for increasing reading speed without a reduction in comprehension or retention. With effort, you can acquire a speed of as much as 1000 words per minute.

PhotoReading   PhotoReading is different from "regular reading" or "speed reading". It is a way to process and understand information than to read it. It claims to process and store the information merely by looking at the page by use, the whole mind rather than only the left or right side of our brain.

In this method, you use speed reading techniques with an additional technique of photoreading to increase your reading speed, comprehension, and retention.

Proofreading   It is a kind of reading for detecting typographical errors in a printed text. One can learn to do it rapidly, and professional proofreaders typically acquire the ability to do so at high rates, faster for some kinds of material than for others.

Structure–Proposition–Evaluation (SPE)   Mortimer Adler popularized this method in his book *How to Read a Book*, mainly for non-fiction treatise, in which one reads a writing in three phases: (1) for the structure of the work, (2) for the logical propositions or progression made, and (3) for evaluation of the merits of the arguments and conclusions. This method demands suspended judgement of the work or its arguments until they are fully understood.

Survey–Question–Read–Recite–Review (SQ3R)   This method involves immaculate and perfect reading, that is, you read so as to teach what is read or be able to explain or apply the knowledge obtained in other domains. In this method, the reader absorbs the information and uses it without having to refer to notes or the book again.

Multiple intelligences-based method   This method draws upon the reader's diverse ways of thinking and knowing to enrich his/her appreciation of the text. Reading is fundamentally a linguistic activity. Most readers use several intelligences, such as auditory, visual, and logical intelligence, while reading, and making a habit of doing so in a more disciplined manner, that is, constantly or after every paragraph, which results in more vivid and memorable experience.

17.9 APPROACHES TO EFFICIENT READING

You should know that forced acceleration may destroy confidence in one's ability to read. The obvious solution, then, is to increase your reading pace as a part of the total improvement of the whole reading process.

In general, *decrease* your speed when you experience certain difficulties (Table 17.3).

Table 17.3   Different Suggested Approaches to Reading (By Decreasing the Speed)

Type of Difficulty	Suggested Approach
Unfamiliar terminology	Try to understand it in the context at that point; otherwise, read on and return to it later.
Difficult sentence and paragraph structure	Slow down enough to enable you to untangle them and get the accurate context for the passage.
Unfamiliar or abstract concepts	Look for applications or examples of your own.
Detailed, technical material	Since it includes complicated directions, statements of difficult principles, technical information be patient and approach the text critically.
Material on which you want detailed retention	Since you want to remember them for a longer duration, read such a material with intensity, observation, and painstaking effort.

Accelerate your pace when you come across certain difficulties, as given in Table 17.4.

Table 17.4   Different Suggested Approaches to Reading (By Accelerating your Pace)

Type of Difficulty	Suggested Approach
Simple material with few ideas which are new to you	Move rapidly over the familiar ones; spend most of your time on the unfamiliar ideas.
Unnecessary examples and illustrations	Since these are included to clarify ideas, move over them rapidly when they are not needed.
Detailed explanation and idea elaboration which you do not need	Since you do not need them, you can move rapidly over them.
Broad, generalized ideas, and ideas which are restatements of previous ones	These can be readily grasped, even with scan techniques.

17.10 TIPS FOR EFFECTIVE READING

Following are some important tips you must keep in mind about reading:

1. Get your eyes checked. Often, very slow reading is related to uncorrected eye defects. Before embarking on speed reading, make sure that you don't have any correctable eye defects.
2. Get rid of the habit of vocalizing words—if you have—as you read. If you sound out words in your throat or whisper them, your reading rate is slowed considerably. You should be able to read most material at least two or three times faster when reading silently than orally.
3. Find out what you want to know, and for that you need to cut through and assess what is to be skipped from reading.
4. Your reading rate should vary according to your reading purpose. To understand the information, for example, skim or scan at a rapid rate.
5. It is always important to stress on qualitative not quantitative reading.
6. Cultivate a positive attitude towards reading. The more you read the more love you develop towards reading.
7. Even if you find reading to be tedious, try to give your task an extended stretch of an hour or so. Don't give up reading in the first 10–15 minutes itself.
8. Learn to underline the key parts/words/concepts for further references.
9. While reading, new ideas and concepts will flash in the mind and the same may be added in the margin of the page.
10. By using different coloured pencils, the importance level of the contents can be underlined for future reading.
11. Try to convert the written information into pictorial format while reading for effective comprehension of contents.
12. Fix some amount of time daily, exclusively for reading, and over a period of time it will become a practice.

RECAPITULATION

✓ There are four skills that are essential for all, namely speaking, writing, listening, and reading skills. However, reading helps you increase your competence in other language skills as well.

✓ Reading speed can be increased up to 1000 words per minute by following certain tips given in the chapter.

✓ For developing effective reading, you need to have the desire to improve, to make conscious efforts to use new techniques and also to have motivation to practise.

✓ Word-by-word reading, slow eye movement, complex words in the text, and lack of concentration are the major hurdles in effective reading.

✓ Reading helps in building confidence besides giving you pleasure and relaxation, as you start reading extended or longer texts.

✓ Vocalization, word-by-word reading, speed reading, SQ3R, SPE, and multiple intelligences-based method are various methods that people use while reading, but you should select the suitable method of reading different types of texts.

✓ As a good reader, you should effectively use skimming, scanning, extensive, and intensive reading skills.

WISEWELL QUIPS

EXERCISES

Concept Review Questions

1. Do you believe in what Francis Bacon has said—*Reading maketh a full man, conference a ready man, and writing an exact man?* Elaborate in 250 words on how important reading is for a professional.

2. What do you mean by effective reading skills? Discuss the major benefits of artful reading?

3. What is the role of speed in the reading process? Discuss the kind of relation that can be observed between speed and understanding.

4. It is said that reading enhances your general ability to use other language skills such as listening, speaking, and writing. Do you agree with the statement? Highlight the importance of reading in about 200 words.

5. Discuss the four basic steps to effective reading. Also suggest ways to increase the speed of reading.

6. What are the major differences between efficient readers and inefficient readers? Discuss and substantiate.

7. Most adults are able to increase their reading rate significantly and rather rapidly without lowering their comprehension. Do you subscribe to the view? Mention the three major factors which are essential for increasing your speed of reading.

8. 'Everybody knows how to read. But not everybody knows how to read skilfully and artfully.' Illustrate the statement and point out the major hurdles in effective reading.

9. What are the major faulty reading habits? Discuss them in detail.

10. Explain the following in about 150 words each:
    (a) Vocalization in reading
    (b) Regression
    (c) Faulty eye movement
    (d) SQ3R

11. Discuss in detail the various methods of reading that are employed by readers.

12. Discuss the different types of reading skills that you need to master for reading different types of texts.

13. When should you increase and decrease your speed of reading in order to develop effective reading? Discuss in detail.

14. Discuss the various tips for inculcating good reading skills.

Reading Comprehension

18.1 INTRODUCTION

Reading comprehension is one of the most commonly assessed skills in many a reputed test meant to select people for further studies or professional positions. In the recent past, in fact, reading comprehension has emerged as one of the most authentic tools to judge a person's suitability for a job or further studies, and it now forms an important part of the preparation on the part of a youngster who intends to attain success in the academic and professional world. It is, however, paradoxical that just as we are beginning to emphasize more on the reading comprehension skills, reading itself, as a skill, has taken a beating in recent times. Lack of reading habits, emanating mainly due to a young mind's overt obsession with technology-driven devices such as the computer, mobile phone, and television, makes it all the more difficult for a large number of students to combat reading comprehension passages which occur almost inevitably in most competitive exams now. That is why, though many brilliant minds around us do very well in other parts of their test, they struggle to answer the questions based on the reading comprehension passages.

18.2 WHAT GOES WRONG AND WHERE?

The question arises—what goes wrong and where? The answer is that many things go wrong at many places. Based on our interaction with students, we have tried to look into the main worries that students have when they confront a reading comprehension passage. We list below all such worries and inquiries, and follow each of those with our suggestion. Read further and understand how many of these worries and enquiries of several students are yours as well. Here we go:

Problem: I read the passage and seem to understand it mostly. When I start attempting the questions, however, I am never confident.

Suggestion: You seem to read perfunctorily. Sometimes, when we perform a task, we pretend to enjoy it as well. However, it may not be the case with casual readers who read in a superfluous manner. You need to read with purpose. Remember, reading with a view to understanding a passage and managing questions based on that is different from just reading for the sake of it.

Problem: When I read, I fail to keep my mind engaged in this activity for a long time. Somehow, the passages I read are boring and tasteless.

Suggestion: This is a symptom of a drifting mind. There is nothing like a 'boring' or 'tasteless' passage. The only thing is that you find them so simply because you are either not familiar with the subject or are not interested in it. A good reader consciously develops interest in a varied reading material. Lack of interest in reading is largely responsible for failure in dealing with reading comprehension passages.

Problem: I want to read but I don't understand many words and expressions in them. What do I do?

Suggestion: For this, you need to improve your vocabulary and read more to give yourself more exposure to how others use language, that is, how they use different words, expressions, etc.

Enquiry: How do I improve my vocabulary? I learn many words but after some time, I forget most of them.

Suggestion: Reading and listening are very essential for improving your word base. For picking up newer words and expressions and retaining them too, you need to not just read a variety of things but also use the words and expressions you have learnt from time to time.

Could you relate to some of these problems that many others face? If you carefully observe the pattern of questions and answers, you will be able to figure out that most of the problems that readers face are related to lack of good reading habits. Look at the instructions given in the next section to understand how to overcome this and other similar obstacles that usually cripple your understanding of a given passage.

18.3 EMPLOYING DIFFERENT READING SKILLS

There are four reading skills, namely skimming, scanning, intensive, and extensive reading skills, all of which help you attempt a reading comprehension passage. Detailed discussion and elaborate tips on how to judiciously employ these skills have been provided in Chapter 17. Let's now develop some other strategies to gain further insight into a given passage as enumerated below.

18.4 UNDERSTANDING THE AUTHOR'S POINT OF VIEW

While approaching a passage, it is also important for us to identify its nature. Broadly speaking, the passages can be *information based* or *point-of-view based*. There are passages in which the author intends to inform the reader about some phenomenon. In such passages, the author does not really have a point of view to prove. In other types of passages, however, he/she may have an opinion or an argument to establish. Understanding the author's opinion helps you answer various *point-of-view based* questions worded in different ways such as:

1. The author's primary purpose in this passage is to…
2. With which of the following statements would the author of the passage be most likely to agree?
3. The author's argument would be most weakened by the discovery of which of the following?
4. Which of the following examples could best be substituted for the author's example of…

5. The author's tone is best described as…
6. The author views his subject with…
7. The author's presentation is best characterized as…

Therefore, understanding the author's point of view really helps in understanding a lot about a given passage. Now the question arises—how do we understand the author's view? Let us go through the following small passage and figure out the author's view or perspective:

Leadership is often confused with managerial skills. It is commonly perceived that a leader leads while a manager just manages. The distinction, however, is not all that easy to maintain all the time…

Observe how the author broaches the topic and prepares to contend the existing view that tends to demarcate leadership from management.

Look at another example to understand how the author's perspective can be understood through the text:

Although myths are often regarded as something apocryphal and misleading, their significance and influence on our psyche cannot certainly be wished away…

Clearly, the author is going to establish the significance of myths in human life and their impact on our psyche in the subsequent section of the passage.

In understanding the author's view, certain linkers also play a significant role. Therefore, look for these linkers and carefully connect them to the point of view, argument, or tone of the passage. Given below are some of these linkers which serve to connect the author's perspective to a given situation:

1. It may not seem worthwhile, **however**, to singularly ignore the advantages of patriarchy…
2. **Regardless** of the controversy that surrounds a celebrity status, it is a drive that is likely to goad a million to its glitter…
3. **Despite/Inspite of the** leader's claim to cut/check the price hike, there seems no let up in the trend…
4. We simply cannot afford to just wait for many more years for the signs of global warming to come out more conspicuously. **In order to** reverse the trend, action must start even before the trend itself is firmly established.
5. **Notwithstanding** the caution exerted by the supremo, the young leader kept on chasing a dreamy ideal.
6. Terrorism seems to have rooted itself in many parts of the world and **as far as** Pakistan is concerned, it has entrenched itself with a terrifying certainty.

Besides these connectives, other linking words such as *but, in fact, in view of, in as much as, since, because, for, regarding, with regard to, in the context of, thus, hence, therefore, yet, moreover, further, as, so that, since that, as far as, besides, apart from, with, along with, as well as, together with, coupled with,* etc. are also to be marked carefully. Some of these such as *therefore, hence, thus,* etc. try to establish the fact in a conclusive way. Some other linkers such as *since, because, as, for,* etc. try to justify the author's stance. Some other connectives such as *apart from, along with, besides,* etc. are used to add ideas or things referred to in the preceding discussion. Whatever may thus be the purpose of the author, connectives play a crucial role in revealing that to us. Hence, while attempting to decode a passage for reading comprehension, we must try to understand what such a linker suggests in the overall tapestry of the author's strategy, style, and scheme in letting his/her reader figure out the point of view.

Connectives, therefore, achieve a lot for the author. They help him/her compare, contrast, augment, substantiate, bear out, establish, nullify, quash, countermand, support, or deflate a view, theory, belief, or perception. For the reader, hence, connectives can serve as vital clues to the stance adopted by the author. Consequently, as a reader, we have to be on the lookout for all such linkers and connecting devices.

Sometimes, clues may present themselves not through connectives but through other linguistic cues which may require a subtle and deft understanding to figure out the author's view. Look at the examples given below and try to establish what the author is trying to suggest:

1. Regrettably, the Kashmir problem has been allowed to feed on itself for far too long.
2. Not surprisingly, children achieve through parenting what they can never achieve through tutoring.
3. To the shock of every conscious citizen, the drainage system of many a city in India is crumbling.
4. Admittedly, improved financial status leads to a general sense of apathy and lack of concern for others in people.
5. Painfully for the whole world, most powerful countries continue to be the most bullying as well.

In all the above examples, there are expressions such as *regrettably*, *not surprisingly*, *to the shock of every…*, *admittedly*, *painfully*, etc. which help the reader establish the author's view. As a reader of passages therefore, we have to be quite alert to such expressions, other than the linkers, to be able to answer the questions based on the author's opinions regarding the issue under the scanner.

18.5 IDENTIFYING THE CENTRAL IDEA

All passages have a central or main idea which becomes very important for us to understand. It is so because out of the questions asked in a reading comprehension exercise, most of these relate to the central idea of the passage. Therefore, attempting a passage without understanding the main or central idea of the passage is like fighting a battle without knowing with whom or for what it is being fought. Let us see how questions related to the central idea of a passage abound in a reading comprehension exercise and are posed to the reader in a variety of ways:

1. The passage is primarily concerned with…
2. Which of the following statements best expresses the main idea of the passage?
3. The author's primary purpose in this passage is to…
4. The author in the passage implies that…
5. It can be inferred from the passage that…
6. With which of the following statements would the author of the passage be most likely to agree?
7. The author's argument would be most weakened by the discovery of which of the following?
8. Which of the following examples could best be substituted for the author's example of…?
9. Which of the following statements best summarizes the main idea of the passage?

Now the question arises—how does one identify the central idea? In the passages that you will be required to read, the author is not usually going to make an obvious statement such as 'My purpose in writing this passage is to…' or 'What I want to say in this passage is…' Authors present their main concerns in a variety of ways and many such ways are often vague, indirect, implicit, subtle, and even obscure. Therefore, identifying the central idea in a passage is not always easy. However, there are a few strategies following which you can figure out the core concern of a passage, some of which are discussed below for your understanding and application.

18.5.1 Tips for Identifying the Central Idea

Following are some important tips for identifying the central idea of any passage:

1. *Never read a passage in a laid-back manner*. Always attack it vigorously so as to scavenge it thoroughly. When you read a passage, keep asking yourself the question—what could be the reason for the author to have written this passage? What is it that the author is trying to tell me here? What could be the main concern of the author in writing this passage? When you keep probing a passage thus, you are likely to push yourself to the core of the argument sooner or later.

2. *A passage—particularly a longer one—is often divided into several paragraphs*. Now each of these paragraphs has a topic sentence which expresses the main concern of that particular paragraph. Many a time, a topic sentence comes right in the beginning of a paragraph and hence introduces the core issue. A topic sentence or an **introducer** is generally followed by some other sentences which are meant to substantiate the topical idea of the paragraph. All such sentences are known as **developers** as they augment, substantiate, compare, contrast, argue, contest, exemplify, or bear out the topical idea already introduced to the reader. A well constructed passage is also likely to have some terminators following such introducers and developers. The **terminators** are likely to summarize and conclude the idea of a paragraph.

   Now, if we carefully observe the structure of a paragraph, we can identify the core issue in each paragraph in a passage. Since we already know the basic idea in each of these paragraphs, we can figure out the core concern of the entire passage as well.

3. *Carefully read the beginning of each paragraph*. As authors frequently present their core concerns in the beginning of a paragraph, beginnings of each of the paragraphs often give us vital clues about the overall idea expressed in the passage.

4. *Also carefully follow the ending of each paragraph* since the author is most likely to wind up his/her main concern in this part of the paragraph. In fact, it is both the beginning and the ending that need to be pursued with intensive reading skills, because most commonly these are the places where the authors intend to share their core concern with the readers.

5. *Predicting and anticipating certain views* to come up and express themselves also helps us move with the central thread of an argument. When you read a passage and look forward to a particular line of thought, it helps you participate in the dialogue that the author intends to establish. This certainly not only enhances your comprehension of the passage but also helps you focus better on it. This is what is also termed as *active, participative* reading, as when you read a passage thus, you do not approach it as a dead text but as a living entity which develops along a particular thought line.

6. *Paying careful attention to each of the linkers* and establishing their function in the overall tapestry also helps you figure out the movement of ideas. Generally, a passage will have one main idea which will be followed by some subsidiary ideas which are linked through various linkers already discussed. Therefore, if you pay special attention to them, you can connect with not just the following and subsidiary idea but also with the preceding and principal argument.

7. We can also understand the central idea of a passage by *thinking about a suitable title for the passage* and making an effort to determine its tone as we read it. If we keep urging

ourselves on with 'The author is trying to denounce/establish/mock at/criticize/appreciate/satirize/justify/explain/inform/instruct/preach/propagate…', we are likely to keep ourselves always in the hunt for the central idea. Similarly, telling yourself, while reading the passage, something like 'The suitable title for this passage could be…' always helps you in searching for the main ideas in a passage.

Understanding the central idea in a passage is, therefore, a culmination of various approaches and strategies which need to be pursued simultaneously by the reader. Though the process may sound a bit complex and demanding, it is both rewarding and enriching.

18.6 INFERRING LEXICAL AND CONTEXTUAL MEANING

While reading a comprehension passage, it is quite important for us to understand the given text both lexically and contextually. Interpreting a text lexically suggests understanding its meaning, while inferring it contextually suggests discovering its underlying meaning in the context. For example, when we come across a remark such as '…that is how the cloud of ignorance obscures our view and does not allow us to transcend…', we are required to look at the expression 'clouds of ignorance' both lexically and contextually. Now, we all know what 'cloud' lexically means, as it is known to be a mass of water vapour that floats in the sky. In the given context, however, it stands for some obstacle or barrier that impedes clarity and obfuscates people. Therefore, it is important for us to understand not just the lexical but also the contextual meaning of the text. In this context, it becomes important for us to take a look at the variety of discourse features employed by the writer while communicating his/her idea to the reader. Employing discourse analysis, in fact, plays an important part in the overall understanding of a given passage.

18.7 EMPLOYING DISCOURSE ANALYSIS

Employing discourse analysis refers to studying the language in use. As we all know, writers employ various techniques in articulating their ideas to the reader. They structure their writings in different styles, use words denotatively as well as connotatively, employ symbolism and imagery by using words in different hues and shades, and adopt a tone or maintain a point of view in a suggestive and implied manner. To be able to understand a given text and answer the questions based on it, hence, requires discovering all such associated features of the text, some of which are discussed here.

Style Writers tend to structure their texts in a particular style or at times mix a couple of them together while writing. Some of the basic styles of writing are informative, descriptive, analytical, narrative, and abstract. When a writer presents the situations as they are and not as they should be, the style adopted is generally informative or descriptive. In choosing to write in such a style, the author focuses more on the factual presentation of a situation rather than on the interpretation of it. Comparatively however, when the author presents a fact and also interprets it, the style becomes analytical or interpretive. At times, the passages that you confront in an exam are far too abstruse. Many a time, it is so because the author choses to write in an abstract manner. The passages that deal with metaphysical, spiritual, or existential issues are at times written in such a manner. When authors, on the other hand, narrate a story, parable, or fable, the style becomes narrative. By keeping in view the style adopted by the author, we can come close to deciphering his/her overall purpose and scheme.

Tone   Just as writing in a particular style helps an author structure his/her ideas appropriately, choosing to write in a particular tone helps him/her express his/her perspective or point of view. For example, at times authors involve themselves in their writings; such writings are mostly *subjective* in tone as they may express the author's ecstasy or dismay, delight or despair, anguish or indignation, appreciation or repugnance, etc. On the contrary, if he/she remains detached, the tone chosen could be seen as *objective*, *detached*, *aloof*, and *matter-of-fact*, etc. At times, authors write to express their reverence or appreciation; so the tone could be *reverential* or *eulogizing*.

In some other passages, we find authors being critical of a person, practice, or phenomenon; so the tone adopted becomes *critical*, *censorious*, *disparaging*, or *belittling*. When authors write to change the world to their ideas, they often mock a particular view, ritual, attitude, or manner-ism and the text acquires a *satirical*, *sarcastic*, *derisive*, *sardonic*, or *mordant* touch. Similarly, authors keen to see their readers laughing often write in a *humorous vein* while others choose to colour their writings in *ironical hues*. Not being able to understand the mood and tone of the author thus amounts to losing sight of the overall sense of the text. Therefore, while reading a passage, it is important for us to keep guessing about the tone, mood, and approach adopted by the writer.

Symbols, imagery, and figures of speech   A literary passage abounds in figures of speech such as simile, metaphor, hyperbole, litotes, synecdoche, metonymy, onomatopoeia, and so on and so forth. Epigrammatic expressions, paradoxes, oxymoronic phrases, sym-bols and imagery are quite frequently employed by authors even while discussing com-mon, everyday affairs. For example, an author may write paradoxically: '...the more you get, the more you need' ...or use an oxymoronic expression such as 'pleasing pain', 'col-ourlessly blue', etc. to express his/her outlook, attitude, or mood. Similarly, a text may be full of symbols and imagery in which the words used may refer to some implicit values, attributes, and characteristics, or to evoke a pictorial image in the mind of readers to cap-ture their attention; induce in them varied emotions such as pity, fear, hope, disgust, eupho-ria, etc; influence them to the ideas, opinions, and thoughts expressed implicitly as well as explicitly.

Employing discourse analysis, therefore, is quite crucial to our understanding of a given text, as missing any such elocutionary, implicit, and insinuating links would tantamount to losing the overall purpose, meaning, sense, and design of a given passage.

18.8 WORKED OUT PASSAGES

Having understood some of the basic features of a passage and the strategies required to com-prehend its contents and answer the questions based on it, we will now go ahead and test our knowledge practically. Here are some passages. Read each of them carefully and answer the questions that follow.

Passage I

What distinguishes humans from animals? For some it is language, for others it is the altruistic willingness to help other members of the species. However, this kind of altruism seems to exist in the animal world as well.

Researchers working with Crisophe Boesch at the Max Planck Institute for Evolutionary Anthropology in Leipzig observed that West African chimpanzees adopt orphaned young, even though they are not related to them. Several animals lavished care on a juvenile for several years. Surprisingly, half of these adoptive parents were male.

This behaviour is thought to be encouraged by the pressure of leopards, with whom the West African chimpanzees share their habitat. The constant threat from the big cats seems to have encouraged cohesion and solidarity within the group. Accordingly, the scientists observed more chimpanzee adoptions in West Africa's Tai National Park than in East Africa.

Wild chimpanzees appear to be more prepared to help than those living in captivity. In zoos, chimpanzees cooperate with other members of the group to only a very limited extent. 'Our observations show that altruism in wild chimpanzees is much more widespread than studies of chimpanzees in zoos would suggest,' concludes Chrisophe Boesch.

Source: 'The Speaking Tree,' *The Times of India*, Sunday, 11 July 2010, p. 1.

Answer the following questions by choosing the right options:

1. Which of the following does the author want to establish by suggesting that animals are altruistic?
    (a) That humans are beginning to behave like animals
    (b) That animals are beginning to behave like animals
    (c) That animals too, like humans, share empathy with fellow creatures
    (d) That humans are not so empathetic to their fellow creatures as animals are

2. While discovering the adoptive streak in animals, what surprises the author is that
    (a) Even chimpanzees adopt orphan juveniles of big cats
    (b) Even male chimpanzees adopt juveniles of other species
    (c) Even big cats adopt orphan juveniles of chimpanzees
    (d) Even female chimpanzees adopt juveniles of other species

3. About the recently discovered altruistic zeal in chimpanzees, all except the following can be inferred from the passage.
    (a) Chimpanzees living in the wild are more altruistic than those in cages.
    (b) Chimpanzees adopt orphans that are even unrelated to them.
    (c) Chimpanzees found in West African forests are more altruistic than those found anywhere else.
    (d) Chimpanzees found in the Tai National Park in West Africa are observed to be more altruistic than those found in East Africa.

4. Which of the following is not a reason for the altruistic behaviour observed in West African chimpanzees?
    (a) The presence of a constant threat from leopards in their environment
    (b) The presence of a natural altruistic willingness to help others
    (c) The presence of a natural sense of competition in animals
    (d) The presence of a natural environment that stimulates such a behaviour in them

5. In the expression 'this behaviour is thought to be encouraged…', the word 'this' refers to which of the following.

(a) That chimpanzees are not much distinguished from humans

(b) That chimpanzees exhibit a sense of altruism existing in them

(c) That chimpanzees do not behave as strictly according to their gender as humans do

(d) That sexual limitations do not stop a male chimpanzee from being altruistic and adoptive in behaviour

Answers with explanation

1. Both (a) and (b) are completely out of tune with the passage; hence ruled out. The author does not compare the extent of empathy exhibited by animals and humans, hence (d) is also not possible. Only (c) sums up the view expressed in the passage; hence the answer.

2. Refer to the last sentence of the second paragraph 'Surprisingly, half of these adoptive parents were male'. Hence, (b).

3. The first sentence of the last paragraph clearly supports (a); the second sentence of the second paragraph establishes (b); option (d) is mentioned towards the end of third paragraph. Therefore, all except (c) can be inferred from the passage. Hence, (c).

4. 'a natural sense of competition', suggested in (c) cannot normally be the reason behind someone's altruism. Hence, (c) is the answer. All the remaining options (a), (b), and (d) can be inferred from the text. Therefore, (c).

5. In the paragraph preceding the statement, the author highlights the sense of altruism found in chimpanzees. Hence, (b).

Passage II

Just as the light sentences awarded in the Bhopal case scandalized the nation, the six-month term given last December to former Haryana police chief, SPS Rathore, for molesting 14-year-old Ruchika Girhotra, provoked an extensive review of the provisions related to sexual offences. In its hastily drafted bill put out for public consultation, the Home Ministry however overlooked the very provision that allowed Rathore to get away with a paltry punishment despite being found guilty of child molestation.

The draft bill failed to address a glaring anomaly in the Indian Penal Code's provisions dealing with sexual assault on under-aged girls. While the rape of a child attracts higher penalty, there is no such distinction when it comes to molestation. Irrespective of the age of the victim, all cases of 'outraging her modesty' (as quaintly put in this provision of Victorian vintage) are punishable under Section 354 IPC with a maximum sentence of two years.

Having left untouched the provision relating to molestation, the home ministry aggravated the anomaly, however unwittingly, by proposing to enhance the punishment under Section 509 IPC for eve-teasing or 'act intended to insult the modesty of a woman', from one year to seven years.

Thus, Chidambaram's ministry ended up sending out a perplexing message to sexual offenders. If they indulge in eve-teasing, they could be put behind bars for as long as seven years. But if they actually molest a woman or a girl, they will not get more than a two-year term. Such a cavalier approach to legislation shows that it is more about politics than about justice.

This is also borne out by the government's prolonged indifference to another gender issue: its international treaty obligation to enact a law on sexual harassment at work places. Though India ratified the convention on the elimination of all forms of discrimination against women (CEDAW) way back in 1993, the government came out with a draft bill only this year.

In between, the Supreme Court sought to fill the gap with a set of directions in the landmark Vishaka judgement of 1997. It is a commentary on its patriarchal society, that despite having a bloated statute book, India does not have legislation on sexual harassment at work places.

Source: 'Fair Play Denied,' *The Times of India: The Crest Edition*, Saturday, 10 July 2010, p. 7.

Answer the following questions by choosing the right options for each of them:

1. The anomaly mentioned in the passage refers to the fact that
   (a) Though there are provisions to punish a rapist, there is no law against molestation.
   (b) There is no difference in the punishment to be awarded to a molester and a rapist.
   (c) There is no difference in the punishment for the rape of a girl and the molestation of a woman.
   (d) There is no difference in the punishment for molesting girl or a woman.

2. By remarking that the draft bill is *more about politics than about justice*, the author intends to suggest that
   (a) The government is more keen on gaining a political advantage by hastily proposing the Bill, than ensuring justice to the victims.
   (b) The government is playing politics by keeping the issue of molestation alive in the media.
   (c) The government is politically motivated to subdue the opposition by proposing the bill in a hurry.
   (d) The government has all the time to play politics but no time to secure justice for the masses.

3. The word 'cavalier' as used in the passage means all of the following except
   (a) Haughty   (b) Rash   (c) Prudent   (d) Frenetic

4. The tone of the passage is
   (a) Approbating   (b) Disparaging   (c) Eulogizing   (d) Captious

5. With which of the following propositions is the author more likely to agree to?
   (a) That the difference in the punishment for rape and molestation be eliminated
   (b) That the difference in the punishment for the molestation of a minor and a major be clearly defined
   (c) That the difference in the punishment for sexual harassment and molestation be equated
   (d) That the victims of molestation and sexual harassment at work places be given adequate compensations

Answers with explanation

1. Refer to the second paragraph. The author clearly emphasizes '…the age of the victim…'. It is this 'glaring anomaly' that the author criticizes in the passage. Hence (d).
2. The entire passage in general, and the third paragraph in particular, suggests that the draft bill has been hastily put up without taking care of the inconsistencies and anomalies, to salvage a political image. Hence (a). The remaining options are either exaggerated or unrelated.
3. Refer to the last sentence of the fourth paragraph. The author obviously finds the step taken by the government to be rash, hasty, and panicky. So, all except 'prudent' (c) can be the adjectives chosen to describe the word 'cavalier' as used in the context. Hence, (c).
4. The author does not appreciate, applaud, or praise the draft bill proposed by the government. Therefore, both 'approbating' and 'eulogizing' can be ruled out. Further, though the passage criticizes the draft bill, he/she does not 'disparage' or insult it, hence, only (d) appropriately suggests the tone adopted in the passage.
5. Throughout the passage, the author has contested the provision that does not take into consideration the age of the victim of molestation. Therefore, he is most likely to agree to (b).

Passage III

Perched languidly near the Konark coast in a state of decrepit grandeur, the Sun Temple may just be sitting atop its own grave. Seemingly on the precipice of extinction after a tenacious battle with the elements, the temple—built in the 13th century as a colossal chariot for the sun god and representative of the pinnacle of ancient Kalinga art—has now presented a great, albeit troubling, riddle before the world's top archaeology experts.

At the heart of a very intricate and equally delicate problem are the sand-filled walls of the Jagmohan, or the front audience hall, that survived the temple's gradual decimation. And as the clock ticks away, the answers, worryingly, haven't come.

Over a hundred years old, faced with the threat of the edifice caving in, experts stuffed the Jagmohan's walls with sand to conserve it. Now, with cracks and damage to the temple apparent even to the naked eye, global archaeology circles are witnessing a raging debate whether or not to extract the sand. If the sand is permitted to remain inside the 130 feet-high Jagmohan, experts fear the structure will suffer irrevocable loss. If it is not, it could disintegrate. Caught in this maze of ifs and buts—and with decay spreading each passing day—archaeologists, conservationists, engineers, scientists, historians, government officials, and other stakeholders are desperately seeking enlightenment. So far, there's been none.

In March this year, over 50 experts from India and abroad converged in Konark to forge a consensus on future conservation strategies for the temple, which was given World Heritage Site status by the Unesco in 1984. They talked, discussed, debated, argued and did everything except arrive at a solution. The range of opinions on offer varied from the scientific to the unscientific, the real to the absurd. Some favoured getting rid of the sand; others opposed it. Some suggested supporting the temple with steel rods. Others felt it was too risky. At the end of it, there was no consensus. And no solution.

Source: Sandeep Mishra, 'Sunset Temple,' *The Times of India: The Crest Edition*, Saturday, 10 July 2010, p. 9.

Answer the following questions by choosing the right options:

1. 'The Jagmohan' in the passage refers to
   (a) The front hall for the audience in the Sun Temple
   (b) The highest wall in the Sun Temple
   (c) The idol kept in the Sun Temple
   (d) The maze in the Sun Temple

2. The Sun Temple in Konark was given the status of world heritage site by the
   (a) UNICEF   (b) OTDC   (c) UGC   (d) UNESCO

3. With the walls of the Jagmohan caving in, the main debate is about
   (a) Whether or not to extract the sand stuffed into the walls
   (b) Whether or not to renovate the walls of the hall
   (c) Whether or not to use steel rods to stabilize the walls
   (d) Whether or not to demolish and restructure the walls

4. By writing that '…the Sun Temple in Konark may be sitting atop its grave', the author wants to suggest that
   (a) The temple was built in the 13th century.
   (b) The temple is in a dilapidated condition.
   (c) The temple has suffered an irrevocable loss in the recent past.
   (d) The temple is built on graves.

5. In the expression, 'so far there has been none', the word 'none' in the passage refers to the fact that

(a) Though there have been scientists, archaeologists, engineers, and historians, there is no political leader to help save the Sun Temple.

(b) Though the temple authorities are seeking government aid desperately, no such aid has come their way.

(c) Though experts from many different walks of life are worried about the decrepit state of the temple, no one is able to find a solution to it.

(d) Though experts such as archeologists, engineers, scientists, and historians are worried about the Jagmohan, they are not bothered about the Sun Temple on the whole.

Answers with explanation

1. Refer to the first sentence of the second paragraph; the correct option is (a).
2. Refer to the first sentence of the fourth paragraph; the correct option is (d).
3. Refer to the second sentence of the third paragraph. Clearly, the 'raging debate' is about 'whether or not to extract the sand.' Therefore, (a).
4. Before using this phrase, the author refers to the 'decrepit grandeur' of the temple. Hence, (b).
5. Refer to the discussion towards the end of the third paragraph. The author clearly highlights that though experts from different walks of life have been trying, no solution is in sight. Hence, (c) can be safely concluded from the passage.

@ For more sample passages, please refer to the Online Resource Centre.

 ## RECAPITULATION

✓ Reading comprehension is a two-fold process as it requires you to understand the passage and reproduce or express the ideas as comprehended.

✓ You need to continuously strive to improve your speed of reading and level of comprehension.

✓ To improve your speed of reading you should try to cultivate the habit of reading larger groups of words and train your eyes to move faster.

✓ Try to get rid of the habit of regression, that is, re-reading the text to ensure clarity of comprehension.

✓ Employ different types of reading skills, such as skimming, scanning, intensive, and extensive, for different purposes.

✓ You should try to figure out the central idea of a passage, the author's point of view, and discourse features of the passage.

✓ You may come across many unfamiliar words of which you do not know the meaning, try to guess these from the given context.

WISEWELL QUIPS

EXERCISES

1. While reading a passage why should you try to understand the author's point of view? Discuss in about 150 words.

2. What are the ways to identify the central idea of a passage? Write a short note in about 200 words.

3. What do you mean by *discourse features* and how well can you figure out the discourse features of a passage?

4. Read the following passages and answer the questions that follow each of them:

Passage I

We see that where there is a process of desire at work there must be the process of isolation through belief because obviously you believe in order to be secure economically, spiritually, and also inwardly. I am not talking of those people who believe for economic reasons, because they are brought up to depend on their jobs and therefore will be Catholics, Hindus—It does not matter what—as long as there is a job for them. We are also not discussing those people who cling to a belief for the sake of convenience. Perhaps with most of us it is equally so. For convenience, we believe in certain things. Brushing aside these economic reasons, we must go more deeply into it. Take the people who believe strongly in anything, economic, social, or spiritual; the process behind it is the psychological desire to be secure, is it not? And then there is the desire to continue. We are not discussing here whether there is or there is not continuity; we are only discussing the urge, the constant impulse to believe. A man of peace, a man who would really understand the whole process of human existence, cannot be bound by a belief, can he? He sees his desire at work as a means to being secure. Please do not go to the other side and say that I am preaching non-religion. That is not my point at all. My point is that as long as we do not understand the process of desire in the form of belief, there must be contention, there must be conflict, there must be sorrow, and man will be against man—which is seen every day. So if I perceive, if I am aware, that this process takes the form of belief, which is an expression of the craving for inward security, then my problem is not that I should believe this or that but that I should free myself from the desire to be secure. Can the mind be free from the desire for security? That is the problem—not what to believe and how much to believe. These are merely expressions of the inward craving to be secure psychologically, to be certain about something, when everything is so uncertain in the world.

Source: J. Krishnamurti, *The First and Last Freedom*, Chennai: Krishnamurti Foundation India, 2001; rpt. 2007, pp. 46–47.

Answer the following questions by choosing the right options.

1. Who, according to the author, would not be bound by a belief?
   (a) The one who believes in the oneness of God
   (b) The one who understands what it means to be dead
   (c) The one who understands the whole process of human existence
   (d) The one who discards all religious theories

2. The author's purpose in this passage is to
   (a) Enlighten the reader about the true meaning of spirituality.
   (b) Enlighten the reader about the true meaning of religion.
   (c) Spread atheism in the world.
   (d) Help the reader understand the process of desires in our beliefs.

3. The author of the passage is likely to agree to all except the following:
   (a) All human beliefs arise from our psychological insecurity.
   (b) Human desires lead to contention and conflict in life.
   (c) People cling to belief also for the sake of convenience.
   (d) All human beliefs are essentially religious in nature.

4. According to the passage, the greatest harm that our belief causes is that
   (a) It leads to our isolation.
   (b) It adds to our insecurity.
   (c) It makes us psychologically complicated.
   (d) It divides people into different religious groups.

5. It can be inferred from the last sentence of the passage that
   (a) Insecurity and uncertainty in human life is the root cause of all human beliefs.
   (b) Human life is based on beliefs because of the variety they offer.
   (c) Beliefs are very important in our day-to-day life.
   (d) Psychologically we are incapable of believing in anything.

Passage II

Once upon a time, it was easy to tell one kind of photograph from another. Photojournalists documented real life while commercial photographers fabricated a happy one. The former relied on the immediacy of the camera's image-making technology; the latter utilised its ability to lie persuasively. There was respect for the former and a certain contempt for the latter. All in all, the two genres of photography were as different as chalk and cheese. Then, in the 1960s, the third genre began to slowly evolve in the West: fine-art photography. These were artists who chose chemicals, silver, and the camera as their tools. Their works were often journalistic, but there was also beauty in them.

Fine-art photography took almost forty more years to come to India. For a long time, the one well-known Indian camera artist was Dayanita Singh. Most gifted photographers, like Raghu Rai, made their reputations through photojournalism. With the exception of a few like Prabuddha Dasgupta, commercial photographers were regarded disdainfully as fabricators of the pretty, soulless reality that a client demanded of them. But staging a photograph is no longer a no-no in Indian fine-art photography. The notions of what makes a photograph artistic are changing.

'When you say you are fine artist, you are saying, "I am making this piece because it's representative of me and who I am",' says photographer Manjari Sharma, a Mumbai girl now based in New York. 'You have no one to hold accountable but yourself. What makes my style fine art is that when I'm making the picture, I'm aiming to fall in love with it myself.' In April this year, Sharma's *My Shower* was showcased in *Burn*, an online journal edited by Magnum photographer David Alan Harvey. *My Shower* is a set of beautifully composed pictures that showcase Sharma's talent for portraits. It began when Sharma noticed that for one hour every day, sunlight poured into her bathroom and transformed it. She started inviting friends to model for her in her bathroom.

As the series developed, Sharma felt as though she was creating a 'personal mythology'. The tight frames have a tender intimacy to them and her models look like they've been gentled into a photograph, rather than captured. Their intensely private moments offer references that range from the purification of a baptism ritual to playful sensuality. *My Shower* is Sharma's first major work and it has won her significant acclaim. It is also a rare example of work that belongs in both the fine art as well as commercial brackets. Bathware manufacturers Grohe commissioned Sharma to create a similar set of images for an upcoming ad campaign. It appears the twain can meet. Sharma said her greatest challenges included packing a 30-person crew into a bathroom and 'getting the models to relax and forget for a second that this is not a Liril commercial'.

Source: Deepanjana Pal, 'War of the Poses,' *The Times of India: The Crest Edition*, Saturday, 26 June 2010, p. 27.

Answer the following questions by choosing the correct options.

1. *Burn*, according to the passage is the name of a
   (a) Photograph   (b) Magazine
   (c) Journal   (d) Book
2. The reason for the contempt for commercial photography was the belief that
   (a) Commercial photography was vulgar.
   (b) Commercial photography was based on lies.
   (c) Commercial photography was not based on real life situations.
   (d) Commercial photography was not as poetic as photojournalism.
3. By stating that '…it appears the twain can meet', the passage suggests that
   (a) Both artistic and commercial interests can be served through fine-art photography.
   (b) Both photojournalism and commercial photography can be merged.
   (c) Both reality and fantasy can be fused to form present day commercial photojournalism.
   (d) Both photojournalists and commercial photographers can share their knowledge.

4. Through the expression 'commercial photographers were regarded disdainfully as fabricators of the pretty, soulless reality that a client demanded of them', it can be inferred that
   (a) People were against commercialization of photography.
   (b) Photography was a western concept.
   (c) Photography was supposed to mirror reality.
   (d) Photojournalism was too dominant a force to be challenged.

5. The ideas in this passage are arranged in which of the following patterns
   (a) Causal pattern
   (b) Chronological pattern
   (c) Problem solution pattern
   (d) Spatial pattern

Passage III

As the former head of the Food and Drug Administration, David A. Kessler confidently took on the powerful American tobacco industry in an effort to regulate the sale of cigarettes. If the tobacco companies had known better, they could have distracted him from his task by sending him cookies. The mere sight of chocolate chip cookies was enough to pulverize Kessler's will power. He fantasized about the moist chunks of chocolate embedded in dough when he left for work. He thought he had defeated desire when he left his home leaving the cookies uneaten. But then he stopped at a café for some coffee, saw a jar of homemade cookies and, unable to resist, had one.

Kessler describes his failed experiment in self-restraint in *The End of Overeating: Taking Control of our Insatiable Appetite*, a highly engaging book that investigates why we are powerless before food. He cites studies that show that the right combination of fat, sugar, and salt—a golden ratio that has the addictive powers of heroin—makes food rewarding. Kessler shows that eating such food makes us want to chase the same sort of fulfilment. Our brains become wired to seek highs in sugar, salt, and fat. And visual stimuli such as attractive packaging, the variety of foods in supermarkets, and delicious sounding descriptions of food spur us, not unlike the lab rats that these studies were conducted on, to consume foods that lead to obesity and eventually heart disease.

Kessler explains in the patiently instructive manner of a school teacher, the vicious cycle that occurs when fatty foods commandeer our lives by chemically altering brain states in such a manner that we actively seek more fat. He suggests that while food companies have not fully understood the science behind the bottomless American appetite, they're doing a stellar job of making food that is highly palatable and unhealthy. Foods are so highly processed, that they become easy to swallow. As a result you never know when you are full. Ingredients are cunningly listed on packages to suggest there's less sugar than there actually is. And just about every flavour can be artificially simulated. So the smoked flavour on your tortilla could well have been cooked up in a food technologist's test tube.

While Kessler's book acts as a health advisory for Americans and the British, it is portentous. As American fast food chains multiply across the world, it won't be long before cultures that still largely follow traditional diets pile on the extra pounds. There are already several studies that show that urban Indians have never been fatter.

Source: Pronoti Datta, 'Eat Right, or Else…', *The Times of India: The Crest Edition*, Saturday, 5 June 2010, p. 25.

Answer the following questions by choosing the correct options.

1. Which of the following does Kessler's book suggest?
   (a) Eating fatty food becomes a habit with us because of its taste.
   (b) Eating fatty food is dangerous to us as it causes cancer.
   (c) Eating fatty food becomes a habit with us for it is swallowed quickly.
   (d) Eating fatty food becomes a habit with us for it governs our brain as drugs do.

2. What adds to the appeal of a fatty food according to the passage?
   (a) Its attractive packaging
   (b) Its wonderful taste
   (c) Its softness
   (d) Its quality

3. By stating that ingredients on fatty foods are 'cunningly listed…', the passage suggests that
   (a) Companies that manufacture fatty foods are operating unlawfully.

(b) Fatty foods promise more hygiene that they can afford to offer.

(c) Fatty foods are far more harmful than their wrappers suggest.

(d) Fatty foods openly declare risk to obesity and hypertension.

4. Which of the following alternatives appropriately suggest the tone of the first paragraph of the passage?

(a) Interactive   (b) Sensational

(c) Captivating   (d) Objective

5. Which of the following do the studies referred to in the concluding part of the passage suggest?

(a) Indians too are fast becoming addicted to fatty foods.

(b) Traditional societies in India are becoming more assertive.

(c) By consuming fatty foods, the Indian urban are getting overweight.

(d) American companies are offering food to the rest of the world.

Passage IV

Targets set by policy makers to slow global warming are too soft to prevent more heatwaves and extreme temperatures in the US within a few years, with grim consequences for human health and farming. A study warned this week.

Although the US and more than 100 other countries agreed in Copenhagen last year to take action to reduce greenhouse gas emissions 'so as to hold the increase in global temperature below two degrees Celsius', a study conducted by Stanford University scientists showed that might not be enough.

Stanford earth sciences professor Noel Diffenbaugh and former post-doc fellow Moetasim Ashfaq wrote in the study, published in *Geophysical Research Letters*, that 'constraining global warming to two degrees Celsius above pre-industrial conditions may not be sufficient to avoid dangerous climate change'.

'In the next 30 years, we could see an increase in heat waves like the one now occurring in the eastern US or the kind that swept across Europe in 2003 that caused tens of thousands of fatalities,' said Diffenbaugh, lead author of the study. 'Those kinds of severe heat events put enormous stress on major crops like corn, soybean,

cotton, and wine grapes, causing a significant reduction in yields,' he said.

Diffenbaugh and Ashfaq used two dozen climate models to project what could happen in the US if carbon dioxide emissions cause temperatures to rise 1.8 degrees Fahrenheit (one degree Celsius) between 2010 and 2039—a likely scenario, according to the UN's International panel on climate change. If that occurs, the mean global temperature in 30 years would be about 2 degrees Celsius hotter than in the pre-industrial era of the 1850s.

Source: 'US to Feel More Heat, More Often in Coming Years: Study,' *The Times of India: Times Global*, Monday, 12 July 2010, p. 20.

Answer the following questions by choosing the correct options.

1. According to the passage, the US and a 100 other countries agreed at Copenhagen to

(a) Fight global warming by cutting petrol consumption.

(b) Fight global warming by keeping the planet temperature below two degree Celsius.

(c) Fight global warming by reducing green house emissions.

(d) Fight global warming by keeping the temperature increase on earth below two degree Celsius.

2. Which of the following is not suggested in the study published in *Geophysical Research Letters* as cited in the passage?

(a) US and other countries may not be able to keep the increase in the earth's temperature below two degree Celsius by the next century.

(b) Heat waves may increase in US in the next thirty years.

(c) Humans will have to face grim consequences on account of global warming.

(d) The measures adopted by nations to fight global warming are not enough.

3. The research reported in the passage has been conducted by

(a) The scientists working in Princeton University.

(b) The earth scientists working in Stanford University.

(c) The scientists working on global warming.

(d) The earth scientists who attended the Copenhagen Conference.

4. Which of the following can be inferred from the first paragraph of the passage?
   (a) In the coming years, global warming is likely to become more intense as the changes seen on the sun are getting virulent by the day.
   (b) The measures promised by the US and the other 100 countries at the Copenhagen are not being followed.
   (c) In the next couple of years, the global warming is going to go soft on earth.
   (d) The measures agreed to be adopted by the US and the other 100 countries at the Copenhagen are too soft to check global warming.

5. With which of the following possibilities of global warming is the author of the passage most likely to disagree?
   (a) Heat waves are likely to cause a large number of fatalities in the US in the years to come.
   (b) The temperature rise on the earth's surface is likely to harm farming in a crucial way.
   (c) The steps proposed at Copenhagen are likely to reverse the pattern of global warming in the next 30 years or so.
   (d) In the next 30 years the earth's surface is likely to get warmer by two degrees Celsius than what it was in the pre-industrial era of 1850s.

Passage V

New scientific breakthroughs have put the spotlight on DHA, docosahexaenoic acid, the most important of the fashionable class of dietary chemicals, the omega-3 fatty acids. It is found most famously in fatty fish and its health benefits include better memory, delay in mental decline and better heart health. 'Fish and no chips', published in last week's print edition of *The Economist* reported a supposed link between diets lacking DHA and increase in the incidence of mind disorders. 'DHA displacement in modern diets by the omega-6 acids in cooking oils such as soya, maize, and rapeseed is a cause of worry. Many researchers think this shift—and the change in brain chemistry that it causes—explains the growth in recent times of depression, manic-depression, memory loss, schizophrenia, and attention-deficit disorder.

There is also a body of data linking omega-3 deficiencies to violent behaviour. Countries whose citizens eat more fish (which is rich in DH) are less prone to depression, suicide, and murder. And a new research by Dr Joseph Hibbeln, a researcher at America's National Institute of Health, shows that low levels of DHA are a risk factor for suicide among American servicemen and women. America's department of defence has taken note. It will soon unveil a programme to supplement the diets of soldiers with omega-3s. The US Food and Drug Administration may change one of its policies, too.

Meanwhile, eating more fish is not the only way out. The oceans are under enough pressure as it is. Hybrid crops with higher levels of DHA could help, too. Till then the best advice is probably 'Nothing in excess'.

Source: 'What the World is Reading: A Summary of Articles from Leading Journals and Magazines', *The Times of India: The Crest Edition*, Saturday, 5 June 2010, p. 27.

Answer the following questions by choosing the correct options.

1. The DHA referred to in the passage is the name of a/an
   (a) Protein   (b) Virus
   (c) Acid   (d) Disease

2. Which of the following, as per the passage, may be caused by lack of DHA in the human body?
   (a) Cancer   (b) Hypertension
   (c) Joint Pains   (d) Schizophrenia

3. Which of the following can be an alternative to DHA as suggested in the passage?
   (a) Omega-6 Oils   (b) Hybrid Crops
   (c) Coffee   (d) Saturated Fats

4. The word *incidence* as used in the passage comes closest to suggest which of the following:
   (a) Prevalence   (b) Predominance
   (c) Occurrence   (d) Happening

5. Which of the following statements cannot be inferred from the passage about DHA?
   (a) It is found in fatty fish.
   (b) Its deficiency can lead to mental disorders.
   (c) It can be replaced by Omega-6 acids.
   (d) Its deficiency can lead to suicide.

@ Refer to the Online Resource Centre for more practice exercises.

CHAPTER 19

The Art of Condensation

Learning Objectives   After reading this chapter, you will be able to

- understand the purpose and significance of the art of condensation
- know the major forms of condensation such as précis, summary, abstract, and synopsis
- develop adequate knowledge of précis writing techniques
- learn tips to condense a given text

19.1 INTRODUCTION

Just as it is difficult to be simple, it is difficult to be precise in what we speak and write. In fact, it is difficult to be elaborate for those who struggle with language in the sense of not having enough words, expressions, and ideas. But for those who have a good number of words at their disposal and have no dearth of expressions and ideas, it is difficult to be precise and brief. Therefore, at times, you see some of the people around you complaining that *it was not possible for them to deliver a two-minute speech though they could fairly well deliver a ten-minute talk!* However, in professional situations, one must be in a position to express oneself not just elaborately but also briefly.

Just to illustrate the view—imagine that you are a marketing executive in your company. The chief of the marketing division has called for a meeting. You have recently read an article in a business magazine which speaks about latest innovations in the field of marketing. You want to discuss some of these innovations in the meeting. Now what would you do? Will you carry the article to the meeting and read it aloud to the other members in the meeting? Or will you take down important ideas, make a summary, and highlight each of the points one by one? Obviously, it is only the second method that would work. Doing so, however, will require some skills of precision and condensation and this is what we propose to help you learn in this chapter.

However, before we understand how to condense, we need to get acquainted with the major forms of condensation that are frequently used by us in our academic and professional life. Given below are the major forms of condensation demanded from and composed by a professional:

- Précis
- Summary
- Abstract
- Synopsis
- Paraphrasing

19.2 PRÉCIS

Among all the major forms of condensation, the précis is most commonly required to be read and written, both by a student and a professional. It is a short and concise account of some text which gives all its important points but none of its details. Since the purpose of a précis is to briefly restate the idea expressed in the original write-up, it does not include any superfluous or illustrative material which may be a part of the original. A précis follows and maintains the view of the author.

19.3 SUMMARY

A summary is often written and included in the reports prepared by professionals. Summaries are also written to briefly present the main findings of a study, a journalistic article, or a geographical survey. Whatever may be the purpose, a summary is quite useful as it presents the entire matter in a nutshell. While writing a summary, the author does not add, develop, or delete any idea. A summary is often shorn of examples and illustrations, and emphasizes the main arguments and conclusions of the original. More often than not, it follows the sequence of the ideas as expressed in the original and detailed work.

19.4 ABSTRACT

Shorter than a summary, an abstract is written to highlight the purpose, scope, and significance of a work. It is often preferred to a summary in technical and specialized forms of communication. Therefore, you often come across an abstract published along with a research article in journals and magazines. In order to understand the features of the abstract and the summary and to figure out the differences between them in detail, please refer to Chapter 23 on *report writing*.

19.5 SYNOPSIS

A synopsis is a condensed and shortened version of an article, research paper, the chapter of a book, a report, or a book itself. It highlights in brief all the essential features of the original document. Normally, a synopsis is required to be submitted to universities when research proposals, dissertations, and theses are proposed to be written by researchers. In a synopsis, the researcher is required to highlight the purpose, scope, and significance of the research. It also includes a reference to the methods adopted for data collection and the research gap that determines the objective of the research.

19.6 PARAPHRASING

Although paraphrasing is not necessarily a condensed form of the original document, it is often believed to be one. Therefore, it is advisable to understand the nature and purpose of a paraphrased text. The purpose of a paraphrase is to reproduce the author's ideas in your own words. So, you may employ as many words as the author has chosen to express himself/herself. Paraphrasing of write-ups is done in order to convey to the reader in simpler terms an idea which, otherwise, appears to be too ambiguous, arcane, philosophical, or poetic to follow. Many a time, you come across a paraphrased text of a classic. It is meant to bring the texts written in different times and languages to readers who can only follow a simpler and familiar version of it.

Despite the differences all such firms of writing, hence, require the original message to be expressed more precisely, briefly, and succinctly. Let's therefore learn the important tips for condensation.

19.7 ART OF CONDENSATION—SOME WORKING PRINCIPLES

In this section, we will look at some principles of the art of condensation.

Be brief and precise   Writing a précis is like reproducing the soul of the matter. Therefore, it needs to be concise, precise, and focused. Normally, the length of the original passage is reduced to a one-third in its précis version.

Be complete   While reproducing an idea, we cannot leave out any of its vital aspects. Therefore, before we launch ourselves into writing a précis, we must carefully read the passage, notice all the important points, and incorporate all of them in our précis. A précis should, in fact, be as complete and comprehensive as the original one, albeit it should be expressed in a less number of words.

Be choosy   Although it is not possible for us to leave out any important idea from the original, it is required that we carefully choose only the material that is an indispensable part of the whole argument. In order to achieve a good précis of the original, we need to discard all the extraneous and superfluous material present in the form of examples, illustrations, instances, quotations, citations, anecdotes, parables, and any other such material that is included in the original to substantiate the basic idea.

Be original   A good précis is both creative and original. Of course, while writing a précis, you are not expected to distort or modify the author's view. You are also not expected to add any idea of your own or leave out some important idea of the author, but at the same time, you are required to express the author's views in your own words. Therefore, try to use your own expressions while rewriting what the author has expressed in the original.

Be coherent   Normally, while writing a précis, we follow the order the author has chosen to arrange his/her ideas. However, since a précis is not a pale imitation of the original, a good précis always has a coherent structure of its own. In any case, it should not look as though some unrelated and disjointed sentences have been yoked together. Remember, the purpose of a précis is to help the reader gather the whole idea in a compact, complete, and coherent way. An incoherent or incomplete imitation of the original would, therefore, be of little worth to the reader.

Be clear   Just like completeness and coherence, clarity too is an important attribute of a well-written précis. At times however, while writing a précis, it is lost as we tend to overemphasize the need to compress the ideas expressed in the original. Since the précis has to serve as a substitute for the original, we cannot afford any type of vagueness to punctuate the reader's comprehension of our précis.

19.8 SEVEN-STEP LADDER TO WRITING AN EFFECTIVE PRÉCIS

Having learnt some essential features and principles that characterize a good précis, let us focus on the process of writing a précis. Writing a précis has some steps and stages. Follow a seven-step ladder, as discussed below, to be able to produce an effective précis.

Read and comprehend   Read the original piece of writing as many times as you require, ensuring that you have understood what the author has expressed in his/her words.

Prepare a skeleton of the main ideas   Having read and understood a passage, identify all main and subordinate ideas and jot them down one by one. This gives you a clear view of all the ideas that are to be incorporated while writing the précis.

Assimilate the essentials   Writing a good precis is to recapture the soul of what the author has said in your own words. For this, you need to not only understand the original passage and jot down its main points, but also assimilate the whole thought embedded in it. To achieve this, you need to focus on each of the points noted down by you and rephrase them in your words. This will help you reshape the overall idea of the original passage in your words without distorting or losing its sense.

Think of a title   Once you have understood the passage, focus on the central idea and think of a suitable title based on it. Thinking of a title and assigning it to a passage is essential as it keeps your thoughts focused on the core of the issue.

Prepare the first draft   While preparing the first draft, remember to neither delete any important idea nor add anything of your own. Focus on the ideas observed and assimilated thus far and try to capture the spirit of the original in as few words as possible.

Review and compare   Having written it once, read your version with a view to observing whether it matches the original. While doing so, ask yourself questions such as—Does my précis capture the essence of the original passage? Does my précis include all the important ideas expressed in the original? Has any idea been unnecessarily added, repeated, or deleted? Does it follow a coherent structure? Does it have clarity and compactness of expression? Does it use linkers and punctuation marks correctly?

At this stage, you can also count the number of words used in your précis. Compare the length of your passage to that of the original. See if you can manage to do away with some more words or add a few more, depending upon whether it sounds redundant or obscure.

Edit and revise   Having reviewed your first effort critically, you can now revise your draft and shape it as the final version of your précis. At this final step of précis writing, incorporate all the alterations, modifications, and changes you thought of while reviewing your first draft.

19.9 WRITING PRÉCIS OF GIVEN PASSAGES

Having learnt the principles of writing a précis, let us practise by working out a few passages given below:

Passage I

A growing number of scientists are going where politicians fear to tread by calling for a wider public debate on the sensitive issue of the global human population, which is set to rise from the present 6.8 billion to perhaps 9 billion by 2050.

Lord Rees, the president of the Royal Society, brought the subject up in his excellent Reith Lectures; Sir David Attenborough has become a champion of those who believe population has been relegated as an environmental issue; and more recently Prof. Aubrey Manning, presenter of the BBC's Earth Story, has stated that the sheer number of humans on the planet is the greatest menace the world faces.

Scientists have a reputation for saying things as they are, not as they should be. Politicians, forever looking for short-term solutions to keep them in office, do not, as a rule, look further than the middle distance. Yet population is one of those over-the-horizon threats without enemies, as Lord Rees put it. It is a disaster in slow motion, and all politicians seem to do is provide the sort of platitudes articulated by Michael Heseltine, who recently fielded a question on Radio 4 by saying that the problems associated with population never turn out to be as bad as predicted—which is probably true if you can enjoy your own Oxfordshire arboretum.

No doubt Heseltine and his fellow politicians who are in favour of doing nothing about population will hardly cite the words of John P. Holdren, President Obama's science adviser, who wrote these words in 1969 when he was a young ecologist: 'If the population control measures are not initiated immediately, and effectively, all the technology man can bring to bear will not fend off the misery to come.'

Misery, what misery? You can, of course, imagine the political class arguing that scientists have consistently got it wrong about overpopulation. But the next 40 years are going to be very different to the previous 40 years, and many scientists fear that there will indeed be extreme misery to come if the world does not take population more seriously.

The facts speak for themselves. The UN estimates that the global population will rise from 6.8 billion in 2009 to 8.3 billion by 2030, with much of the increase in the poorest countries, notably sub-Saharan Africa, which is set for a 51 per cent increase in the same period—four times that of the UK.

World food production will have to increase by 50 per cent to meet rising demand; water availability will have to increase by 30 per cent; and global energy demands by 50 per cent. Politicians may think that science and technology will provide what is needed, as it has done in the past at a cost to the environment, but many scientists are not so sure.

Holdren himself came up with a simple equation to try to quantify the sustainability of a given population level: $I = P \times A \times T$—where I represents the impact on the environment, P the population size, A the affluence or level of consumption per head, and T the technology that determines how efficiently resources are used. The equation simply says that the impact on the environment is a factor of the number of people, how much they consume, and how efficient the consumption is.

It is a crude equation, but the aim was to show how we can limit the impact on the environment by intervening at any of these three levels. What is clear is that a continuing increase in human numbers makes everything else we do to reach sustainability far more difficult. As David Attenborough says: 'I've never seen a problem that wouldn't be easier to solve with fewer people, or harder, and ultimately impossible, with more.'

Just in case anyone thinks that this is just a problem for poor countries in Africa, they should read last week's report on population by the 'Forum for the Future'. Official statistics show that the UK population grew by 2 million between 2001 and 2008, its fastest rate of growth since the post-war 'baby boom'. Over the period from 2008 to 2033, the British population is set to grow from 61.4 m to 71.6 m, with 2029 being the year when there will be 70 million people living officially in Britain—there could be another 1 million or so living here illegally.

That means we will have to accommodate another city the size of Bristol every year. About two thirds of this projected increase is expected to be either directly or indirectly due to future migration, according to 'Forum for the Future'. Politicians should be forced to debate this issue, rather than relying on people being cowed by suggestions that to do so is somehow pandering to illiberal, xenophobic, and racist elements who have hijacked the subject for their own nefarious ends. (No. of words: 810)

Source: http://www.independent.co.uk/news/science/steve-connor-we-need-a-global-debate-on-population-1999538.html, accessed on 12 December 2010.

Supposing you are required to prepare a précis of this passage, what would you do? Of course, you would read it carefully. Then, you would mark the important ideas and prepare the skeleton of the passage. Therefore, let us have a look at the important ideas of the passage as given above:

1. Population increase is one of the most serious problems of the world today.
2. Studies conducted in this field suggest that the world population is further likely to increase by a few more billions.
3. According to many environment scientists, ecologists, researchers, and other intellectuals, the rise in population will lead to several unpleasant consequences.
4. Most significantly, more number of people will require more water, more food, and more energy resources to survive.

5. Much of this increase in population is likely to take place in the poorest countries of the world.
6. However, it is not just the poor countries but also the developed ones which are set to see an increase in their population.
7. Ecologists and environmentalists the world over fear that this increase will have a direct bearing on the environment as the population size, their affluence level, the appropriate consumption of resources, and the effective utilization of technology by people are inter-linked phenomena.
8. However, with the increase in population, the other three factors also become difficult to control and effectively balance.
9. Therefore, many scientists and other intellectuals regard population growth to become the greatest menace in the years to come.
10. Despite the gravity of the situation, politicians have been trying to downplay the population menace, owing to their parochial and personal objectives.
11. However, they need to be brought round to discuss the issue rather than appeasing some equally parochial groups who divide the world on small issues for their personal gains.

If we conscientiously work on the given passage therefore, this is how one of the condensed versions may look like:

Population increase is one of the most serious problems of the world today. Studies conducted in this area suggest that the world population is further likely to increase by a few more billions. According to many environment scientists, ecologists, researchers, and other intellectuals, the rise in population will lead to several unpleasant consequences. Most significantly, the increasing number of people will require more water, more food, and more energy resources to survive. Much of the population increase is likely to take place in the poorest countries of the world. However, it is not just the poor countries but also the developed ones which are set to see an increase in their population. Ecologists and environmentalists the world over fear that this increase will have a direct bearing on the environment as the size of population, their affluence level, the appropriate consumption of resources, and the effective utilization of technology by people are interlinked phenomena. Therefore, all concerned intellectuals regard population growth to emerge as the greatest menace in the years to come. Despite the gravity of the situation, however, politicians have been trying to downplay the population menace owing to their parochial and narcissistic objectives. Keeping in view the seriousness of the matter though, they need to be brought round to discuss the issue rather than appeasing some equally parochial groups who divide the world on small issues for their personal gains.
Suggested title: Menace of Population
No. of words: 249

Passage II

Stilettos are an apt metaphor. They give the illusion of heights reached. Their name derives from the stealthy, deadly knife favoured by the Mafia. They are associated with those who can walk the talk, and talk their way through even when they have no walk. But they are also precarious. It takes a brave woman to slip into them. And a braver one to kick them off—not for a night of being bad, but for good. Stilettos are never put away voluntarily, only with a reluctant sigh. This is the nature of these sexy shoes. And of the profession that struts the ramp in them.

Viveka Babajee's short life and sad end is chilling testimony to the fact that stilettos flatter to deceive. She's not the first victim. She won't be the last. Ugliness is built into the beauty trap.

(Contd)

Why shouldn't the media have gone to town on this story? It is couturier-made for the Velcro-effect on eyeballs. With its lowdown on high life, the shadows behind the spotlight, the tarnish after the glitter, it packed a double whammy of a punch, pandering to both our salivating voyeurism and our prim sanctimoniousness. Clearly a model's life and a model life are two very different entities. We lap up one while paying lip service to the other.

It has taken yet another suicide to reveal the dark, tortured, hateful world behind the insouciant sashay down the ramp, lit by a million megawatts of adoration—and envy. And it has forced its besotted followers to face up to the cruel question: should the glamorous model believe she is a genuine high-end product entitled to all its aura, or is she just a knock-off? 'Between the idea and the reality… falls the Shadow.' Which is why T.S. Eliot's next line does not follow. There is little 'Shanti, Shanti, Shanti', only an unsettling disillusionment. And as in the case of Viveka Babajee, a fatal disintegration.

Liberalization's laissez-faire has unleashed a revolution quite unknown in the 70s when Jenie Naoroji's swish of models, led by the ethereal Shai, glided smilingly and barefoot down the ramps, barefoot instead of today's arrogant stomp. True, the profession was always sequined by glamour, but in its early days it only flirted timorously with the entertainment, and it certainly wasn't souped up by celebrity. More significant, the big-bucking broncos of global brands, lassoing the market and upping the ante for the cowboys of Indian industry, hadn't yet arrived to make this into the high-stakes game it is today. Nor had the Big Bad Wolf of Bollywood gatecrashed into everybody's party. Modelling was always about dog-eats-dog-and-everybody-bitches, but as Brand India got into stride by the turn of the millennium, a nuclear-grade mutation seems to have hit this world of canines in corsets.

India-with-a-designer-label has opened up undreamt-of opportunities for every young girl to turn her face into her fortune, and her body into her wealth manager. But it has also sharpened the competition into a killing machine. It has savaged an industry notorious for its disposable culture. Reduce? You must. Recycle and Re-use? Never! Careers increasingly have a shorter life-span than mascara wands, and there is no fairy godmother to stop their returns to rags. A godfather would be better, but he's chasing the taller girl in the shorter skirt who is thinner and younger.

The former model and Page 3 habitué Queenie, recalls, 'Back in the 1980s and early 90s, we were a small, loyal, close-knit fraternity, more like a family. Today, it is all about fierce competition and back-biting. You get sucked and drawn deeper into it, like quicksand. Not just models, today's entire younger generation is under terrible pressure to be part of the glam crowd, they are exposed to booze, drugs, and competitiveness earlier and in a greater degree than ever before.'

The Queenie Bee astutely puts her bejeweled finger on the latest Culprit No. 1: Bollywood, which seems to be responsible for more damage than we feared. 'Today, the concept of a supermodel doesn't exist. Modelling has become an open-ended industry with Bollywood on one and a sea of faces on the other. There is no longer any sense of professional identity.'

The society columnist, Simi Chandoke, endorses this concern. 'Modeling in itself is no longer enough; there is the additional pressure to make it big in Bollywood too. Because B. Town is now the object of worship, and the fashion frat has been elbowed out of that niche. This wasn't the case a decade and a half ago. Back then, models were more respected. But today the tables have turned and everything from brand endorsements to shows falls in the lap of film stars.'

Chandoke goes on to the hazardous fall-out. 'Once ousted from the scene, the girls have no education or job to fall back on, and when they don't find an anchor in a financially stable man, they break and turn to drugs and drinking. Short-lived romances and flings and keeping the wrong company are also part of the culture. It's a vicious circle. Bollywood, on the other hand, is more grounded and surrounded by family and a strong circle of friends. There is less scope for loneliness.' (No. of words: 866)

Source: Bachi Karkaria, Nicole Dastur, and Diya Banerjee, 'Not a Model Life',
The Times of India: The Crest Edition, 3 July 2010, p. 2.

Let us try to understand the main points in the above passage:

1. Modelling is a glamorous profession but not without an ugly side to it.
2. A model's real life is far from being what it seems.
3. Liberalization has changed the face of modelling.
4. Today, models have to perform in trying circumstances.
5. With competition increasing, models have shorter careers but a lot at stake.
6. In order to achieve success, models try to use their physical beauty.
7. The promoters however are always interested in new models.
8. Because of ruthless competition around, the profession of modelling has lost its sense of fraternity that characterized it in the 1980s and 90s.
9. Moreover, because of the entry of film stars into this profession, modelling has lost the distinctive identity and respect the profession once enjoyed.
10. Rather than a profession in itself, modelling is now seen as a gateway to filmdom.
11. Since modelling is both extremely demanding and also short-lived a profession, it does not allow the aspiring models to prepare for any alternative.
12. The models with inadequate educational competence find themselves nowhere once out of modelling.
13. Lack of alternatives and a sense of rejection makes them resort to drugs, drinking, and unstable relationships.
14. The immanence of suicides in modelling not only establishes its cruel sordidness but also brings to the fore our own hypocrisy.

Précis   The much envied profession of modelling actually conceals beneath its apparent charm a world of darkness, hate, torture, exploitation, and loneliness. Though once a respectable profession, modelling in a post-liberalization world is marked with ruthless competition, loss of identity, and a sense of alienation. Today, models have to perform in trying circumstances. With competition increasing, models have shorter careers but a lot at stake. In order to achieve success, models try to use their physical beauty. The promoters however are always interested in new models. Because of ruthless competition around, the profession of modelling has also lost its sense of fraternity that characterized it in the 1980s and 1990s. Moreover, because of the entry of film stars into this profession, modelling has lost its distinctive identity as well. That is why the profession no longer has the respect it once enjoyed, as most of the products earlier advertised only by models are also being advertised by Bollywood stars. Today, models enter this profession with an aspiration to become a Bollywood star one day. Since modelling is both extremely demanding and also ephemeral, it does not allow models to prepare for any alternative. The models with inadequate educational competence find themselves nowhere once out of modelling. Lack of alternatives and a sense of rejection makes them resort to drugs, drinking, and unstable relationships. The immanence of suicides in modelling not only establishes its cruel sordidness but also brings to the fore our own hypocrisy, as by watching a glamorous model on the romp we seek to gratify our senses and at her fall express our shock at the absence of morality in the world of modelling.

Suggested title (Besides the one suggested in the article):
- *Modelling: Problems and Perspectives*
- *Modelling: Not a Model Ramp to Stomp*

No. of words: 275

Passage III

The right kind of education consists in understanding the child as he is without imposing upon him an ideal of what we think he should be. To enclose him in the framework of an ideal is to encourage him to conform, which breeds fear and produces in him a constant conflict between what he is and what he should be; and all inward conflicts have their outward manifestations in society. Ideals are an actual hindrance to our understanding of the child and to the child's understanding of himself.

A parent who really desires to understand his child does not look at him through the screen of an ideal. If he loves the child, he observes him, he studies his tendencies, his moods, and peculiarities. It is only when one feels no love for the child that one imposes upon him an ideal, for then one's ambitions are trying to fulfill themselves in him, wanting him to become this or that. If one loves, not the ideal, but the child, then there is a possibility of helping him to understand himself as he is.

If a child tells lies, for example, of what value is it to put before him the ideal of truth? One has to find out why he is telling lies. To help the child, one has to take time to study and observe him, which demands patience, love, and care; but when one has no love, no understanding, then one forces the child into a pattern of action which we call an ideal.

Ideals are a convenient escape, and the teacher who follows them is incapable of understanding his students and dealing with them intelligently; for him, the future ideal, the what should be, is far more important than the present child. The pursuit of an ideal excludes love, and without love no human problem can be solved. If the teacher is of the right kind, he will not depend on a method, but will study each individual pupil. In our relationship with children and young people, we are not dealing with mechanical devices that can be quickly repaired, but with living beings who are impressionable, volatile, sensitive, afraid, affectionate; and to deal with them, we have to have great understanding, the strength of patience and love. When we lack these, we look to quick and easy remedies and hope for marvellous and automatic results. If we are unaware, mechanical in our attitudes and actions, we fight shy of any demand upon us that is disturbing and that cannot be met by an automatic response, and this is one of our major difficulties in education.

The child is the result of both the past and the present and is therefore already conditioned. If we transmit our background to the child, we perpetuate both his and our own conditioning. There is radical transformation only when we understand our own conditioning and are free of it. To discuss what should be the right kind of education while we ourselves are conditioned is utterly futile.

While the children are young, we must of course protect them from physical harm and prevent them from feeling physically insecure. But unfortunately we do not stop there; we want to shape their ways of thinking and feeling, we want to mould them in accordance with our own cravings and intentions. We seek to fulfil ourselves in our children, to perpetuate ourselves through them. We build walls around them, condition them by our beliefs and ideologies, fears and hopes—and then we cry and pray when they are killed or maimed in wars, or otherwise made to suffer by the experiences of life. (No. of words: 609)

Source: J. Krishnamurty, 'The Right Kind of Education', *Education and the Significance of Life*, London: Victor Gollancz Ltd, 1966, pp. 26–28.

Let us try to understand the main points in the given passage:

1. The right kind of education for a child cannot be without love, care, and understanding.
2. We try to condition the child to our ideals mostly through education.
3. We make efforts in making the child fall in consonance with our own desires and aspirations.
4. Demanding conformity thus, we force a child to fit into a frame desired and devised by us.
5. Conformity however instills in the child fears, conflicts, and confusions.
6. True education, thus, is not governed with a tendency to conform a child to our ideals.

7. It requires a great deal of time, patience, and effort on the part of the parent and teacher.
8. Moreover, only those who not conditioned to chase certain ideals can intelligently and freely understand the child and help him get true education in life.

Précis   Imparting true education to a child is one of our major challenges. Mostly, in the name of education, we tend to condition a child to our expectations, ideals, desires, and wishes. Education induced thus, demands conformity which hardly helps the child, for such coercion only induces in him a sense of insecurity, fear, conflict, and confusion. Through such kind of education, we can only perpetuate ourselves, which is of little worth to the child in understanding himself or the life around him. True education for the child is the one based on love, care, and an understanding of him as an individual. This requires a great deal of patience, love, and creativity on the part of the parent and the teacher. More significantly, those who aspire to teach true education, must themselves be truly educated. Therefore, the educator has to free himself from all the conditioning, and then focus on the child as he is and not as he should be according to some ideology, hope, fear, or expectation. However, it does not usually happen as we try to automate a child to a particular system, impose on him our own wishes, thoughts, beliefs, and aspirations and lament the loss incurred by them on our account.

Suggested title (Besides the one suggested in the write-up):

- *True Education*
- *Education: The Process of Deconditioning*

No. of words: 206

Passage IV

Ever wondered what keeps people happy all the time? An underlining factor in every happy person's life is that he/she tends to have supportive relationships. The American author Lois Wyse said, 'A good friend is a connection to life—a tie to the past, a road to the future, the key to sanity in a totally insane world.'

Social networking sites are one of the best ways to stay connected with friends and family. They have changed the way people socialize. Rahul Kulkarni, Product Manager, Google India, says, 'Being socially connected has a definite positive effect on your well-being.' Namrata Aswani, PR consultant, says, 'I feel connected to the world—it's so far, yet so near. You can share photos and videos with a click. Chat with your old friends and the happiness just flows. It is a relief from a boring day.'

A recent study shows that social media is now our favourite online pastime, beating email by a wide margin. According to research, if you have a supportive social network, you may add years to your life. Says Dr Harish Shetty, Mumbai-based psychiatrist, 'Friends provide emotional nourishment and tangible social support. The feeling of having access to someone during those dark hours accelerates hope. Hopelessness leads to loss of confidence.'

Many who blame the world for not having friends are those who don't trust anyone with their feelings or confidences. Good emotional contact causes networks to be built. Even studies suggest that we need close relationships that involve understanding and caring.

Prahasitha, senior faculty at One World Academy, Chennai, feels, 'Our work and the way we live is so structured that it isolates people from each other. Social networking sites fulfill a primal need in all of us.'

Social networks seem to be the same as conventional networks but bigger and more casual. They may not match up to the physical presence of a friend, but are reliable during those dark, depressing times.

Source: Norbert Rego, 'Make Friends, Live Longer!', *The Times of India: Times Life,*
3 October 2010, p. 4.

Précis Supportive relationships in life add to our happiness. Studies reveal that staying connected to family and friends has a positive effect on a person's sense of well-being. In modern times, when people lead a busy and isolated life, social networking emerges as one of the best forms of socialization. It can help people stay in touch with one another despite physical distance, through mails, chatting, and sharing photographs online. The relationships developed through social networking, though appear to be casual and may not be as effective as the physical presence of a person, can help people during sadness and depression.

Suggested title:

Social Networking: Happiness unlimited?

(No. of words: 101)

Passage V

For many centuries, India embraced a unique culture—a culture that invested most of its resources in meeting the needs of the soul rather than fulfilling cravings of the flesh. The sages, who were the founding fathers of this great culture, stressed that the world we live in has its source in a higher reality which is divine, eternal, and intrinsically beautiful. This reality is the Lord of the Universe, the Lord of Life. It is the primordial pool of intelligence, existence, and bliss. In this ever expanding universe, everything blossoms because the world is intrinsically connected to a source of boundless nurture. Knowledge of this truth grants us freedom from bondage, and ignorance of it robs us of the gift of a joyful life.

All living beings have their rightful place here on earth, and we all have the right to enjoy its bounties. But as soon as we attempt to possess them while denying others their share, we invite misery into our lives. Work hard and collect worldly possessions if you wish, but consume judiciously so that you yourself do not get consumed. And do not waste your energy preventing others from collecting the objects of their desire. There is no need to place yourself in the position of a priest or a policeman. Simply set an example. That will send a message a million times clearer and more powerful than all the messages broadcast by all preachers and policemen combined. This is the message of the sages, the message from which our Vedic culture evolved: the world consisting of numberless forms is rooted in a single reality, one that is all-pervading, beginingless, and endless. This understanding was the backbone of the lifestyle embraced by the people in Vedic times.

The founding fathers of Indian culture also introduced a social system that rewarded the use of one's own conscience. Their social, moral, and ethical laws rested on the basic principles of spirituality—life is sacred; hurting and harming anyone is a spiritual offence; therefore, do not do to others what you do not want others to do to you. Complying with the voice of one's conscience was the driving force behind all moral action. According to Vedic sages, refraining from hurting others out of fear of punishment is the hallmark of a low-grade person. Attending to one's own well-being and the well-being of others constituted the core of spirituality. As this core principle of spirituality infuses everything with lasting life, the sages called it dharma, the law of sustainability. They also called it *sanatana dharma*, the perennial or original dharma of mankind.

This law is eternal and unalterable. Regardless of who you are and what you wish to become, practising this dharma is an absolute necessity. The law of sustainability communicated by dharma constitutes the principles of duty, for such principles are charged with power to lift the human spirit from a base existence to a key role as a member of civil society. Without the principles of sustainability as a foundation, we cannot define right and wrong, good and bad, just and unjust. Unless we embrace the fundamental principles of dharma—sustainability in its broadest and purest sense—we cannot

(Contd)

prevent our personal tastes and interests, our whims and ambitions from colliding with those of others. And unless we see things in the light of this higher dharma, we cannot decide whose desires, goals, and ambitions are valid and whose are not. It is only in the light of dharma that we can do justice to ourselves without being unjust to others. It is the understanding of this dharma, therefore, that will bring the fight for justice to an end, for this dharma sheds light on our relationship with ourselves, with others, and with the world of nature.

Source: Pandit Rajmani Tigunait, 'Dharma for All the Ages', *The Times of India: The Speaking Tree, 26 September 2010, p. 4.*

Précis What characterizes ancient Indian culture is its emphasis on addressing the needs of the soul rather than those of the flesh. To Indian sages, the absolute reality was divine, eternal, and intrinsically beautiful. This reality was believed to be all-pervading, beginningless, and endless. That we are a part of this single reality was to have true sense of knowledge which set people free from bondage, while an absence of it brought misery and pain into one's life. In this sense of boundless togetherness, all living beings were supposed to have their rightful existence on earth and any form of unnatural possession, coercion, or usurpation was not advocated. In this form of religion, conscience had a great role to play. It was believed that one's conscience was a part of the divine force and following one's conscience was a sacred way of life. Since life itself was considered sacred, hurting and harming others was considered sinful. Listening to the inner voice prevented people from doing anything irreligious or sinful. What thus constituted the core of our ancient religion was the dharma or sustainability which emphasized on attending to the well-being of all. This religious sustainability hence formed the core of our civil society and emphasized on a sense of religion, which rather than focusing on the selfish and parochial choices, ambitions, and desires brought into view an all-encompassing sense of universality and commonality of existence.

Suggested title:
True Religion
(No. of words: 234)

RECAPITULATION

✓ Many a time, in our academic and professional life, we are required to put the ideas expressed by others in a condensed form. The art of condensation, therefore, becomes an important part in our writing skills.

✓ There are four major forms of condensation, namely précis, summary, abstract, and synopsis.

✓ Précis is the form of condensation that is most commonly used.

✓ For writing a good précis, one needs to observe the principles of exactness, conciseness, completeness, coherence, and clarity.

✓ While preparing a précis, examples, anecdotes, quotes, and references are not included. Only their gist is assimilated in the text.

✓ Writing a good précis requires careful reading and comprehension of the passage, taking down all the important points, arranging them in a coherent order, maintaining clarity and conciseness in expressions, and above all assimilating the essence of the original without getting into the lengthy details and substantiation of key issues.

✓ A good précis has a suitable title and reduces the length of the original piece of writing to its one third.

WISEWELL QUIPS

EXERCISES

1. Define a 'précis' and discuss its characteristic features.

2. 'Writing a précis does not mean resorting to a pale imitation of the original, but involving yourself in a creative process.' Do you subscribe to the view? Discuss and substantiate.

3. What is the seven-step ladder to writing an effective précis? Discuss and illustrate with appropriate examples.

4. Distinguish each of the following in about 200 words:

   (a) Summary   (b) Abstract

   (c) Précis   (d) Synopsis

5. 'The art of condensation is an essential element in our writing skills.' Discuss and substantiate.

6. What are the necessary tenets for writing an effective précis? Discuss and exemplify.

7. Condense each of the following small passages retaining the main idea and using a minimum number of words:

   (i) When one does not really understand the true purpose of life, one is not in a position to really figure out how to focus on the actions to be carried out. That's what happens to most of us. We live and keep involving ourselves in actions which may not necessarily define us. Such a life is like a rudderless drift in a dark, befuddling ocean where you sail and sail, and still don't know where to shore up.

   (ii) 'We pine for what is not there,' says Keats and thus, like an enlightened soul, captures the cause behind perennial human suffering. Throughout our lives, we keep chasing a falling star; running after a goal, the completion of which should give us a sense of fulfilment and achievement. However, the misery lies in the fact that what is chased and achieved becomes immediately tasteless and redundant. No longer interested in what we possess, we hurtle ourselves into achieving what we don't have.

   (iii) Despite all our claims to be able to cobble up an educated and intellectual society, it seems we are hardly inching closer to one. There could be various reasons, ways, and instances to vouchsafe a view like this. One of the quickest ways to figure out the pulse of the nation: it won't be a bad idea to watch the most favoured and popular serials that swarm hundreds of channels on your television screen. Look at the lurid content, the slipshod presentation, the melodrama, the abundance of gimmicks, vulgarity, and cheapness that characterize most of these programmes. And then you realize, with a gasp of sigh, that we are hardly headed towards any awareness, education, or enlightenment.

   (iv) One of the dubious distinctions of our society surely is to be one of the most corrupt countries in the world. Corruption today has become an integral part of our system. If you want to get anything done, you need to bribe a person. It doesn't matter whether the matter is right or wrong, small or special; you need to grease a bureaucratic palm in order to get a file moving off its blocks. For those who cannot afford to fall in line in with such unsaid expectations, there hardly is a ray of hope. After all, their wishes may soar but the files won't register a budge unless sufficiently winged on currency notes.

(v) Adults are scared of death, as children are scared of darkness, so says Bacon, the legendary Renaissance English essayist. It is, however, just not children who are scared of darkness. After all how many of us can claim that darkness does not frighten us? Imagine yourself being caught in the darkness of night with no one around you! Even if you are trapped within your own house that you have lived in for years and years together, you don't really enjoy seeing your large, lonely shadow against your own walls. Similarly, not many of us feel like walking alone in long streets with no light around. And even if the grown ups choose to watch late night horror movies in big theatres, a realization that they have to cross a long, unknown, dark street all alone after the show is over, can give them a cold shiver down their spines.

(vi) Superstition is one of the peculiar features of human life. Normally, the word is conceived, viewed, and interpreted in ancient terms and with the rise of the scientific spirit and a general sense of awareness all around, one would imagine that its tentacles on us are ebbing away. A closer look into the behaviour of even the most educated, affluent, and respected though hardly endorses such surmises. On the contrary, there is a spurt of superstitious beliefs all around us. No marriages are solemnized today without the matching of 'gunas' of the girl and the boy. Nobody is prepared to believe that it is a 'kundli' generated through an engineered software after all that is actually calling the shots. We would love to see 'the hand of God,' in all the things that happen around us. If, by chance, the Gods fail to win matches for us, it is an octopus or a parakeet that can decide the matter.

(vii) Whether science is a friend or a foe has been one of the raging debates for many decades now. Gradually, however, the debate is giving way to the belief that it does not matter whether it is a friend or a foe. We all now understand that it is not just difficult but almost impossible for us to conceive of a world without the machines, motors, and computers that do not just make our life much more comfortable but almost define our existence. At the same time, we are also aware of the damage that the proliferation and advancement of science and technology has caused to the environment. The real issue for us today, therefore, is not whether to see science as a friend or a foe but to be able to survive without science and still not leave it.

(viii) Looking at the variegated shades and hues of Indian cinema in Hindi, one wonders what happened to the stream of 'parallel cinema' that characterized many movies made in the 1970s and early 1980s. Those were the days when you could see directors such as Shyam Benegal, Sai Paranjpe, Aparna Sen, Basu Chatterjee, Basu Bhatacharya, Gulzar, Saeed Mirza, Ketan Mehta, and the likes of them raising pertinent socio-political issues, capturing realities in an intense, artistic manner, and attempting to redefine the codes and ethos that always torment the creative souls. In retrospect, they seemed much like a ruthless continuation of cinema as a means of expression of self conceived reality and propagation of ideas, an experiment that was almost poetically presented in the 1950s and 60s by movie makers such as Bimal Roy, Hrishikesh Mukherjee, Gurudutt, Mehboob, K. Asif, and Raj Kapoor.

(ix) Whether humanity would survive global warming or not is only a matter of speculation for us. The depleting water resources, the rising temperature, the melting glaciers and strange climatic phenomena recorded in the last couple of decades however, does not augur well. Besides, the burgeoning population, the staggering amount of inventions, and growing consumerism, have only been adding many more twists and turns to the plot that now looks increasingly tragic. Of course, predictions have always been made about the catastrophe striking us, and since most of these have not materialized, we tend to believe that what is doing rounds in the media is just an unfounded and exaggerated projection of the whole situation. We may choose the position we may like to but we must know that though ignoring a rumour is judicious, ignoring reality is no wisdom.

(x) One of the toughest things to do today is to be a child. These days, children are expected to be mature, disciplined, focused, sensible, assiduous, competitive, and above all successful. So, they are supposed to be all, but not a child. Childhood today is devoid of innocence; they cannot expose their ignorance in the quizzes they have to participate in. They cannot score poor marks in exams because they were charmed by a particular sport and followed it both in their congested streets and on a television screen. Whether willing or not, they all have to live upto one single dream of their parents—they have to be successful. One really wonders what has happened to all our intelligence. By making a child chase his tail in a rat race, we are hardly making him successful. Though in an effort to do so, we ourselves are being hopelessly petulant and childish.

8. Read the following passages and write a précis for each of them. Also assign a suitable title to every passage and write the number of words you have used in making a précis of the original:

Passage I

It was the cold midnight of 2 and 3 December 1984. The people of Bhopal were sleeping in the warmth of quilts as the mercury dipped, with the sounds of police vehicles assuring them 'all is well'. Suddenly, they started coughing and sneezing incessantly.

They wondered what was wrong. Those who came out of their homes discovered that hell had broken loose. People were running in panic. Sunil Singh, a grocer, recalls the nightmare: bodies strewn all over, and the air irritating the eyes, throats, and lungs. 'I was unable to see properly, but I ran until I collapsed in a pond meant for pigs and buffaloes,' he says.

The next morning, the people of Bhopal woke up to the news of the world's worst industrial disaster. About 8,000 people had perished and 5 lakh were left with multi-systemic injuries, as over 40 tonnes of deadly methyl isocyanate gas leaked from the multinational Union Carbide India Ltd's (UCIL) plant.

Twenty-five years on, the survivors struggle with cancers, pain, breathlessness, and other problems. Every month, about 30 people die because of the after effects of the gas. Children, who were born after the tragedy, too, are exposed to the toxic waste.

The environment watchdog Greenpeace recently declared the defunct UCIL plant as a 'global toxic hotspot'. It found heavy concentrations of carcinogenic chemicals and heavy metals such as mercury, which have been found at 20,000 to 60 lakh times the permissible levels.

Over 40 per cent of the women who were exposed to the gas have had abortions. Also, many women have not found grooms because of the fear of deformed babies. Femida, a survivor, was just four during the mishap. Unmarried and scarred, she wanders in slums, and often tears up her clothes in a fit of madness.

Activist Rachna Dhingra, who runs the Sambhavana clinic, says that many women who were toddlers at the time of the mishap are now suffering from menstrual problems. Nazma (name changed) was just a year old at the time of the mishap. Now, her menstrual cycle is erratic. And she is yet to get married. 'I have been suffering for the last 10 years,' says Nazma. 'I do not know when I would become a normal woman.' (No. of words: 374)
Source: Deepak Tiwari, 'Murder at Midnight,' *The Week*, 31 January 2010, p. 45.

Passage II

In a saturated marriage market, the matrimonial bio-data—that business-like document brandished by *aunties* and local matchmakers—is worth its weight in gold. But Gaurav Agarwal knew that M/28/Wheatish/Garment Exporter could not sum up the chutzpah he wanted to sell. So, he decided to hand out VCDs instead.

Those who dare to press 'play' can see Agarwal in an embroidered shirt, thick-rimmed glasses and goatee, tapping away at his laptop, strumming the guitar, doing a salsa shimmy and showing off his beatboxing skills for eight long minutes. Then, in Bollywood style, he cruises down the streets of Lucknow, wind blowing through his hair, as he confesses what he is looking for in a partner; someone who 'maintains her figure, is from the same caste, and works from home'. Later there is a sneak peek into his room and even an introduction to his parents for prospective brides. 'Nowadays, if you want to get married, you need to advertise yourself', he says, 'I thought, why not do it with a *dhamaka*?'

Agarwal is not the only one trying to cut through the clutter of dull matrimonial biodata with a 'dhamaka'. From video clips to PowerPoint presentations to mini feature films, several prospective brides and grooms are making sure their pitch is memorable. So you have eligible bachelors cooing at pups, in the hope that bachelorettes will sense their sensitivity; prospective brides walking down imaginary ramps in their living rooms to show their 'open-mindedness', and girls putting their multifacetedness on display by smashing a tennis ball over the net in one shot and demurely making rotis in the next.

Meghna Chitalia, a wedding planner who has recently added making new age biodata to her work profile, had one Mumbai girl change 11 outfits in her matrimonial commercial. 'She got into a sari, salwar kameez, mini skirt, and jeans and sported different looks. I guess she wanted to show that she could be traditional and modern too,' she says. In another case, the planner helped a bride make a short feature film about her life. The video, interspersed with grabs from home videos of her childhood, interviews with members of her family and accounts of her favourite hobbies, were woven into a storyline. In yet another case, a prospective groom actually did a spoof of Amitabh Bachchan in *Deewar*.

Excessive as it may seem to many, Chitalia says the idea is to make a lasting first impression. 'You have a few minutes to say everything about yourself and the idea is to say it in the best way possible,' she says, adding that a professional video shoot could set a client back by anything between ₹15,000 and ₹1 lakh. Not everybody outsources the video profiles though often they are handycam jobs or a mix of the amateur and the professional, like in the case of singer Vikram Sachdev who chose to begin with a clip of herself participating in a musical reality show. (No. of words: 490)

Source: Mansi Choksi, 'The Shaadi Screen Test', *The Times of India: The Crest Edition*, 29 May 2010, p. 13.

Passage III

Millennium jokes at the turn of the century were a dime a dozen, but what drew the loudest laugh from a friend was a channel's choice for a New Year Eve special movie: *The Titanic*. 'They are welcoming the new century on a sinking note.' He guffawed, little knowing that the movie, the accident it was based on, and his one-liner, all were going to be a metaphor for the 21st century. For when American Airlines Flight 11 and United Airlines Flight 175 flew into the twin towers, it was but a morbid parallel to the Titanic crashing into the iceberg. After that day, the world was never the same again.

It is a decade the world wants to forget, especially the west—marred as it has been with bloodshed, extremism, bankruptcy, and unrest. The party for the 21st century, it seemed, had ended even before it began. But the story was slightly different for India; indeed it seemed as if someone forgot to tell us that bad times were afoot. Like the lone drummer marching to his own beat, India plugged in her iPod and swayed gaily, sometimes stumbling, but always managing to get back on her two feet, through the decade that ends in a few days.

It was the decade of *desi*-ism, the era when we finally came into our own. The early promise of liberalization led to the high that only money and success give, and with these two guests, every other good time followed. It was okay to be Indian, heck it was cool to be us!

In literature, we learnt to look beyond Salman Rushdie and Vikram Seth, and Bollywood with all its over the top movies was fun once again. In soaps we left behind *The Bold and the Beautiful* for *Kyunki Saas Bhi Kabhi Bahu Thi*, and when hitting the dance floor we preferred to jive to *Desi Girl* rather than *Single Ladies*.

India, at least for the Indians, was shining as one of us got his daughter married at the Versailles and another pocketed British car giants. We lived well, ate, drank, drove, and wore the best of what the world had, worked hard, partied harder, and looked better than we had ever done before. Our women were taking over the work place and the bar, but it was also becoming important to have someone escort them home while our men were trying to decide whether they were metro or retrosexuals.

Where there is light, there is also darkness, and our chickens did start coming home to roost with the Naxal problems raising their heads. In politics, the more the things changed, the more they remained the same; but what did not budge from its decades old position was the distrust and anger our politicians aroused in us.

Pakistan, we reduced to a mere footnote, both in our history and present times, but sometimes at our own peril. We now look at China, wondering if we dare take it on openly.

Amid all this upheaval, a small section of India discovered its conscience and made itself heard, time and again. It is still a trickle, this awareness that the world does consist of others besides you and me but it has all the makings of becoming a gushing river one day. We still continued to subjugate our women even more so because now they asserted themselves but we also realized that 'our head in the sand' approach to alternate sexuality did indeed belong to the Victorian era. And while our torrid love affair with cricket continues, we learnt to appreciate other forms of sport, too, from tennis to shooting to golf and even hockey, even though that love affair lasted only till *Chak De India* was in the theatres.

It is a decade in which we laid the foundation for India to be a completely different nation from the one that we inherited. We tried to make her cooler and hipper though we did not always succeed. The next 10 years will give us plenty of opportunities to take this effort further, learn from our mistakes, and perhaps one day finally arrive at an India which we can be proud to have created.

(No. of words: 702)
Source: Nikita Doval, 'AD 2019', *The Week*, 3 January 2010, p. 25.

Passage IV

The Indo-Pak border in Punjab is foreboding and romantic, untouched for most parts, except by the footfalls of BSF men and a sparse population. The Rodanwala Khurd village here is serene and pristine.

The landscape stretches endlessly, with fields of knee-high paddy or wheat crop, punctuated by tracts of fresh earth, and the border fencing broken only by iron gates every half a kilometer or so and electrified at night. The fence has made all the difference in the lives of the villagers. 'We are sandwiched between the gun-wielding Pakistani Rangers and our own BSF men because we farm our land beyond the fence,' says former sarpanch, Daljinder Singh.

At the ceremonial retreat, 4 km away, Pakistani and Indian soldiers lower their national flags in marching style that is a barometer of Indo-Pak relations. Anger on the soldiers' faces and ferocious thumping of boots indicate all is not well. If the gates are banged shut and bolted, it means Pakistan is not taking India's complaints seriously. In good times, a deadpan expression pervades the ceremony and visitors are allowed close to the Zero Line between the gates. There is a healthy competition about which side has more viewers to cheer and whose patriotic songs are louder. Fridays go to the team across the fence, and Sundays are ours.

Some villagers have benefited from this spectacle that gives the patriotic types goose pimples. Farmer Puran Singh's relatives have tea stalls near the venue. A couple of people plying taxis between Amritsar and Attari are doing well thanks to tourists. 'But most tourists hire taxis in Amritsar. And most of our *khokas* (kiosks) were dismantled when the government undertook development of the area,' says Puran. He feels the ceremony has not benefitted people in the area.

But Rodanwala has its own show twice a week. The Delhi–Lahore Samjhauta Express halts here for around four hours. Customs officials carry out checks, and two heavy iron gates open to allow the train across the border. Children throng to see the train. 'We have only one government primary school. Most people prefer to send their children to private schools in Attari as private education is better and English is taught,' says farmer Sukha Singh.

A good school and hospital remain a dream. But sarpanch Manjit Kaur dare not demand it, as the border is a potential war zone and, hence, to be kept without buildings that could obstruct view of enemy movement. 'Even in terms of agriculture, we cannot go beyond paddy and wheat because of the height. We cannot have trees. We cannot grow fodder for cattle, as it would grow into a thicket that could be a hiding place for the enemy,' says Daljinder.

The border fence passes through 500 acres belonging to 120 farmers in Rodanwala. They require border fence gate passes to go to their lands across the fence. 'Getting this pass is difficult and depends on the whim of the BSF commandant. They ask us to apply again, and we have to run to wherever they tell us, including Jalandhar. Even with a pass, the guards won't let us cross on a day when the weather is bad, or some reasons we don't know,' says Sukha. 'The power supply is at night, but we cannot irrigate our land. If our tube-well pump gets spoilt, the electrician cannot enter,' he says. For Puran, wild boars from Pakistan ruin his wheat crop. 'One farmer was killed by wild boar,' he says, 'We cannot chase them into Pakistan as we may be shot.'

Farming holds no charm for the sons of farmers. Gurpej Singh, 19, has cleared his Class 12 and dreams of joining the BSF. 'Promises are made that the BSF will recruit us, but nothing happens. Money seems to work, and we don't have that. Lots of my friends are like me, doing nothing,' he says.

Recently, women from farmer families with gate passes have been allowed to work in their fields across the fence as the BSF has posted women along the fence. But the lot of most women is pathetic. A 'polio nurse' who comes every fortnight is their only access to health service. The dispensary building, which has been there for 15 years, has no doctors or staff. When Paramjeet Kaur developed labour pains in the middle of the night in December, a villager who owns a taxi drove her to hospital in Attari, from where the doctor referred her to Amritsar, 30 km away.

But Paramjeet is luckier than people in the Dera Baba Nanak sector, where Pak rockets landed recently. 'In Dera Baba Nanak, the Ravi river is the border. There is no fence. When it floods, it devastates our village,' says Daljinder. 'And if there is shelling on the border, our movements are restricted. When we see unusual troop movement, we abandon the villages and flee to relatives in Amritsar, Jalandhar or Ludhiana.

(No. of words: 815)

Source: Vijaya Pushkarna, 'Pristine, Yet Perilous,' *The Week*, 7 March 2010, pp. 18–19.

Passage V

An insomniac city breeds nightbirds, who have cocked a snook at nature by liberating their body clocks from the confines of the circadian rhythm. They thus work nights in brightly illuminated offices or party till day-break in the garish neon of night-clubs, sleep by day, and snack whenever it suits their fancy.

Tejal Rajyagor, 25, for instance, lives in Mumbai, but keeps GMT timings. She works for a UK-based company that has outsourced offices to the subcontinent, starting her working day at 1.30 p.m. IST, which is 8 a.m. in London. 'Office hours are till 11.30 p.m., but work generally spills over to 1.30 a.m. So by the time I reach home, it's usually 3 a.m.' The area outside her office, which has many BPOs, comes alive post midnight, when employees step out for an 'evening' snack. 'There are chaiwalas, the idli-vada man, traffic zipping past. You would never know it was so late in the day unless you checked your watch,' Tejal says.

Tejal earns big money. But she is aware that she is paying a heavy price. 'Socially, I don't exist to the outside world, as I have barely attended a do in the last two years. Worse, I feel so disconnected, and cannot get my body back to Indian timings on weekends,' she says. Already reed thin, she has lost over five kilos in the past few months because of erratic meals. 'Most colleagues have the opposite problem, they are piling up weight, many having notched up to 10 extra kilos.'

Not so long ago, only a skeletal staff worked the graveyard shift—nurses, resident doctors, journalists, railway and airlines staff, cops, and some factory workers. With darkness no longer a restricting factor, the after-sunset workforce has burgeoned. While on the one hand, the business process outsourcing industry emerged, on the other, people are working later and later even in traditional sectors.

There aren't many India-specific studies done on the effects of the new work culture. But health experts point out that there are reasons why humans were created as diurnal beings, and tampering with this rhythm could open the door to a host of health troubles.

Manvir Bhatia, Chairperson of the Department of Sleep Medicine at Delhi's Ganga Ram Hospital, notes that hormones such as Growth Hormone (GH) and melatonin spurt at night. 'In adults, GH is associated with the body's repair mechanism. Its deficiency leads to the fatigue and low immunity that we see in those who stay awake at night.' A day's sleep cannot compensate for a night's rest. 'Melatonin is secreted after the onset of sleep, in the dark. It has a vast role in anti-ageing, and cancer prevention. The traditional concept of beauty sleep has a scientific basis here.' Cortisol, the stress hormone, too, is related to the circadian rhythm. Its disruption is manifested in the increased stress levels, high blood pressure, and obesity.

While India, where manpower is not a shortage, may be in denial of these studies, Denmark has already taken the lead. It is paying compensation to women who developed breast cancer after working in late night shifts, after a World Health Organization study concluded that altering sleep patterns suppressed the production of melatonin, thus escalating the risk of cancer. So far 40 women have got compensation.

'The effects build up over a period of time. They are not immediately evident,' says Bhatia. 'I'm seeing memory loss in younger people, in their 30s and 40s. There's also an increase in the Type A personalities hyperstimulated people, for which one reason could be exposure to a lot of light for long hours.'

It is not just about staying awake beyond traditional bedtime. the quality of darkness is as important. A US study showed that infants who slept in rooms lit by night lamps had increased chances of developing myopia later.

A decade ago, Mumbai resident Ashwin Jajal went to the Bombay High Court with the complaint that the blinking neon hoardings at Marine Drive, which invaded his fifth floor apartment, were hurting his health and sensibility, not allowing him a good night's rest. A subsequent state government committee found that these lights were not just bad for the eyes, but affected sleep, appetite, and caused psychological problems. The court ordered the lights be switched off after 11 p.m. A small battle was won in the war against light pollution. (No. of words: 729)

Source: Rekha Dixit, 'Let There be Night', *The Week*, 17 January 2010, pp. 18–20.

@ **Please refer to the Online Resource Centre for more exercises on precis writing.**

Paragraph Writing

Either write something worth reading or do something worth writing.

–Ben Franklin

20.1 INTRODUCTION

Take a look at the paragraph that follows:

Rabindranath Tagore was a great poet. He is known for the lyrical quality in his poems. Rabindranath Tagore was a multifaceted individual. His poems are wonderful for their musical appeal. Tagore was born in Bengal. His whole life was dedicated to the service of society and literature. Tagore, however, was not just a poet. In his poetry, we observe not just melody but also the commitment to change the social taboos. Tagore was a poet who had great associations with other great Indians. Even today, his songs are sung and heard by many. In fact, you cannot imagine Bengal and its culture without imagining Tagore. Mahatma Gandhi was one of those Indians with whom Tagore shared a lot of his intellectual perspectives. When Tagore died, he left millions of hearts crying behind.

Can you figure out what is wrong with this passage? Would you describe it as a well written passage? If you do so, you must really be appreciated for your patience. This is so, because most of the readers of this passage are not going to go beyond the third or the fourth line, as the passage does not really make much sense. Why? Actually, the passage lacks continuity, focus, coherence, and unity. When we read this passage, it is difficult to keep track of the thought process of the writer.

The passage starts by introducing *Tagore as a great poet*. It quickly adds that *Tagore's poetry is known for its lyrical quality*. This opening makes the reader subconsciously anticipate something more about Tagore's poetry. What we are, however, told is that *Tagore was a multifaceted individual*. Somewhat confused, we look for some details about Tagore's versatility, but are told that his *poems are wonderful for their musical appeal*. With some effort, we start feeling that after all, the passage is about Tagore the poet, and the lyrical and musical quality of his poetry. However, some kind of an introductory idea—*Tagore was born in Bengal* confronts us. This is followed by another introducer—*his entire life was dedicated to the service of society and literature*. Then we are made to read a transitional sentence—*Tagore, however, was not just a poet*. Our concentration waning by now, we expect the passage to finally achieve some continuity but the writer refuses to oblige us! But we are made to read something about Tagore's poetry: *In his poetry, we observe not just melody but also the commitment to change the social taboos*.

It is by this time that the patience of readers starts deserting them. Like all other readers, we too feel like screaming for something to be told with some continuity. The tired readers are then offered a transition as the sentence introduces another idea that Tagore was the one who had great associations with other great Indians. With some hope, we start looking for the relevant details about some other great Indians, but are assaulted with another inconsequential sentence: *Even today, his songs are sung and heard by many*.

As though the writer has not hurt us enough, the paragraph continues to drift into incoherence in telling the reader about Tagore's contribution to Bengali culture—*In fact, you cannot imagine Bengal and its culture without imagining Tagore*. But then it changes gear all too suddenly and returns to pick the thought squandered earlier—*Mahatma Gandhi was one of those Indians with whom Tagore shared a lot of his intellectual perspectives*. It seems the writer prefers to write the way he/she thinks and feels and we are sorry to say that the writer does not seem to be doing anything in a coherent manner.

Finally, when the writer ends the passage on a supposedly emotive note—*When Tagore died, he left millions of hearts crying behind*—we heave a sigh of relief. Such passages can test the tolerance of the readers.

Though it is impossible to really figure out what the writer actually intended to communicate to us by writing such a hopeless passage, we can make a valiant attempt to carve out some method of this madness:

> Born in Bengal, Rabindranath Tagore was a great poet. He is known for the lyrical quality of his poems. His poems are wonderful for their musical appeal. Tagore, however, was not just a poet. He was a multifaceted genius whose whole life was dedicated to the service of society and literature. Therefore, in his poetry, we observe not just melody but also a commitment to change the social taboos. Because of his intellectual eminence, Tagore enjoyed a great association with other great Indians. Mahatma Gandhi was one of those Indians with whom Tagore shared a lot of his intellectual perspectives. When Tagore died after years of service to the domain of knowledge, intellect, art, music, and culture he left millions of crying hearts behind. Tagore's influence on Bengali culture is so significant that even today one cannot imagine Bengal and its culture without remembering Tagore.

We admit that rescuing a passage that has been **wrecked** so hopelessly is an exacting task. Even then, you probably can figure out that the second version of the passage is decidedly better than the first one. It starts with an introducer: *Born in Bengal, Rabindranath Tagore was* *a great poet*. Having introduced the idea that Tagore was a great poet, the passage sets out to highlight the lyrical and musical quality of his poetry—*He is known for the lyrical quality of his poems. His poems are wonderful for their musical appeal*. Following this, we are introduced to the other side of the poet's personality—*Tagore, however, was not just a poet. He was a multifaceted genius whose whole life was dedicated to the service of society and literature*. A developer that follows—*Therefore, in his poetry, we observe not just melody but also a commitment to change the social taboos*—augments and substantiates the idea proposed through the transition.

Emphasizing Tagore's intellectual prowess further, the next sentence in the paragraph cites his association with other great Indians of his time: *Because of his intellectual eminence; Tagore enjoyed a great association with other great Indians*. The next sentence immediately illustrates the idea by referring to the fact that *Mahatma Gandhi was one of those Indians with whom Tagore shared a lot of his intellectual perspectives*.

Honestly speaking, the paragraph needs further development as the idea of Tagore's genius, as a poet, lyricist, intellectual, and social reformer, tangentially referred to the in the passage, are not satisfactorily developed. But then we have not been given any other details on these aspects of Tagore's versatility. In the want of that, the passage makes its way out by putting together both the terminators—*When Tagore died after years of service to the domain of knowledge, intellect, art, music and culture he left millions of crying hearts behind. Tagore's influence on Bengali culture is so significant that even today one cannot imagine Bengal and its culture without imagining Tagore*.

Though we could not completely redeem the original passage, the two versions of the passage throw up some vital points, necessary to write well-structured and hence well-understood paragraphs. Our tryst with the passage written as atrociously as in the original attempt, makes us construct a loose definition about what a paragraph is; as we can say that a *paragraph is a group of sentences that introduces, presents, develops, and winds up one main idea on a topic*. It is an argument or a stand-alone piece of writing that usually has one controlling idea. Ideally speaking, if you have more than one main idea to communicate, you need to write as many paragraphs. Put to this test, the original paragraph again falls short of our expectations as it tries to yoke together many differing ideas all at one place. Therefore, a paragraph has a group of sentences rather than a group of ideas. The idea in a paragraph should actually be one, and the different types of structural devices should help it come out clearly. Though the paragraph cited originally presents an unusual example of messy structure, the mistake committed is not all that unusual. In fact, to some extent, quite a few of us tend to lose the structural sense of the paragraph. By and large, the entire problem arises due to the fact that people start writing without understanding the structure of a paragraph. Therefore, it is quite vital for us to understand the essential components of a paragraph and how they appear, so that while writing we can ensure their presence in the paragraphs that we write.

Let us first study what constitutes the structure of a paragraph.

20.2 STRUCTURE OF A PARAGRAPH

In a broad way, a paragraph can be divided into three major parts:

- Topic sentence/introducer
- Supporting details/developers
- The concluding sentence/terminator

Having acquainted ourselves with these three different types of sentences used to construct a paragraph, let us study them in greater detail. Read the following paragraph and figure out these three different types of sentences:

Beauty Lies in the Eyes of the Beholder

The concept of beauty is an ambiguous phenomenon. Some find beauty to be a physical attribute while others regard it as a matter of intellectual comprehension. For an epicurean, beauty never filters underneath human skin whereas for a grave thinker it is always skin-deep. It can always be argued as to which of these two perspectives is worth assimilation in life. I feel we cannot choose one and discard the other altogether. This is so because we are neither only a physical entity nor are we just a brooding sage. We are both—the mind and the body. So, if we say we can turn a blind eye to physical beauty, we end up deceiving ourselves. On the other hand, if we focus obsessively only on physical beauty, we are unable to comprehend the other, and at times far more subtle, vistas of beauty. Hence, there is no harm in believing the dictum that beauty lies in the eye of the beholder, for, after all, it gives you an opportunity to appreciate the beauty of all that your eye beholds and mind observes.

The first sentence introduces the central idea of the passage. It is, therefore, the *topic sentence* or the *introducer* in the passage. The second to the ninth sentences constitute the main body of the passage, and hence are the *developers* in the paragraph. The last sentence winds up the discussion, and hence is the *terminator*.

Thus, a good paragraph generally has a beginning, a middle, and an ending. Since it is a group of sentences, they can always be distinguished as *introducers*, *developers*, and *terminators* in a well-structured paragraph.

20.2.1 Topic Sentence

The sentence that introduces the main idea in a paragraph is called the *topic sentence*. Mostly, it appears in the beginning of the paragraph but sometimes it does appear towards the end of the paragraph. At times, however, the topic sentence can also be seen hidden somewhere in the middle of the passage. Since the purpose of the *topic sentence* is to emphasize the main idea of the passage, it is not generally suggested or practised to keep the *topic sentence* hidden under the debris of other details. This provides the core idea(s) that run throughout the paragraph like an underlying thread. The topic sentence guides the readers and lets them know what it is all about. So it performs two major functions. They are as follows:

Structural   It describes the shape of the argument.

Interpretive   It offers a conclusion or reaction or feeling.

Structural topic sentences   Structural topic sentences can look like the following openings:

1. There are three main reasons for the high inflation rate in Indian economy at present.
2. Positive thinking has several benefits.
3. Meditation, which is an intensely personal and spiritual experience, leads to three major important results.
4. There are various causes for underemployment in urban areas.
5. Distance education in the past one decade has had the following results.

These examples demonstrate how the structural topic sentences guide the readers to anticipate and move with the rest of the paragraph that unfolds as outlined by such introducers. Using structural topic sentences like these will help you follow your argument easily, as long as you link your ideas together.

Interpretive topic sentences　An interpretive topic sentence not only introduces the reader to the main idea but also acquaints the reader to the author's perspective on the issue. The interpretive topic sentence, hence, becomes more valuable than the structural topic sentence. The structural topic sentence helps the reader follow the rest, but it does not tell us a lot about the topic. The interpretive topic sentence, on the other hand, allows the writer to freely express his/her interpretation of the data, and makes an effort to convince the reader at the same time. In order to use interpretive topic sentences effectively, you can use some of the following strategies:

1. You can use descriptive words such as **high, low, widespread, limited, half**, etc.
2. You can interpret/conclude using words such as **suitable, beneficial, unsuccessful, serious**, etc.
3. You can even give your opinion, for example, **shocking or disturbing**, if you want your reader to share your perspective on the issue.

Some examples:

1. Many communicable **diseases of man are known to be** caused by micro-organisms. Some of these micro-organisms are…
2. In education, **girl children drop out earlier than boys**. Girls' enrolment is just 61 per cent, compared to…
3. A recently released report by the Ministry of Human Resource Development (MHRD) shows a **nation-wide decline** in school dropout rates. In Maharashtra too, the number of students…
4. **Almost 49 per cent of the children** fail to complete primary level education. In a recent study in 11 districts of Rajasthan, it was found that…
5. Access to basic services is **extremely limited**. It was found that …

Since both structural and imperative topic sentences introduce the main idea or the central argument of the passage, they both are known as introducers. The purpose of an introducer is to place in perspective the central idea in a paragraph. Coming usually at the beginning of the passage, introducers serve as a signpost to the central idea of the entire passage.

An introducer generally lays the foundation for the rest of the argument to follow. It raises hopes and makes promises that the remaining sentences in a paragraph are required to fulfil.

The sentences that aim at fulfilling the promise made by the introducers are called *developers*. Just as the function of introducers is to introduce and emphatically place the central idea in a passage, the job of developers is to substantiate, augment, and authenticate the claims made by the introducers. Read the following passage and understand how the developers drive home the idea floated by the introducers:

What will happen? Absolutely nothing. The media will lose interest as soon as something still more grisly takes place, and this shocking case will soon be forgotten. Perhaps, the husband who threatened the young mother will get picked up for questioning. Perhaps, his family members will also get hauled in. Then what? The thing is, even the police will be on the side of the husband, even though the law dictates that he be considered a co-accused for instigating the twin murders. Nothing of the sort will happen.

Source: Shobha De, 'The Twin Tragedies', *The Week*, 25 October 2009, p. 82.

The introducers at the beginning of the paragraph pose a rhetorical question—*What will happen? Absolutely nothing.* Notice how the remaining paragraph tries to illustrate how and why the writer feels that nothing will happen. The developer that follows—*The media will lose interest as soon as something still more grisly takes place...*—illustrates how nothing much should be expected to take place in the case under review.

The other developers that follow—*Perhaps, the husband who threatened the younger mother will get picked up for questioning* and *Perhaps, his family members will also get hauled in*—concede the idea that something might perfunctorily will be done and that a couple of token steps would be taken, but ultimately nothing much would matter. This is how the properly constructed developers establish the idea introduced by the topic sentence in a paragraph.

The third category of sentences is known as *terminators*. The purpose of terminators in a paragraph is to wind up the discussion in a manner that is fulfilling and satisfies the reader in a psychological way. Though coming at the end of a paragraph, the importance of terminators in a paragraph can never be underestimated. They leave on the reader the final impression about the crux of the entire paragraph. Look at how the following paragraph leaves the desired impact on the reader:

> From a demure wife to a jet-setting socialite having a tumultuous relationship with a Hollywood star, Jemima Khan, former girlfriend of Hugh Grant, is now getting ready to tell her side of the story of the years she spent in Pakistan, married to Imran Khan. Jemima, who now lives in London with her two sons after her divorce from the cricketer, has reportedly been paid a handsome amount for spilling the beans on what was undoubtedly the most controversial coupling of their times. Hope the book lives up to that juicy past.
>
> Source: 'Confessions of a Socialite', *The Week*, 25 October 2009, p. 96.

See how the terminator in the paragraph—*Hope the book lives up to that juicy past*—drives home the point in an intimate manner. Emphatic though abrupt, the terminator in the above paragraph actually induces in the reader a desire to discover how in her prospective book, the author, Jemima Khan, is likely to unfold her side of the story. This is how even a small terminator becomes quite crucial in the overall development of a paragraph.

Thus, it is important for us to understand what constitutes a paragraph. Since we have seen that a paragraph includes introducers, developers, and terminators, it is important for us to use them judiciously in order to maintain a sense of structure in a paragraph.

PRACTICE TEST 20.1

Read the following sentences and identify the introducers, developers, and terminators:

1. Further, it is also important for a manager to be punctual and methodical.
2. Man is born free but he finds himself in chains everywhere.
3. To sum up, suffice it is to say that corruption has corroded our country.
4. Although there are many ways to achieve success, none of them leads to happiness.
5. But it is not possible for us to forget the sacrifice of our forefathers.
6. Poetry developed as an art form much later than drama that has its roots in religious conventions.
7. Therefore, all we can do is just sit silently and pray.
8. Surprisingly enough, the police kept aloof throughout the incident.
9. Meanwhile, his business had started taking a downward slide.
10. Of course, there is no point in overlooking the advantages of a democratic governance.

20.3 CONSTRUCTION OF A PARAGRAPH

Just as it is important to understand the structure of a paragraph, it is also very important for us to learn ways to construct a good, emphatic, and effective paragraph. Let us learn how to use different strategies with the help of which paragraphs are constructed effectively.

20.3.1 Narrative Description

Look at the following paragraph and figure out the technique employed:

> The long, steep road that lead to Sholagar Thotti—a village atop the Sathyamangalam hills—is strewn with stories of fear, pain, suffering, and ignominy. Our guide Jeeva Jothi has a tale to tell at every turn. 'This was where Veerapan kidnapped actor Rajkumar,' he says, pointing out a bungalow in Thalavadi village. 'Veerapan is supposed to have frequented this,' he says, pointing to a tea shop at Hasanur.
>
> *Source:* Kavitha Murlidharan, 'Weary Victims', *The Week*, 25 October 2009, p. 58.

This small paragraph is quite catchy. Can you guess what makes it interesting? It actually is the narrative description of the passage that makes us read this. In this paragraph, the writer ventures to tell a story to the reader so that the message is communicated in an engaging manner.

20.3.2 Comparisons and Contrasts

We can construct a whole lot of paragraphs with this device. For developing paragraphs on these lines, two similar things are compared or two dissimilar things are contrasted. The purpose for employing this technique is to make the entire argument forceful and emphatic.

For instance, look at the paragraph given below. See how the speaker brings to the fore the significance of changed perspective that characterizes a teacher's life in modern times:

> It has been the tradition of our country that man is not measured merely in terms of money, and this is particularly true of scholars who impart education to others. In ancient times, a student used to live with his teacher or the Acharaya and, just like the latter's own children, become a member of the family. The guru had to bear the entire burden of maintaining the pupil and in lieu of fees, the pupil would gather wood from the forests or take the cows for grazing. Till comparatively recent times, it used to be considered improper to charge any fees for imparting education. But times have now changed. India cannot keep herself isolated from the rest of the world. Not only cotton, wheat, and maize but everything else of value is measured in terms of money. We should neither feel surprised nor should we blame any one of those who have adopted the educational profession as a career, also measure their worth in terms of money. Such a development was inevitable, for, when social recognition is based on money, teachers could not but fail to come under its influence. Teachers have, therefore, begun to make demands for more money. Previously, the students used to discharge their obligations towards their teacher by serving him while they were his students and later, when they had completed their education and entered family life, by imparting education to others.
>
> *Source:* Rajendra Prasad, Speech at the opening ceremony of Engineering Block, Birla Vidya Vihar, Pilani, cited in Deepvali Debroy, compiler, *Famous Faces, Famous Speeches*, New Delhi: Vikas Publishing House, pp. 174–175.

By saying that in ancient times it was the student who used to help the teacher in the times of the latter's need, the speaker puts the entire complexity in a proper perspective. The comparisons drawn by the speaker help him sound authentic and unbiased. Thus, through comparisons and contrasts, we can carry conviction and authenticate our perspectives in an objective and emphatic manner.

20.3.3 Sustained Analogy

Using analogies is another way to draft emphatic passages. Comparisons and contrasts are carried out also through analogies. However, in its impact and appeal, this approach is more figurative and literary. Unlike simple comparisons and contrasts, analogies are used to compare things which are generally not from the same class. When such comparisons are used extensively, the device is termed as *sustained analogy*. Look at the paragraph given below how through a sustained analogy, the author compares life to a journey:

> Life is a journey. Just like a journey—that starts with a single step—our life also starts with a single breath or cry. During our journey, we come across new people, places, and experiences. Some of these are bitter, others sweet, and some others bitter-sweet. Sometimes we are made to tread a jerky, bumpy road and at times we keep marching ahead on a silken, smooth road! In journeys, it is common to confront deserts, oceans, and mountains. The one on a journey, however, cannot call it quits. Once committed to the journey, one has to keep moving. So stop not till your journey is over.

20.3.4 Cause and Effect

Cause and effect is an important device with which we construct paragraphs on a variety of topics. Through this method, the paragraph attempts to establish a relationship between certain events and the reasons behind them. While using this method, authors are able to convince their readers in a scientific and logical manner. Read the following paragraph and see how this device is used effectively:

> The effect of guilt in a person's life can easily be observed throughout his/her life. Those who are constantly gnawed by a deep-seated guilt often blame themselves for all the problems around them. Such people are never optimistic or excited about anything in their life. Most of the time they are sad and gloomy. They keep lamenting their actions of the past, and their present too keeps drifting away from them. It leads to multiplication of guilt, as the guilt of lost opportunities gets combined with the mistakes committed in the past. This vicious circle of guilt hence never allows its victims to succeed or be happy in life. It is so because guilt saps all enthusiasm, energy, and an urge to survive or excel. The result is that a guilt-ridden individual tires quickly. Not being able to enjoy life or carry out their responsibilities in full measure, such people retreat into an apathetic, dull, listless condition and are even prepared to bring their life to an end.

By relating the feelings of guilt to an apathetic, dull, and listless condition, the author provides to us the psychological perspective through which one can observe why the underperformers in life lead a life of little activity as they constantly wriggle under the weight of a constantly tormenting sense of guilt.

20.3.5 Quotations and Paraphrasing

Quoting authorities is an excellent way to develop a paragraph. Read the following paragraph:

> Isn't it surprising that despite so many schemes announced by the government, nothing much changes in the life of the poor? Probably it isn't. Knowing how deep-rooted corruption has become in our country, it really seems the only natural outcome of all government policies. Modern day capitalism and the contiguous narcissism have added to the culture of corruption in our country. People feel like accumulating money; by hook or crook. Understandably therefore, the gap between the poor and the rich is widening. Sometimes, I feel like reminding all the corrupt people about Gandhi who once said, 'Recall the face of the poorest and the weakest man whom you may have seen, and ask yourself, if the step you contemplate is going to be of any use to him. Will he gain anything by it? Will it restore him to a control over his (her) own life and destiny? In other words, will it lead to swaraj for the hungry and spiritually starving millions? Then you will find your doubts and your self melting away.'

By quoting Gandhiji, the passage drives home the point that a life of purpose is basically the one that is not obsessively self-centred. It is not essential that the quotations are cited only from eminent personalities. Even common people are quoted to substantiate a point of view. When the words of the commoners are used, it is known as *peer testimony*.

Look how a paragraph on a similar issue can authenticate itself even by quoting common people:

> Isn't it surprising that despite so many schemes announced by the government, nothing much changes in the life of the poor? Probably it isn't. Knowing how deep-rooted corruption has become in our country, it really seems the only natural outcome of all government policies. Modern day capitalism and the contiguous narcissism have added to the culture of corruption in our country. People feel like accumulating money, by hook or by crook. Understandably, therefore, the gap between the poor and the rich is widening. Corrupt people have no time to listen to story of a poor peasant who says, 'I have two children who are given food on alternate days. I cannot feed both of them everyday. It is really painful to see them hungry. But what can I do? I don't have enough to make both ends meet everyday.'

20.3.6 Enumeration

At times, we list a series of ideas in order to substantiate the topic sentence. This device, used quite often to construct a paragraph, is known as *enumeration*.

Read the following example to see how enumeration leads to not just authentication but also coherence in a paragraph:

> Despite all the growth and development registered in post-independent India, our country continues to be tormented by a large number of social evils. Some of the most disturbing ills prevalent in our society are casteism, communalism, corruption, dowry system, hooliganism, untouchability, intoxication, and child labour. It is difficult to say which of these are worse than the rest of them. Though all such ills and problems can be categorized as bad, worse, or worst, the impact that they leave on the social milieu is always only worst.

By enumerating the evils in the society, the writer is able to highlight a number of problems that beset our lives even amidst the times of all-round growth and development. Look at the following paragraph and see how through enumeration, the author builds up the paragraph:

> No rules are required for you to lose your life on the road. But if you want to be safe on the road, go by the following: While driving your vehicle on the road: keep to your left; always stick to your lane; never try to overtake from the wrong side; drive within the prescribed speed-limit; stop at every red-light; use dippers at night; use your seat belt while driving a four-wheeler; and wear your helmet while riding a two-wheeler.

20.3.7 Definition

Another way to develop a paragraph is to use *definition*. This method for developing a paragraph is particularly employed in those situations where the author intends to take up some topic, term, issue, or argument in a particular way. By taking the reader through the definition of a particular word, the author is able to prepare his/her readers to follow the intended line of argument or thought. See how the author in the following passage attempts to define the term 'insomnia' for a specific purpose:

> The term *insomnia*, derived from the Latin root *somn* refers to the *chronic and habitual inability to fall asleep or remain asleep for an adequate length of time*. Though insomnia can strike people of any age group, old people are more prone to it. However, owing to a stressful life and unhealthy food habits, a large number of young people today suffer from it. In a mechanical, alienated, frantic, and guilt-ridden life, there are other sleep-related disorders, as some people become *somnambulists*, that is, they begin to walk in their sleep. Almost all patients of *insomnia* feel *somnolent*, that is, sleepy or drowsy during their waking hours and some others become *amnesiac*, that is, they start losing their memory.

By defining the term 'insomnia' and other related words, the author is able to make things clear to his/her reader. Look how the speaker, by introducing a little known expression 'anorexia nervosa', through its dictionary meaning, immediately establishes a rapport with the listeners in the following extract of a speech:

> Friends, today we are going to talk about the rise of a very peculiar health disorder in young girls. In medical terms it is known as *anorexia nervosa* which stands for a psychological disorder, mainly afflicting adolescent girls and young women, characterized by a significant decrease in body weight, deliberately induced by refusal to eat because of an obsessive drive to lose weight. In this age of glamour, an urge to look smart and presentable is very common and we see a large number of young girls trying to shed their weight at a frantic pace. Therefore, a frequent food evasion or massive cuts in food intake often becomes a way of life for them. While doing so, however, little do they seem to realize that by indulging in such weight loss obsessions, they are likely to fall victim to a debilitating health disorder known as anorexia nervosa.

Having already defined the term 'anorexia nervosa', the speaker makes it convenient for him/her to refer to the term as and when required in the subsequent part of the speech, without causing ambiguity or confusion in the mind of the listeners.

Reading further, take a look at how by defining the multi-layered term 'personality', the author, in the following paragraph, prepares the reader to approach the subject of discussion in an appropriate way:

Personality is a term with many varying connotations, depending on the context of usage. It is a term that may be used to denote a celebrity (a public personality or figure), one's character or temperament, or the way one comes across to others (he or she has a good personality). In medical and psychological parlance, however, personality is used to denote 'those characteristics of a person that account for consistent patterns of feelings, thinking, and behaving'; unique and enduring patterns of behaviour and emotional response, which make us distinct individuals.

Source: Ennapadam S. Krishnamoorthy, 'The Inside Man', *The Hindu*, (Sunday, Weekly Edition) 29 November 2009, p. 1.

See how the author takes the readers through the general understanding of the term *personality* before he/she makes them figure out the specific sense in which he/she probably intends to take up the term and discuss it in the ensuing part of the write-up.

20.3.8 Testimony

Giving testimony is another way to develop a paragraph. Testimony can be of two types—*peer testimony* and *expert testimony*. Just as *peer testimony*, cited earlier in the discussion, expert testimony also lends credibility to the author's opinion. See how in the following passage, the author uses expert testimony, to drive his/her ideas home:

In our country, we say that we should transform our nature, grow from the slavish, unregenerate condition of ignorance, to a state of wisdom. The growth or the transition from the one to the other constitutes the goal of the religious quest. From the disruption of being we must rise to the articulation of being. The Buddha said exactly the same thing. We are sunk in suffering and ignorance, and our goal is to grow into enlightenment. According to Ezekiel, 'Thus saith the Lord God—I will put a new spirit within you; and I will take the stony heart of their flesh and will give them an heart of flesh.' For the Jews, 'the spirit of man is the candle of the Lord.' Speaking of the mystery religions of Greece, Aristotle observes, 'The initiated do not learn anything so much as feel certain emotions and are put in a certain frame of mind.' To live one must first die to his old life. Orpheus believed that the soul was 'the son of the starry heaven', that its dwelling in a body is a form of original sin, its earthly life was a source of corruption and its natural aim was to transcend this life. This view is at the heart of Plato's idealism.

Source: S. Radhakrishnan, *Towards a New World*, Delhi: Vision Books, 1983, p. 65.

Through a series of quotations, the writer attempts to put the problem in the proper perspective. Since the topic at hand is controversial, the author of this passage uses expert testimony extensively to make his reader understand the variegated nuances of the subject under perusal.

20.3.9 Facts, Figures, Instances, and Examples

Facts and figures compose a very important part of many a professional write up. Since facts, figures, and examples can carry conviction and illustrate the point, authors generally prefer them for developing paragraphs. It is done so in order to create an immediate and intimate rapport with the reader. Read the following paragraph and see how it is constructed through facts and figures:

> The Indian woman continues to live on the periphery. And this, despite the fact that she is now more self-reliant, more conspicuous in social circles, and well rooted in the professional world. Though she earns well, this does not ensure equality for the Indian woman. According to a survey conducted by an NGO based in Mumbai, it is suggested that more than 50 per cent newly married males are uncomfortable with the idea of not having a son born to them, while 65 per cent resented the idea of their wives getting a job and financial independence at their expense. Not only this, nearly 35 per cent of them felt that it is better to have a less educated girl for a wife as she would be easy to be tamed. So much so, that nearly 15 per cent of men were not against using violence against their women, for, *if nothing else works, this does!*

By citing factual data from the findings of an NGO research, the above paragraph builds up the entire edifice in a convincing manner. The figures that follow the topic sentence clearly illustrate the central idea that even today the Indian woman continues to lead a life of secondary existence in the patriarchal Indian social structure.

Given below is one more example which suggests how with facts, we can lend credence to discussion and analysis:

> The sales dropped to 63% in Delhi while it grew to 79% in Bahadurgarh. Similar trends are observed in other parts of the country. As can be seen from the above rectilinear graph, Chandigarh reported a drop of about 13% whereas in Mohali, the sales registered a growth of 15% as compared to last year. It is felt that the demand for our product is picking up in the upcoming areas whereas it is somewhat on a decline in the cities already developed.

Look how again the examples in the following paragraph construct it well:

> It is generally believed that Americans are very good speakers of English. Their articulation, however, does not really seem to support such beliefs. In fact, observed closely, it seems that Americans use sloppy articulation for quite a few expressions. For example, *I did not* for them becomes *I dint*, whereas the true contracted form for *I did not* should be *I didn't*. Similarly, *you ought to* in American English sounds like *you oughta* or *you otta* and *you have to* becomes *you hafta*. The forceful and intense *yes* in British English is always a sloppy *yeah* for them and *I don't know* sounds like *I dunno*.

By citing relevant examples, the author of this paragraph tries to bust the myth that Americans are very good speakers of English.

See further how in another paragraph, facts and figures highlight the gravity of the situation with regard to climate change:

> Climate change is for real and its alarming impact is evident all around us. What else could it be when the proportion of extreme category hurricanes worldwide rises from 20 per cent in the 1970s to

35 per cent in the 1990s; when a heat wave in Europe can kill more than 30,000 people; and when the glaciers are fast receding. In fact, the entire ecosystem is deteriorating, resulting in new diseases with newer virus mutants. Bee colonies are disappearing and there has been a marked change in cropping patterns in many developed countries.

Source: Rajiv Vij, 'Change, for Climate's Sake', *Sunday Times*, 13 December 2009, p. 19.

By citing statistics, percentages, and other facts, the author tries to illustrate the topic sentence that *climate change is for real and its alarming impact is evident all around us.*

20.3.10 Episodes

Like facts, figures, instances, and examples, episodes also help the construction of a paragraph. At times, situations require writers to talk about various episodes in order to drive home their ideas in a convincing and emphatic manner. Look at the following paragraph and note how through episodic illustrations, the paragraph is constructed in a neat and convincing manner:

I trembled as I boarded the train the next time. Could this be my last ride in life? Will I too die, blown away by some bomb? I thought and thought. Should I get off the train? Catch an auto and reach office? But how long can I do that? With a monthly income of a couple of thousand bucks, how can I afford to be that extravagant? No, nothing would happen. Otherwise, all these fellow passengers of mine would not have boarded the train. They must have sensed that after the bomb blast last week, everything was normal. But then, everything is always normal before it becomes abnormal. After all, before a blast, there is no blast. And no fear. No anxiety. But how can we say that? There was fear and anxiety on every face and all eyes. Like me, all faces around me seemed to be thinking—would this be their last ride? I shuddered how a blast in your city rips your confidence apart. How it reduces you to a nervous wreck who has no option but to move on with his daily life! Even if it takes him to death!

By using the episode in a narrative form, the paragraph builds up the environment of threat, fear, anxiety, and panic that lurks beneath the apparent calmness and normalcy after the incident of a bomb blast in the city. While establishing the fear and nervousness of a person by citing an episode, the author is able to illustrate the central idea of the paragraph which is pushed to the end of the construction. As we can make out, the central idea in the paragraph is to depict how an incident of a bomb blast in the city induces fear and anxiety in the minds of the people. However, before we are made to read the central idea: *how a blast in your city rips your confidence apart! How it reduces you to a nervous wreck who has no option but to move on with his daily life!*, we are made to read the thought process of a traveller who goes through great anxiety while boarding a city train after a tragic incident of a blast that has recently occurred in the city, shattering the commoners' assurance and composure while performing their day-to-day activities.

Thus by employing examples, facts, figures, analogies, enumeration, comparisons, contrasts, definition, causes, effects, etc., paragraphs on different issues and topics are constructed. Many a time, you will find authors combining a variety of such devices in order to structure a paragraph. Therefore, it is important to learn to use these strategies as effectively and appropriately as possible.

Just as it is important to understand how to build a paragraph through these devices, it is also of crucial importance to learn how to provide transitions in a paragraph.

20.3.11 Using Transitions and Connecting Devices

Transitions are the expressions which connect different ideas expressed in a paragraph. Without learning to use them accurately, it is impossible to develop an impressive paragraph. For instance, read the paragraph given below and decide whether the transitions provided are apt and relevant:

> Those who believe in God seem to be aware of His presence around them. Moreover, skeptics suggest that it is one thing to believe in God and quite another to be aware of His presence. They feel that but it is quite easy to keep oneself in illusion, ignorance, and darkness, and quite difficult to be aware, awake, and alive. Nevertheless, they feel that God is not a matter of belief and God is a power, a source of energy of which we need to be conscious about.

The second sentence starts with 'Moreover'. Do you think it is appropriately placed? *Moreover* is additive in nature. It adds to the preceding argument. When we read the paragraph, it becomes obvious that the statement contradicts the claims made in the first statement. Therefore, we need *however* and not *moreover*. Similarly, the choice of *and* in place of *but* further in the same sentence is also misplaced. In the third sentence, we need a concessional connective *though* and not *but* which is used to introduce a contrastive idea. Further in the sentence *and* is again wrongly placed and we need either *yet* or just a comma to contradict the preceding idea. Even in the last sentence *nevertheless* is quite misplaced as we need something that reinforces the continuing argument.

Look at the revised version of the paragraph and see how the transitions are properly used:

> Those who believe in God seem to be aware of His presence around them. **However**, skeptics suggest that it is one thing to believe in God **but** quite another to be aware of His presence. They feel that **though** it is quite easy to keep oneself in illusion, ignorance, and darkness, **yet** it is quite difficult to be aware, awake, and alive. **In fact**, they feel that God is not a matter of belief **as** God is a power, a source of energy of which we need to be conscious about.

In fact, different types of transitional words and phrases are meant for different purposes. Look at the table given below; see the transitional and connecting devices listed; and observe the purpose for which they are used in a sentence:

Transitional and Connecting Devices	Purpose
• Therefore, consequently, as a result	• Establish cause and effect
• While, meanwhile, in the meantime, simultaneously, even as	• Suggest simultaneous actions
• For instance, for example, again, such as, specifically, especially, to illustrate	• Cite examples and illustrations
• However, in contrast, on the other hand, but, though, although, nevertheless, notwithstanding, despite, inspite of, yet, surprisingly however, alternatively	• Contrast the preceding idea

(Contd)

Transitional and Connecting Devices	Purpose
• Finally, to conclude, in conclusion, thus, so, on the whole, therefore	• Wind up an idea
• And, moreover, more importantly, over and above, furthermore, in addition to, further, besides, last but not the least, again, first of all, secondly, finally	• Used for adding ideas
• To sum up, to summarize, in a nutshell, in brief, in short	• Summarize the foregoing discussion
• In the same way/fashion/manner, similarly, likewise, by the same logic/account/token	• Suggest similarity; add similar ideas
• Because, since, as, for, on account of, due to the fact that, for that reason, because of, thanks to/thanks largely to	• Used for highlighting the cause and reason
• For the same reason, to emphasize, most significantly/importantly, first and foremost, not to mention/forget, in fact, indeed, by all means, undoubtedly, to be sure	• Emphasize the point in discussion
• To clarify, conversely, namely, in other words, that is to say, to put it in simpler terms, to rephrase it	• Used to clarify the preceding argument
• Without doubt, clearly, obviously, evidently, clearly enough, in no uncertain terms, of course, needless to say	• Used for establishing a point of view
• Having said that, even though, there is no denying the fact that, admittedly, to be honest, may be, possibly	• Used for conceding an idea
• Accordingly, purposefully, for this reason, for this purpose, so that, in order to	• Highlight the purpose for which an argument is made

PRACTICE TEST 20.2

Use the following transitional/connecting devices in sentences of your own:

As regards, furthermore, nevertheless, even if, even though, as soon as, for, not only but also, as soon as, not to forget, notwithstanding the fact that, whereas, for instance, whereupon, in other words

PRACTICE TEST 20.3

Read the paragraph given below and provide transitional/connecting devices wherever required and develop on the 'idea':

Indian cricket team has achieved remarkable milestones under Mahendra Singh Dhoni's captaincy. Indian team has won the limited over cricket World Cup in 2011. In 2007 itself, the team had won the inaugural T20 World Cup.

Team India has also won the Champions' Trophy, the Triangular Series and the Asia Cup. Dhoni's captaincy has been instrumental in bringing about an unprecedented sense of commitment and passion for the task at hand. There is a seamy side to this rosy tale. Dhoni has not been able to infuse in the team similar courage when it comes to playing test cricket abroad. Indian team has not won much in any of the countries Australia, England, New Zealand and South Africa.

PRACTICE TEST 20.4

Given below are a beginning and an ending of a paragraph. Develop the remaining part of the paragraph in about 150 words:

Eyes are considered the most significant organ of the human body ... Hence, use your eyes well to win the audience.

20.3.12 Extended Definitions

As the title suggests, an *extended definition* is written to explain a complex term used in a book, research paper, or a report. In fact, some of the terms may be so crucial as well as complex that without defining them or placing them in the context in which we expect our readers to do, it becomes difficult for us to establish proper communication with them. An extended definition often includes the literal meaning of the term used with all its characteristic features. Although it may be written in a variety of ways, the basic purpose of an extended definition however is to define, explicate, and establish the basic features of a term before the writer ventures to take it up further.

An extended definition often appears in the beginning of a work, often immediately follow-ing the term when it is introduced for the first time. For example, a research article on 'ethical relativism' would require the author to first define the term suggesting that 'ethical relativism is a methodical principle of interpreting mortality based on the assumption that moral ideas and standards are mere conventions' and then only build the discussion further. Similarly, a report on 'spectroscopy' needs to first establish that 'spectroscopy is a branch of analysis devoted to identifying elements and compounds and elucidating atomic and molecular structures' before proceeding further with the rest of the discussion. When required to use extended definitions in your writings, keep in mind the following tips:

1. Use extended definitions for the terms generally perceived to be complex, intriguing, or multifaceted.
2. Whenever required to give extended definitions, provide them when the term is used for the first time in the discussion.
3. Rely on definitions given in the encyclopaedia, dictionaries, or standard books. In case the definition is picked up from the Net, try to cross-check it with a standard source before using it in your work.
4. Give all the characteristic features of the term besides providing the definition. In ency-clopaedia and dictionaries, the definitions given are at times too compact to be understood well by the reader. Therefore, it is only when one such a term is explained with the help of its characteristic features that the readers are able to perceive it in the sense it is required to be seen. Moreover, one can always suggest a deviation from the commonly perceived

understanding of a term. In this case also, characterizing a particular term in which it is required to be observed by the reader helps us achieve a common frame of reference with the reader.

20.4 FEATURES OF A PARAGRAPH

In the following section, let us learn the key features of a good paragraph with the help of a few examples.

20.4.1 Unity

Unity in a paragraph stands for the togetherness of ideas. Ideally, a paragraph should have one central idea—outlined through the topic sentence—and the subordinating ideas which help the main idea to come to the fore. This means that if the writer finds it important to introduce another equally important idea, he/she should switch over to another paragraph. This sometimes does not happen and hence it affects the unity of a paragraph.

Read the following paragraph and figure out whether it maintains unity in the ideas expressed:

> Reading books is a great hobby. Coming from great minds, books are generally the storehouse of wisdom. Reading books helps us grow intellectually. Reading gives us ideas and ideas change the world. Reading is also helpful in making us more articulate and spontaneous in expressions. These days, however, the habit of reading books is declining. Children today prefer watching television to reading books. Television charms them with striking images. Somebody has correctly called it an *idiot box*. After all, what is there to watch on television? In the name of entertainment, it offers us loud and meaningless programmes. The so-called family serials are horrible to sit through, and idiotic masala movies are repeated endlessly. On television, you can't even watch the news; for more than the information, what you get is sensationalism.

Though the passage starts quite well, it loses unity of thought immediately after the first few sentences. The paragraph starts with the idea that reading books is a great hobby. The sentences that follow try to establish why the author feels so. It keeps building on the same theme until the author finds television to be responsible for the declining reading habits. From this point onwards, the paragraph loses the original track altogether and starts discussing only television, highlighting the weaknesses of this medium. Even if the details related to television are to be tagged along with the main idea, the author cannot afford to singularly leave out the thematic core of the paragraph, which has something essentially to do with books and reading habits. Since the passage focuses far too exclusively on television and its drawbacks, it tends to lose the central idea altogether. Read the revised version of the paragraph to know how to maintain the main argument.

> Reading books is a great hobby. Coming from great minds, books are generally the storehouse of wisdom. Reading books helps us grow intellectually. Reading gives us ideas and ideas change the world. Reading is also helpful in making us more articulate and spontaneous in expressions. These days, however, the habit of reading books is declining. Children today prefer watching television to reading books. Television charms them with striking images. So, unlike books, it captures the attention

of young minds quite immediately. Books require reading, which requires concentration and effort. Alternatively, television offers them entertainment, though loud and meaningless, without demanding any effort from them in return. Moreover, books generally don't have to offer to their readers the sensational stuff that the television so readily supplies in order to capture the attention of young, innocent minds.

PRACTICE TEST 20.5

Read the following paragraph, try to make out where it tends to lose the unity of thought and expression; and rewrite it making it effective:

A life without ambition is a like a train without an engine. Just as it is the engine that drives the train, it is human beings' ambition that drives their lives. By all means, ambition occupies a central position in human efforts. Some theorists refuse to believe that ambition has crucial significance in human life. They feel that if we want to be happy, we must have no ambition whatsoever. They often quote the *Geeta* that urges us to concentrate on our work without bothering about the result. Of course, the *Geeta* is the most authentic work on spirituality, teaching us invaluable things. But it is not possible in life to just do our work and forget about the result of the work that we keep doing. Who would like to do a job that doesn't give them money or status or satisfaction? We are humans and we need something to keep us moving. It is really not possible to agree to the suggestion of the theorists who give no credit to ambition in human life.

20.4.2 Coherence

Maintaining *coherence* in a paragraph is different from maintaining *unity* in it. Look at the following paragraph and find out what it lacks:

I really appreciate the idea of arranging a trip to Srinagar for our school children. Certainly, it is a wonderful idea. Srinagar is a beautiful place and our students must be able to see that before they leave the campus. Since the journey is a long one, we can arrange a good, luxury bus for our students. They can watch mesmerizing sights on the way. The whole trip may consume some ten days or so. With their exams slated in March next year, there is nothing much to worry about losing precious time. And seeing a beautiful place like Srinagar is no wastage of time whatsoever. Staying our students should also not be a problem. We have good hotels in Srinagar. We can book one good one. After all, our students belong to well-off families and no student would mind sparing a couple of thousands for a trip to Srinagar. With sporadic violence and militancy, it is no longer a safe place but. Though March is far off, don't you feel that they will miss the crucial ten days on their run up to the board exams?

As we read this paragraph, we can figure out that the author seems to review the proposal whether to send the children of his/her school to Srinagar. In that sense, though loosely, the paragraph does hang onto the same idea and hence can be said to have maintained unity of idea. What it grossly lacks is coherence. In the beginning, the author appreciates the idea of taking the school children to Srinagar. Towards the end however, he/she seems to be arguing against the idea of taking students to Srinagar because of *sporadic violence* and *board exams*. In between, the author talks about the various arrangements and facilities required for the trip and does not seem to be bothered on that account.

Nothing, however, seems to move in a coherent manner in the above paragraph. If the author intends to argue against the idea of taking the students to Srinagar, it seems absurd on his/her part to begin with all accolades such as: *I really appreciate the idea of arranging a trip to Srinagar for our school children. Certainly, it is a wonderful idea.* Not only this, he/she also finds it almost mandatory on his/her part to take children to a beautiful place like Srinagar as he/she writes *Srinagar is a beautiful place and our students must be able to see that before they leave the campus.* Further into the passage, the author starts moving from one idea to the other mentioning *good, luxury bus, mesmerizing sights, good hotels,* and the *well-off families of students who will not mind sparing a couple of thousands.* In a way, the author seems to be quite excited at the idea of arranging a trip to Srinagar *before they leave the campus.*

While discussing all this, the author does not focus much on the continuity of thought and hence the passage does not seem to progress coherently from one point to the other. It is, however, towards the end that the paragraph scrambles into complete incoherence as the author changes his/her perspective quite abruptly on the issue: *With sporadic violence and militancy, it is no longer a safe place but. Though March is far off, don't you feel that they will miss the crucial ten days on their run up to the board exams?*

Look at the revised version of the paragraph and make out how the same idea can be expressed without lapsing into incoherence.

> I appreciate the idea of arranging a trip to Srinagar for our school children. Srinagar is a beautiful place and it would be nice if our students are able to see the mesmerizing sights of the place before they leave the school campus. However, with sporadic violence and militancy, it is no longer a safe place for such adventures. Moreover, for the students appearing for the board exams next March, a ten-day trip might be a little too long. Regarding other arrangements, however, there doesn't seem to be much to worry about. For the trip, a luxury bus can be arranged and students can be lodged in a good hotel there. Since most of our students come from well-off families, they would be able to bear the cost of expenses the proposed trip is likely to incur.

PRACTICE TEST 20.6

Improve the paragraph given below making it more coherent:

One of the most challenging issues of urban living is to find suitable gifts for suitable people on suitable occasions. Suitable occasions exist in plenty and suitable people too are quite in multitude. I don't mean to say that suitable gifts are in any dearth. How can they be? They are far too in abundance. In the world of consumerism, they cannot be expected to be in any dearth! The only thing in dearth is your ability to choose the suitable gift that suits your pocket too. That seems to be found wanting. On the last days of the month when mostly it looks like a flat toothpaste. The situation requires you to squeeze it with all your might for you to manage to look presentable to others with polished teeth and a plastic smile on your face. All this for you to be able say suitable words to suitable people on suitable occasions with suitable gifts in your hand!

20.4.3 Expansion and Emphasis

Alongside maintaining coherence and unity, it is also required that the idea that is introduced in a sentence is properly expanded and emphasized. Some paragraphs fail to click with the reader, simply because the idea that is generated in the passage is not taken to its logical conclusion. Read the following passage and see what it actually lacks—unity, coherence, or expansion.

Compared with the FBI, which threw light on the anti-India terror designs of the LeT, Indian security officials have made little headway regarding 26/11. Security officials here are yet to back up the charge that Col. Saadat Ullah, who allegedly belongs to the Pakistan army, was party to the 26/11 conspiracy. There are incredible intelligence inputs from the Russian Anti-narcotics Bureau that Dawood had used drug money to fund 26/11. It is reported that he contributes close to $1 billion to Pakistan's ISI.

The paragraph fails to achieve its purpose as it is supposedly written to highlight how compared with the FBI, the security system in India has not been able to make much progress in investigating the 26/11 terror attacks. However, what we are made to read is just one such evidence of these claims. Without adequate details, the paragraph limps to a feeble ending for it lacks expansion and emphasis.

Read the full version of the same paragraph and see how successfully it emphasizes the point:

Compared with the FBI, which threw light on the anti-terror designs of the LeT, Indian security officials have made little headway regarding 26/11. Security officials are yet to back up the charge that Col. Saadat Ullah, who allegedly belongs to the Pakistan army, was party to the 26/11 conspiracy. On the other hand, the Mumbai Crime Branch has not been able to find evidence of the local support the terrorists had in executing the attacks flawlessly. It has been unsuccessful in arriving at a pattern that vaguely suggests the role of Dawood Ibrahim. There were credible intelligence inputs from the Russian Anti-narcotics Bureau that Dawood had used drug money to fund 26/11. It is reported that he contributes close to $1 billion to Pakistan's ISI.

Source: Dnyanesh Jathar and Anupam Dasgupta, 'Still in the Dark', *The Week*, 29 November 2009, p. 37.

Sometimes it is not the lack of expansion but the absence of emphasis that keeps a paragraph from achieving its intended purpose. Since a paragraph is written to put across certain ideas, it should be written in a convincing and emphatic manner. Sometimes, however, the paragraph fails to clearly articulate the purpose for which it is essentially constructed.

Read the following paragraph and try to figure out where it falters:

Although seven dossiers with quantum evidence of LeT involvement were handed over to the Pakistan government, it chose to turn a blind eye to most of the points. The 26/11 probe at the grassroots has not made much headway, apart from the hastily stitched up charge sheet filed by the crime branch. It lists in detail the crimes committed by the group of 10, material evidence gleaned, and fragments of conversations between the terrorists and LeT handlers in Pakistan.

The passage has no ambiguity but it lacks the clarity of purpose. The ending of the passage leaves something to be desired. We feel like asking for more from the author. When we read the passage in full, we get what we want:

> Although seven dossiers with quantum evidence of LeT involvement were handed over to the Pakistan government, it chose to turn a blind eye to most of the points. The 26/11 probe at the grassroots has not made much headway, apart from the hastily stitched up charge sheet filed by the crime branch. It lists in detail the crimes committed by the group of 10, material evidence gleaned and fragments of conversations between the terrorists and LeT handlers in Pakistan. Naturally, questions are being raised about the Mumbai Police's competence in handling an investigation of this proportion and complexity. So, could the NIA be asked to take over the probe now?
>
> *Source:* Dnyanesh Jathar and Anupam Dasgupta, 'Still in the Dark', *The Week*, 29 November 2009, p. 37.

The last two sentences clearly define the purpose for which the author has constructed this paragraph. In fact, if we chop off the last two sentences of the paragraph cited above, we will stand to lose much of what the author is trying to put across through this piece of writing. It is, therefore, very important to expand and emphasize the ideas that are required to be highlighted.

At times, it is just a single sentence that makes all the difference. Look at the following instance:

> Global warming refers to a gradual warming of the earth. Though many scientists believe that global warming is a natural phenomenon, most of us are aware of the fact that many human actions such as emitting carbon dioxide and other gases into the atmosphere, cutting down trees, excessive consumption of water, petrol and other natural resources, besides unplanned, unnatural, and mechanical development have contributed heavily to the triggering or escalation of the process. The results are for all of us to both witness and withstand. August 2008 was the hottest month since weather records have been kept. The ten hottest years in recorded history have occurred since 1970. There is an increased number of instances of natural fury in the form of floods, droughts, tsunamis, and cyclonic storms. The number of patients suffering from skin cancer is increasing by the day. All this and a lot more establish the pattern of global warming for sure.

The paragraph does an admirable job in establishing a relationship between global warming and the human practices that cause it. However, it fails to strike an emphatic message simply because it does not reinforce the idea for which it is constructed in the first place.

Take a look at the passage once more and now form an opinion about it:

> Global warming refers to a gradual warming of the earth. Though many scientists believe that global warming is a natural phenomenon, most of us are aware of the fact that many human actions such as emitting carbon dioxide and other gases into the atmosphere, cutting down trees, excessive consumption of water, petrol and other natural resources, besides unplanned, unnatural, and mechanical development have contributed heavily to the triggering or escalation of the process. The results are for all of us to both witness and withstand. August 2008 was the hottest month since weather

records have been kept. The ten hottest years in recorded history have occurred since 1970. There is an increased number of instances of natural fury in the form of floods, droughts, tsunamis, and cyclonic storms. The number of patients suffering from skin cancer is increasing by the day. All this and a lot more establish the pattern of global warming for sure. Now the question arises, can we do something to stop this? The answer is YES, we can; and therefore, we should.

The only difference in these two versions of the same paragraph is the last part that attempts to inspire the reader into action. It is only because of the last two sentences in the paragraph that the author registers a change in the tone and makes the passage look like a part of some persuasive and instructive piece of speech or writing. Without this ending in place, the remaining passage, of course, would seem as though the author intends to share a piece of bad news with the reader. Hence, many a time, it is the proper expansion and emphasis of the ideas intended that secures the real purpose.

Thus, it is extremely important for us to keep in mind the significance of unity, coherence, expansion, and emphasis while we venture to construct effective paragraphs.

20.5 DESCRIPTIVE WRITING TECHNIQUES

Broadly speaking, paragraph writing can be divided into four kinds—expository, narrative, persuasive, and descriptive.

Expository Paragraphs are used for defining and introducing different concepts and ideas to the reader. For example, when you say 'Democracy is a form of governance in which people's urge for freedom and equality is respected', the statement is expository and defining in nature. If a paragraph is built on these lines, it would be known as an expository piece of writing.

Narrative Writing technique is very commonly employed by creative writers. It is essentially a story telling technique. For example, it is quite common to read something like 'When I entered the room, it seemed unusually quiet. They were all sitting there. Not speaking. Just looking down, with their heads buried in their knees. More than the surprise, it gave me a shock, and a wave of trepidation swept through me.'

Persuasive Writing is very commonly employed by marketing people. It is the heart and soul of write-ups such as sales letters and business proposals. In such writings, you often come across expressions such as 'All that you have to spare is a meagre amount of ₹500 per month to buy a life-long pleasure that comes of travelling and seeing unseen places. Just pay peanuts and let us do the rest. For almost nothing, the company takes care of all your travel programmes; gets the booking of your travel tickets done through our agent; fixes hotels for you and arranges tour guides…' Of course, a writing like this is done in order to promote the sales or services of a product, scheme, or proposal.

Descriptive Of all the writing techniques, descriptive writing is the one that is most commonly employed by us. It is quite extensively used for describing an idea, object, process, procedure, event, product, features, functions, etc. Regardless of the profession, all of us have to use descriptive writing in order to make an idea, object, process, event, feature, or function known to others. Since we all have to employ descriptive writing techniques, it won't be out of context for us to learn them in some detail.

Given below are the tips for making your descriptive writing effective:

1. Create a picture of the object/person/place/thing in your mind.
2. Memorize the object of description.
3. Visualize its features intimately.
4. Employ memory and imagination.
5. Create an intimate image in your mind.
6. Provide verbal structure to the object imagined.
7. Make your description interesting and innovative.
8. Use vivid and lucid language to communicate the idea.
9. Choose words which strike clear, unambiguous images in the mind of the reader.
10. Be intimate and warm in your approach.
11. Use catchy, informal phrases to capture the attention of the reader.
12. Use humour and wit if possible.

20.5.1 Examples

In this section, you will see a few descriptions.

Description of a Pet   If you intend to pay us a visit; don't hesitate doing it; at least not because of the fact that we have a dog. That too, a female one; the one you generally call a bitch. For more than anyone of us, she will be the one to receive you with open arms. Like any other dog, our Genie reacts to a doorbell call in a swift way. The moment a knock is heard or the call bell is rung, Genie's ears spring out of their drooping confines. Her eyes open and her neck arches upward. In a split second, she props herself on her all fours. In anticipation, her tail begins to wag. She knows not who might be at the door—a friend or foe, a family member or a stranger. To Genie, it doesn't matter who calls on us, at what time, or for what purpose. She is there to receive all those who come knocking. Though she cannot open the door herself, she makes up for the loss the moment anyone of us lets the latch go off its hook and the door opens. For without wasting even a single moment, she raises her mighty—and potentially threatening figure, to those who see her for the first time—to all those who use their sweet will to come to us. While she does so, her arms rest lovingly on the chest of the caller, her tongue licks the person wherever he/she can afford her to let her do so, and her tail, swirling like a perpetually moving broom as a non-stop sign of welcome to the guest, sweeps the floor beneath her at a frantic pace!

Description of a Toaster   One of the most compact and the best performing oven-toaster-grill (OTG) continues to perform life long for the satisfaction of its owner. The OTG is fitted with special and powerful sheathed heaters made of chrome nickel steel tube for fast, even cooking. Special pilot lamps are provided to give an indication of Upper Lower Heater. The OTG also has two level fixed shelf supports along with a fork. A special toughened see-through glass front window allows the owner to actually observe the food being cooked. The thermostatically controlled oven uniformly heats up to 300 degree centigrade to suit the individual requirement of each item to be cooked. With its unique features and extraordinary design, the appliance is sleek in look and functional in purpose. Compact and graceful, the OTG is specially designed to save fuel, time, and space.

See how with some changes a descriptive paragraph can be transformed into a persuasive paragraph:

> One of the most compact and the best performing oven-toaster-grill (OTG), the bake-all oven continues to perform life long for the satisfaction of its owner. The bake-all oven is fitted with special and powerful sheathed heaters made of chrome nickel steel tube for fast, even cooking. Special pilot lamps are provided to give you an indication of upper lower Heater. The bake-all oven also has two level fixed shelf supports along with a fork. A special toughened see-through glass front window allows you to actually observe the food being cooked. The thermostatically controlled oven uniformly heats up to 300 degree centigrade to suit the individual requirement of each item you would like to cook. With its unique features and extraordinary design, this lovely appliance which can be kept on your dining table, provides you with a variety of 5-Star menus. Sleek and functional, your bake-all oven is an ideal saver of fuel, time, and space. Literally, the OTG warms up your lifestyle.

Apart from writing descriptive paragraphs, you would be often required to write argumentative and analytical paragraphs. Therefore, it is desirable that these two types be discussed in some detail.

20.6 ARGUMENTATIVE PARAGRAPH

An *argumentative paragraph* argues against the view that is generally established. Authors writing such paragraphs normally choose to maintain a forceful and emphatic tone to contend the view that is normally taken for granted.

Look at the following paragraph and observe the style and tone adopted by the author:

> *Loyalty* is one of the most complex and debatable issues in human life. Generally regarded as a great attribute, loyalty stands as a firm evidence of a person's socially approved demeanour. In delicate phenomena like love, friendship, and marriage, it is seen as a single most important yardstick to judge someone's trustworthiness and dependence. This, however, is not the only side of the coin. At times, loyalty becomes a crippling factor in our life. It stops people from realizing their true potential. Particularly in the life of a creative and imaginative individual, maintaining loyalty to a system seems like an unnecessary appendage. This is so because creative souls tend to think beyond the existing frame. History is replete with instances where the creative people's imagination has challenged the existing codes and have preferred listening to their inner urge rather than sticking to a patterned existence. Though disapproved initially, such rebellion is often seen as an achievement and a hallmark of a person's courage. Regardless of the fact whether such 'disloyalties' are appreciated or disparaged, it is such rebellion alone that makes human life so creative, interesting, and enigmatic. Aware of the immense possibilities in not adhering to a loyal social system, creative souls often see *loyalty* only as a disruptive force that unnecessarily intends to act as an impediment in the march of a soul's innermost urge to realize its true potential.

See how carefully the author builds up the thesis to base his/her argument. As we can see that the paragraph is about 'Is loyalty all too often a destructive force, rather than a virtue?' he/she prepares to contend the traditional view that loyalty is always a virtue. However, rather than achieving this objective in a hurry he/she concedes the view that loyalty does matter in certain aspects of human behaviour before moving on to establish the paramount importance of a non-conformist, uncompromising, and creative urge in human endeavours.

Therefore, while writing an argumentative paragraph, which finally forms a part of some issue or argumentative essay, you need to agree first so as to disagree later on. This is important as a style in which the author jumps to his/her beliefs without paying any attention to the counter-view is not generally appreciated. In fact, if you do so, you are likely to sound prejudiced and parochial in your view.

Hence, allow the argument to develop naturally out of the discussion rather than forcing it straightaway in a paragraph.

In short, keep in mind the following while writing an argumentative paragraph:

1. State the established/opposing/counter view in the beginning.
2. Highlight its possible advantages, if any.
3. Introduce your view logically.
4. Give proper examples to substantiate the details.
5. Sound convincing and forceful.
6. Avoid sweeping statements and hasty generalizations.
7. Don't sound derogatory and insulting.

PRACTICE TEST 20.7

Given below is a paragraph written to argue against the view 'Money cannot Buy Happiness'. Trace what is wrong with it and rewrite it to make it sound like a well-written argumentative paragraph:

After all what is there in life without money? I don't think anyone will move even a single inch without the help of movement of the golden wheel of rupee. Money is all important in all walks of life and gives you everything that you can think of getting in this world; it gives you power; it gives you status and position; it gives you respect and recognition; it gives you pleasure and enjoyment. Take money away from human affairs and all his affairs cease to be. You are anything and everything with money. Without it, you are nothing. With money you can buy your comforts; move round the world; become educated and important in life; stay healthy and fit; get popularity and fame and what not? As regards its buying happiness is concerned, tell me what do you mean by happiness? Does poverty give that? Are you any happier when you don't have money and feel hungry, embarrassed, low, defeated, and humiliated in a world that thrives on the power of money? If you are not happy when you are comfortable, recognized, powerful, and respected, you will never be. Then it is not just money but nothing else as well that can give you happiness.

20.7 ANALYTICAL PARAGRAPH

An *analytical paragraph* analyses a situation with the help of facts, figures, and information and tries to draw inferences on the basis of these.

Read the example below and observe how the author analyses the issue of *gender disparity* with the help of facts and figures:

In a highly stratified society like India, there are numerous layers of differentiations apart from those concerning caste and class. Gender is now recognized as a more pervasive and distinct category of social stratification. The literacy rate among the tribals is not only low but also shows a high level of gender disparity. During 1971, female literacy among tribals was 4.85 per cent at the all-India level and only 0.49 per cent in Rajasthan. By 1981, it had increased to 8.05 per cent at the all-India level and 1.2 per cent in Rajasthan. Despite massive efforts by government and non-government agencies, it was still 19 per cent at the all-India level and just 4.42 per cent in Rajasthan in 1991. The states of Andhra Pradesh with 8.68 per cent and Rajasthan with 4.42 per cent have remained at the bottom of the tribal female literacy table. On the other hand, states like Mizoram (78.74%), Nagaland (54.51%), Sikkim (50.37%), and Kerala (51.07%) have more than 50 per cent literacy among the tribal female population. It is significant that Andhra Pradesh, which has a lower tribal literacy than Rajasthan, has higher literacy among the tribal female population.

Read another example of an analytical paragraph. In the paragraph given below, the author analyses the situation without using any statistics and figures:

Adverse weather (or the threat of adverse weather) can cause an accident in many different ways. Weather information is generally a prediction but not always accurate. Weather information is provided to flight crews at dispatch and in flight also, but is not always timely. Flight crew decisions based on available information are taken, but not always made in accordance with prescribed procedures. There is no clear way, and indeed no practical need, to separate the entirely environmental factors from the truly operational ones.

In a succinct way, the author of the above passage deftly analyses the situation. Therefore, in an analytical paragraph, it is not mandatory for us to be elaborate all the time. A tidy and precise analysis should be preferred to the detailed statistical data if situation so warrants.

In short, remember to keep the following points in mind while writing an analytical paragraph:

1. Present the situation as it is and also as it should be.
2. Interpret the data or situation with interpretation, comparison, and contrasts.
3. Be elaborate while interpreting data and succinct while analysing a situation without it.
4. Don't sound equivocal and far too philosophical in tone and style.
5. Present both sides of the coin.
6. Choose direct and emphatic word order.

PRACTICE TEST 20.8

Construct an analytical paragraph on *terrorism* with the help of the outline provided below:

Terrorism, an expression of violent dissent—dissatisfied minority group—seeks its course through coercion—creating a sense of fear—reckless killings—hijacking and blowing of aircraft—Government's indulgence—cross-border terrorism—weakens the country.

 RECAPITULATION

✓ A paragraph is a group of sentences that focuses the expansion and emphasis of an idea, view, or perspective with clarity, unity, coherence, and precision.

✓ It constitutes various sentences such as introducers, developers, and terminators.

✓ Introducers are also known as topic sentences, which are further broadly categorized as structural topic sentences and interpretive topic sentences.

✓ Various techniques are used to construct a paragraph. Some of these are known as narrative description, comparisons and contrasts, sustained analogy, cause and effect, quotations and para-phrasing, enumeration, definition, testimony, facts, figures, instances, examples, etc.

✓ A good paragraph always has features such as unity, coherence, expansion, and emphasis.

✓ Descriptive writings are used to describe a person, place, object, building, procedure, system, etc.

✓ Descriptive writings are aimed at creating mental pictures, focusing on a single idea.

✓ Powerful images, interesting language, catchy sentences, humour and wit, etc. make descriptions lively and arresting.

✓ Argumentative essays are written to argue in favour of or against an argument, point of view, perspective, etc. They are written in a forceful and emphatic style.

✓ Analytical essays are written to analyse a situation with the help of facts and figures. Analysis of a given situation or state of affairs is carried out through interpretations and inferences drawn on the bases of given facts, data, and statistics.

WISEWELL QUIPS

EXERCISES

I. Write a paragraph in about 250 words on each of the following topics:

1. Religion is more important than science.
2. Art lies in concealing art.
3. A borderless world is just a dream.
4. Global warming: A threat to our planet.
5. One may smile and smile and still be a villain.
6. Peace cannot be attained through violent means.
7. Science is both man's greatest friend and worst enemy.
8. The solution always lies in the problem.
9. Child is the father of man.
10. Bhopal: A forgotten tragedy.
11. The rich always exploit the poor.
12. Without education man is just a beast.
13. Beauty needs no ornament.
14. Deforestation vs forestation.
15. Terrorism: A global menace.

ANSWER KEY

Practice Tests

20.1 **1.** Developer   **2.** Introducer   **3.** Terminator   **4.** Developer
     5. Developer   **6.** Introducer   **7.** Terminator   **8.** Developer
     9. Developer   **10.** Developer

20.2   **1. Hence**, it is not possible for us to reconcile to our fate all that tamely.

    2. Furthermore, you can pick up from this book how to use water colours effectively.

    3. Nevertheless, nothing can finally be said about the patients who have heart problems.

    4. Even if we go by the data, it is hard to justify the steps proposed by the committee.

    5. However, it is not possible for India to allow America to intervene in all its matters.

    6. Although they claim so much, nothing substantial has yet been achieved in this matter.

    7. He never tried to be an entrepreneur **for** he was not sure of his management skills.

    8. But in all probability, recession is likely to worsen.

    9. As they approached the house, they saw the thief running away.

    10. Cricket has brought us many laurels. Winning from the 1983 World Cup to the T20 World Cup in 2007, and the ICC World Cup 2011, it has won us moments of pride and glory, **not to forget** of course the many victories registered against archrivals like Pakistan, Australia and Sri Lanka.

    11. Notwithstanding lofty claims by the government, child labour in India is not likely to vanish any time sooner.

    12. By all means the king wanted his daughter to laugh.

    13. For instance, it is Hamlet who needs to act but he doesn't and not King Lear who does.

    14. After all, there is no way by which you can make the unwilling souls listen to you.

    15. In other words, it is our thinking that makes us happy and not our situation.

20.3   Indian cricket team has achieved remarkable milestones under Mahendra Singh Dhoni's captaincy. With Dhoni at the helm of affairs, Indian team has won the limited over cricket World Cup in 2011. **Before that**, in 2007 itself, the team had won the inaugural T20 World Cup. **Besides these**, team India has also won the Champions' Trophy, the Triangular Series and the Asia Cup. Dhoni's captaincy has also been instrumental in bringing about an unprecedented sense of commitment and passion for the task at hand. There, **however**, is a seamy side to this rosy tale. **Despite** his attainments in the limited over cricket Dhoni has not been able to infuse in the team similar courage when it comes to playing test cricket abroad. **Consequently**, Indian team has not won much in any of the countries such as Australia, England, New Zealand and South Africa.

20.4   Eyes are considered the most significant organ of the human body. Placed close to our brain, they are the focal point of our consciousness. It is said that eyes are the windows of the soul. There is no emotion and no expression that a human eye cannot transmit effectively. Therefore, in terms of non-verbal communication, eyes acquire a place of crucial importance. If as a speaker your eyes seem to be dull and blank, it is not really possible for you to create a positive impact on your listeners. In fact, it is impossible to keep dull eyes and reveal intensity through your voice or other body organs. Imagine a face that is bright but has dull eyes. You can't, because it is simply not possible. On the other hand, it is quite natural to have bright eyes and a lively face. It is so because when your eyes shine, your face expresses the positivity that is so crucial while making a speech. Hence, use your eyes well to win the audience.

20.5   A life without ambition is a like a train without an engine. Just as it is the engine that drives the train, it is human beings' ambition that drives their lives. So, by all means, ambition occupies a central position in human efforts. Some theorists, however, refuse to believe that ambition has crucial significance in human life. Quoting the spiritual text, the *Geeta*, they feel that if we want to be happy we must have no ambition whatsoever. There is no denying the fact that the *Geeta*, is the most authentic book on spirituality, and what it teaches us is invaluable. For common mortals, however, it is not possible to involve themselves in action without desiring a result. Since ordinary people are normally goaded while they are able to see the result in sight, it is not possible for us to forgo human ambition.

20.6   One of the most challenging issues of urban living is to find suitable gifts for suitable people on suitable occasions. And suitable occasions exist in plenty and suitable people too are quite in multitude. Of course, I don't mean to say that there is any dearth of suitable gifts. How can they be? In fact, they are far too in abundance. The only thing in dearth of is your ability to choose the suitable gift that suits your pocket too. Often, only that seems to be found wanting. Particularly, on the last days of the month when it looks like a flat toothpaste. In such situations, you are required to squeeze it with all your might to manage to look presentable to others, with polished teeth and a plastic smile on your face. All this for you to be able say suitable words to suitable people on suitable occasions with suitable gifts in your hand!

20.7 I agree with the view that money is not all that one needs to have in life. I also agree that it cannot buy happiness for us. But it can buy all the things that can be bought. I don't think anyone will move even a single inch without the help of the movement of the golden wheel of the rupee. Money is important in all walks of life and gives you everything that you can think of getting in this world; it gives you power; it gives you status and position; it gives you respect and recognition; it gives you pleasure and enjoyment. With money you can buy your comforts, move round the world, become educated and important in life, stay healthy and fit, get popularity and fame, and what not? As regards its buying happiness is concerned, tell me what do you mean by happiness? Does poverty give that? Are you any happier when you don't have money and feel hungry, embarrassed, low, defeated, and humiliated in a world that thrives on the power of money? If you are not happy when you are comfortable, recognized, powerful, and respected, you will never be. Then it is not just money but nothing else as well that can give you happiness. In fact, it is wrong to blame money for not being able to buy us happiness. Happiness is a matter of inner experience and it depends much on our attitude and way of thinking. So, if we are not happy, we are to blame, not money.

20.8 Terrorism is an expression of violent dissent. The dissatisfied minority groups, whose views are not entertained by the majority, resort to it at times. Terrorism seeks to achieve its course through coercion, the motto being—the end justifies the means. Creating a sense of fear by reckless killings and wounding of innocent citizens, wilful destruction of public and private property, hijacking and blowing up of aircraft, and kidnapping and assassinating political personalities, it has now acquired international dimensions. Intelligence reports reveal informal contact between terrorist groups of different countries for the purpose of financing, gun running, guerilla training, and providing shelter to the terrorists. Many governments also indulge in the act of terrorism directly or indirectly, by dispatching miscreants across borders, aiding, abetting, or providing logistical support to the terrorists of other states. This type of cross-border terrorism is employed either to weaken another enemy state across the border or to promote certain other political goals.

Essay Writing

- figure out the relevance and importance of essay writing
- understand the characteristic features of an essay
- learn about the different stages in the writing of an essay
- develop techniques required to construct an effective essay
- overcome the common pitfalls in the task of essay writing

21.1 INTRODUCTION

Vasla wants to become an IAS officer. What drives her is not just the glamour of power and position that civil services officers enjoy in our country, but also her belief that an IAS officer has a real opportunity to dedicate his/her life for a good common cause. Despite such a noble intent, Vasla has decided not to take the civil services examination this year, as she does not feel confident about cracking the English section of the test and finds herself on a slippery ground particularly in the essay writing part.

Vasla is not the only girl in the world who dreads the prospect of writing an essay because thousands of students regard the task of writing an essay an onerous one and find themselves unsure of taking a test that evaluates them on this particular skill. Now, the question arises—why is it so? The answer is that essay writing is a challenging task as when we are asked to write an essay quite a few of our skills are being put to scrutiny. Therefore, first of all, let us understand the multiplicity of the task of essay writing.

As we all know, essay is a written composition in which the author shares his/her knowledge about a certain topic, reveals to the reader his/her perspective on the issue being discussed, and offers criticism and comments on the situation. The writing of an essay, therefore, requires the author to display not only his/her knowledge of the subject but also the maturity of vision, clarity of thought, and felicity of expression. Besides these, the author should also be able to weave together the different parts of an idea into a thread of unity. The task, though difficult, is not impossible to achieve and turns in hefty returns to the aspirant who commits himself/herself to the arduous but exciting prospect of writing an essay. However, before taking up the task of writing an essay, let us see the different types of essays we generally come across.

21.2 TYPES OF ESSAYS

The word 'essay' comes from the French expression 'essai', which means an effort or a verbal sketch which reveals the author's perspective on a given subject. Certain distinct features of different types of essays will be discussed in the following sections.

21.2.1 Argumentative Essays

An essay that is written to contend an established view is argumentative in nature and is known as *argumentative essay*. Read, for example, the beginning of some such essays:

> Cancer is generally regarded as a disease of severe physiological disorder. Most of us believe that the malignant growth of tumour is caused by the chaotic and aggressive disorder in the human metabolism which leads to an aggressive growth of dead cells in our body. Cancer thus is essentially seen as a disease rooted in a physical disorder. The recent studies, however, suggest that cancer can be rooted in our attitude and can be linked to the way we think, feel, and perceive the world around us…

By linking cancer to a psychological bent, attitude, and mindset, the author in the above extract of the essay contends the established view.

In an argumentative essay therefore, the author is often keen to challenge the established notion. Because of this, such essays are also known as *point-of-view essays*. While writing an essay of this type, we need to establish the argument that is reason-based and not governed by our subjective opinions or emotions. In such an essay, it is always helpful to state the rationale behind the existing idea before suggesting the alternative view.

21.2.2 Analytical Essays

An *analytical essay* often reviews a book, movie, topic, situation, or a given text by bringing to the fore its subtle nuances. Take a look at the following extract from one such essay:

> Set in the turbulence of partition times, the novel brings to the fore the lurking sense of insecurity and incertitude that ticks the characters in the story. As the plot develops, the initial calm suggesting harmony and peace gives way to discord and desperation that sets in the people of both the communities. The novel is remarkable for its ruthless yet objective depiction of reality as lack of political will and administrative commitment lead to aggravation of the situation. The minute details with which the novel observes the sense of restlessness and nervous anxiety that sets in the environment are suggestive of the author's psychological penetration into the working of the mind.

While writing an analytical essay, we need to carefully observe the finer aspects of a work of art, situation, text, book, or topic and highlight all its subtleties. In an analytical essay, the data and material collected play an important role as they often form the basis of an analysis. Read the extract taken from such an essay:

In terms of sex ratio, both Haryana and Punjab seem to fare rather poorly. The last census had shown both these states to have the worst sex ratio in the country. The picture appears really skewed in the 0–6 age group. The census record showed that the child sex ratio (0–6 years) in the state was a dismal 819 in Haryana, while the national average stood at 927. Barring a few districts like Gurgaon and Faridabad, the rest of the district in the state of Haryana show a pathetic situation with regard to sex ratio. And this is the situation even after a whole lot of different schemes launched by the state government to stop female foeticide.

21.2.3 Descriptive Essays

A *descriptive essay* is written to get the reader the specific and concrete details of a situation or an object. In descriptive essays, the author primarily harps on his/her senses to help the reader visualize, feel, or enjoy the object of description. Subjectively written at times, descriptive essays are quite often a reflection of the author's personality. Take a look at a part of one such essay:

The scene at the airport is so very special. You don't generally see so many people presenting themselves in such a disciplined way. So you see people standing in a long queue without trying to jump it; you see them patiently waiting for their baggage to be weighed and checked. You also see them moving through the security check-ups without grumbling or frowning. So much so that you can see some people maintaining their sense of humour even while being 'felt' by the security personnel. One thing that keeps airports strikingly different from other public places is the fact that the number of people going around with a smile on their face far exceeds than at any other similar terminus such as a railway station or a bus stand. Now, every-one is aware of the pleasant smile of the airhostesses and stewards, but it is only at an airport that even a bearded man behind a window or the one with a bushy moustache at a security point smiles at you before they let you go off to embrace the huge, silver bird waiting for you to take you to cloud nine.

21.2.4 Expository Essays

Unlike an argumentative essay, an *expository essay* is meant to explain a topic without giving the author's opinion. It is essentially designed to convey a piece of information with the reader so that he/she comes to know about a situation, topic, fact, or state. The tone of an expository essay is often detached, objective, and matter of fact as rather than establishing the author's point of view, it is meant to impart to the reader the information and knowledge that the author possesses. Take a look at how the following expository paragraph takes off on a discussion on health insurance:

Health insurance refers to a system for the advance financing of medical expenses through contribution. When proposed as a public policy by a government, people have the facility to pay their contribution or taxes into a common fund to pay for all or part of health services specified in an insurance policy or law. The key features of health insurance are advance payment of premiums or taxes, pooling of funds, and eligibility for benefits. Known as public health insurance, this form of health insurance may apply to a limited or comprehensive range of medical services and may provide for full or partial payment of the costs of specific services. Benefits under such a scheme may range from the right to certain medical services or reimbursement of the insured for specified medical costs. Unlike public health insurance which is run by a government, private health insurance is organized and administered by an

(Contd)

insurance company or other private agencies. It makes provision for accumulation of funds by the regular and systematic contributions made by the policy holders. Offered with a wide range of flexibilities, private policy holders are required to pay for a certain period of time towards health insurance and are offered reimbursement or cash-free facilities as per the terms and conditions initially agreed upon.

21.2.5 Reflective/Philosophical Essays

A *reflective* or *philosophical essay* is meant to discuss a profound and deep issue. In such essays, the authors discuss universal human issues, such as life, death, love, faith, truth, etc. Since the subject matter of a philosophical essay is universal, the authors rise above the immediate and mundane, universalizing the personal. Take a look at the following extract that can form a part of one such essay:

Peace is not simply the absence of war. A truly peaceful society is one in which everyone can maximize their potential and build fulfilling lives from threats to their dignity. A transformation in the inner life of a single individual can spur and encourage similar changes in others, and as this extends into society, it generates a powerful vortex for peace that can steadily shape the direction of events. The collective impact of 'ordinary citizens', awakened and empowered, can propel humankind toward the twin goals of genuine disarmament and a flourishing culture of peace.

Source: 'A New Era of the People: Forging a Global Network of Robust Individuals',
'The Speaking Tree', *The Times of India,* Sunday, 24 October 2010, p. 7.

21.3 CHARACTERISTIC FEATURES OF AN ESSAY

Though different types of essays can be written in a variety of ways, following are some of the characteristic features of a well written essay:

1. A good essay is the result of a *careful planning and selection of material.* Since an essay relates to a specific situation, problem, or fact, it selects the matter that is required to be selected and rejects what is redundant. Hence, good essays are never produced abruptly but sculpted with careful consideration and thought.
2. A good essay is *comprehensive in its approach and vision.* A well written essay highlights all the aspects related to the issue under discussion by highlighting the various aspects of the problem or issue.
3. Though an essay is a *reflection of the author's perspective* and at times also throws light on his/her personality, it is considered most mature and relevant when written in an objective and detached manner.
4. A good essay is normally *well balanced* and not lopsided. A well crafted essay strikes a balance in its different parts and a good writer of essays gives due importance to each of its various parts.
5. *Coherence* is another feature of a good essay as the different parts of an essay are well coalesced into one another. A good passage, rather than focusing on any individual aspect of the problem, always creates the impact of one organic whole on the reader.
6. A good essay reflects *consistency and logical sequence of ideas* in a composed and controlled manner. In a well written essay, exaggerations and hyperboles have no role to play.

7. A good essay is always written *without ambiguities, verbal juggleries*, and equivocations. The style of a good essay is therefore direct, simple, vigorous, and lucid.

8. Just like the other components of an essay, its title is also chosen very carefully.

21.4 STAGES IN ESSAY WRITING

Following are the different stages in writing of an essay:

Collecting the material   A good essay is the result of a careful research. Therefore, before writing an essay on some topic, we need to collect the relevant material required to be studied so that the essay we compose is authentic, substantial, and convincing. Without proper data and understanding of the relevant details, we are likely to produce a perspective that would seem subjective, trivial, and prejudiced.

Defining the scope   Since an essay is always specific and to the point, it is important for us to define the scope of our presentation of the idea in question. While defining the scope, we need to look carefully at the title of the essay. For example, when we have to write on 'The Problem of Terrorism in Modern World', we need to give a worldwide view of the perspective and while discussing 'The Problem of Terrorism in India', we need to restrict our scope and limit it to India alone.

Making an outline   An outline of an essay relates to its skeletal form. It consists of the main and sub-points of our essay. An outline is always helpful as it keeps us focused and systematic in taking up the various issues involved in an essay.

Making the first draft   Since essay is a detailed composition, it is generally best produced after revision and editing. It is always handy to jot down the sequence in which we plan to take up the different ideas that are going to form the fulcrum of our essay. Without the main points in our mind or in front of us, it is not possible to maintain coherence and unity in the ideas expressed.

Revising and editing   While revising our essay, we must pay careful attention to maintaining logical development of the idea throughout our effort. Besides the content and the style, the natural evolution of a thesis is what makes an essay eminently readable. The style chosen has to be compact, direct, and shorn of bombastic words, clichés and jargonistic structure. Remember that any ambiguity in style and substance is likely to erode the impact of the overall presentation and hence it must be carefully avoided.

21.5 COMPONENTS COMPRISING AN ESSAY

An essay can be divided into three distinct parts:

(a) Introduction

(b) Main body—development of an idea

(c) Conclusion

Take a look at the introduction of an essay and judge whether you find it effective:

> When it comes to analysing a prospect such as 'India's Progress: A Myth or Reality?', it seems the word *progress* is not good enough. It is so because *progress* means something positive and good but in many ways our country is either not progressing or just becoming worse. Take for instance, the case of growing fundamentalism among several sections of the society and also the gap between the rich and the poor that seems to widen up all the time.

The introduction of the essay appears too abrupt to be impressive. Rather than introducing, the author here seems to be discussing the idea and starts giving the different aspects of the issue straightaway. Such beginnings do not allow the reader to feel settled with the topic. Consider another beginning that alternatively introduces the idea:

> *Progress* is a multi-dimensional term and is generally confused with other similar terms such as 'evolution', 'growth', and 'development'. In the past, it has often been suggested that our country, in the name of development, has only registered a lopsided development. To be able to evaluate whether India has not just *evolved* and *developed* but has also *progressed*, we need to take into consideration the quantitative and qualitative aspects of the country's development over the years.

With this introduction, the essay seems to have begun well. The author tries to put the word *progress* in its right perspective and takes the reader further into the discussion. Therefore, we need to avoid introductions that plunge into the discussion right in the beginning itself. However, a hasty discussion is not the only way in which an introduction is spoiled. At times, the introductions written in essays have been found to be too long, irrelevant, flashy, or abstruse to be appropriate. Therefore, while writing the introductory part of the essay, keep in mind the following points:

Tips for writing effective introduction

1. Keep your introduction brief and effective.
2. Avoid starting abruptly or too philosophically.
3. Define or explain the title in a precise, specific way.
4. Use quotations, dictionary meanings, statements, or sayings to introduce the reader to the main idea.
5. Don't take sides on an issue or sound prejudiced in your approach.
6. Avoid jargons, clichés, and bombastic beginnings.

Once the beginning is set, we can move on to the main body of the essay. It is in this part of the discussion that we compare and contrast, challenge and question, reveal and establish, and hence, bring into view the different nuances of the main idea. Quite often, this part of the essay runs into several paragraphs with each of those consisting of a topic sentence, a set of developers, and a terminator. To be able to come out well in this part of the essay, we need to keep in mind the following ideas:

Tips for developing main body

1. Evaluate all the possible aspects of a problem, topic, or issue.
2. Give due importance to each aspect; don't appear prejudiced or biased in your approach.
3. Relate all your ideas to one another.
4. Connect this part to the hopes raised or promises made in the introduction.
5. Avoid a miscellany of too long or too short paragraphs in this part of the essay; let there be a semblance of equality in the length and size of different paragraphs.
6. It is in this section that the author attempts to convince the reader of his/her perspective on the issue; therefore, analyse the different aspects of the problem exhaustively and leave nothing to chance.
7. Use supporting material to augment and develop ideas. For carrying conviction, use brief or extended examples, facts, comparison, contrasts, expert testimony, and other such devices to make the text look comprehensive and authentic.
8. Let the main body lead the reader automatically to the conclusion of the essay.

Let us take a look at the main body of an essay on the value of sports in life.

The physical benefits of sports are well known. Participation in sports builds the stamina and makes the player strong. The sports such as swimming, football, hockey, volleyball, tennis, badminton, and basketball that require running, stretching, bending, and constant physical movement help us build up resistance and reflexes. Physical activity helps us develop immunity; keep the body fit and fine; take up physical strain as and when required. It is only out of physically fit people that we get physically fit countrymen who can take our country to the ladder of success, growth, and development besides getting us the brave soldiers who protect our national territories and international boundaries.

Sports help us not just physically but psychologically as well. They help us maintain mental and emotional balance. Since sports require competition and constant effort to win, they lead us to a positive attitude. Participating in sports activities leads to purging mind of unwanted emotions. It gives us a sense of well-being and help us maintain a positive mindset. As it is aptly said, *a healthy mind lives in a healthy body*, participating in games and sports helps us attain both a healthy mind and a healthy body. Even in terms of social adjustment, sports and games play a crucial role. One who has participated in sports and games knows the importance of team spirit, discipline, cooperation, fairness, and cheerfulness and displays these traits both in his/her personal and professional activities. Such people respond to challenges of life with a spirit of competitiveness and enthusiasm and finally achieve their goals both at the personal and professional front.

Realizing this magical importance of sports, human civilizations have always assigned to games and sports vital importance. Historically, sports have swayed our imagination since time immemorial. Long ago, the Greeks realized the importance of sports in life and started a festival every fourth year that included contests of sports, music, and literature. It was these ancient games which were revived in 1896 in Athens and are now known as the prestigious Olympic Games, which involve thousands of players in numerous sports events every fourth year. Besides these, there are World Cups in sports such as football, hockey, and cricket. Table tennis, badminton, volleyball, and basketball players keep competing in other premier championships. The Commonwealth Games and Asian Games give opportunities to players from different nations to prove their mettle in a variety of sporting and athletic events.

Just as the introductory part or the main body of the essay, the conclusion of an essay is also quite crucial. In fact, many a time a poor conclusion can adversely affect the overall impact of an essay. In order to come up with a good conclusion, bear in mind the following suggestions:

Tips for developing a good conclusion.

1. A conclusion is meant to reinforce the idea already illustrated and established in the main body; avoid therefore developing any new idea in the concluding part of the essay.
2. Avoid feeble endings; pack it with a punch of force and vigour.
3. An unrelated or irrelevant ending makes an essay look ludicrous at times; hence, let the conclusion naturally emerge out of the discussion.
4. Keep your conclusion crisp and in cohesion with the other parts of the essay.

Take a look at the conclusion of an essay on *Superstitions in Society*.

In essence it seems that lack of education and superstitions are deeply intertwined. The poorer a society is in terms of education, the greater the prevalence of superstitions in its circles. It is only through education that we can purge a society of its superstitions. As individuals in a society become truly educated, they develop scientific temperament; are able to see through false notions and fake beliefs imposed on them through rituals and customs and hence, can steer clear of them. Therefore, in order to keep our society away from the canker of superstitions, we need to make constant efforts in providing education to as many people as possible, broaden their knowledge, and make them see the futility of the mental cobwebs that assault them in the form of superstitions. The task seems challenging but is not impossible to achieve as the darkness of superstitions can certainly be eliminated by spreading the light of awareness and knowledge through education.

21.6 ESSAY WRITING—GUIDING PRINCIPLES

Having understood the nature and features of an essay, let us now learn how to write effective essays. Here are some of the principles following which we can develop essays of different types.

21.6.1 Work Hard on the Introduction

Just as is the case with other compositions, a good beginning is quite crucial to writing an essay. A good, imaginative start can help us capture the attention of the reader. For example, let us understand which of the two beginnings given below appears to be more appropriate on the issue 'Freedom of Press: An Indian Perspective':

> Press in India is not free. Even after so many years of independence, press has not been able to assert itself. Though there are reports that question the policies of the government and attack the ills of political and bureaucratic system every now and then, we are yet to see a press that is fully independent of its view and can interfere directly into government's policies and play a constructive role in the nation building.

An introduction of this sort is certainly not appropriate. It hardly introduces and in fact seems to end the discussion even before the discussion gets started. Rather than introducing, this piece of writing seems to wind up a discussion and hence need to be avoided. Take a look at the revised version of the same:

> Freedom of press is often equated with the freedom of a nation. It is believed that the extent of freedom of press determines the extent of freedom that people of a particular nation enjoy. Freedom of press is crucial to the growth, development, and progress of a nation. In India, press has the constitutional freedom to express its voice. Representing the voice of people, the press in India has always played a significant role in expressing the views of the masses on matters of importance and general concern. Despite having achieved that, press in India has not been able to assert its prowess as a tool of change. A closer look at the variegated nuances of freedom of press in India brings to the fore a very complex and intriguing scenario for us to observe.

21.6.2 Make the Main Body Look Authentic and Unified

Developing the idea broached in the introduction, the main body of an essay actually is its most expanded part. Since it constitutes the main argument of the whole idea, the main body may also be seen as the most important part of an essay. Running into a couple of or several paragraphs, the main body is expected to look authentic and unified. Without an authentic and compact main body, an essay is likely to sound hollow and trivial. In order to achieve unity and provide authenticity to this part, authors tend to divide the main body of their essay into different thought units and construct a separate paragraph on each of them. Before writing this part of the essay, it is quite useful to prepare a small outline by dividing the main idea into sub-topics. For instance, an outline such as the one that follows might be designed before we launch ourselves into writing the main body of our essay on a topic such as the one that follows:

The Problem of Brain Drain

- *Causes of Brain Drain*
  Lucrative Jobs
  Better Opportunities and Life Style
  Freedom of Work
  Lack of Corruption
  Better Facilities and Working Conditions
  Charm of Foreign Tag
  Enhanced Social Status

- *Recent Trends*
  Statistics in the Past Two Decades
  Future Projections
  Suggested Remedies
  Better Work Environment
  Freedom for Creativity and Innovation
  Improved Infrastructure and Facilities

Once we are sure of *what* to write, we can start thinking of *how* to write that. As already suggested, the main body of an essay constitutes different paragraphs, each of which has its own *introducer*, a couple of *developers*, and a *terminator*. Every paragraph deals with a separate sub-topic within the broad range of the principal or main idea. In order to lend credence to the thought, writers often use devices such as comparison, contrast, analogy, examples, instances, statistics, quotations, enumerations, definitions, etc. Using strategies such as these provides conviction and authenticity to the content of the essay and it is difficult to conceive of a topic that can be developed into a cogent and substantial essay without employing such techniques. For further details on how to build effective paragraphs using these strategies, please refer to Chapter 20 on *paragraph writing*.

Just as it is important to provide conviction, it is also equally important to make an essay coherent and well-knit. A loosely constructed essay is generally the one that does not convey the central idea with concentration and effect and where the sub-topics are fragmented arbitrarily and the different parts of the essay seem disjointed from one another. Such an essay leaves on the reader a poor impact and fails to convince him/her about the views of the writer of the essay. Therefore, we must pay special attention to put the different parts of an essay in consonance with one another. Linkers and connectives help in putting different ideas in accordance with one another. Take a look at how by using linkers, different ideas in an essay on a topic such as the one cited below can be presented:

Media: Uses and Abuses

- *Though* media is often accused of using unfair means or sensationalism, its advantages cannot be overlooked.
- *Besides* popularizing vulgarity and buffoonery, media can also be accused of having changed the cultural ethos of modern generations.
- *Moreover*, by reporting about the corruption cases, media brings to the fore the ugliness of the system that often remains hidden from our view.
- *However*, rather than highlighting reality, most of the new channels today focus on showing what can keep their audience hooked to their programmes through alluring advertisements, catchy cover stories, sensational sting operations, and violent crime sequences.
- *Further*, the different forms of media should be alerted and made more conscious about their constructive role in the society.

A detailed discussion regarding using linkers, connectives, and transitions has been given in Chapter 18 on *reading comprehension* besides the chapters on grammar in the book. Since connectives provide order and consistency to an idea, it is important to use them appropriately.

Let us understand through an example how the lack of coherence can spoil the impact of the main body of an essay:

Corruption in India

The roots of corruption in our country have become so entrenched that it seems difficult and even impossible to uproot it. In a post-independent India, corruption has made heavy inroads in all aspects of our life and has permeated the social, political, economic, and religious fabric of our society. The pains of the masses are further increased by corruption in government offices. Moreover, there are smugglers which go on increasing the tentacles of corruption in our country.

As far as trade in India is concerned, there too corruption rules the roost. Corrupt traders keep hoarding the essential commodities and create artificial shortage of a commodity. Because of this, the commodity becomes rare to find anywhere and then the corrupt hoarders increase the price and accumulate huge amount of money. Not only these but black marketers also introduce spurious goods to replace the original items and thus innocent people end up buying the fake item at an exorbitant rate.

The above part lacks both coherence and unity. The first paragraph starts by introducing corruption in a post-independent India but then suddenly jumps to the idea that '…the pains of the masses are further increased by corruption in government offices'. This illustrates lack of coherence in the organization of the paragraph as there seems no consistent development of a principal idea that a good paragraph aspires to achieve. The structure seems flawed also in terms of unity as the second paragraph starts abruptly with the idea 'As far as trade in India is concerned, there too corruption rules the roost…'.

Take a look at how a revised version of the same seems more compact, unified, and coherent:

The roots of corruption in our country have become so entrenched that it seems difficult and even impossible to uproot it. In a post-independent India, corruption has made heavy inroads into all aspects of our life and has permeated the social, political, economic, and religious fabric of our society. For decades now, corruption has reflected itself in a variety of ways. The problems of black marketing, bribery, scams, and adulteration raise their ugly face only because of corruption in our society. Moreover, other problems such as poverty, inflation, lack of governance, and lawlessness can also be seen as other ramifications of this hydra-headed problem.

For example, if we look into the nuances of trade in India, corruption rules the roost. Corrupt traders keep hoarding the essential commodities and create artificial shortage of a commodity. Because of this artificially created scarcity, the essential commodities disappear from market. With the rise in demand, the corrupt hoarders increase the price and accumulate huge amount of money. The masses have to pay the increased price for buying the essential household commodities. This really affects their monthly budget which, for millions and millions in India, is always tenuous and dwindling. Not only these but black marketers also introduce spurious goods to replace the original items and thus the poor, innocent people end up buying the fake item at an exorbitant rate.

As can be seen in the revised version, first the different forms of corruption are highlighted in one paragraph and the one that follows takes up one particular form of corruption and explores its impact on commoners. It is evident therefore that maintaining coherence and unity in an essay becomes one of its important features and the essay that lacks in this aspect fails to create the right kind of impact on the reader. The concept of coherence and unity has also been discussed in Chapter 20 on *paragraph writing*.

21.6.3 Keep the Conclusion Short and Effective

Though conclusion of an essay comes at the end of it, its importance cannot be undermined. In fact, quite a few essays fail to click with the reader because of a poor and ineffective conclusion. Take a look at the following paragraph that comes at the end of an essay on corruption in India:

> Recently, even health and educational institutes seem to have been involved in widespread corruption. There have been instances of fake degrees, illegal approvals, and unlawful admissions in various institutes across the country. In our country, corruption has made every illegal activity seem fairly common; criminals can commit a heinous crime and can roam around freely; smugglers can freely sneak through barriers; engineers can make bridges that can collapse any day; transporters can run buses on roads without permit; employees can be selected, promoted, or transferred by greasing the palms of those concerned; doctors can sell kidneys of their patients; teachers can change grades and marks they have awarded to students. In these ways and many more, corruption is rampant in our country. Seeing this, one can only wonder what sustains our country despite this widespread wave of corruption that sweeps almost all aspects of our social, political, economic, and religious life.

A conclusion like this may seem relevant but it hardly seems impressive and forceful. A paragraph like this should be placed in the main body of the essay. Moreover, the concluding sentence expresses a despair and helplessness which is bound to create a pessimistic impact on the reader. Consider an alternative conclusion for the essay on the same topic:

> Observed thus, it is apparent that corruption in India is one of the most challenging problems that our country faces. It is also clear from the discussion above that unless corrupt practices are stopped in our country, millions among us will be forced to lead a substandard, dissatisfied, and miserable life. However, since corruption emanates from loss of values and ethics, we must make untiring efforts in reviving them in all aspects of our life. Individually, each of us needs to set a high standard of personal conduct and judge ourselves scrupulously. Since a society is made up of individuals only, if the individual changes, so will the society and the country with the passage of time.

While writing the conclusion for an essay, keep in mind the following points:

1. The conclusion of an essay should be short, crisp, and effective.
2. While writing a conclusion, do not start elaborating a particular point.
3. No new ideas should be added in a conclusion.
4. The conclusion should essentially be in consonance with the discussion.
5. The conclusion should not be pessimistic and gloomy.
6. In conclusions, writers are expected to give their own ideas; therefore, avoid quotations from other sources while making final statements.
7. Give convincing ideas in conclusions; avoid making sweeping statements.

21.6.4 Write in an Effective Style

Just as keeping an essay well-structured is important, so is writing its contents in an effective and appropriate way. The style of an essay is, in fact, as important as its contents. Therefore, writing in a style that is direct, emphatic, and elegant becomes crucial to the overall impact of an essay. At times, essays are spoiled because of a style that is circumlocutory, bombastic, and ornate. We should bear in mind however that using unnecessary frills and verbose expressions is likely to create a negative impact on the reader. Given below are some of the principles for composing an essay in an effective style:

Avoid ostentatious and showy beginnings   Since essays are supposed to be written on serious issues of relevance and significance, the style chosen for these should also be sober and graceful. While writing an essay, we need to avoid ostentatious or shocking beginnings. Though we can use a statement which is arresting and interesting for the reader, a conversational or anecdotal opening is generally avoided in essays. Take a look at the following example and see how ridiculous the beginning seems for the essay on ragging in colleges.

> A friend of mine is scared today. He applied for admission to a college of repute in our city. Is he scared of the possibility of not getting a seat in the prestigious institute? Has he not managed admission there? Well, the answer is yes, he has! Then why should he be scared? It is so because the college is notorious for its reputation for ragging. The guys and girls getting into the first year of their degree programme are invariably in for a 'treat' and it is that 'treat' which gives my friend a scary feeling.

The beginning such as the above may suit a speech but does not go well with the tone required for an essay. Consider alternatively beginning the essay as suggested below:

> Though started as a healthy convention of introducing the juniors to their seniors in academic institutes, *ragging* today has come to mean something really unpleasant and undesirable. A by-product of modernization, ragging has come to convey constant harassment, pressure, and torture of juniors by their seniors in colleges, institutes, and universities. Every other day, we get to know some frightening story through some report in a newspaper or a cover story on a television news channel in which a youth is reported to have committed suicide following constant torture inflicted in the name of ragging by the seniors. So much so that ragging is now regarded as an offence that is punishable by law.

At times, openings are rendered ineffective not because they sound unreal and artificial but also because they sound too passive and ineffectual. Written in a laid-back manner, some of the essays start on a note that does not stimulate the imagination of the reader at all. Take a look at one such beginning and observe how dull it sounds:

> It is commonly believed that the right place of a woman is within the four walls of her house. There is a general perception in the society that a man should earn the livelihood for the family and the woman should look after the household affairs. It is suggested that mixing these roles leads to chaos and confusion in the society and leads to family discords, insecurity in children, unnecessary ego clashes between husband and wife, and also an increasing consumerism which is a direct offshoot of woman's economic independence.

A beginning such as the one cited above seems far from being impressive as the writer uses a series of laid-back expressions such as 'it is believed…', 'there is a general perception…', and 'it is suggested that…'. This makes the passage take off on a drab and passive note. Consider the following opening which can suitably begin the essay in a more emphatic manner:

> One of the important debates of our times has certainly been the rightful place of a woman. Ever since woman has stepped outside the proverbial *four walls* of her home, views and counter-views have started doing rounds on whether she is required to stay back, look after the household affair, and rear up children or play a more constructive and meaningful role by using her knowledge, education, and skills at the professional front as well. Traditionalists blame the economic freedom of woman for discords in the family, rise in consumerism and insecurity of children in families where both husband and wife engage themselves in a clash of egos. Before delving deep into these issues and supporting or discarding such views however, it would be appropriate to take a look at the shaping factors that have given rise to such changes in the society.

Use examples, statistics, and quotations sparingly   Though examples, quotations, and statistics are required to be used in essays, their use must not be over-exercised. It is so because essays are essentially lengthy discourses and are required to present the author's views on a subject. Therefore, rather than harping excessively on statistics, examples, and quotations, we must restrain our usage of such devices and give our perspectives on the issues. Nevertheless, using examples, statistics, and quotations sparingly helps the reader feel convinced about the views of the author. An essay without such supporting material would seem rather flimsy and too personal, whereas the one that primarily rests on these would appear unoriginal and borrowed.

Choose elaborate sentence structure   Since essays are indicative of a scholarly pursuit, they need to be written in an impressive style. Therefore, using a jerky and choppy sentence structure shows the writer of an essay in poor light. Take a look at the following extract of an essay on the importance of co-education.

> The strict puritans won't have it. They would believe their morality being questioned. Girls should not study with boys, they believe. It spoils both boys and girls. In their view concentrating in girls' presence is hard for boys. Similarly for girls learning without being nervous or scared, is not possible in boys' company. Hence those against it, oppose co-education. They believe studying together would lead to several problems. It would lead to illicit affair. It will lead to poor academic performance. It would lead to hasty marriages. It would lead to early exposure to sexual life. Therefore, fastidious puritans emphasize segregation of boys and girls during their school and higher education.

As can be observed, the style of the extract is far from being impressive. The writer uses a prose that stumbles, stops, and resumes every now and then. The jerky, choppy prose tends to sound not only repetitive but also immature and unpolished. See how the revised version of the same extract brings maturity and focus in expression:

> Puritans generally oppose the idea of co-education. In their view, educating boys and girls together leads to several problems such as distraction from studies, poor academic performance, premature exposure to the opposite sex, immature love affairs, and hasty love marriages. Fearing such offshoots of co-education, puritans suggest educating boys and girls separately during their schooling and higher education.

Choose common, familiar words   At times, essays tend to lose their appeal as writers choose to describe their views in a bombastic and pompous manner. Choice of long and unfamiliar words over simple everyday expressions leads to ambiguity in style and at times suggests artificiality

Reading pompous prose gives the reader an impression of vanity and unnecessary pretensions and fails to convince the reader of the writer's views. While writing an essay, we must bear in mind that a piece of writing that is not understood can hardly be appreciated. Therefore, there is no point in being pompous and artificial. Prefer writing familiar and common words instead of artificial, bombastic expressions. This builds immediate rapport with the reader and convinces him/her of the writer's views on a given subject. Take a look at how unusual expressions, circumlocution, and flowery style of writing creates a befuddling impact on the reader:

> Sartre once said that when two people interact, there should be subject-to-subject correspondence between them. Groping around, what one generally excavates is a subject and object relation that seems to characterize the nuances of their mutual relationship. It means that governed by an imperceptible emotion of ravaging hatred and revulsion coupled with a spree to dominate and coerce, man wishes to manoeuvre and manipulate all the time so that all others in his existential vicinity should seem sufficiently dwarfed and thereby he be declared a winner. All human growth, development, and achievement therefore seem to have been built on the debris of countless corpses not seemingly unlike him. The difference between the vanquished and the victorious seems to be minimum as all men venture into dislodging the others and getting themselves perched on the seat of success. Digging their graves thus would be unearthing their shrouded gambits—some camouflaged in grins of achievement and some others masked in smirks of failure.

The paragraph above makes a proper start by quoting Sartre's view. Soon afterwards however, it starts losing itself in the labyrinth of ideas as from one abstruse sentence we keep moving on to the other of the same type. The deliberate use of unfamiliar and uncommon words, and verbose and roundabout expressions render the content in the passage all the more confusing. A consciously carved style of this type seems belaboured and pompous affecting the overall understanding of the passage. See how the revised version of the same passage communicates the same idea with ease and clarity:

> Sartre once said that when two people interact, there should be subject-to-subject correspondence between them. What is observed in human relations however is a subject and object relation. It means that when two people share a relationship, they tend to dominate each other. It seems that because of a desire to dominate and control, we try to manipulate and manoeuvre others. This desire to coerce others can also be attributed to a deep-seated hatred or disliking for the others. In any case, in order to achieve greatness or victorious status, we want to crush many others into a vanquished lot. Seen thus, all our development and achievement seems to be a sad tale of fierce struggle among human beings that quite a few lose and only a few others win.

Avoid being unnecessarily equivocal or obscure   Obscurity is one of the avoidable traits of many an essay that writers of philosophical and metaphysical bent tend to write. Since philosophical indeterminism often leads to obscurity and confusion, an essay addressing such profound issues needs to be written in a style that is lucid and transparent. Even if the subject matter is profound and eludes direct interpretation, one needs to strive for clarity and lucidity in one's expressions. Writing in a deliberately profound manner and dropping unresolved equivocations and leaving the reader baffled with metaphysical overtones is likely to have an adverse effect on the readability of an essay. A large number of essays are not read by readers simply because they sound far too arcane, abstruse, and obscure in their choice of vocabulary, syntax, or sentence construction. Take a look at how obscurity affects the expression in the following extract from an essay:

> The deliberate, purposeful embroidering of God's immanence in man's life fills in the fathomless vacuum, the scorching sense of emptiness he feels himself being confronted with once the thought of insubstantiality of all his achievements, futility of all his endeavours, and incertitude of all his plans confronts him with all the darker, gloomy shades of reality. Snubbed into an existential predicament and assured of nothing else except the triviality and mortality of his own existence, man gropes for an assurance and permanence that the existence of his creator, God would ascribe to him. Deeply disquieted about his own sense of transience and a sloppy, ephemeral existence, it is in His presence that man seeks permanence and continuity of life. God is man's assurance to himself that even when he reduced to ashes or dust, he would continue to exist, if not here on this land of mortality then in the sweet continuity of heaven. That seems like a great snubbing of nature, the powerful but unruly, unintelligent daughter of God who most foolishly seeks to destroy her father's most loved and a chosen creature on this planet by drawing curtains on his physical existence. How else otherwise would you vouchsafe of pontificate such madness on the part of nature? If nature were not insane, she would have shown more reverence for the most gifted creature of God. Then she would not have destroyed man. She would have preserved him, as a representative species of God Himself. So an angry man, deeply annoyed at having been treated in the same way like any other creature, any other modest and mortal creature on the earth, decides to not only discard nature and cling to God but also pledges to destroy his destroyer, nature.

A write-up such as this is bound to leave a reader confused and bewildered. The writer expresses a complex idea in a style that is equally mystifying and confusing. Take a look at the revised version of the text to understand what the passage actually intends to communicate:

> It seems that it is a deep-seated insecurity of human beings that turns them against nature and sets them in their pursuit of devising ways that assign to them a sense of permanence and continuity. Aware of their short-lived existence, humans tend to seek immortality by creating an image of God which gives them a sense of assurance that even after death they will continue to exist in heaven.

Thus, by observing the principles of clarity, coherence, and consistency, we can write effective essays.

RECAPITULATION

✓ Essays that form an important part of competitive examinations can be written on a variety of issues such as social, political, religious, economic, academic, and other issues of general interest or significance.

✓ There are different types of essays such as argumentative, analytical, descriptive, expository, and reflective essays.

✓ Writing a good essay requires careful planning, selection of appropriate material, and in-depth research.

✓ A good essay has consistency, logic, and sequence of the idea explored by the author.

✓ A good essay expresses the comprehensive vision of the writer. Though it expresses the author's perspective on an issue, it is not expected to be lop-sided and unbalanced; it is shorn of ambiguities and equivocations.

✓ Structurally, an essay can be divided into three parts—introduction, main body, and the conclusion.

✓ While writing an essay, we need to start on an appropriate note, make the main body of the essay unified and consistent, and keep the conclusion short and effective.

✓ Since an essay is written on a matter of seminal importance, avoid flashy beginnings, leave out bombastic and obscure expressions, choose familiar and common expressions, and construct the matter in an elaborate yet lucid prose.

EXERCISES

I. Write an essay in about 300 words on each of the following topics:

1. Generation gap
2. Population explosion
3. Value of discipline in life
4. The great Indian dream
5. Indian cinema: An escape from reality?
6. Man vs machine
7. Global warming
8. The menace of corruption
9. Rituals and religion
10. Science and spirituality
11. Exodus from villages in India
12. The plight of slum dwellers
13. Can we achieve peace through atoms?
14. Has dowry system ended in India?
15. Do we need a revamp in our education system?
16. Are we losing our culture?
17. Is woman's lib just a fib?
18. Do we deserve democracy?
19. Should capital punishment be abolished?
20. Are elections free and fair in India?
21. Black is black, white is white
22. The child is the father of man
23. All that glitters is not gold
24. Reading habits of modern youth
25. Life in a hostel
26. Parenting
27. Creativity
28. Sense of humour
29. Success and failure
30. Uninvited guests
31. Technology is changing the way we think
32. Poaching of animals
33. Time management: a perennial problem
34. English in the digital era
35. The rise of Narendra Modi in Indian politics
36. Is fundamentalism on a rise all over the world?
37. Are we still a colonised nation?
38. Is media playing its role?
39. Journalism: art or business?
40. The big fat Indian marriages

Business Letters and Résumés

Learning Objectives   After reading this chapter, you will be able to

- understand the various elements of business letters
- learn the different layouts of a letter, such as indented layout, semi-block layout, and full block layout
- acquire various specific features of effective letter writing
- learn how to write different types of business letters

22.1 INTRODUCTION

A letter is the most ancient form of communication with those who are separated by distance. Correspondence has been playing a vital role in both social and business worlds. This helps the writer keep in touch with others. Though with the emergence of new technology and increasing use of emails and SMS, it is assumed that letters have become outdated, this is not the case. In fact, in the matters of high importance, letter writing continues to be a preferred tool of communication. Importantly, business letters create an impression about the organization and hence it is necessary to learn the art of writing good letters.

A *business letter* is a formal written document (as compared to a personal letter) through which companies try to correspond with their customers, suppliers, bankers, shareholders, and others. They are sometimes called *snail-mail* (in contrast to an email, which is faster). Business letters are written for various purposes such as informing, congratulating, requesting, ordering, enquiring, complaining, making an adjustment, applying for a job, and selling a product.

22.2 IMPORTANCE

Business letters are important for the following reasons:

1. Business letters help organizations in strengthening their rapport with customers, stakeholders, suppliers, etc.
2. They can be filed for future references and they serve as an important repository of information.
3. They help in conveying information that is confidential or complex.
4. Letters help companies reach the organizations, clients, shareholders, and others who are geographically in distant places.

5. They help companies to know the problems in their products, services, and deliveries. Thus, mutual exchange of information helps in filling the gaps and eventually helps in the smooth functioning of the organization and contributes in its growth.

22.3 ELEMENTS OF STRUCTURE

Since business letters are forms of formal writing, they are written in a distinct format. They have a margin of at least one inch on all four edges and are written on 8½″ × 11″ (or metric equivalent) unlined stationery.

Now, let us get familiar with different elements that appear in business letters and also learn their proper sequence of appearance.

Standard Elements	Additional Elements
Letterhead and date	Addressee notation
Inside address	Attention line
Salutation	Subject line
The body	Reference initials
Complimentary close	Enclosure notation
Signature block	Copy notation
	Mailing notation
	Postscript

Letterhead This contains the return address (usually two or three lines) with the organization's name, full address, email, telephone, and fax numbers. Sometimes, it may be necessary to include a line after the address and before the date for a phone number, fax number, email (also e-mail) address, or something similar.

Often a line is skipped between the address and date. This should always be done if the heading is next to the left margin. Letterheads are in fact designed with a lot of creativity and imagination, as they create the first glance brand image of the organization. Given below is a sample letterhead:

U and V Medicos Ltd
70-72 Vidya Nagar, New Delhi 111031

Phone: (0111) 2442460   Fax: (0111) 2442473   http//www.uandvmedicos.com

25 October 2010

However, when you write your job application letter, you cannot have the company's letterhead; you can write your address first and then the receiver's address with one line space in between. These addresses should be well aligned with the left hand margin as shown below:

Snigdha Mathews
245, Civil Lines
Mirja Ismail Road
Jaipur 302004

The Managing Director
Torrent Pharmaceutical Ltd
Ahmedabad – 380009

The date on which you are writing the letter should be mentioned in one of the following ways:

October 25, 2010 or 25 October 2010

The inside address   This is the address you are sending your letter to. Make it as complete as possible so that the letter reaches to the right person in time. Before writing the recipient's address, leave one line space. Include titles and names if you know them.

This is always written on the left margin. If an 8½″ × 11″ paper is folded in thirds to fit in a standard 9″ business envelope, the inside address can appear through the window in the envelope.

An inside address also helps the recipient route the letter properly. Skip another line after the inside address before the greeting. Now let us go through the examples given below:

Dr Judith Briganja
Head
Department of Biotechnology
Agricultural University, Hisar

Mr Deepak Gilhotra
Assistant Manager
Amul Dairy Products
Gujarat

While addressing a firm, 'Messrs' is used before the name and in this case 'The' is not used before the name. For example:

Messrs M.B. Sons
33/2, Cristal Palm
Tughlak Road, New Delhi

In case of 'limited/incorporated' company, it is better to write the receiver's name and designation and if you do not know the name of the person, it is best to write the designation of the officer because in these companies they require to have people in different positions. For example:

The General Manager
Shine Gems Limited
Jawahar Lal Nehru Marg
Meerut – 250006

Mr Pankaj Pitwal
The Chief Accounts Officer
Kitchenware Limited
Ashoka Marg
Mumbai – 400006

Attention   An attention line refers the letter to the person or department in charge of the situation covered. The word *Attention* is followed by the name of the individual or department. Do not abbreviate the word *Attention* or follow it with a colon.

The attention line is placed two spaces below the last line of the name and address of the addressee, either flush with the left margin of the letter or in the centre of the page. When paragraphs are indented, the attention line is placed in the centre of the page. For example:

The General Manager
Shine Gems Limited

Jawahar Lal Nehru Marg
Meerut – 250006

    Attention Mr Dilip Dewan

Salutation   This is nothing but greeting. The greeting in a business letter is always formal. It normally begins with the word 'Dear' and always includes the person's last name. For example,

Dear Professor Chakraborty

It normally has a title. Use a first name only if the title is unclear—for example, you are writing to someone named 'Steller' or 'Soumya' but do not know whether the person is male or female, it is better to address as

Dear Steller   or   Dear Soumya

And if your letter is addressed to a head of an organization or firm whose name is not known, it is advisable to use

Dear Sir/Madam

Nowadays, the salutation in a business letter usually ends in a colon.

The body   The body is written as text. A business letter is never handwritten. Depending on the letter style you choose, paragraphs may be indented. Regardless of the format, skip a line between paragraphs.

Skip a line between the greeting and the body and also between the body and the close.

The complimentary close   This short, polite closing ends with a comma. It is either at the left margin or its left edge is in the centre, depending on the business letter style that you use. It begins at the same column the heading does.

The block style is becoming more widely used because there is no indenting to bother with in the whole letter.

Signature line   Skip two lines (unless you have unusually wide or narrow lines) and type out the name to be signed. This customarily includes a middle initial, but does not have to. Women may indicate how they wish to be addressed by placing *Miss*, *Mrs*, *Ms* or similar title in parentheses before their name.

The signature line may include a second line for a title, if appropriate. The signature should start directly above the first letter of the signature line in the space between the close and the signature line. Use blue or black ink.

Identification initials   The initials of the typist appear left-justified two spaces below the signature block.

Enclosure notation   It is located with the identification initials or in place of them with the notation *enc., encl., enclosures* (3), or 3 *encs.*

Copy notation   Left justify two lines below identification initials with the notation *cc: full name* or *initials* or *designation of people* who are to get the copy of the letter.

Postscript   It is included two spaces below the last text on the page. It is written as *P.S.* and then a short sentence. Never use the postscript to add something that was forgotten during the writing of the letter. Instead, rewrite the letter.

22.4 LAYOUT

There are various layouts, using which information is given in letters. However, for our professional purpose let us know the major ones. They are as follows:

- Full block layout
- Semi-block layout
- Simplified layout

Full block layout Nowadays, this letter layout is the most popular layout because it is very attractive, easy to read, and very simple to draft. It saves time because indentation is not required. Therefore, in this chapter, this particular letter layout has been used in sample letters. While using this layout you should keep the following points in mind:

1. All elements except the letterhead are aligned to the left margin.
2. It follows open punctuation except in the cases of salutation (:) and message.

A full block layout will look like as follows:

Letter Head

Ref. No.
Date
Inside address

————
————

Sub:
Salutation
Main body

Complimentary close
Signature
Enclosure

Semi-block layout This layout has become outdated. The heading, complimentary close, and signature block are aligned vertically with the right margin. The rest of the elements are left aligned. Moreover, the salient feature of this style is that each paragraph of the message begins a few spaces away from the margin.

Simplified layout This layout follows the following principles:

1. Omits salutation
2. Often includes a subject line in capital letters
3. Omits complimentary close

This is often true that very few customers ever get to see the company office or a branch office. However, what customers see is correspondence issued or sent by the company or firm. An untidy or ungrammatical letter gives the immediate impression that the company's product or service is equally flawed. On the other hand, upon receiving a well-drafted, handsomely spaced, and well-organized letter, a customer unconsciously assumes the sender of the letter to be an up-to-date, well-organized, and successful business house.

It is an established fact that letter writing occupies at least one-third of all office work, and good letter writing may display your capability and establish your credibility. They help you work quickly and effectively with a growth in your career. To be able to achieve this, however, one needs an effective business letter writing style. The following section illustratively highlights some of the important aspects of an effective letter writing style.

22.5 BUSINESS LETTERS—ELEMENTS OF STYLE

Don't talk like a machine   Essentially, a business letter is more personal than a business report or a technical proposal. Since a letter is a communication between two individuals, it has elements of warmth and human touch that makes it especially interesting and reflective of the writer's personality. Therefore, writing a business letter mechanically can suggest the mechanical way in which things are carried out in an organization. Write in a style that makes the reader feel being addressed personally.

Incorrect Usage	Correct Usage
• The policy requires the person concerned to fill in and return the enclosed form within a fortnight to facilitate the processing at the company's end.	• We request you to fill in and return the enclosed form within a fortnight so that we can process it further.
• It is regretted that the goods sent by the company did not reach the buyers in time.	• We regret that the goods sent by us did not reach you in time.
• The firm would be glad to present a demonstration of the washing machine for the customer's benefit.	• We would be glad to present a demonstration of the washing machine at the time and venue convenient to you.
• Consequent upon our due consideration of your request, we regret to inform you that the termination of your said account cannot be forestalled.	• We have carefully considered your request but regret to inform you that the termination of the said account cannot be avoided.

Don't Write Letters in a Language that Others cannot Understand!

Display a 'You' approach   An important strategy to get your message across is to write the letter from the point of the view of the reader. At times, letters fail to communicate the message because the writers writing them seem obsessed with the idea of highlighting their priorities, choices, and predilections. Such a smug approach, however, harms the prospect of business communication as no one would like to read something which rather than helping the reader, helps the writer. Therefore, follow a reader-oriented approach as suggested in the following examples:

Incorrect Usage	Correct Usage
• We are glad we can now send our VCD players. We also look forward to continuing to receive orders from our customers in future as well.	• You will be glad to know that your VCD players are being sent shortly. We look forward to continuing to receive your orders in future as well.
• We are happy to receive your request for…	• Thank you very much for your enquiry/request/order for…
• Let us know whether we can extend Dr Chawla's services to your company. We are now informed of the latest advances in nano-technology. Dr Chawla's expertise enhances considerably the extent of our consulting services.	• Kindly let us know whether you require Dr Chawla's services. Dr Chawla is a well-known expert in the field of nanotechnology and is aware of the latest advances in this emerging field of study. We are confident that Dr Chawla's guidance and expertise can provide further impetus to your company's drive in this area of research.

Be courteous and considerate Since letters are essentially human in touch and appeal, it is expected that we show courtesy and consideration towards the reader. Some professionals do not pay much attention to this aspect in their communication and their letters sound quite drab, dry, or blunt. However, courtesy and consideration are an essential aspect of an effective communication and we need to express this in our business letters, as illustrated below:

Incorrect Usage	Correct Usage
• You have paid no attention to our complaint.	• Kindly look into our complaint. (OR) This is to remind you that our complaint is yet to be looked into at your end.
• We cannot grant you credit.	• We have carefully considered your application for a credit. However, as per our company policies, we are not in a position to grant you a credit of `…. However, you can use our short-term investment plan which can entitle you for a term loan at the end of the third year.
• We cannot do anything about your problem.	• We have carefully looked into the matter and wish to help you. However, owing to our other professional preoccupation, it is at present not possible for us to entertain your problem.
• Your application for the post of Purchase Manager has been rejected.	• This has reference to your application for the post of Purchase Manager in our company. Having carefully considered your profile, we regret to inform you that right now we don't have a suitable position that can utilize your qualification and experience. However, we have kept your application and curriculum vitae on our files and will inform you as soon as any such need arises in future.

Remember, courtesy and consideration lead to the goodwill of the organization. Hence, even if we have to write more to earn it, we should not see it as an expensive bargain. After all, it is better to be indirect if being direct suggests rudeness and lack of consideration to the reader.

Don't blame the reader   Business letters deal with a variety of professional situations. Many such situations are unpleasant and leave people fuming, grumbling, and complaining. As a golden rule, we should not dash off a letter when we are furious and agitated. Even after having received a letter that is full of accusations, complaints, and charges, we need to keep our cool. If we are to blame for any problem, openly and candidly confess your mistake. In case we are needlessly blamed, explain in a calm way the reader where the fault lies and how it can be avoided in future. In no way we should find faults with the reader or blame him/her for any mistake. Take a look at how by avoiding hurt, we can help ourselves as well as the reader:

Incorrect Usage	Correct Usage
• You have not mentioned the colour and design of the mobile phone set you wish to buy.	• A folder comprising all the designs and colours of the mobile phone sets produced by our company is being sent again. Please tick the colour and the design of the mobile phone set you wish to choose and mail it back to us. We'll immediately get back to you with our quotations as soon as we understand the specifications of your choice.
• This problem would not have occurred if the food processor had been operated according to the guidelines in the instruction manual sent to you with the product.	• We are sorry to learn that the plastic lid of your food processor has developed cracks. This problem usually occurs when the appliance is used without the lid being tightly fixed on the jar. Please refer to the Instruction Manual (pp. 33–34) to see how to fix all such accessories before using the appliance. This will help you avoid any future damage to the food processor.

Avoid being negative   Many a time, a good message loses its impact because the tone and verbal structure employed by the writer is negative. However, what matters in business letters is the fact that one needs to express positivity of attitude and mindset. Hence, avoiding a negative approach and replacing negative verbal structures with those which sound positive or neutral is certainly worth practising. Take a look at how the message can be communicated without sounding negative or dejected:

Incorrect Usage	Correct Usage
• None of your cheques have been received by us.	• We are yet to receive the payment for the 20 laptops sent to you earlier this month.
• Despite our repeated reminders, you have not yet looked into our complaint.	• We regret to inform you that despite repeated reminders, we are yet to hear anything about the complaint we registered with you. (OR) Please note that despite our several reminders, we are still waiting anxiously to hear from your end.

Be natural and precise   A very common problem with business letters is that many of them are written in a style that sounds artificial, unnatural, and full of technical jargon. An unnatural and artificial style leads to verbosity in expressions while jargon leads to ambiguity in the message. We need to understand that professionals have to deal with a large number of reading assignments and rather than receiving an unnecessarily long-winded, confusing, or verbose letter, they would appreciate a communication which is precise, natural, and effective. Therefore, avoid writing sentences which are characterized by jargon, redundant expressions, or artificiality of some sort. Take a look at the following sentences to see how unnaturalness, jargon, and verbosity spoil the impact of the message and how once devoid of these, the message shines up:

Incorrect Usage	Correct Usage
• I am sorry to have to point out that we do not have these goods in stock at the present moment of time.	• We are sorry to inform you that due to limited stock, we cannot honour your order for the next three weeks.
• It will be necessary for you to fill in and complete the enclosed form and return the same to us before we can proceed with a consideration of your request.	• Please fill in the enclosed form and return the same to us so that we can consider your request.
• I have immense pleasure in informing to you the fact that…	• You will be glad to know that…

Be simple and specific Writing and reading business letters involve time, money, and energy on the part of the professionals who are involved in the process. Any message that is indirect, vague, confusing, and complicated is, therefore, likely to vex the reader's nerves. Lack of clarity in a business letter sometimes leads to unnecessary confusion and controversy. At times, professionals are seen debating over or speculating about the actual meaning of the contents of a business letter. All this can easily be avoided by writing in a simple, direct, and specific style. Therefore, while writing a business letter, ensure that the letter is correct, complete, and clear in every respect. Take a look at how maintaining clarity and specificity in letters helps us achieve effectiveness in our business letters:

Incorrect Usage	Correct Usage
• The payment towards the said consignment will be returned in a short period of time.	• Thanks for sending the goods in time. The payment cheque will reach you by 25th November.
• Our company offers substantial discount on paying three consecutive loan instalments in time.	• Pay three consecutive instalments in time and avail our 3 per cent discount scheme.
• Once we have a communication from you to this effect, we will revert to you in a short span of time.	• We will get back to you within three days from hearing from you in this regard.
• A cheque containing the stipulated amount of money has already been dispatched to you.	• A cheque for ₹52000 has already been dispatched to you.

Carefully distinguish between 'I' and 'We' Personal pronouns (*I*, *we*, and *you*) are important in business letters. While writing business letters, it is perfectly appropriate to refer to yourself as *I* and to the reader as *you*. Be careful, however, when you use the pronoun *we* in a business letter that is written on company stationery, since it commits your company to what you have written. When stating your opinion, use *I*; when presenting company policy, use *we*. Take a look at how by maintaining a proper distinction between *I* and *we*, we can maintain clarity in business letters:

Incorrect Usage	Correct Usage
• I am targeting a growth of 20 per cent in customer base expansion.	• Our company is/we are targeting a 20 per cent growth in our customer base expansion.
• Since we don't make any special discounts to our overseas customers, we would suggest that you utilize a local warehouse facility.	• Since we do not offer any special discounts to overseas customers, I would suggest that you utilize a local warehouse facility.
• I am prepared to sign up the contract once the terms and conditions are clearly defined by you and me.	• We are prepared to sign the contract after the terms and conditions are mutually finalized.

Judiciously use the active and passive voice   One way to achieve a clear style is also to minimize your use of the passive voice. Often, it not only makes your writing dull but can also be ambiguous or overly impersonal. However, passive voice is sometimes necessary; particularly when the action is more important than the agent. Using passive voice also helps us take away the bluntness and crudity that crops up in a direct expression; particularly the one with a bad news or accusation. Thus, both the active and the passive voice have their own use and should be chosen carefully. When we want the message to be direct, clear, and precise, we should prefer active voice. However, when the message has a negative tinge to it and would sound blunt and rude in the active voice, we should use passive voice. Take a look at how the judicious use of active and passive voice can help us compose effective business letters.

Prefer using passive voice for expressions such as these:

Incorrect Usage	Correct Usage
• You have not yet informed me.	• I have not yet been informed.
• We have decided to stop trade with them.	• It has been decided to stop trade with them.
• The chairman conducted the meeting in a routine way.	• The meeting was conducted in a routine way.
• You have not yet paid the balance amount.	• The balance amount is yet to be paid by you.
• The manager announced that DA will be enhanced from the next financial year.	• It was announced that the DA will be enhanced from the next financial year.
• It seems you did not wrap the scenery properly before sending it to us.	• It seems that the scenery was not wrapped properly before being sent to us.

Prefer using active voice for expressions such as these:

Incorrect Usage	Correct Usage
• We have been thankfully in the receipt of your letter dated…	• Thank you very much for your letter dated…
• An outstanding growth rate has been achieved by our company in the last five years or so.	• Our company has achieved an outstanding growth rate in the last five years or so.
• It has already been proposed through our earlier communication that a canteen be installed in your premises to avoid absenteeism.	• We have already suggested opening a canteen in the office premises to avoid absenteeism.
• Some new schemes involving hefty incentives to our workforce are likely to be introduced by us very soon.	• We are likely to introduce some new schemes which will offer attractive incentives to our workforce.

Avoid using clichés and jargon   Using clichés and jargon, particularly in the beginning of a letter or at the end of it, makes the letter sound dull, routine, and repetitive. Since both the beginning and the ending of a letter require strong, positive, and emphatic structure, using some empty claptrap either in the beginning or towards the end of a letter is likely to make it sound weak and ineffective. Given below are some such expressions which need to be avoided:

1. The writer begs to acknowledge the receipt of…
2. The favour of your early reply will be appreciated by us.
3. Enclosed please find herewith…
4. Awaiting the favour of your early reply.
5. Assuring you of our best possible services and attention at all times.
6. I have the pleasure of informing you…
7. Please be good enough to advise us further on this.

22.6 TYPES OF BUSINESS LETTERS

There are different types of business letters, which are as follows:

- Acknowledgement letter
- Goodwill letter
- Letter of recommendation
- Credit and collection letter
- Appreciation letter
- Inquiry letter
- Sales letter
- Claim letter
- Request letter
- Adjustment letter

Let us take a look at the most common types of business letters.

22.6.1 Acknowledgement Letter

This type of letter is written when you want to acknowledge someone for his/her help or support when you were in trouble or you required that person's monetary help or guidance for the completion of some task. For example, as the convener of the conference you sought help from various departments of your organization and even outside the organization. An acknowledgement letter speaks volumes about the gratitude that you have. Thus, this letter can be used to thank for something you have received from someone, which has been of great help to you. In your professional life, you would require to write this type of letter quite often. An example of this type of letter follows:

Footsteps Incorporations Ltd
12, Winners Enclave, Jaipur
Phone: 0141-27354661-69; Fax: 0141-27354660

RJP/T/29

1 October 2014

Ms Sakshi Gupta
Event Manager
340, Nehru Place
New Delhi – 110019

Dear Ms Gupta

We write this to appreciate the committed support and help we received from you in organizing a three-day workshop on *Advertising: Possibilities and Perspectives* from 28–30 September, 2010, in Jaipur. Throughout the event, your suggestions and guidance helped us organize the workshop in a systematic and methodical way. Everybody appreciated your commitment, positive approach, and professionalism with which you managed the event. What particularly stood out was your cheerful disposition and team spirit that helped the other members in the team use their acumen and ingenuity in trying circumstances.

We look forward to more such associations in the time to come.

Thanks and regards

Yours sincerely

Alok Rastogi
Coordinator

22.6.2 Letter of Recommendation

This type of letter is written to recommend a person for a job position or admission in a higher degree or a specialized kind of study programme. The letter simply states the positive aspects of the applicant's personality, required skills, and how he/she would be an asset to the organization. Sometimes, a letter of recommendation is even used for promoting a person in an organization. An example of this type of letter follows. As already suggested, we normally structure letters according to the full block letters format. The body of the letter is given below:

Dear Sir

Thank you very much for your letter dated 2 March 2015. I welcome the opportunity to support Ms Aprajita Ghosh's application for the post of Assistant Marketing Officer in your organization.

Ms Aprajita Ghosh has been one of the most diligent and talented students of our college. Though her CGPA (9.7) itself speaks highly of her intelligence, there are other aspects of her personality that make her a distinguished student of this institute. Besides doing well exceptionally in her studies, Ms Ghosh has successfully developed herself as a versatile and multifaceted individual during her stay on the campus.

(Contd)

To enumerate a few of her outstanding achievements, Ms Ghosh has been the Student Coordinator for a creative cell Education Beyond Boundaries (EBB), which is formed to stimulate the creative imagination of students and faculty. Some of the common activities of the EBB have been presenting a research paper; discussing a movie, book, painting or any other work of art; commemorating a special day or event; organizing open poetry reading sessions; arranging group and panel discussions on seminal issues, etc. Being the Professor-in-Charge for EBB, I always found Ms Ghosh to be participative, communicative, and committed to the task. Besides this, she also has displayed her leadership qualities from time to time. Last year, she helped our department in the organization of an International Seminar by looking after the media reporting about the event. Not only this, Ms Ghosh is also the Sponsorship Coordinator for the TECHFEAST, an annual cultural programme of the institute that is organized independently by our students. A faculty colleague has recently informed me that Ms Ghosh has done exceedingly well in a study-oriented project on marketing strategies she undertook in his guidance.

In these aspects, I find Ms Ghosh's attitude, temperament, and sense of commitment not only satisfactory but also exemplary in certain respects. I thus, have great delight in recommending her to work for your organization. I am quite sure that upon her appointment, Ms Ghosh will perform her duties most diligently, efficiently, and reliably.

Yours sincerely

Dr Sudhir Saxena

Head, Management Dept

22.6.3 Appreciation Letter

All human beings feel good when they are appreciated for their good work. Realising its importance, an appreciation letter is written to appreciate someone's work in the organization. This type of letter is written by a superior to his/her junior. An organization can also write an appreciation letter to another organization, thanking the client for doing business with them. This type of correspondence certainly helps in strengthening the bond between two individuals or organizations. Read the example of this type of letter given below:

Dear Sir

Thank you very much for your letter dated 25 October 2014 enquiring about Mr Ajitender Chauhan's credentials, conduct, and performance as the Chief Security Officer in our organization.

It gives me pleasure in introducing Mr Ajitender Chauhan as a dedicated, hard working, and innovative professional who has admirably served our organization in the past four years. All these years Mr Chauhan has proved himself to be a very efficient security officer. When he joined our organization, the incidents of thefts had been on a rise. Besides this, the traffic rules were never implemented in the campus; helmets were not in vogue; and teenagers rode their bikes at a breakneck speed inside the campus. Mr Chauhan saw to it that the rules were not just framed but also followed punctiliously. Immediately after his joining, Mr Chauhan introduced the concept of check-posts in the campus area; he recruited security guards and put them on night patrolling. Mr Chauhan also suggested using modern technology to help the security develop a refreshingly robust look. Now we have latest security equipment in our security office; our guards communicate through a wireless network; they have enough vehicles and weapons to deal with any eventuality. Besides this, teenagers are no longer allowed to drive vehicles and the regulations regarding wearing helmets, maintaining speed

(Contd)

limit, using seat belts are also vigorously followed. An efficient twenty-four hour vigil has kept the thieves and other miscreants at bay. During his regime, no major theft, robbery, or accident have been reported in the campus. People living in the campus are a fearless lot today; so much so that they can at times leave their vehicles and houses unlocked. This speaks volumes about Mr Chauhan's vision, efficiency, and administrative acumen. Besides all these qualities, Mr Chauhan is a competent communicator and possesses good interpersonal skills.

I believe in having Mr Ajitender Chauhan as Coordinator, Security and Campus Maintenance, you will have a very competent, imaginative, communicative, and efficient person for your organization.

Yours truly

Manager (HR)

22.6.4 Acceptance Letter

Although you will often accept a job offer in person, it is a good gesture and a wise practice to formalize it with a letter. Begin your acceptance letter by thanking whoever has sent you the job offer, and then make it clear that you have decided to accept it. Express how much you look forward to filling this new position and mention one or two aspects of the job you will especially enjoy. The main text of such a letter is given below:

Dear Sir

Thank you very much for your letter dated 24 October 2014 offering me the post of Marketing Executive in your esteemed organization. I am indeed delighted in accepting this post on the terms stated in the appointment letter and confirm that I can commence my work from 1 November 2014.

I can assure you that I shall do everything I can to make my association with your organization productive, meaningful, and rewarding. Particularly I look forward to making a solid contribution to the project Vista 21st Century that should provide me an opportunity to use my ideas and expertise for the growth and development of the organization. Besides this, I also look forward to making a constructive contribution to the International Conference the organization is planning to organize in the month of October 2015.

I am earnestly looking forward to an exciting association in the time to come.

Yours sincerely

Anubhav Sood

22.6.5 Apology Letter

In the professional world, an apology letter is written for a failure in delivering the desired results. If the person has taken up a task and he/she fails to meet the target, then an apology is generally offered. This letter is also written if someone happens to have inflicted undue or undesirable inconvenience to somebody. This type of letter helps in patching or alleviating and thereby saving the writer from spoiling the relation. A sincere apology can go a long way towards winning back a lost professional linkage. A few occasions that warrant an apology letter are missing an appointment, missing a deadline, cancelling and postponing an appointment, causing inconvenience to someone, and owning up responsibility for someone else's offensive or unacceptable behaviour. While writing such a letter, we need to express our genuine sense of apology, express concern and unhappiness at the inconvenience thus caused to the reader, and promise to make up for the loss at the earliest opportunity. The main text of such a letter is given below:

Dear Sir

We are sorry to learn that the trainees sent to you from our consultancy firm for the summer training have not really been able to live up to the expectations of your organization. It is clear from your letter that they have not showed enough interest in completing the projects assigned to them. You have also mentioned that they have been casual in their outlook and have also been found wanting in terms of interpersonal skills and discipline. We deeply regret and apologize for letting you down in your hopes, aspirations, and expectations you might have had from our trainees. We have sincerely looked into the various factors that might have led to this unpleasant scenario and would like to share some of our observations with you in this regard.

It seems that the trainees undergoing summer training in your organization have been asked to work on projects which either do not interest them or do not match their profile. We have individually spoken to each of the trainees and have been given more or less the same feedback. Since the projects assigned to them do not stir their imagination or engage them sufficiently, almost all of them find the whole industrial exposure to be of little worth. A common feeling has also been lack of response from the staff of the organization that does not seem to involve our trainees in their professional assignments—something that might have them engaged and provided them an invaluable exposure to the nuances of the professional world. Lack of transportation and lodging facility at your end has also resulted in the absenteeism reported in your letter. As the group sent to you mostly comprises girls, they find it difficult to stay back in the late hours especially since they have to rely on a public transport to reach back their hostel which is a good 25 km away from the organization's campus.

Having said that, there is no denying the fact that once assigned a work or given a deadline, our trainees must have committed themselves to the task of living up to the reputation of our institute and the expectations of your organization to the best of their ability and calibre. They have been cautioned and have also been asked to seek an appointment with you and apologize for the lapse in the display of professional commitment and perseverance on their part. Most likely they will try and make up for the loss of prestige by working extra hard and meeting all the deadlines during the remaining period of their training at your end.

Thank you very much for giving an opportunity to improve our mechanism.

Yours sincerely

Asst Dean

Industrial Training Programme

22.6.6 Complaint Letter

A complaint letter is written to tell someone that an error has occurred and that needs to be corrected as soon as possible. In business world, there are numerous situations which warrant a complaint or claim letter as quite a few things go wrong several times. For instance, wrong billing of goods/services is done; wrong goods are dispatched; customers are overcharged for the goods sent to them; and at times they receive the goods in a damaged condition. Read some of the excerpts of complaint letters:

1. On 21 October 2014 we placed an order with your firm for 120 ultra-super long-life batteries of 1 KV and 60 of 5 KV. The consignment arrived yesterday but contained only 100 batteries of 1 KV and 6 of 5 KV.
2. The bedsheets that we ordered have been received through consignment number 206/1233. We regret to inform you that the texture, colour, and print are not as per the specification of our order.

While writing a letter of this type, maintain a poised and calm tone. Admittedly, we all feel angry when we have to suffer because of someone else's mistake. Even then there is no point in blaming, accusing, and being angry while drafting such a claim or complaint letter. Hence, the tone of complaint letters should not be aggressive or insulting, as this would annoy the reader and not encourage them to solve the problem. In addition, questions such as 'Why can't you get this right?' should not be included. Follow the following steps for writing a letter of this type:

1. Inform about the problem in a clear, precise way.
2. Refer to the order/invoice number to avoid ambiguity in communication.
3. Avoid being rude, angry, or humiliating in your tone.
4. Suggest a solution to help the other rectify the problem.

A text of such a letter is given below:

Dear Sir

This is to acknowledge the receipt of consignment containing 100 copies of Sidney Sheldon's *Best Laid Plans*, last evening. On its arrival apparently, something seemed amiss about the consignment as a large part of it seemed to have been drenched quite noticeably. With a view to retrieve the books in order as quickly as possible, our reception staff immediately opened the consignment. On opening the parcel, however, we realized that the damage was far more extensive than initially observed.

In fact, we have found most of the books to be in a bad shape. As many as 56 of them have been received in puffy and bloated condition. The remaining books too seem to have either lost their cover or a part of them comes off the moment you turn pages in them. Some of the books in the lot have developed cracks in between the page and a few of them seem to have been frayed in the corners. Of the entire lot, we could retrieve only ten books which we can keep in our book racks for sale without embarrassing ourselves while confronting our customers.

Since Sheldon's best seller has been in quite a demand recently, we were quite optimistic about the sale of the book in good number. However, since now we are left with only a handful of those copies which can be sold, it seems we will have to manage without our own best laid plans.

We are returning a parcel containing 90 copies of the Sheldon's *Best Laid Plans*. Kindly send another 100 copies of the novel as soon as possible. Please also send with the consignment 30 copies of Ken Follett's *The Third Twin*.

You are further requested to send the fresh invoice covering the cost of the new order after crediting our account with the invoiced value of the returned copies including reimbursement for the postage cost of ₹476 incurred in returning the damaged books to your store.

Keeping in view the vacation time that is just round the corner, we request you expedite the order at the earliest. While sending the parcel however, please ensure a safe arrival of the consignment.

We look forward to your early reply.

Vikram Sodhi

Manager

22.6.7 Adjustment Letter

Dealing with a complaint is usually not a very happy thing to do. Most of us feel bad at being found wanting in something. Professionalism, however, requires us to deal with complaints and claims in a manner that suggests maturity and clarity of approach. Viewed thus, receiving a complaint can be a blessing in disguise as it may help us review our services and products and bring about an improvement wherever required. A letter that deals with a complaint and claim letter is termed as an adjustment letter. While composing an adjustment letter, keep in mind the following points:

1. Acknowledge the complaint immediately. Remember that the person on the other side has already suffered; expecting him/her to wait any further for your reply could seriously jeopardize your relationship with the customer. Even if you cannot give a full reply, send an interim response assuring him/her further action at your earliest possible.

2. Generally, customers make complaints only when they cannot cope with the things at their end. Thus, handling complaints requires sympathy and consideration as the chances are that the customer is right in his/her claims.

3. Once you understand that the fault lies at your end, it would be graceful on your part to admit your fault, express regret, and promise to rectify the error.

4. Even if the complaint is baseless or the adjustment the claimant seeks is unreasonable, avoid being rude. Politely point out where the fault lies and suggest alternatives in an inoffensive manner.

5. Do not blame others to save your skin; after all by working in an organization you have to share the responsibility for the actions of a co-worker.

6. Thank the customer for bringing the matter to your knowledge.

The full text of an adjustment letter dealing with the complaint cited earlier is given below as a way of illustration:

Dear Mr Sodhi

We are sorry to learn from your letter dated 3 April 2014 about the difficulties you have faced in receiving the consignment containing Sheldon's *Best Laid Plans* in bad shape. This has caused us a great deal of concern and we are thankful to you for bringing the matter to our notice.

Having looked into the matter, we understand that the consignment containing books was damaged due to careless handling by the transporter. It seems that the said consignment was loaded in a lorry that had no awning to protect it from rains during transit. Sending our best laid plans haywire, it just rained when the vehicle was about to reach Jaipur. However, this hardly justifies the sense of complacency on the transporter's side that led to this inconvenience to you and financial loss to us.

We have acted swiftly on your complaint. The erring transporter has been blacklisted by our company and we are sending you a fresh consignment containing another 100 copies of Sheldon's *Best Laid Plans* and 30 copies of Ken Follett's *The Third Twin*.

Further, your account has already been credited with the invoiced value of the 90 books returned to us including the postage expenses incurred in returning us the earlier consignment. A fresh invoice towards the cost of 130 books now being sent is enclosed.

We again apologize for the inconvenience this has caused you and your customers and look forward to a continued association in the times to come.

Yours truly

Ashok Arora

Sr Sales Manager

22.6.8 Inquiry Letter

A letter of inquiry is written to enquire about a product or service. While writing an inquiry letter, keep in mind the following:

1. State clearly and precisely what information you require—a catalogue, some general information, samples, price lists, quotations, etc.
2. Ask about the time period the supplier is likely to take in facilitating the order.
3. Seek clarification regarding the mode of payment, discount offer, credit facility, if any.
4. Keep your inquiry brief and to the point.

At times, enquiries are also written to find out the status of an order already placed. Sometimes we may like to know when we will get our scholarship for the year, receive our original documents since we have paid all the instalments of our car/house, or when our thesis viva voce will be held. In all such enquiries, the matter written should be concise, complete, and yet brief.

22.6.9 Permission Letter

Permission letters are widely used to send and receive messages of request, or to seek or grant permission for utilizing the resources, facilities or services. In the business world, you may request the management to grant you the authority to do something or take decisions on authority's behalf. The amount of details you need to provide in a permission letter will depend on what you are seeking and from whom you are seeking it. Depending on the need, hierarchy, and relationship, the tone, the length and the language employed in a permission letter differ. For instance, if you ask a colleague for permission to share an article, the letter may be very short and informal. However, while requesting for loans, asking for facilities for conducting experiments, or seeking permission from some author for using his/her copyrighted material, the letter will have to be written elaborately and worded in a tone explaining the reason for asking for the specific favour. Some of the expressions commonly used while writing this type of a letter are, *Could you possibly...* or *I would be grateful if you could..., Kindly help us organize..., It would help us a long way in assessing..., We look forward to your kind approval in this regard...Will your organization be in a position to allow us...*, etc.

An example of a permission letter is given below:

Birla Institute of Technology and Science
Pilani - 333031 (Rajasthan) India

10 November 2014

The Dean - Administration
BITS Pilani
Rajasthan

Subject: Request for approval of travel expenses for the UGC major research project (Transforming Language Learning with New Technologies for Enhancing Employability of College Students from Rural Background)

Dear Sir

Since I, along with my project fellow, need to go to Delhi for giving a Mid-Term presentation for the above stated UGC project, I request you to approve travel (by air-conditioned taxi), stay and incidental expenses as per actual.

(Contd)

The expenses may kindly be approved under Travel/ Field Work for which I have received 37, 500/-.

I thank you and look forward to your consideration in this regard.

Sincerely,

Dr. Pushp Lata

A/C -3028

Associate Professor

Department of Humanities and Social Sciences

22.6.10 Invitation Letter

A letter of invitation is sent by an individual or an organization to invite someone to attend an event. An invitation letter can be formal or personal. A personal invitation letter is usually designed to invite someone to attend a social event, such as a birthday party, an engagement ceremony, a wedding party, an anniversary dinner, etc. On the other hand, a formal invitation letter is usually designed to fulfil some organizational or official interest.

The main text of such a letter is given below:

Dear Prof Pathak,

You will be delighted to know that we are organizing a two-day "ELT@I International Conference on Interfacing Language, Culture and Technology" (ELT@I Rajasthan) on 8-9 October 2014 at MNIT, Jaipur. The conference would provide a suitable platform to the academic professionals and research scholars to initiate and facilitate discussions among the academicians and the educational planners on the existing practices and emerging challenges in English language teaching.

We wish to invite you as the Chief Guest for the valedictory function of the conference. It would be an honour and privilege to have you with us. We shall provide you airfare for economy class from Allahabad to Jaipur and back. A university car will be arranged to receive you at the Jaipur airport. We shall also be glad to care of your hospitality and comfortable stay in Jaipur.

The brochure providing details of the conference is attached for your perusal. Kindly accord your consent at the earliest so that the preparation may be made accordingly at our end. Needless to say that your Valedictory Address will inspire hosts of academicians actively devoted to ELT.

For further details of the Conference, please visit our website: http://discovery.mnitjaipur.ac.in/ELTAI

We earnestly look forward to your inspirational presence on this occasion.

With Warm regards.

22.6.11 Rejection Letter

Rejection letters are written for declining somebody's request. It could be declining a donation, turning down a request for some favour or facility, or for denying a candidate a job offer. Not all the companies or organizations write rejection letters to candidates who fail to sail through the recruitment process. However, by writing one, we can make a difference. It leaves us in favourable light, mollifies the unpleasantness or sharpness of refusal, and also eliminates the follow-up enquiries from the anxious candidates.

A text of such a letter is given below:

I take this opportunity to thank you for the interest you have shown for the post of Customer Care Executive. However, we sincerely regret that we cannot offer you employment with our organization at this time.

We would like to keep your application in our files for a period of 120 days in case a suitable position falls vacant. In the event of an appropriate available position, we will give your application a strong consideration. If your address or contact number changes within this period of time, kindly inform us about that.

if you wish to obtain more detailed feedback regarding your application, you may contact us through email or telephone.

Thank you once again for applying.

Wishing you all the best for your future endeavours,

22.6.12 Order Letter

This letter, as the name suggests, is used for ordering products. This letter can be used as a legal document to show the transaction between the customer and the vendor. While sending an order letter, accuracy and clarity needs to be ensured. An order letter needs to include the following points:

- An accurate and full description of the goods required
- Quantities of goods required
- Prices already agreed on between the customer and the supplier
- Specifications of the goods ordered
- Details regarding delivery requirements—place, date, mode of transport
- Details regarding terms of payment already agreed on
- A concluding sentence urging the suppliers to send the material urgently/well in time

22.6.13 Application Letters and Cover Letters

It is a popular misconception that application letters and cover letters are essentially the same. However, both these kinds of letters are different. The letter of application is just like a sales letter in which you market your skills, abilities, and knowledge. On the other hand, a cover letter is primarily a document of transmittal as it identifies that an item is being sent. It includes the information like the person to whom it is being sent and the reason for its being sent. It, in fact, serves a permanent record of the transmittal for both the writer and the reader.

Application letters Whenever you write an application letter for a job or internship, you have to compete with all the other candidates who apply for that job position. The receiver of the letter or your audience is a professional who screens and hires job applicants. This recruiter looks through hundreds of other applications. Therefore, the immediate objective of your application letter and accompanying résumé is to attract this person's attention.

Since your immediate goal is to obtain an interview by writing your application letter and résumé, it is better to keep the following points in mind:

1. Catch the reader's attention favourably.
2. Convince the reader that you are a qualified candidate for the job.
3. Make a request for an interview.

The job application letter should include the following elements of information:

1. Mention the job position for which you wish to apply and let the recipient know how you came to know about it.
2. Sum up your qualifications for the job specifically talking about your work experience, activities that show your leadership skills, and your educational background.
3. To call attention to your strengths as a candidate, state your objective directly at the beginning of the letter.
4. Include the information that is not included in your résumé but is pertinent to the job.
5. Give references of people who can support your credentials and qualification for the post.
6. Emphasize the qualification that the prospective employer would like to seek in you.

See the following example:

> I am seeking a position as a manager in your centre. In such a management position, I can use my master's degree in Information Systems and my experience as a Programmer/Analyst to address business challenges in data processing.

If you are interested in the post that has been advertised, you are required to refer to the advertisement right in the beginning of your letter. For example:

> With reference to your advertisement No. HT/Rect/10/22 dated 7 May 2014 in *The Times of India*, I wish to offer my candidature for the position of Sales Manager in your company.
> (OR)
> I was interested to see your advertisement in today's *Hindustan Times* inviting application for the post of Systems Analyst in your reputed organization. I wish to be considered for this position.

If you have been referred to a company by one of its employees, a career counsellor, a professor, or someone else, mention that before stating your job objective. For example:

> During the recent conference on Artificial Intelligence held at IIT, Roorkee, one of your research scholars, Pooja Shastri, informed me of a possible opening for a manager in your data centre. My extensive background in programming and my master's degree in Information Systems make me highly qualified for the position.

At times, we apply to an organization hoping for a possible opening in an organization. While drafting such unsolicited applications, we need to be persuasive in approach, clear in our objective, and convincing in our expression.

Given below is the full text of an unsolicited job application letter:

> Dear Sir
>
> For the past five years I have been working as a Media Officer with Skylet Pvt. Ltd, New Delhi. I am now looking for a change of employment which would widen my horizon, utilize my creative potential to

(Contd)

the full, and at the same time improve my prospects. It has occurred to me that a growing and promising organization such as yours might be interested in using my calibre and services.

After my masters in Journalism and Media Studies from JNU, New Delhi, I pursued my PhD from BHU, Banaras, and was awarded the doctorate degree for my research work and thesis in the field of Advertising. Since then I have been associated with Skylet Pvt. Ltd, New Delhi, and currently head the cell that drafts advertisements for professional organization. I particularly love providing to an advertisement the socio-linguistic hue that makes it strike a chord with the viewer immediately. In the past, we have designed and marketed some of the most popular, catchy, and decent advertisements. Quite a number of the advertisements designed by us have had a successful run on the small screen, in newspapers, and magazines. A list all such advertisements is provided in my curriculum vitae, which is enclosed for your perusal.

At a personal level, I get along with people quite comfortably and thoroughly enjoy living up to the tough challenge of being creative, innovative, and yet methodical in meeting the deadlines. Presently, I head a group of 8 talented people and they all seem to enjoy working with me. I have provided references of people who are well aware of my expertise and behaviour in the professional world.

I shall be pleased to provide any further information you may need and look forward to be given an opportunity for a personal interaction.

Yours truly

Vijay Kajla

22.6.14 Sales Letters

Of all the types of business letters, sales letters are the most interesting and unique in their approach and appeal. Sales letters are written to advertise and promote a product. A good sales letter is able to achieve the following objectives.

Catching the reader's attention   The most immediate purpose of a sales letter is to capture the attention of the reader. For this, the beginning of the letter should be so captivating that the reader should not be able to put it down without reading. The beginning can be made catchy with the help of a quotation, by telling an anecdote, by asking an intriguing and interesting question, by making an appeal to the reader's vanity, pride, comfort, health, and economy, or by using statistics that startle the reader.

Creating a desire   Having aroused the interest of the reader, the next part of the sales letter strives to sustain it. For this the letter has to point the benefits, features, and advantages of the product. In order to gain favour from the reader, we should be able to stress the highlights of the product from the prospective customer's point of view. Instead of making an exaggerated claim, a good sales letter focuses on the outstanding features of the product and creates a desire in the reader to buy it.

Carrying conviction   Having created a desire in the reader's heart for the product, the next step is to convince him/her of the authenticity of our claims. A good sales letter achieves it in a number of ways—by arranging free demonstration for the potential customer; by providing a guarantee; by making the reader read through the favourable comments and statements by other users of the product; by showing statistics in favour of the product; by enclosing literature that presents the product in favourable light, etc. By carrying conviction thus, this part of the sales letter legitimatizes the claims made earlier about the features and performance of the product.

Inducing action   The closing paragraph of a sales letter is designed to persuade the reader to take the action. The desired action could vary—we may want our reader to call our sales branch, ask for a demonstration, call the reception to know further about the product, or send an order. To achieve this, the final part of the letter too needs to be strong enough. Generally, in this part of the sales letter we are required to make the offer tempting by making special offers, by attaching easy-to-fill-in-pro forma or tear-off slips or by facilitating action at the end of the potential customer.

Thus, a sales letter creates the impact that is achieved by an effective advertisement. The outcome of a good sales letter is the same as that of a good advertisement—both of them convert their target audience into potential customers. Since the purpose of a sales letter is to persuade the reader to buy their product, the style of a sales letter should be persuasive and emphatic.

An example of a sales letter is cited below:

Dear Sir

Are you tired of your old printer that has acted up right from the moment it was installed? Haven't you had enough of your printer's jamming the paper, rumpling the fresh sheets of white paper into unrecognizable lump, and creating a low rumble before it obliges to print a paper at a pace that embarrasses even the proverbial tortoise? The IMM Technologies bring to you its new product—Prifax 6400. Prifax 6400 is the new arrival in the IMM family, India, and has already been successfully selling in the USA, Germany, and France.

Prifax 6400 has a unique inbuilt technology of printing and faxing. Yes—both simultaneously by one machine! For years together your Prifax 6400 works for you. It operates even when the power is low, the paper is crumpled, the number of copies to be produced is in hundreds, or the time at your disposal is minimum. Prifax 6400 has been designed keeping in mind the emphasis in corporate sector and academic institutions on speed, reliability, cost saving, and virtually no wastage of sheets. This printer-cum-fax machine has a laser 6400 pixel printer technology, and is combined with 32 MB for buffer memory and inbuilt colour to black and white printing and back-to-back printing technology. That ensures an additional cost saving of the cartridge besides carrying out the printing task at a staggering pace. Prifax 6400 not only handles the printing job but also handles the fax job efficiently with the same reliability and neatness. This is easily done as it is directly connected to your phone line with an internal buffer of 120 GB for the fax. It is the only printer-cum-fax machine available in the market which not only gives you a package of printer–fax but also has an additional feature of being connected to a satellite phone.

Prifax 6400 comes in three bright colours: black, white, and grey with a superior metallic finishing body and a power backup inbuilt mode. This machine is available at an unbeatable price of ₹7200 with an introductory discount of ₹500 and a free-service-for-a-year scheme in the presentation of this coupon/letter on the sales desk or a reply of this letter with the coupon enclosed. So place your order today and enjoy the facility of both printer and a fax machine for the price of one.

Yours truly

Amitabh Srivastava

Sales Manager

22.7 RÉSUMÉ PREPARATION

In the process of getting a job, résumés play a very vital role. A good résumé can get us short-listed and a bad one can lead to the rejection of our claim to a position. Therefore, preparing a résumé is an important skill that we need to develop at an early stage. Preparing a good résumé or curriculum vitae (CV) requires good imagination, creativity, ingenuity, and consistent effort in keeping it updated, comprehensive, and effective. Following are some of the most crucial elements of a résumé in brief:

- Appearance
- Personal information
- Career/Professional objective
- Education/Academic qualification
- Work experience/Professional skills
- Activities and achievements/Special interests and aptitudes
- Awards and honours
- Memberships
- References

22.7.1 Types of Résumé

There are various types of résumés that you can prepare while applying for a job. However, the selection of the type will be determined by the kind of job profile the company offers and the kind of skills and work experience you have. These types offer you *different ways to organize your details* you want your employer to know.

Chronological   A chronological résumé gives your work history in the chronological order, that is, in the sequence of occurrence. Nowadays, a reverse chronological order is in practice, since employers prefer this type as they get to know what you are doing now and where and when you have worked in different organizations.

Functional   This type of résumé mainly focuses on your skills and experience. If you keep changing your career quite often and there are gaps in your employment history, it is advisable to prepare this type of résumé.

Hybrid/Combination   Mostly this type of résumé is prepared. In this type, you list your skills and experience history first and work/employment history next. While drafting your résumé, you need to be strategic since you can highlight the skills that are relevant for that particular job profile.

Based on the kind of channel used, résumés are of two types:

- Paper-copy/Traditional print résumés
- Electronic/Scannable résumés

Paper-copy/Traditional print   As many offices are going paperless, résumé tradition has also changed. More recruiters are looking for electronic résumés rather than paper ones. But some smaller companies still want a hard copy/traditional print résumé, so it is important to know how to get your résumé through the door. While drafting your résumé, keep the following points in mind.

1. Keep the résumé short as recruiters do not have time to go through a bulky one.
2. Take care of font size and adequate margin.
3. Use one type of heading—serif or sans serif.

Serif and non-serif are the two important typeset designs which are available under 'font' in your computer. Serifs are the small finishing strokes on the ends of characters, whereas sans serif fonts do not have these finishing strokes (Fig. 22.1).

While sending a paper-copy résumé, keep the following tips in mind:

1. Keep the copies crisp, clean, and dark.
2. While drafting the résumé keep busy readers in mind.
3. Tabular form is easily readable and helps the reader form a quick opinion.
4. While editing do triple check.

Fig. 22.1 Examples of Serif and Sans Serif Letters

Electronic/Scannable résumés An electronic résumé is called a scannable résumé. It is a plain text either in ASCII (American Standard Code for Information Interchange) or in HTML (Hyper Text Mark-up Language) document. It is submitted along with a job application. The applicant should use key words to provide the recruiter or employer information regarding his/her key skills, work experience, qualification, etc. You must know that many companies process all their incoming résumés electronically by using an automated applicant tracking system, that is, ATS. As and when the company wants employees for them, they use data mining technique to search through the data base created for this purpose. Résumés are scanned by the Optical Character Recognition (OCR) software. In this process, the résumés that match the maximum score for the relevant keywords and desired skills are printed and the candidates are called for an interview.

Tips for scannable/electronic résumés

1. Use standard fonts which have distinct letters.
2. Use various techniques to draw the attention of the recruiters. Some of these are boldface, capitalization, indentation, etc.
3. Make it computer friendly.
4. Do not use underlining and fancy scripts.
5. Never use any lines, boxes, or graphics.
6. Provide important information in the beginning and use technical words reflecting your core competencies and skills.
7. Use industry buzzwords and common technical acronyms.
8. Use standard font size (10–12) and avoid columns.
9. Don't worry about length.
10. Use descriptive nouns and noun phrases (surveyor, programmer, manager, six years' experience).
11. Post a high-quality print quality without folding the paper.

Non-traditional Résumés Due to the scarcity of time, organizations sometimes require non-traditional résumés. The purpose is to go through the profile of the candidate quickly to ascertain his/her suitability for the available position. Such résumés are prepared keeping in view the need and focus of the organization. While preparing non-traditional résumés, we should remember the following points:

- Leave out detailed descriptions of our past accomplishments and responsibilities.
- Focus on revealing the qualifications, skills and capabilities related to the position advertised.
- Let your résumés be communicative, easy to handle and crisp.

Let's discuss some of the non-traditional résumés:

Video Résumé In video résumés, the prospective candidate has to speak in front of the camera, highlighting his/her qualifications, abilities, skills and accomplishments. This requires meticulous planning with the text, use of expressions, tone, pitch and pauses. In order to do it well, we also need to be aware of the concepts such as lighting, framing, editing and scripting.

Visual CV At times enterprising job hunters create their own websites to augment their credentials and achievements. However, one needs to be crisp and catchy while designing such websites. A website with easy operative access through minimum links along with strong visual and linguistic appeal may help us create good impact on prospective recruiters, while a clumsy and longwinded website may disorient and discourage a prospective employer from approaching you.

In the past few years, Prezi, the free, online presentation tool has become so popular for the creation of online résumés that a new term – Prezumes has been assigned to it. This online template helps people design and format their résumés in an attractive manner. If used sensibly, this online tool can help job seekers showcase their skills, abilities and qualifications in an effective way.

Regardless of the method, tool or model used to draft a résumé, it must bring out effectively a candidate's qualifications, skills, and abilities to perform a specific job.

22.7.2 Important Features of a Selling Résumé

Here are some important features of a selling résumé:

1. It creates crucial first impression.
2. A selling résumé will always have catchy appearance and contents.
3. It is well-organized, properly written, and presented with an apt layout.
4. It is free of errors.
5. Its purpose is to persuade that you have abilities, skills, and personal qualities that the employer is looking for.
6. Both hard copy and scanable résumé could be attractive and serve the purpose, provided the details are presented well.
7. A good résumé is always accompanied by a well-drafted cover letter.

Résumé contents Now let us know the contents of a résumé. They are as follows:

Identification

- Name
- Phone
- Address
- Email ID

Career objective You should always draft a career objective for yourself based on your skills and professional aspirations. Do not copy and paste what others have written in their résumés. Given below are three career objectives and a comparison is given for your better understanding.

Flawed Seeking employment in a business environment offering an opportunity for my professional growth and aspirations

Good Achieving excellence as a computer programmer

Better To market financial planning programme and provide financial counselling to assure positive client relations

Education

Degree	University/College	Year	CGPA/Division

Employment/work experience   Always provide in reverse chronological order.

Student Assistant

University of …….. 2000–present prepared and processed ……….

Sales Associate

Or

Give in a tabular form

S. No.	Designation	Company	Pay scale

Professional skills (Related course work) technical knowledge

Managed a retail design studio producing over …….

Hired, trained, and supervised …….

Provided training ………

Professional affiliations (membership, etc.)

Membership of various professional organizations/societies, etc.

Activities and interests

Co-curricular (brief)

Any other special interests

Awards/Honours/Achievements

Academic/non-academic

Professional

References

2 from university/earlier organization

1 from reputed person

Sample Résumé I

Ganesha Nilayam,
#9, III Street, South Sector,
Adambakkam, Chennai – 600088
Phone 044-22444386
raj2004@yahoo.co.in

Rajagopal Vaideeswaran

Education

- Graduation
  Institution: Birla Institute of Technology and Science (BITS) Pilani
  Degree: M Sc (Tech) Information Systems
  Year of Graduation: 2004
  CGPA: 8.31 (Till 6th semester)

(Contd)

(Contd)

- XII
  School: G.K. Shetty Hindu Vidyalaya
  Board: Tamil Nadu State Board
  Year of Passing: 2000
  Percentage: 96.67%
- X
  School: G.K. Shetty Hindu Vidyalaya
  Board: Tamil Nadu State Board
  Year of Passing: 1998
  Percentage: 90%

Software Skills

LANGUAGES KNOWN	:	C, Java, Perl.
SCRIPTING	:	Shell programming (Unix), HTML, XML
OPERATING SYSTEMS	:	Unix, Linux, Windows 95/98
ASSEMBLY	:	MASM (8086)
DATABASES	:	SQL (Oracle), PL/SQL, MySQL

Projects Completed

- PS-1 Project Title: Website Development Using Flash
  Description: It involved development of an online demo on the working of fire for Wels Secutrons Ltd
  Team size: 4
- Title: Design and Implementation of File System for an Experimental Operating System
  Description: In this project we developed a file system emulation of DOS, which could perform the functionalities of Creation, Read, Write, Modify, and Delete operations on Files. We used 8086 Assembly programming using MASM (Microsoft Assembler).
  Team size: 2
- Title: Paint program in 8086
  Description: I simulated the MS Paint using 8086 Assembly Programming using MASM.
- Title: Developing an Automated Library Management System
  Description: This project was done as a part of the course Object Oriented Programming and automatic Library Management was done using JDBC Programming using Java as front end and MS Access as back end.
  Team Size: 3

Currently Doing

- Title: Application of Bayesian Networks in Bioinformatics
  Description: This project aims at reconstructing Phylogenetic Networks from Phylogenetic Trees using Bayesian Networks.
  Team Size: 2
- Title: Design and Development of Desktop Utilities for BITS Linux Operating System.
  Description: This project aims at developing new desktop utilities such as lockscreen, switch user, shortcut keys, pseudo user profile, etc., in GNOME desktop by tweaking GNOME source code, for the BITS distribution of Linux named 'BITS LINUX'
  Team Size: 3

Electives Completed

- Real Time Systems
- Data Communications and Networking

(Contd)

(Contd)

Currently Doing

- Machine Learning
- Introduction to Bioinformatics
- Effective Public Speaking

Awards and Achievements

- Successfully completed National Himalayan Trekking expedition in 1999 as a part of Duke of Edinburgh Scheme arranged by Youth Hostels Association of India
- Won first place in Oratorical Contest held during National Science Day Celebrations 1999
- Won Merit Position in Painting Competition in All India Schools Festival organized by United Schools Organisation of India in 1997
- Won Certificate of Honour (Gold Medal) for painting Young Envoys International in INTRART – 1993

Personal Information

Name	:	Rajagopal Vaideeswaran
Father's name	:	P.R. Vaideeswaran
Date of Birth	:	01-01-1983
Sex	:	Male
Marital status	:	Unmarried
Nationality	:	Indian
Languages	:	English, Tamil, Hindi

Extracurricular Activities

- School Pupil Leader in the year 1999–2000
- Conducted programming competitions during APOGEE 2003.
- Zonal Runners-up in Table Tennis Zonal Championship
- Active Blood Donor

I declare that the above mentioned details are true to the best of my knowledge and belief.

(V. RAJAGOPAL)

Sample Résumé II

Form for on-campus placement

Company code: ☐ ☐ ☐

Jaipur Institute of Technology & Science, Jhotwara, Jaipur (Rajasthan)

1. (a) Name: _____

    (b) ID.No.: ☐ ☐ ☐ ☐ ☐ ☐ ☐ ☐ ☐

2. Degree Programme: _____

3. (a) Date of Birth: _____ (b) Age: years _____ (c) Sex (M/F): _____
    (dd/mm/yy)

4. JITS Hostel Address: (a) Hostel: _____

    (b) Room No.: _____

    (c) Hostel Ph one No.: 0141 _____

5. Email ID: (a) JITS Email ID:_____

    (b) Alternate Email ID: _____

6. Permanent Address: _____

(*Contd*)

7. Educational Qualification:

Degree/Examination	Board/University	Year of Passing	CGPA (Max. 10.00) (or Marks %)
12th Standard			
Integrated First Degree or its equivalent: I Year			
II Year			
III Year			
IV Year			
V Year			
Higher Degree: I Semester	JITS, Jaipur		
II Semester			
III Semester			

8. Month and year of completing the present programme: _____
9. Details of Practice Training (PT)/Thesis/Dissertation:
   PT I at:
   PT I Project Title:
   PT II at:
   PT II Project Title:
   Thesis/Dissertation Title:
10. Projects completed/currently doing:
11. Elective courses: (a) completed   (b) currently registered
12. Extra curricular activities:
13. Any other relevant information:

Date: _____   _____
                                                                        Signature

RECAPITULATION

✓ Business letters form an important part of professional communication and are written to deal with a large number of business situations, such as enquiries, complaints, claims, promotions, advertisements, reminders, payment collections, acknowledgements, appreciations, apologies, recommendations, or applications for jobs, tender, or contracts.

✓ The structure of a business letter is more or less defined—it has a reference number, a date line, the inside address, the subject, the body of the letter, the complimentary close, the name of the signatory with designation, enclosure list and notations regarding copies to be circulated.

✓ Business letters are messengers of the organization. Therefore, they need to be written in a style that is clear, cordial, warm, reader-oriented, and professionally appropriate.

✓ For composing different types of business letters, different strategies are to be adopted so that the letter serves the intended purpose and is not misunderstood.

✓ Résumés should necessarily include details such as career objective, key skills, work experience, academic qualification, honours, awards and distinctions, and reference of professionals who can support the credentials of the person applying for the job.

WISEWELL QUIPS

EXERCISES

Inventing the necessary details, write the following letters in full block form:

1. Assuming yourself to be the Purchase Officer of Aradhana Opticals, 24, Vijay Marg, Patna, write a complaint to Ageless Glasses, 121, Paharganj, New Delhi, reporting that the three of the six consignments containing glasses have been received in a damaged condition. Ask for the replacement of the damaged goods and seek compensation for the postage charges incurred.

2. Assume that as the Sales Officer, Cozy Mattresses, Ajmer, you have received a complaint from a local dealer complaining that the two dozen mattresses sent to them have serious defects. Write an adjustment letter refusing or accepting the claim. Provide suitable details for your acceptance or refusal of the claim.

3. Assume that as the head of the department, you have received a request for writing a recommendation letter for a former student of yours who intends to pursue his postgraduate degree from a university in Australia. Your recommendation letter should highlight the student's strengths, achievements, and suitability for the course he intends to pursue.

4. Imagine that your company Gracious Foods Enterprises, Pune, has decided to enter the catering business and is planning to open its outlets in some of the major cities of the country such as Delhi, Mumbai, Calcutta, Pune, Chennai, Bengaluru, Jaipur, Chandigarh, Lucknow, etc. Assuming yourself to be the Marketing Head of the company, prepare a sales and promotion letter to be sent to the public to promote and publicize the company's food outlets.

5. Assuming that your organization has recently conducted a week-long training programme for newly recruited junior level managers. The training programme was aimed at improving the communication and interpersonal skills of the young personnel. The programme included expert lectures by resource persons including Prof. Ajay Kashyap of the Institute of English Language and Communication Studies, New Delhi. Assuming yourself to be the

Coordinator of the programme, draft a letter of acknowledgement appreciating the cooperation and expertise received from the resource persons during the programme.

6. As the Head of the Production Unit of New Age Vision, Jaipur, you have received a complaint from the Graam Sevak of Ladanpur village—a remote hamlet in Rajasthan—who has complained about the unruly, rude, and offensive manner in which some of the crew members of your company conducted themselves during their stay in the village while shooting for a documentary entitled 'Vision Village: Vision India'. Draft an apology letter expressing your regret and assuring action from your side.

7. Assuming yourself to be the Controller, Software Operations, prepare an appreciation letter for Mr Sunil Bhatti who has been working in your company as a Software Engineer. Mr Bhatti has applied for the position of Senior Software Engineer in Relics Solutions, Bengaluru and they have written to you to vouchsafe Mr Bhatti's credentials. Draft an appropriate response to this effect.

8. Assuming that you have the required qualifications and skills for the posts advertised, draft job application letters in response to the following classifieds:

(a) World Megasoft Pvt. Ltd, Mumbai, seeks talented and enthusiastic Software Engineers and Software Architects for its new branch office in Gurgaon. The candidates having 3–5 years of experience possessing expertise in SDP IN Services, Provisioning, Charging/Billing, Fraud Management, Revenue Assurance, NMS, SIP, IMS, IPTV may apply. The candidates are expected to have required competence in Java, XML, RMI, C++, Visual Basic, Cold Fusion, BO, Crystal Reports, Webfocus, Oracle, DBA, PL/SQL, and Unix/Solaris. The candidates must have excellent communication abilities and outstanding customer-facing skills along with software architecture. Salary and perks in commensuration with the qualification, experience, knowledge, and skills of the candidates. To view the detailed job descriptions and submitting your on-line application, please log on to http:/www.megasoft. com/in/megasoft/recruitments within ten days.

(b) Urgently wanted Competent Programmers at AhimsaCircle.com PHP Programmers for its Jodhpur office. The programmers recruited would be responsible for developing multiple bespoke e-commerce website using Magento, Joomla, Drupal. Candidates must possess either of the following qualifications: a graduate in B.Tech/B.E. Computers BCA, PG—M.Tech—Computers/MCA. Besides the necessary technical skills, the candidates must possess excellent communication and interpersonal skills. Salary no bar for deserving candidates. To apply for the following positions, please register on the company's website and then upload your résumé and covering letter with your complete profile.

(c) Zeniture Finances Ltd—a company in import/export, financial services, and development—seeks application from young, energetic, qualified persons for the following positions: Front Office Executives, Marketing Sales Executives, Import/Export Executives, Receptionist. Incumbents must possess good command of English language and have a pleasant disposition besides good academic record and educational qualification up to graduation/postgraduation, preferably from commerce/management stream. Interested candidates may submit their applications along with their detailed résumés on the company's website http://www.zeniturefina.com

9. With winter approaching, you intend to purchase electric heaters for your office at Safdarjang Road, New Delhi. Assuming yourself to be the Maintenance Officer of Ubiquitor Technologies Pvt. Ltd, New Delhi, write a letter of enquiry to Westworld Appliances and Equipment, Mumbai, seeking information about the availability and price of the product.

10. Write a formal letter of invitation to the MLA of your constituency requesting him/her to be the chief guest on the inaugural function of the newly formed cooperative society of your locality. Assume yourself to be the secretary of the cooperative society. Invent necessary details.

11. Global Electronics, Gurgaon, Haryana, has recently launched a new film projector. Assuming yourself to be the Sales Manager of this company, write a sales

letter to the principals of all the colleges in your town to promote the new product. Point out its features and facilities.

12. Imagine yourself to be the Sales Manager of High Tech Action Computers, Bengaluru. Your company has recently launched a new low cost laptop 003 CP in the market. In order to promote the sale of this model, draft a sales letter to be sent to colleges, universities, and other organizations.

13. Innumerable mobiles have inundated the Indian markets; however, the land phones have their own advantages. Keeping in mind the competition with the mobiles, BSNL has launched land phones with extra facilities so as to cater to the requirements of the potential customers. Now as a Sales Manager of BSNL, write a sales letter in consonance with the AIDA technique.

14. Zap International is launching a new mobile set with latest configurations next month. Assuming yourself to be the Area Sales Manager, write a sales letter inventing necessary details to be sent to all the stockists/wholesalers of your area for promoting its sale.

15. Catmoss Electronics ordered for twenty-five computers from HCL. But on arrival of the consignment, the purchase manager found complaints in at least twelve of them while verifying their quality. In order to keep its reputation, the manager wants to ensure that everything is in order.
    (a) As Purchase Manager of Catmoss Electronics write a complaint letter to HCL, Bombay, suggesting the adjustment you seek.
    (b) As Sales Manager of HCL, write the reply for the complaint letter expressing regret and showing your readiness to do the needful to retain the customers.

16. Keeping in view the elements of effective letter writing style, rewrite the following expressions and extracts taken from various business letters.
    (a) We simply cannot entertain your request for the replacement of electric geysers supplied to you three months back.
    (b) As suggested earlier, our company is likely to offer substantial discount to those who return their investment files within a certain limit of time.
    (c) Enclosed please find herewith the required report for your information and necessary action.
    (d) We beg to request your kind attention to the fact that had washing machine been operated as per the instructions provided in the instructions manual sent to you at the time of delivery, the occurrence of the event could have been forestalled.
    (e) The company takes great delight in informing all their shareholders that a decision to the effect of offering an additional dividend has been evaluated, considered, finalized, and approved of in the last meeting of the company's board of governors.
    (f) With reference to your application for the post of Production Manager (Milk Products), please note that we have evaluated your competence, skills, knowledge, and temperament and have found you wanting in most of those. So, we regret to decline to accede to the offer of your services to be utilized at our end at the present moment of time.
    (g) Consequent upon being in the receipt of a numerous complaints for a host of our regular customers, it is to be brought to your notice that the highlighters sent to our stationery store have performed abysmally, much of course to our surprise, as the product sent this time is nowhere comparable to the kind of quality your reputed company has prided itself in offering to its clientele.
    (h) Since we are not responsible for the loss or damage of goods during transit, we are not to blame for whatever went wrong with the consignment sent to you and hence cannot accept your claim.
    (i) With reference to your application No. PRTC/1208/1336 dated 28-09-2009, please note that it has been rejected by our company.
    (j) Our product is simply the best in the market. It is absolutely immaculate and ultimate in its performance. After all, the other food processors in the market are a pale imitation of our innovations and just as the copy can never replace the original in any other sphere, in terms of providing you kitchen appliances, no other company can take the crown away from us. So simply choose the best and forget the rest.
    (k) We beg to state that we are interested in getting waterproof garments for our delivery boys who deliver food door to door on order without

stopping even in rainy seasons and not stopping even in cities that are known for their endless rains. Hence, we wish to know whether you can supply us 100 pairs of garments for our boys so that they can carry out orders without getting drenched in the rain whenever required.

(l) When we opened the consignment, we got the shock of our life. The entire crockery items seemed to have been reduced to rubble. There were broken bits and pieces of what were once the handsome glasses, jars, cups, plates, and trays—all shattered beyond a hope for any redemption or recovery. We are very sure that your packing staff seemed to have done an awful job. We are returning the consignment and request you to take your packing staff to the task besides returning our compensations.

(m) When we heard from your boss that a young guy would be sent for managing the event for us, we felt really disappointed and were not looking forward to your assistance or expertise in arranging our seminar. In fact, I had asked my subordinates not to expect anything from a young chap. However, with your admirable communication skills and excellent managerial qualities, you not only challenged but also changed our perspectives. The way you organized the event really impressed us. We really are now thankful to you for all your contribution in making the event a huge success and would request your expertise as and when an occasion so warrants.

(n) We are extremely delighted to have Atul Varshney as our Quality Control Adviser. In the past one decade or so, Mr Varshney has added admirably to the standards of our products. Mr Varshney possesses excellent communication skills and his expertise and acumen are simply incomparable. We are quite sure that by asking to utilize Mr Varhsney's expertise and services, your company has taken a definite stride forward in ensuring excellence in quality control of your product.

(o) We are dismayed in learning from you that trainees sent for improving their communication skills and sent to you for a communication development programme displayed a perfunctory and casual attitude towards improving their communication skills. It certainly is a matter of grave concern to us and we have forwarded your performance report to our senior management for further necessary action. The trainees had been sent because we sensed they needed improvement in their communication skills. But what can you do with those people who are least inclined to learn anything in their life? Anyhow, it seems the management is likely to take a firm action in this regard. Thanks for letting us know the truth that really matters in such situations.

ANNEXURE 22.1

Checklist for a Business Letter

Having drafted a business letter, answer the following questions before sending it to the reader:

Q.1. Is the letter written in a courteous, polite, and warm manner? Yes ( ) No ( )
Q.2. Does it have all the information required by the prospective reader for taking the desired action? Yes ( ) No ( )
Q.3. Does it have a proper reference no., date, and subject line? Yes ( ) No ( )
Q.4. Does the letter begin with a proper salutation? Yes ( ) No ( )
Q.5. Does the letter have an appropriate tone suited to the occasion and subject matter? Yes ( ) No ( )
Q.6. Does the letter end on an appropriate complimentary close? Yes ( ) No ( )
Q.7. Does the layout of the letter consistently maintain the block, semi-block, or fully block form? Yes ( ) No ( )
Q.8. Is the letter divided in different paragraphs for taking up different ideas with the reader? Yes ( ) No ( )
Q.9. Does the letter give the complete information about the company's name, address, contact numbers, website, etc., in its header at the top? Yes ( ) No ( )
Q.10. Does the letter leave on the reader a good impression about you and your organization? Yes ( ) No ( )

Business Reports

Learning Objectives   After reading this chapter, you will be able to

- know what a report is and how important it is in the professional arena
- understand various aspects of a business report including its features and different types
- learn how to collect data required for reports
- learn how to write various types of reports effectively
- understand the structure of a business report
- discern and develop an effective style for writing reports

23.1 INTRODUCTION

Report is an important form of business communication. Generating and analysing reports is a routine task in every professional's day-to-day activities. It is in fact hard to think of an organization where, as a professional, you would not be required to write some sort of a report. Whenever there is a decline in production or sales, frequent strikes in a company or a fire breaks out in the factory, the authorities ask for reports which consist of the data related to the problem, its interpretation, and the findings arising out of such an analysis. Reports are thus written to analyse a situation, to offer an alternative method of operation, to study the growth rate of a company, to observe the trends in socio-political-psychological changes happening around us in all walks of life and so on and so forth.

Since reports acquire such inevitability in the professional world, it is mandatory for us to understand in detail how they are written, what their structure is, and what makes them effective. However, first of all, let us understand the definition, meaning, features, and significance of reports in professional situations.

23.2 DEFINITION

In general terms, a report is an account of or a statement about something that happened in the past. The word 'report' has originated from the Latin word 'reportare'. Etymologically 'report' means 'to carry back' because 're' means *back* and 'portare' means *to carry*. In other words, it is a description of some event or situation that has already happened. Since the very nature of the report requires it to be written or talked about, a report is a document that is written to be carried back to someone who requires it. The comprehensive definition could be as follows:

A business report is a formal communication written for a specific purpose, conveying authentic information to a well-defined audience in a completely impartial and objective manner. Written in a conventional or usable form, it describes the procedures followed in the collection and examination of data, analyses the facts collected, derives conclusions from them, and gives recommendations, if necessary.

23.3 SALIENT FEATURES

On the basis of the above-stated definition, we can easily trace the characteristic features of a business report, as follows.

A formal piece of writing A report is a formal piece of writing. It is not a document where one expresses his/her ideas and feelings freely the way they come to him/her. A report is essentially written in accordance with certain rules and norms. The facts and ideas are recorded, analysed, and sequenced in a particular way. The elements to be inducted in a report follow a certain pattern.

A factual account A report is a factual account of data or information. Essentially, every report is a collection of data for the intended reader(s) who will make efficient use of it. The facts contained in a report may be an account of something that has already happened or something latest, an account of any new information, any plan for a course of action, etc. Facts should always be accurate and complete and arranged in a way to project clear meaning.

Written with a specific purpose A report is always written with a specific purpose. In fact, it originates with a need, desire, or purpose, either to inform or to analyse. These are always written to help the intended reader(s) to keep track of information or to take important decisions or actions.

Written in an organized manner Since reports are based on facts, they have an organized structure. Generally reports follow a conventional or usable form. Hence, while writing a report, a proper planning and presentation of data is quite important.

Written for a specific audience To keep abreast of current information, authorities need data or information. They demand the relevant information from a person who has this information. Thus, the person who generates a report is aware of its primary audience. Since the subject matter is related to the reader, it is interpreted for his/her awareness and future use. It helps the authority to take a sound decision, to find solutions to the existing problems, and remain ahead of others. There are various types of reports, but each report is audience specific.

Written in an objective manner Reports are always written in an objective manner when a collection of facts is to be communicated. Not many shades of meaning may be bestowed to the report since it is written in an impartial and objective style. Data is analysed in relation to the problem. The facts are presented the way they are; no scope is left for personal evaluation.

Includes only relevant information A report includes only essential information. Therefore, redundant information is not generally a part of its structure. It consists of the information that helps readers save their time and make them understand what exactly they require to understand. In other words, reports give readers exactly what they want—neither an iota more nor an iota less.

23.4 SIGNIFICANCE

As already suggested, reports are an indispensable part of professional communication. Broadly speaking, reports help professionals achieve the following objectives:

1. Reports help professionals plan, acquire, execute, organize, coordinate, manage, and evaluate business activities in an effective way.
2. Reports facilitate the flow of information to ensure smooth execution of tasks so as to meet the challenges successfully.
3. Sometimes they serve as a record of facts where information is organised and recorded for the readers' benefit. Thus, reports also serve as a repository of information.
4. Reports enable the authorities to take timely decisions. They may also be used for further analysis.
5. They can be helpful in creating awareness among shareholders and other investors when reports are sent to them regarding the market position of the company from time to time.

23.5 TYPES

Different types of reports are written for different purposes. Figure 23.1 clearly shows the classification of reports.

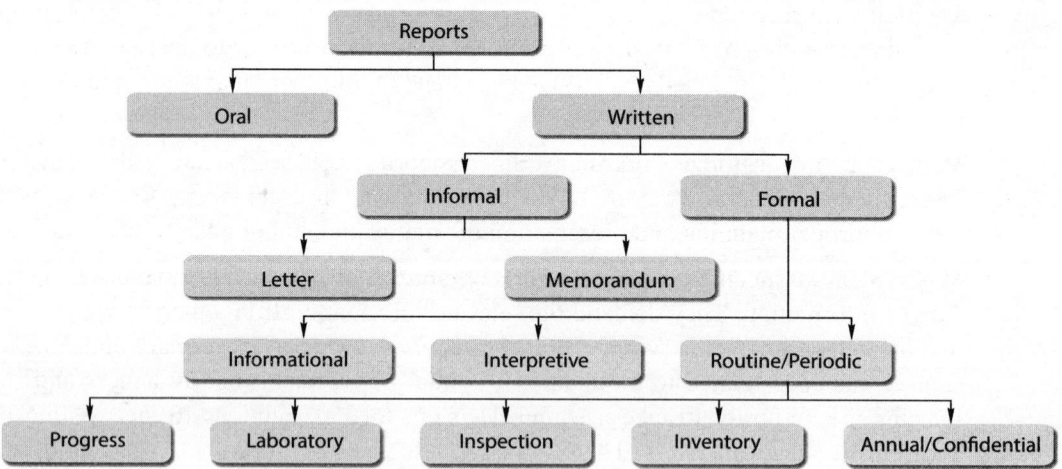

Fig. 23.1 Classification of Reports

Mostly reports are written and presented to the authorities but at times, due to paucity of time, the writer of the report is asked to present the whole report orally in a nutshell. Going further on this, let us understand the major differences between the two types of reports:

Difference between oral and written reports
The obvious distinction is that the oral report is spoken and you can imply and suggest by using nuances of your voice, whereas written reports are devoid of such insinuations.

The table given below briefly sums up the difference between oral and written reports:

Table 23.1 Oral vs Written Reports

Oral Reports	Written Reports
• Spoken	• Written
• Presented face to face	• Not necessarily
• Easy for the speaker since preparation is already done and difficult for listener	• Easy for the reader to take his/her own time in grasping the facts but difficult for the writer since he/she has to gather facts, analyse them, and draw conclusions
• Ephemeral in nature	• Permanent record of information
• Immediate clarification is possible	• In the absence of face-to-face communication, immediate clarification is not possible
• Less accurate and reliable	• More accurate and reliable
• Informal	• Formal

After discussing the differences between oral and written reports, let us look at other classifications of reports based on their degree of formality—formal and informal reports.

23.5.1 Informal Reports

The ultimate purpose of an informal report is to fulfil an immediate requirement which does not require an elaborate analysis or presentation. So, these are shorter than formal reports. For instance, if a managing director wishes to know the current status of production or performance of employees in a particular department, the information can be presented in an informal report. Informal reports are also written by using the same format as that of a letter or a memorandum. The following part discusses these two types of reports in detail.

Letter reports   The business letter format is very important for communicating formally in or outside an organization. Whether it is the findings from a major R&D project or a list of individuals recommended for merit bonuses, or an explanatory report for a legal proceeding, writing a report in the letter format is a regular task. Take a look at the layout of a letter report:

Letter Head

Date

Inside address

————————

————————

Sub:

Salutation

Main body

   • **Introduction:** Purpose, context, background

- **Findings**: Supporting text with topic headings, analysis, illustrations, etc.
- **Conclusion**: Major findings and expected action.

Complimentary close

Signature

Enclosure

@ Refer to the Online Resource Centre for some tips for writing letter reports.

Memo reports   Memo is also called an inter-office memorandum. It is a prescribed form which is used to send important information within an organization. It is used to convey information regarding routine business matters, such as sending information from one department to another, announcing a change in policy matters, intimating the minor administrative changes done, etc. A report that deals with a minor problem or provides facts of routine nature, using this inter-office memorandum format, is called a *memo report*.

A memo report should give an account of what has been done and is required to describe the important findings and also their significance. Since it is meant for circulation within the organization, it is more informal in style as compared to letter reports. As far as the length of the report is concerned, it is generally not more than three typed pages. Moreover, the main body of the report consists of a few relevant headings such as findings, recommendations, etc. Look at the format of a memo report.

Groomwell Finishing School, New Delhi

To:

From:

Date:

Subject:

Introductory paragraph: reference, authorization, objective

Main text: such as findings or details of conference, significance

Concluding Paragraph: formal closing

Signature

@ Refer to the Online Resource Centre for sample letter and memo reports. Also read 23.15 Report, Sample I.

Differences between letter report and memo report   Though letter and memo reports are both informal in style, there are a few major differences between the two, which are given in Table 23.2.

Table 23.2 Letter Report vs Memo Report

Criteria	Letter Report	Memo Report
Size	Longer than memo, i.e., comprises 4–5 pages	Shorter than letter report, i.e., usually one or two pages
Format	Uses letter layout	Uses inter-office memorandum format

(Contd)

(Contd)

Criteria	Letter Report	Memo Report
Style	Less informal	More informal
Purpose	Both for internal and external communication	Only for internal communication
Content	Letter reports provide more details, arguments, evidences and possibly one important and one relevant illustration.	Memo reports are precise and specific

Though informal reports such as memo and letter reports are written quite occasionally, it is the formal reports which are more frequently used and play a crucial role in professional situations. Let us now discuss the formal reports.

23.5.2 Formal Reports

Formal reports have the seriousness of purpose and content presenting elaborate data which cannot be informally presented. There are various types of formal reports. The pattern of these reports is set according to established procedures or conventions. They are broadly classified under three categories:

1. Routine/Periodic reports   2. Informational reports   3. Interpretive reports

Routine/periodic reports   Routine reports are also known as *periodic reports*. These reports are submitted annually, quarterly, monthly, weekly, or at any other prescribed intervals in the routine of business affairs. Usually, in these reports, some columns are given in a pro forma where some ticks are to be put or if anything is to be written, it is very briefly written. Such reports present the collected data and facts in their original form and sometimes they include brief recommendations. For example, the confidential reports on employees may include whether an employee should be promoted or not; a periodic report on the progress of project may include a brief recommendation for more funds or an extension of the duration of a project, etc. Routine reports can be classified into the following five types:

- Progress report
- Laboratory report
- Inspection report
- Inventory report
- Annual confidential report on employees

Progress report   A progress report informs the reader about the status or the progress of a particular project undertaken by an organization during a specific period. It gives an account of the various stages of the project in chronological order along with the details of the work completed till date and the amount of work yet to be completed. If the task is investigative, then simple information of what you did is not sufficient; you are also expected to summarize whatever you have found out. It is appropriate that you evaluate the work completed to determine the progress of the project (we are ahead or behind the schedule; we are under or over budget). It may also make a mention of special problems that may arise during the course of work progress. These reports are also written on the construction work of a factory building or dams for water supply schemes or on the construction of a huge institute premises, etc. Read 23.15 Report Sample II.

Laboratory report   Laboratory reports are those reports which are written by scientists and students of engineering, science, psychology, biology, and those who work regularly in laboratories.

These scientists and students record their experiments in a pro forma mentioning various details related to the experiment. This provides a step-by-step account of the process to be followed in completing a certain task. If the process includes the operation of mechanical devices such as computers or laboratory equipment, then these devices are usually described before the process is explained. As a matter of fact, various findings are to be put in a logical order. This also consists of a brief summary of the findings so that it becomes comprehensible when others read it or wish to carry out further research in the same field. This pro forma necessarily includes the name of the experiment, the apparatus used, the procedure followed, and the findings and conclusions. Read the sample report III from 23.15.

Inspection report   Inspection reports are the routine reports that are compiled only after a thorough investigation of objects or products. These types of reports are used by organizations either (a) to see whether the product is functioning properly or needs some repair, or (b) to see whether the quality of the piece of equipment is up to the standard or not. This report helps in checking the quality of products in a systematic manner and ensures the right standard and smooth functioning of equipment. Such reports are normally submitted to the executive heads of the maintenance, production, or sales departments. A report of this type is written in a prescribed pro forma where there are several columns indicating the options regarding the quality and operation of each and every part of a particular product. The purpose of such reports is to keep a check on the mishandling and poor production of equipment. Read the sample report IV from 23.15.

Inventory report   Often this type of report is computer-generated and requires the user to put in the relevant data regarding the stock on hand. These reports are submitted at regular intervals, may be weekly or monthly or annually on the amounts and kinds of items in stock, stock-out ratios and projected needs and order dates for supplies. These reports include statistical details including the number, amount, and type of material required. Thus, these reports keep an account of the material available, its consumption, proper maintenance, etc. These are written by the person who checks the stocks and fills in all the details in a prescribed form and signs it. Read the given sample report V from 23.15.

Annual confidential report of employees   As the name suggests, these reports are submitted annually by the controlling officers about their subordinates. It evaluates their work performance and behaviour in their respective departments. For example, the senior executives of production, sales, marketing, and quality control departments submit a report on the performance of their subordinates annually. That is the reason why all the related officials keep a record of the performance and conduct of their employees. These reports determine a professional's appraisal. Important decisions such as promotion, demotion, transfer, or termination of a contract are normally based on the study of such reports. It is not easy to prepare the pro forma of this report because there are certain human attributes which are hard to assess. Therefore, in this report, sometimes questions are included and some space is given for providing short explanation. Read the given sample report VII from 23.15.

Informational reports   The term 'informational' is generic for any report whose primary purpose is to convey information. It entails all the details related to the subject under discussion. These reports develop an understanding of the aims, objectives, organization, policies, regulations, procedures, problems, and future outlook of a company. Here the data collected is presented in an organized form and the situation is presented as it is and not as it should be. These reports mainly convey information so as to serve various purposes such as making a discussion, determining a course of action and coordinating the operation of the organization. In this type of report, neither data analysis is done nor are recommendations provided. Attention should be given to present the material in a proper way. Each part should prepare the reader for what is to come, and the discussion of one point should be complete before another point is taken up. In any institute or college, a report on how many students got registered in various disciplines would be an informational report unless it analyses the question 'why so?'.

Thus, for writing an effective informative report, what you have to do is to collect data, arrange it in a proper order, and present it in a manner that is suitable for business communication.

Interpretive reports Interpretive reports are also known as *analytical* or *investigative reports*. These types of reports help the readers analyse, interpret, and evaluate facts and ideas. An interpretive report differs from an informational report, as it analyses and interprets the data obtained and arrives at some conclusions and recommendations and hence leads the reader to some course of action. The major emphasis is on the analysis of the results of an investigation or the proposed solution to the problem. The report writer analyses and interprets the data in such a way that it naturally leads to some suggestions which may be of great importance when implemented. The writers of such reports try their level best to present their findings and recommendations in the most convincing and persuasive way. The report submitted by the director or secretary to the shareholders is an example of an interpretive report.

An interpretive report that abounds in recommendations is also at times termed as *recommendation/recommendatory report*. Interpretive reports, owing to their nature, are more expansive and elaborate. These reports vary widely in scope and subject matter, but they are always associated with some sort of business activities such as personnel, marketing, sales, accounting, advertising, productions, equipment, plant location and distribution; and invariably look for solutions to a problem or suggestions for improving the existing conditions.

Thus, an interpretive report is a piece of a formal business communication which helps the organization in taking timely decisions to solve a problem, to launch a new product, to reduce the gap and create awareness among its employees as well as shareholders.

23.6 USE OF GRAPHIC AIDS/ILLUSTRATIONS

Effective business report writers stress on the layout of reports because a neatly furnished report not only enhances the appearance but also enables them to communicate effectively. Since most of the reports include complex, voluminous data, you can tell the story better by including graphic aids (e.g., tables, charts, diagrams, etc.). Graphic aids supplement the text, help you communicate the report content, give emphasis to key points of the coverage, and make the report more interesting and readable. Above all, they help you give it a professional flair. Therefore, it is advisable to know the answer to some of the following relevant questions with regard to the use of graphic aids/illustrations:

How to use graphic aids/illustrations?

- Illustrations should be neat, accurate, and self-contained.
- Contents should be closely related to the text.
- They should be explained and placed as close to the first reference as possible.
- Size of the illustration should be big enough to be clearly visible.
- If you photocopy or directly copy from another source, you must give proper credit to your source.
- Illustrations should be numbered and captioned as suggested below:

    1. Tables–Roman numerals at the top   2. Figures–Arabic numerals at the bottom

Keep the following tips in mind while incorporating graphics into a report:

1. Place the graphic aids as close to the related text as possible, usually following the paragraph in which it is introduced.
2. Always refer to a graphic aid in the report text (e.g., 'Figure 1 illustrates').
3. Give a title and number for tables and charts. Tables are usually numbered separately; the term *figure* may be used for graphic aids other than tables (e.g., Table I, Figure 1, Figure 2, Table II, Figure 3). Or, a numbering scheme identifying the type of figure may be used (Table I, Chart 1, Map 1, etc.).
4. Table numbers and their captions (titles) are given at the top of the tables in capital Roman numerals, whereas figure numbers are given at the bottom of the figure along with the caption. Look at the examples given below.
5. Graphics are usually included in the 'Discussion' section of the report. Generally, you require to include your results graphically as well as in written form.

Table II
Creative Writing Mid Semester Grades
I Semester 2009–2010

Grade	Male Students	Female Students	Total
A	7	10	17
B	27	27	54
C	43	46	89
D	13	16	29
F	5	6	11
Total	95	105	200

Note: Based on Simulated Data

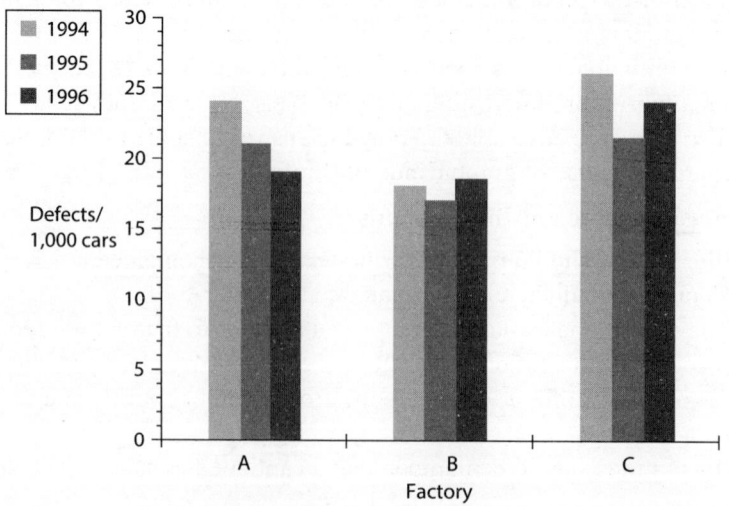

Fig. 23.1 Comparative Bar Graph Showing Defects in Cars in A, B, C Factories

Now let us look at the different types of graphic aids that can be used in reports.

Figure 23.2 shows the various types of graphics.

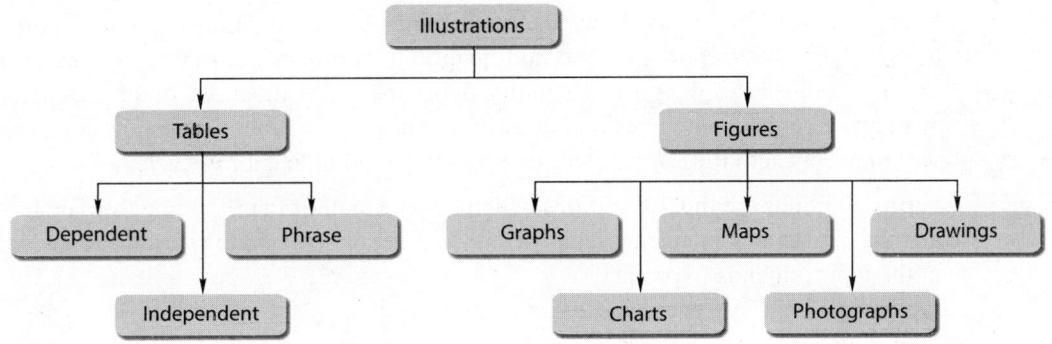

Fig. 23.2 Various types of Graphics

Let us look at an example each of independent table and phrase table.

Independent table Independent table is the one that includes rows and columns giving the complete statistical information related to the issue under discussion. Independent tables are used for helping the reader understand the trend or pattern without necessarily going through the preceding or the following text. Table 23.3 gives an example of an independent table.

Table 23.3 Trend of Admissions in Various Academic Programmes at Oxbridge College 2005–2015

S.No.	Programme	Year	
		2005	**2015**
1.	B.A	110	100
2.	B.Sc.	35	80
3.	B.Com	70	90
4.	Journalism	20	50

Dependent table Dependent table is the one which is closely associated with the text preceding and following it. It cannot be interpreted effectively independent of the text.

Phrase table Phrase table is a table which consists of rows and columns and instead of giving data in figures, provides information by using words and phrases. Table 23.4 gives an example of a phrase table.

Photographs Sometimes you may be required to provide a photograph of a machine or site or building or equipment to illustrate a realistic and accurate view of your subject. It makes your verbal expression more powerful and assertive.

Table 23.4 Details of Availability and Durability of Essential Goods

Goods	Durability	Nature/metal	Availability
Toys	Breakable	Plastic	Abundant
Utensils	Long lasting	Steel	Scarce
Flower Pot	Long lasting	Brass	Scarce

Maps   Maps are the other wonderful aids which help you depict geographical and spatial distribution. For example, the spatial location of a produce, building, route, highways, minerals, etc. can easily be presented by showing maps.

Charts   Tables and graphs show quantitative analysis and relationship among various heads, whereas charts in a report give the non-quantitative information. For example, an *organizational chart* illustrates the positions, units, departments, or functions of an organization or the way they are connected to one another. A flow chart, on the other hand, shows pictorially how a series of events/activities/operations are arranged to complete a full cycle.

Graphs   Graphs are the illustrations which help the author in presenting the data in a creative, colourful, and catchy manner. Graphs help the reader understand the various trends and patterns that the report intends to project.

Graphs are of various types. These are as follows:

- Rectilinear
- Semi-log
- Multiple line
- Pictorial
- Bar
- Scatter
- Pie
- Surface

By this time, you must have understood the basic nature of a report and its most characteristic features. An important aspect of reports, like all professional tasks, is its planning and preparation, both of which are essential in letting your report fall in consonance with its desired tenets. Let us look at the stages of planning and preparation required for writing a formal report.

23.7 PREPARATION AND PLANNING

It is said *to fail to prepare is to prepare to fail*. And this holds good for reports. The importance of preparation and planning in writing a report cannot be overemphasized. Often, however, we simply tend to ignore this aspect or dismiss it while writing a report, as we consider it too mechanical and not really worthwhile to be followed diligently. As a result, we happen to plough too quickly into the writing process itself and end up failing to let the reports realize their true objective. Before you start writing a professional report, you must do the following:

1. Set your objective.
2. Assess your audience.
3. Decide what information you will need and collect data.
4. Prepare your skeletal framework, that is, form an outline.

Collectively, these activities constitute the planning stage of report writing, and the amount of time and thought you spend on them will make a *vast* difference to the effectiveness of all the work that will follow.

23.8 DATA COLLECTION

The next step after defining your objective, scope, and purpose is to collect relevant data. Recognizing the sources of information and collecting information become the primary tasks at this stage. Formal writing cannot depend on imagination and assumptions. It requires solid proof. This marks the importance of data collection in report writing.

23.8.1 Sources

The following sources can be approached for data collection:

(a) Encyclopaedias   (b) textbooks   (c) office records

(d) files   (e) journals   (f) handbooks
(g) manuals   (h) government publications   (i) Internet
(j) magazines   (k) newspapers   (l) computer databases

23.8.2 Methods of Data Collection

The following methods can be used for data collection:

- Personal observation
- Personal interview
- Telephonic interview
- Mail questionnaire

Personal observation Some reports will be based on your personal observations. For example, you may be required to write a report on an experiment you conducted in the laboratory, a job you performed, or an event to which you were an eye-witness. In all these situations, you rely on your sensory perceptions, and hence you should develop careful observation. This method has the following merits and limitations.

Advantages	Disadvantages
• First-hand information	• Time consuming
• Very reliable	• Not necessarily convincing for others

Telephonic interview Another method of collecting data is interview—personal or telephonic. If the information you seek is of a routine nature and only brief answers are required from a small number of people, then you can contact them on telephone and save time on travelling. But often, you may not get as effective a feedback on the telephone as is possible through personal interviews. Following are some of the merits and demerits of this method of data collection.

Advantages	Disadvantages
• Quickest of survey techniques	• No detailed data available
• Low refusal rate	• Observation eliminated
• Memory factor eliminated	• Limited information
• Low cost occasionally	• Little time for orientation and reaction
• High returns	• Respondents' antagonistic
• Approach and questions standardized	• Not essentially representative
• Mostly reliable in matters of costly and time-consuming routine research	• Low credibility

Personal interview While collecting data by conducting personal interviews, you should be shrewd, observant, and sensitive to the reaction of the person you are interviewing. You must be quick to readjust your approach and attitude to suit each case. You need to first secure a person's attention, excite his interest, and establish a rapport so that he/she responds and cooperates in giving you the information you want. And this depends on your knack of handling people. First you must do your own homework well. While deciding on the questions to be asked during the interview, break your discussion into significant components for easy handling and frame the questions accordingly. The set of questions that you prepare for this purpose is called an *interview sheet*. Before we discuss the guidelines for planning and conducting personal interviews, let us discuss the advantages and disadvantages of this method.

Advantages	Disadvantages
• Flexible	• Limited coverage
• Direct information	• Costly and time-consuming
• Orientation possible	• Prone to discussion
• Non-verbal interpretation possible	• Subjective information
• Least obscure	• Given to chance and failure
• First-hand impressions	• May influence responses
• Questions can be repeated or rephrased	• Respondents may not respond to personal or embarrassing questions
• Useful in market survey	

Tips for planning and conducting personal interviews   Here are some important tips for planning and conducting a personal interview:

1. Get a prior appointment from the person whom you plan to interview and inform him/her about its purpose.
2. Prepare an interview sheet at least with ten possible questions relevant to your report survey.
3. Think of your opening and closing and also visualize the kind of expressions you plan to use for making the environment conducive at that time.
4. Be clear about your purpose of collecting the data.
5. Dress well and reach the venue on time.
6. Carry a pen and a notepad for taking down the answers.
7. If you wish to record the interview, seek permission beforehand.
8. After reaching the venue, greet your interviewee warmly, orient him/her to the interview briefly and then begin your interview.
9. Exhibit your zeal and interest in the answers by listening actively.
10. At times of confusion, do not get annoyed; rather seek clarification in a strategic manner.
11. Do not digress, and bring the interviewee back to the topic if he/she gets digressed.
12. Avoid asking embarrassing or personal questions.
13. Do not get involved in unnecessary or heated arguments.
14. Since you are seeking a favour from the interviewee, always assume a subordinate position.
15. Do not interrupt your respondent unnecessarily.
16. Thank your interviewee for sparing time and always keep the lines of communication open as you may require some more information later.

Mail questionnaire   When a wide geographical coverage is required to be covered and a large number of people have to be contacted, the most efficient and convenient method is to collect data through mail questionnaires. Before going further, let us take a look at the relative merits and demerits of this method.

Advantages	Disadvantages
• Cheapest method	• No clarification possible
• Covers wide area	• Not necessarily correct
• Specific, accurate, processable	• Illiterates can't answer
• Covers large number of respondents	• Unwilling souls won't answer
• Most scientific and reliable	• Time consuming
• Reduces hesitation	• Not face to face
	• High refusal rate

Tips for preparing a questionnaire

1. Basic requirements
   (a) Good mailing list   (b) Good covering letter   (c) Interested respondents
2. Tips for framing a good questionnaire
   (a) Set objectives for every question you ask
   (b) Be pointed, short, and clear
   (c) Figure out possible answers
   (d) Phrase the questions clearly; let there be no room for ambiguity
   (e) Avoid leading questions
   (f) Avoid delicate questions
   (g) Avoid long and complicated questions
   (h) Arrange the questions logically
   (i) Don't ask questions that require lengthy answers
   (j) Leave enough blank space for answering
   (k) Don't repeat the questions
3. Additional tips
   (a) Enclose the covering letter with a self-addressed, stamped envelope
   (b) Assume confidentiality
   (c) Ensure action and be courteous

Sample Questionnaire

Ministry of Health
Government of India
Occupational Hazards Faced by Information
Technology Professionals
Questionnaire
PART–A

Please fill in the details:
Name:
Age: Sex:
Marital Status:
Name of the organization you work for:
No. of years have you worked for it:
Designation:
Hours of work:

(Contd)

(*Contd*)

PART–B

Please tick your choice, unless mentioned otherwise.

1. For how many years have you been in the IT field?
   () Less than a year   () Less than 5 years   () 5–10 years   () More than 10 years

2. Why did you choose this field?
   () Parental pressure   () Lure of money   () Glamour   () Genuine interest
   () There was nothing else to do

3. What are your qualifications with respect to this field?
   () B.E./M.E.   () BCA/MCA   () Diploma   () Any other (please specify)

4. How do you rate yourself as an IT professional?
   () Excellent   () Good   () Above average () Average

5. As regards your job-satisfaction, you are
   () Very satisfied   () Satisfied   () Dissatisfied

6. If your answer is (c), why are you dissatisfied?
   () Competition   () Security threat () Work pressure () Any other (please specify)

7. Is the work pressure very high?
   () Yes   () No

8. If it is so, why do you think it is?
   () New demands   () Meeting of project deadlines
   () Learning new skills   () Any other (please specify)

9. How many hours of work per week do you put in?
   () Less than 50 hours () 50–60 hours   () 60–70 hours   () More than 70 hours

10. Do you work on weekends?
    () Yes   () No

11. For how long do you sit in front of a computer per day?
    () Less than 1 hour () 1–5 hours   () 5–8 hours   () More than 8 hours

12. How frequently does your job require traveling?
    () Once a month   () Twice a month
    () Every week   () Twice a week
    () More than two times a week

13. How do you rate the atmosphere in your office?
    () Hostile   () No change   () Friendly   () Motivating

PART–C

(Can tick multiple options)

14. At the end of the day, you suffer from
    () Stress   () Fatigue
    () Pain in joints   () Watery eyes
    () Numb fingers   () Any other (please specify) _____

15. How often do you go to a doctor?
    () Once a week   () Once a month
    () Every two month   () Any other (please specify) _____

16. Have you considered (or are you) consulting a psychiatrist?
    () Yes   () No

(*Contd*)

(Contd)

17. Do you constantly feel the urge to get away from work?
    ( ) Yes   ( ) No
18. Is your profession affecting your family life?
    ( ) Yes   ( ) No
19. How many hours a day do you spend with your family?
    ( ) 4–5 hours ( ) 3–4 hours   ( ) 1–2 hours   ( ) Less than 1 hour
20. How many social outings do you have every month?
    ( ) 2–5   ( ) 5–10   ( ) More than 10   ( ) Do not go out at all
21. Do you think you have enough leisure and recreation time?
    ( ) Yes   ( ) No
22. Have you ever thought of changing jobs due to ill-health or lack of recreation?
    ( ) Yes   ( ) No
23. If yes, how often (please specify)?
24. Does your salary include allowance for medical check-up and medical insurance?
    ( ) Yes   ( ) No
25. If yes, please mention the amount: _____

26. Any other relevant details (please specify): _____

23.9 ANALYSING AND ORGANIZING THE DATA

Once data collection is done and notes have been taken, the next step for the report writer is to analyse the data. As a report writer therefore, you should determine more valuable information from the data collected. Revise your questions and data; add new items to it if necessary. If the information does not fit into your purpose, do not use it.

The next step is to organize the data collected. An old but highly workable adage is that *a writer should tell the audience what he or she is going to do, then do it, and then tell the audience what has been done*. This can be conveniently divided into the more typically labelled parts as the introduction, body, and conclusion of a report.

A formal outline is necessary if your report is very long, because the brief informal jottings are not detailed enough to hold together a 20-page report. An outline is necessary in a formal written report because it finally helps us form the table of contents of a report.

While deciding on the sequence of ideas, save the most important statements for the first and last positions, as they receive more stress than anything else.

23.10 WRITING AND REVISING

Keep the audience in mind and prepare a rough draft with everything relevant. In the first review, concentrate on blocks of ideas. In the second review, concentrate on finer details, matters of syntax and diction, and write the final report in a specific form using a style sheet or manual.

23.11 PREPARING AN OUTLINE

Outline is an arrangement of words, phrases, or sentences which indicates the nature and sequence of topics and sub-topics to be discussed in the report. Since thoughts do not always come in the order in which we like to put them in writing, an outline is required. It gives a framework and indicates the pattern of our report. Given below is a sample outline prepared for a report on *Leadership: Now and Then*.

1. Introduction
    1.1 Historical Background
    1.2 Interpretation of Leadership
2. Evolution of Leadership in History
    2.1 Industrial Revolution
    2.2 Leaders in History
    2.3 Leadership Eras
3. Theories on Leadership
4. Leadership Styles
5. Contexts of Leadership
    5.1 Leadership in Organizations
    5.2 Leadership and Management
    5.3 Leadership by a Group
6. Performance
7. Conclusion

23.12 STRUCTURE OF FORMAL REPORTS

As we have learnt that reports vary widely in length, structure, purpose, style, etc., there are various parts in a report that need to be well signposted in long formal reports and some of them are not essential in shorter reports. The various elements in a report can be mainly classified as follows:

1. Front matter   2. Main body   3. Back matter

Let us now discuss each of these parts in detail.

23.12.1 Front Matter

The front matter can include the following:

- Cover
- Preface
- Title page
- Acknowledgements
- Frontispiece
- Table of contents
- Copyright notice
- List of illustrations
- Forwarding letter
- Abstract (or summary or executive summary)

Cover   The formal reports customarily have a binding or hard cover. This hard cover of the report contains important information like the *title of the report, report number* if any, *name of the report writer and organization*, and *date*. Usually, it is of white or neutral colour (soft). While formatting the cover of your report, you should spread it well in the whole space, but do not crowd it with too much of information or by using an extravagantly bigger font. Always keep proper space between the different items of information so as to make them distinct from each other. While writing the title, use either all capital letters or the first letter of each key word should be written in capitals. You should compose the title well by using key words. Thus, the cover should have a label that provides enough information to the reader about the report without looking inside it. An example of a cover is given below:

> **A Report**
> **on**
> **Fatal Road Accidents during 2004–2014**
> By
> Sriram Raichuri
> Senior Traffic Inspector
> **Department of Traffic and Signal Control, New Delhi**
> 28 November 2014

Title page This is the first right hand page of the report. A title page contains additional information along with what is given on the cover of the report. It tells the reader about the *subtitle (optional) of the report, the name and designation of the authority for whom the report is written*, and *approvals if any*. But remember not to crowd the title page with unnecessary information. Let us see a sample title page:

> **A Report**   **R. No. B239/2014**
> **on**
> **Fatal Road Accidents during 2004–2014: A Case Study of New Delhi**
> Prepared
> for
> Ujjawal Pratap Singh
> Director
> Traffic Police, New Delhi
> By
> Sriram Raichuri
> Senior Traffic Inspector
>
> Approved by
> Abhishek Sayal
> Deputy Director
> Traffic and Signal Control
> **Department of Traffic and Signal Control, New Delhi**
> 28 November 2014

Frontispiece Frontispiece is nothing but the window display of the whole report as it consists of artistic drawings, pictures, photographs, or maps. The sole purpose of giving a frontispiece is to ignite curiosity among the readers about the contents of the report. However, these are not normally used in technical reports.

Copyright notice It is always placed on the inside of the title page. It actually indicates who has the legal rights of the document—whether it is with the publisher or with the author. This is the very reason that there is a small note at the bottom of every copyright notice. A sample is as follows:

Forwarding letter Forwarding letters forward a report to the primary recipient of the report. Forwarding letters can be 'covering' and 'introductory' type. The former serves as a record of transmission of the report. It is not bound with the report, as it does not contain any important information. Introductory covering letters often refer to specific parts of the report, and may repeat particularly important points such as purpose, scope, limitations, conclusions, and recommendations. Since an introductory covering letter contains an appreciable amount of information, it is usually bound into the report, immediately after the title page.

Preface The preface of the report introduces the report to its readers. In fact, the preface tells the readers broadly what the report is all about. Since the last paragraph of a preface includes acknowledgements, it is not necessary to have separate acknowledgements in the report.

Acknowledgements This is an element of the report that appears before the abstract of the report. It mentions the names of those people and sources that have helped the report writer to accomplish the task. However, this is not a list. It is properly crafted since it not only includes the names of people but also informs the readers how those persons helped the writer to carry out his/her work within the stipulated time. Let us learn a few more important tips for drafting good acknowledgements for your report:

1. First categorize the persons whom you want to acknowledge according to courtesy sake, real help, emotional help, and technical help.
2. Bring variation in your expressions while expressing your gratitude.
3. Always mention the reason due to which you intend to express your gratitude for the person.

Let us look at a sample of an acknowledgement.

Acknowledgements

I would like to express my deep sense of gratitude to Prof. B.N. Jain, Vice Chancellor, Birla Institute of Technology & Science, Pilani, for giving me an opportunity to carry out the project on 'Cross-cultural Communication: A Study from the Business Perspective of the Indian and American Cultures', an area of my interest.

I would also like to thank the Instruction Division for providing me this opportunity to work.

I also thank Dr Devika Rathi, my project instructor, for her guidance without which this report would not have been possible. I would also like to acknowledge the help and support of Mitesh Singh, whose inputs helped me know a lot about intercultural communication. I would also like to acknowledge Mr Rakesh Kumar, employed with a leading hedge firm in USA, whose inputs were of immense help in preparing this report.

(Contd)

(Contd)

I owe a lot to my parents for being my constant motivation to work hard. Above all, I thank Virender and Sushila, my friends, who have been instrumental in the successful compilation and presentation of this report.

I also thank Mr Rakesh Saini, the binder for giving it an elegant shape.

Table of contents   A table of contents (ToC) is an essential element in a long formal report as it has several sections and subsections. However, it is not needed if the report runs into four to five pages. Since ToC of a report includes the chapter headings and subheadings and their respective page numbers, it becomes easy for the reader to locate specific information from a report. The discussion section of the report is normally broken down into 2 or 3 levels to give the reader a clear picture of the report's contents. It serves as a guide to the report. It does not include a title page or table of contents. The decimal numbering system (DNS) is used while preparing the ToC. See the example given below:

Table of Contents

List of illustrations   When more than twelve illustrations (maps, charts, graphs, etc.) are used in a report, a list of illustrations is prepared like a ToC. In this list, page numbers are also provided so that a reader may quickly locate a specific illustration that he/she wishes to refer to.

Abstract   The abstract is a brief write-up which tells the reader what the report is all about and its accomplishments. An abstract captures the essence of the entire report, based on which the reader decides whether to go ahead with reading the report or not. It is very short in length, not generally exceeding 250 words. The abstract should contain relevant information such as main purpose, main design point, methodology, and some eye-catching results to show the significance of the report. Read the sample abstract given below on the topic *Exploratory Research on Risk Taking Attitudes in Relation to Approaches in Innovation of Effective Business Leaders*.

Sample Abstract

This report is based on the profile study of ten top most business leaders. It analyses the various factors such as risk-taking abilities, inventiveness, adaptability, resilience, and business acumen among these leaders. It also suggests an existence of strong correlation between effective risk-taking with strategy-based innovation. The report serves as an effective introduction to cross-factor analysis of effective leadership. It has implication for HR in a situation when they are finding top leaders to lead a firm towards innovation, grow its revenues, or turn around the company in crisis.

Executive summary   An executive summary presents the entire report in a nutshell telling the readers about what is to be expected in the report and what the report explores. You can specify the following points in your executive summary:

- Objectives of the report
- Major findings
- Overview of subject matter
- Recommendations
- Methods of analysis

An executive summary to the report is like a window to the building. It allows its readers to get to know how significant and relevant the study is for him/her. Since it includes major recommendations as well, it enables the authorities to take decisions to solve the problem under discussion.

Differences between abstract and executive summary are explained below:

Abstract	Executive Summary
Very brief, i.e., not more than 300 words	More elaborate than an abstract
Relevant only when the report is meant for people from the same domain knowledge	Meant for readers also other than subject experts; hence, can relate to all
Presents only the essence of the report	Presents the entire report in a nutshell
Cannot help in taking decisions	Can help in taking quick decisions if required
Does not include illustrations	May include one or two illustrations

23.12.2 Main Body

The main body consists of the following parts:

- Introduction
- Conclusion
- Discussion
- Recommendations

As mentioned earlier, this is the part of the report that talks about the actual contents of the report. Now let us know in detail how each of the four different parts listed above can be written effectively.

Introduction   A good introduction to the report is an effective curtain raiser to its contents, significance, methods adopted, etc. It attempts to give the readers a clear picture of the problem and get their attention. Ideally this section should provide the following elements of information.

Background information   What is the setting of the problem? To answer this question, the introduction of the report necessarily includes the historical or technical background that has given rise to the problem. It throws light on the exact state of affairs and the factors responsible for the situation in brief.

Problem statement   You also need to specify what exactly the problem is that you are trying to solve. This is called the *problem statement*.

Reference to authorization   It usually includes the person requesting the report and the reference to the letter or memo or meeting or telephonic conversation held on a specific date by which the person was asked to submit a report and he/she thus becomes the writer of the report. This is known as *reference to authorization*.

Purpose and scope   The background of the problem logically leads to the purpose for which the report is being prepared. Sometimes in government offices, the author of a report is given certain guidelines which he/she is supposed to follow verbatim. These are called *terms of reference*. Thus, when you state the points or issues that you would cover, indicate the scope of investigation.

Significance   Since reports are formal in nature, they serve a specific task. While writing the introduction, you are required to specify the scope of the study, its limitation, and also its significance.

Methodology   While writing an introduction, you are required to state the method or the sources that you have used for collecting the facts and analysing them for the report.

Procedure   In the introduction, you need to include the procedure adopted for analysing the data.

Basic principles or theories involved   Whenever problems are identified, certain principles or theories are used for finding their solutions. Therefore, the introduction of the report should refer to those principles or theories.

Summary of findings   What are the main results? A well-written introduction also includes a brief summary of the major findings of the report.

General plan of work   Towards the end of the introduction, you should also provide the general plan of work talking about the different chapters and the varied issues discussed in them. Here is a sample introduction.

Sample Introduction

Topic: A Report on the Impact of Culture and Communication on Global Business Growth
The end of the Cold War marked a significant breakthrough in international relations. Major countries like the United States, Russia, the United Kingdom, and their likes, realized the need to cross borders in the fields of business, commerce, politics and also to ensure a leading position globally. And that called for communication

(Contd)

(Contd)

between representatives of different countries on different platforms in different fields with an intensity never felt before. Since the early 90's, we have seen developed nations and fast developing countries such as the BRIC nations, focusing on and giving in their best efforts to ensure effective and productive cross-cultural communication. Post the liberalization reforms introduced by the P.V. Narsimha Rao government in 1991, India has emerged as a competitive international player in the world's business scenario. However, the competition is very stiff and there are quite a few countries like China and Brazil which are giving us a very tough fight for the top spot. Time thus is ripe for us to work on our cross-cultural communication strategies and training to ensure that the dream of becoming a global superpower is soon realized. It was this reason that inspired me to take up this topic for the project work. This study looks into the intricate details of cross-cultural business communication, efficient cross-cultural communication strategies and ways to improve our present understanding of the topic.

Indian and American cultures are representative of two very different cultures, broadly classified as the Eastern and the Western cultures. The two show contrasting features when compared from different perspectives on different parameters. The application of one's cross-cultural understanding can lead to brilliant results in all fields of negotiation and interaction, especially in the business arena. Cross-cultural business communication, therefore, is a rapidly growing field of interest. It stresses on the need to include business, its related variables and an efficient business strategy in the cross-cultural communication base.

When communicating with someone from a different culture, we can therefore expect cultural differences to have an influence. Cultural differences stem from our differing perceptions, which in turn determine how we communicate with people of other cultures. By understanding how people perceive the world, their values and beliefs, we can better understand their perspective and can anticipate potential cross-cultural misunderstandings.

For this report, a questionnaire was drafted and sent to 100 professionals working in multinational companies and 83 were received back. These duly filled questionnaires helped in procuring the details regarding cross-culture communication barriers. The data was compiled using SPSS tool.

In this report, after the introduction, the second chapter highlights the general barriers in cross-culture communication and the third chapter brings to the fore the various cross-culture communication barriers particularly in business. Chapter four focuses on making a comparison of various cultural practices in business in India and America. Finally conclusions are given in chapter five.

Discussion The discussion part of the report may be divided into different chapters as the case may be. These chapters may delve into different aspects of the problem. The organization of the report here is problem-specific. It may include various illustrations to substantiate the discussion. The results are usually presented as tables and graphs. While explaining the tables and graphs, you have to explain them as completely as possible. Identify trends in the data and try to prove what you want to establish. While describing a table, you have to describe every row/column. And similarly while describing a graph, you are required to describe the x/y axes. If necessary, you have to consider the use of log-axes. Thus, this section will typically show pie charts, bar graphs, and other statistics relevant to your research. Through these graphic aids, the discussion part of the report actually comprises the whole data, presents it in a lucid format and offers interpretation from various perspectives. It is here that all the comparisons, contrasts, parallels, and arguments are brought into sharp focus which helps the author interpret the situation and also leads the reader to conclusions and inferences that stem from such discussions. So keep the following points in mind:

1. Divide your text into several sections/chapters.
2. Organize the matter into topics and sub-topics.

3. Include substantial matter under each topic/sub-topic.
4. Provide detailed analysis/interpretation.
5. Do not insert recommendations in the discussion section.
6. Include illustrations for making the analysis clear to the reader.

Conclusion   In the conclusion section of your report, you can briefly recapitulate the problem studied, the approach adopted, and the results based on the analysis done in the discussion part of your report. It can be written either in points or developed in separate paragraphs. Without introducing new ideas or offering elaborate discussion on the facts already established in the report, the conclusion section draws inferences in a crisp and tidy manner.

Recommendations   Recommendations state the actions that according to the writer are required to be taken based on the findings and conclusions. Here you will suggest solutions, ideas, or recommendations. This section will include answers to all the questions, such as What is to be done? Why? How? Where? When? etc.

23.12.3 Back Matter

The back matter of a report would include the following:

• Appendices   • Bibliography   • Index   • List of references   • Glossary

After the main body is over, you may have to include a few more elements which may find a place in a full-length report. Let us look at the various parts that make up the back matter.

Appendices   This section is for reference and you can add additional charts and graphs, sample questionnaires, worked-out calculations, site plans and so on. These elements are placed in the appendices keeping in mind that these may impede the flow of information if they are placed within the text. While making a decision on the material that is to be pushed to the appendix part of the report, keep the following points in mind:

1. The appendix should contain material closely related to the topic but not absolutely essential to the overall understanding of the report.
2. Provide a proper cross-reference in the text, for example, 'see Appendix A for the questionnaire sent to the respondents' or for detailed calculation 'see Appendix B'.
3. Experimental results, detailed calculations, statistical data, comparative analysis, sample questionnaires, sample pro forma, etc. can be included in the appendix.

Documentation   Documentation is an indispensable part of any professional writing such as reports, proposals, and research papers. It includes how the various sources such as books, anthologies, newspapers, journals, etc., which have helped you expand your understanding of the problem or topic, are to be cited. For this purpose, you are required to prepare references, bibliography, and sometimes footnotes. There are four major documentation styles; MLA (Modern Language Association) style, APA (American Psychological Association) style, Chicago style, and IEEE style. In the following section, let us understand the differences between references, footnotes, and bibliography.

List of references   This section of the report includes the bibliographical details of the books and sources from which certain ideas, facts, and data have been borrowed as it is. These sources are mentioned so that the readers may get to know the specific location of the idea or data if they wish to refer it further. This is done by citing such sources in the main text and listing them in the alphabetical order or in the order of citation at the end of the report.

Footnotes   If the number of references is small, they are mentioned in the footnotes at the bottom of the page and a star or asterisk mark is used in the text to draw the reader's attention.

Bibliography   Bibliography is an alphabetically arranged list of all the sources, such as websites, newspapers, books, magazines, journals, documentaries, manuals, reports, movies, etc. consulted by the author for writing the report. It also includes the sources not visited by the author but are deemed to be useful for such studies. A bibliography differs from the list of references in the following respects:

Differences between Bibliography and List of References

Bibliography	List of References
It lists all the works consulted for ideas/information.	It always helps the reader know the specific location of an idea or a piece of information.
It contains works recommended for further study.	No such function.
Entries are necessarily made in alphabetical order.	Entries may be given either in the alphabetical order or in the order of citations.
The writer starts preparing the bibliography before writing the report and continues to add entries till the writing task is over.	More conveniently prepared while the report is being written.
It may be 'annotated' or 'select.'	Only lists all the works cited in the report.

Glossary   A glossary is a list of technical words and terms which appear in the text of the report. In a glossary, these words and phrases are arranged alphabetically. Their meaning and explanation are also provided.

Index   This is the list of various topics, sub-topics, and various other important aspects which have been discussed inside the main text but could not find a place in the table of contents. Thus, it serves as a quick guide for the reader to locate any specific idea or concept given in the main text somewhere.

23.13 STYLE OF REPORTS

An effectively written report is one that follows a logical flow of ideas and is cohesive. It contains proper links between and within its sentences, paragraphs, and sections which make it easy for the reader to follow the message it conveys. Moreover, it uses language to keep track of the report's purpose and the reader's needs. Therefore, before you sit down to compose your report, think about the recipient of your document. What are you trying to say to this person? Another question to ponder at this stage is—how to say whatever you are trying to say, effectively? In fact, for compiling an effective report, both the organization of the contents and its language are crucial. For maintaining a clear, concise, and objective style for writing a report, follow the tips discussed in the following subsections.

23.13.1 Provide Complete and Accurate Information

The documents which provide incomplete information are likely to be inaccurate in some aspect or the other. Consequently, the recipient of the document interprets the data differently from the way the writer intended. Take a look at the following examples:

Original The sales force will meet at 6 pm on Wednesday.
Revised The sales force will meet at 6 pm on 30 June 2015, Wednesday in the Senate hall.
Original This year the firm has incurred high profits.
Revised This year the firm has incurred 4.8% more profit as compared to the last year.

23.13.2 Use Plain, Familiar, and Concrete Words

Always use plain, familiar words while writing a report. If you use the words which are commonly used in day-to-day life, your readers will not get baffled by the language and will be able to appreciate the message you have conveyed in the document. Avoid ostentation and flowery language as far as you can. Also, prefer using concrete words rather than abstract words to maintain clarity and precision in the report. See the following examples:

Original In all probability, we are likely to launch a mobile set that is bound to have a resounding repercussion on the way mobiles are designed.
Revised Our new mobile set is likely to change the existing mobile designs.
Original The committee will *decipher* the factors responsible for lack of *synergy* between the two departments.
Revised The committee will find out the reasons responsible for the lack of cooperation between the two departments.
Original The *ramifications* of the experiment done in the laboratory should be *corroborated*.
Revised The results of the lab experiment should be verified.

23.13.3 Avoid Wordiness and Redundancy

After you have decided on the meaning you are trying to convey, work on putting it in a concise language. Be brief, whenever possible. Avoid wordiness and unnecessarily long words. When you use more number of words to convey the same meaning, it is called *redundancy*. Strive for clarity in your writing and avoid vagueness. See the following examples:

Original Tsunami had the effect of a destructive, disparaging and caustic impact on the manufacturing plant.
Revised Tsunami destroyed the manufacturing plant.
Original Keeping in mind the objective of facilitating and having smooth operation for timely processing of rebate requests, it appears that more employees are required.
Revised To fulfil the rebate requests quickly, we need six more employees.

23.13.4 Judiciously Use Active and Passive Voice

In general, active voice is preferred to express ideas in reports. The unnecessary use of passive voice sometimes makes us write more than required. However, passive voice is preferred in order to avoid curtness and give prominence to the action rather than the agent.

Sometimes you will be called upon to write a report describing a corporate disaster that occurred because someone made a mistake. Passive voice can be used to describe the mistake without directly placing blame on the person who made the error which led to the disaster. In this situation, it would be tactless, to say the least, to use active voice boldly. See the following examples:

Original Because Mr Thareja, Accounts Officer, forgot to include the correct budget projections with the bid, we lost the client.

Revised The correct budget was accidentally left out of the bid, which led to the loss of the client.

Original You have not yet informed me about the latest marketing strategies you are likely to follow from this month onwards.

Revised I have not yet been intimated about the latest marketing strategies you are likely to follow.

Thus, there are times when judicious use of passive voice can enhance the subtlety and tactfulness in your writing.

23.13.5 Follow Emphatic Word Order

Since a report is a concise presentation of facts, an emphatic word order is required to arrest the attention of the reader. For a reader, what comes in the beginning and at the end has an extra emphasis in meaning and idea. Therefore, try and supply the most important part of your message either in the beginning or at the end of a sentence and let it not be lost somewhere in the middle of a sentence. See the following examples:

Original When he returned to the production unit, the Production Manager tried to implement the strategies he had learnt about in the training programme attended in the previous month.

Revised On his return from the training programme, the Production Manager attempted to implement strategies recently learnt.

Original In the months to come, a proper order and mechanism to tackle the problem of absenteeism would be in place.

Revised To tackle the problem of absenteeism, a proper mechanism would be in place shortly.

Original Individual preferences in the decision to allocate the division of work among employees would be considered before making a final decision on this.

Revised Before allocating the division of work among employees, individual preferences would be considered.

23.13.6 Maintain Parallelism in Writing

While writing, remember to follow parallel grammatical structures. Grammatically, a gerund is parallel to a gerund while a to-infinitive is parallel to a to-infinitive. Look at the following examples and learn to use parallel grammatical structure as suggested:

Original Reading in the conference hall may be accommodated to some extent but to cook there seems weird and strange.

Revised Reading in the conference hall may be accommodated to some extent but cooking there seems weird and strange.

Original While you speak to inform, remember to use slides for carrying conviction.

Revised While speaking to inform, remember using slides for carrying conviction.

Original The Manager asked for improving the marketing strategy and to introduce attractive schemes for enticing more customers.

Revised The Manager asked for improving the marketing strategy and introducing schemes to entice more customers.

23.13.7 Prefer Using Verbs to Long Nouns

Psychologically, it is easier for us to remember actions than objects. In language, action is often described through verbs while objects are denoted through nouns. Preferring verbs to nouns, particularly longer ones, helps you express yourself in an emphatic, clear, and forceful way. Consider the following examples to understand how the use of verbs in the revised versions makes the idea more lucid and appropriate:

Original After a careful revision of the suggestions made, a consideration for their implementations is likely to follow.

Revised We are likely to implement the suggestions after carefully considering and revising them.

Original Please make an approbation of the recommendations only after a meticulous observation thereof.

Revised Please approve the recommendations only after meticulously observing them.

Original A new procedure has achieved its full realization. All the members of the staff are requested to make a utilization of the new procedure for obtaining short-leaves during office hours.

Revised A new procedure for obtaining short leaves during office hours has been implemented. All the members of the staff are requested to follow that.

23.13.8 Carefully Use Acronyms/Abbreviations

The difference between an *acronym* and an *abbreviation* is that an acronym is a shortened phrase and an abbreviation is a shortened word such as Mrs, Mr, Pvt., etc. According to answers. com, this difference may be true, but it is a debatable point. More succinctly, an acronym is an abbreviation for a series of words, and which is *pronounceable*. An abbreviation is usually the first letter of each of a series of words *without considerations whether it is pronounceable*. An example of an abbreviation is 'FDIC'. An example of an acronym is 'AFLAC' (as uttered by the company's mascot duck). Other examples of acronym are NATO, AIDS, BRIC, SAARC, LASER, UNESCO, etc. So, abbreviation is a broader term. All acronyms are abbreviations but vice-versa is not true. However, you need to understand that in official documents, we should not use abbreviations and acronyms generously. If you are using, define them in full parenthetically at the first place and later use them. Besides, if you are using abbreviations which have more than one expansion, define them in the glossary.

@   Please refer to the Online Resource Centre for a list of commonly used abbreviations and acronyms.

23.13.9 Avoid Clichés

Just as overusing a machine makes it groan, overusing an expression makes it sound clichéd and hackneyed. Therefore, while writing a report, try to avoid expressions which sound drab and repetitive. See the following examples:

Original We assure you that we would spare no stone unturned in helping the company achieve its objectives.

Revised We are committed to doing our best in order to help the company achieves its objectives.

Original Last but not the least let me thank the management.

Revised Finally, I would also like to thank the management.

Besides clichés, avoid overusing jargons and foreign words. Doing this unnecessarily confuses the reader and leads to ambiguity which certainly is not a desirable trait.

23.13.10 Avoid Circumlocution

Since report writing is all about being concise and precise, beating about the bush and not arriving at the point of discussion is almost like a criminal offence you can commit as an author of your report. Remember that more number of words in professional situations means more money, time, and energy; and an extravagant use of any of these is certainly suggestive of an amateurish and unprofessional approach towards work. Therefore, avoid circumlocutory expressions and be pointed and straight while writing a report. Look at the following examples to understand the point better:

Original In the given circumstances, it would hardly augur well if the company decides to retrench its tried and trusted employees especially when suitable replacements of theirs would not be easy to come by. Hence, our suggestion is to forestall the process of termination of employees, particularly those who are quite tried and trusted and would be hard to replace.

Revised We don't suggest retrenchment of tried and trusted employees for it's generally difficult to replace them.

Original All in the present moment of time, mobile industry has witnessed a mad scramble for more customer orientation with all the major players in the field readily making frequent announcements and thereby securing their clientele as the inducements they offer that are far too luring for the average customer in India to put a resistance to.

Revised At present all the major mobile companies seem trying to attract customers and secure clientele by making announcements and offering inducements difficult to be resisted by an average customer in India.

23.13.11 Maintain Unity in Paragraphs

Though paragraph writing techniques have been taken up in a separate chapter in the text, it is worthwhile to keep in mind that finally all your report is structured in words, phrases, sentences, and paragraphs. Structuring a paragraph needs practice, revision, and concentration. Often a report is criticized for not being coherent as the paragraphs don't progress systematically. At this stage, it is important to understand that you write separate paragraphs to deal with distinct ideas in your mind. Therefore, if you learn to maintain unity within a paragraph, you can express your ideas in a coherent, lucid way. Consider the following examples to observe how lack of unity in a paragraph can lead to ambiguity and confusion:

Original We are not of the view that you should start launching a new product in the market. We are not able to compete with other players in the market. Our company policies for marketing and sales are also being revised. Moreover, the company has not been able to do well in the north India where the customers prefer a new product with attractive schemes. Before launching a new product, we must understand how other products have fared in the recent past. Otherwise, taking a hasty decision would adversely affect our reputation in case we launch a product and it fails to click.

Comments: The author seems to struggle with his/her multiple ideas. It is difficult to ascertain what really bothers the author. He/she jumps from one idea to another without being able to touch where the fault lies in launching a new product. A possible way to revise the paragraph is given below.

Revised We don't suggest launching a new product in the market. It is so because we are yet to firm up our marketing and sales policies. Reportedly, our products have not fared well, particularly in north India. Before launching a new product, we also need to understand what customers expect from us. Taking a hasty decision at this stage would only adversely harm our company's reputation.

Original Of all the management styles, inspiring leadership has no parallel. Management itself is a complex phenomenon. You need to work with people. And people are no machines. They feel, they think, they mind. Machines respond the same way to a treatment; people don't. There are other forms of management. Some managers tend to mechanically carry out their task. This is a drab and unimaginative way of managing people. It hardly helps. Some others try to coerce their subordinates. They are known as despots. And they are despised. The best form of management of course relates to the inspirational, democratic and empathetic way of managing people.

Comments Another example of a flawed construction. The passage moves from one idea to another without a proper direction and coherence in ideas. See how the revised version helps us assume a better shape of the author's perspective:

Revised Managing people is far more challenging than handling machines because human beings tend to respond to situations, policies and treatments in diverse ways. There are quite a few management styles in practice. Some of these are mechanical and lack imagination; some others are too despotic to be appreciated. The best form of management appears to be the one that inspires, empathizes and sounds democratic.

23.13.12 Avoid Punctuation and Grammatical Errors

Grammatical and punctuation errors can jolt your credibility severely in the professional world. Therefore, you need to be correct in terms of grammar and usage in your writings, including formal reports. In the grammar section of the text, you have already been instructed how to avoid grammatical and punctuation errors. Make good use of these instructions and try to make your report as error-free as possible. Many a time, errors are caused simply because of insufficient proofreading of the document. However, before you release a formal report, especially the one that will be seen by the management or will go outside the firm, you must proofread your document.

23.14 PREPARING A CHECKLIST

After preparing the final draft of your report, ask the following set of questions to ascertain that your report is complete and fulfils the desired purpose:

1. Does it have a title which clearly reflects the subject of the report?
2. Does it have a properly formatted cover and title page?
3. Does it have the acknowledgements and abstract?
4. Has the table of contents been prepared using decimal Arabic numbering?
5. Does it clearly state the purpose, scope, and limitation of the report?
6. Have you given a proper historical background of the problem?
7. Have you given a clear reference to authorization?
8. Does your introduction clearly entail the method and procedure for collecting data?
9. Have you carefully analysed and interpreted the data obtained for your report?
10. Have you included adequate illustrations in a professional and methodical manner for facilitating your readers' understanding of the problem?
11. Have you made proper inferences based on the analysis of the data?
12. Do all the figures and tables have proper numbering and captions?
13. Does your report provide clear recommendations to improve the situation under discussion?
14. Have you stated the answers to all the 6 'wh' questions that your recommendation should necessarily answer so as to enable your reader to take timely decision?
15. Does it have a smooth flow of information?

16. Is it free from typographical and formatting errors?
17. Finally, does it really cater to the needs of the primary reader?

23.15 SAMPLE REPORTS

In this section, a few sample reports for the following types are given.

Read them carefully and follow the finer aspects of their presentation, layout, and content:

- Memo report
- Inspection report
- Interpretive report
- Progress report
- Inventory report
- Laboratory report
- Annual confidential report

Sample Report I—Memo Report

Mandelia Institute of Technology and Research, Timarpur

Inter-office Memorandum

Date: September 30, 2014

From: Mohan Lal Sukharia, Unit Chief, Women Development Centre

To: Firoj Ahamad Abbas, Director, Research Programs

Subject: Commendation of Anjani Kumari Singh regarding her performance on Women Development Project (WDP)

The purpose of this report is to officially commend Anjani Kumari Singh for her exceptional contribution throughout her assignment to the Women Development Project (WDP).

As you know, Anjani Kumari Singh has been working on a special assignment with the WDP team for the past two years. Now that this WDP plan has been completed, I wanted to make sure that she gets some recognition for her significant and exceptional contributions to the project.

As a senior research fellow, Anjani's role in the project was pivotal for its timely and successful completion. It was Anjani who worked for long hours, numerous nights and weekends doing the field survey and data analysis with her small team of three junior researchers. The quality of Anjani's written work was also exceptional. Her regression analysis summaries were always very well written and rarely required revision.

Anjani was also outstanding as a colleague and a project team member. Her upbeat enthusiasm and keen interest in the project was infectious, and she in fact motivated the entire project team.

In closing, I would like to say that I have worked with many research fellows over the years and have never come across one as professional and productive as Anjani Kumari Singh was on the WDP. I believe that the organization as a whole should recognize her exceptional contribution to a major project.

Please let me know if you have any questions or comments.

Mohan Lal Sukharia
Unit Chief, Women Development Centre

Sample Report II—Progress Report

Apex Builders, Delhi

Progress Report

Date: _____

Name of the work or project _____

(Contd)

(Contd)

Total work to be completed _____

Work completed till date _____

Work to be completed _____

Possible date for completion _____

Suggestions   (a) If more funds, why and how much _____

               (b) Extension of duration _____

               (c) Increase in skilled/ unskilled labour _____

Remarks, if any _____

_____

Signature

Designation of the reporting officer

Sample Report III—Laboratory Report

ABC Technologies Pvt Ltd

Faridabad, Haryana

Laboratory Report

Date: _____

Name of the Investigator _____

Position _____

Name of the other Investigators of the team (if any) _____

Heading _____

Experiment No. _____

Date of experiment _____

Statement of purpose _____

Apparatus used _____

Method or procedure followed _____

Detailed calculations/Observations _____

Conclusions _____

_____

Signature

Sample Report IV—Inspection Report

Promax Multimedia Works Ltd

Sitapur

Inspection Report on Stereo

Date: _____   Model: _____

(Contd)

Number: _____   Type: _____
                                      Serial No: _____

_____
Call Registration Number _____
Customer's Name & Address _____
Call Reported _____
Call Assigned _____
Call status _____ closed/pending for spares/repair

_____

Note: Please put a tick √ against the relevant item.

(a) Visual inspection   Case: broken/scratched/ normal
                         Column Speakers: worn out/ sticky/dusty/ normal
(b) Condition of wires: broken/normal
(c) Switches and keys: broken/normal
(d) CD/DVD case: broken/scratched/ normal
(e) Mechanical movement: Stopped/obstructed run/intermittent/noisy/smooth
(f) Amplifier: Playback: dead/poor/distorted/humming/normal
(g) Recording   (i) Direct: nil/poor/distorted/normal
                   (ii) Recording from Other source: dead/noisy/poor/normal
                   (iii) Microphone: dead/noisy/poor/normal
(h) Frequency Response: high frequency missing/low frequency missing/speaker booming/jarring/normal

Signature

Supervisor's name

Sample Report V—Inventory Report

Bintex Glass Works
Bijapur
Inventory Report

Sl. No.	Item	Stock on Hand 31st March 2001
1.	White paper (foolscap size) plain	10 reams
2.	White paper (foolscap size) ruled	5 reams
3.	Letterheads	500
4.	A4 paper	4 reams
5.	Cartridges	14
6.	Carbon paper	2 boxes
7.	File covers	250

(Contd)

(Contd)

Sl. No.	Item	Stock on Hand 31st March 2001
8.	Envelopes: Large Small	1500 600
9.	Pencils: black	3 doz
10.	Erasers	2 doz
11.	Blank rewritable CDs	6 doz
12.	Clips	6 packets
13.	Stapler	4 packets
14.	Pens Black Blue Red	6 doz 10 doz 6 doz
15.	Pen drives 1 GB 2 GB	2 doz 2 doz

Date _____   Checked by _____
Place _____   Signature _____

Sample Report VI—Annual Confidential Report

Baba Engineering Works, Lalgarh
Annual Assessment Form 2014 - 2015

Name of the Employee _____
Educational Qualification _____
Current Designation _____
Date of Appointment _____
Current Responsibilities _____

Note: Please assess each item by putting a tick √ in the appropriate column.

		Excellent	Good	Average	Below average
(a)	Appearance Smartness Tidiness				
(b)	Character Dedication Honesty				

(Contd)

(Contd)

		Excellent	Good	Average	Below average
(c)	Motivation Work habit Initiative Innovativeness				
(d)	Interpersonal Relationships: With superiors With colleagues With subordinates				
(e)	Technical knowledge Computer knowledge				
(f)	Performance Diligence Consistency				
(g)	Communication Skills: Written Oral				
(h)	Is he/she fit for promotion to the next higher post?				
(i)	Special Achievements				
(j)	Other Remarks				

Date: _____

Signature and Designation
of the Reporting Officer

Sample Report VII—Interpretive Report

R. NUMBER: rp12/2014

A Report on Sale of HMT Watches

Prepared for
Mr B. Gupta
Managing Director, HMT, Delhi

By
Ms. Krishna Gulyani
Marketing Deputy Manager

Approved by
Mr Jeff Willam
Chief Marketing Manager

**HINDUS MACHINE TOOLS PVT. LTD. Magadhpur
October, 2014**

(Contd)

(Contd)

Acknowledgements

I express my deep sense of gratitude to Mr Rahul Saxena, Managing Director, Hindus Machine Tools, who permitted me to do a project which interested me.

I would like to express my thanks to Mr P. Verma, A.C.M., Tiban Watches, Mr Arun Sen, A.C.M., Citigen Watches, for providing us with the necessary information about their company market.

I am also grateful to all the people who have spent their valuable time in answering the questionnaire and interview. This report would not have been sufficiently meaningful without their responses.

I am indebted to Mr Gaurav Kapoor, Manager, Marketing Department, for providing the sales figures and making me understand various factors involved in this project.

At the home front, I acknowledge the great support of my parents and encouragement received from my friends during the project.

I would also like to thank Mr Rakesh for word-processing the manuscript with speed and efficiency.

Ms Krishna Gulyani

Abstract

The watches produced by HMT in spite of having the recognition of ISO 9000, are unable to compete with the current market. The report deals with the declining sales of HMT watches during 2010 – 2014 and identifying the possible reasons responsible for this situation. It also discusses how the company lags behind in introducing new models and lacks proper measures in facing competition from other companies. It also provides suggestions following which HMT can improve its sales and compete in the current market situation.

Chapter One

Introduction

The Hindus Machine Tools (HMT), one of the most reputed industrial units of Magadhpur, was set up with a motto of promoting small-scale as well as large-scale industries. It manufactures watches and machinery and has earned fame for its quality products. Increase in profits flowed in till 2010, when the gradual decline in the market value for HMT products started.

As the marketing manager of HMT, North Zone area, I have been asked through the letter number 249/MD/2014 by the Managing Director of the Unit to study the matter and suggest measures to improve the situation. This report attempts to give a bird's eye view on the market value of the watches produced during this period. Relevant data involving market competition from other reputed companies have been collected from records and through personal interview of the concerned authorities. Public response has also been gathered through questionnaires and personal interview of around 1000 people.

Since the report is based on the responses of consumers and authorities the results might be inaccurate in case of fluctuation in the supply of raw materials and overall market.

Recent trends in market have been analysed through available records whereas interviews with authorities helped in gaining an insight into marketing and feature enhancement strategies adopted by different companies. Survey reports tell about consumer behaviour, their choices, and also reasons behind the decline in sales of HMT watches. In the second chapter a comparative study of different designs and models of HMT watches has been done to find out which class of watches has faced maximum decline in sales. The third chapter deals with current sales policy, reasons for its decline, the fourth chapter provides the conclusions and the fifth and last chapter highlights the suggestions using which the sales can be increased to revive the company's profit and market value of its products.

Chapter Two

An Overview of Sales during 2010–2014

Ever since its inception, the Hindus Machine Tools (HMT) is renowned for its quality watches. It is a matter of pride to state that even now HMT's sales are influenced by its age-old reputation. However, in the recent years, there have been challenges to its monopoly.

(Contd)

(Contd)

As per the ratings of 'BUSINESS INDIA', February, 99 issue, HMT had 85% of the total watch market in Delhi in 2000–2005, whereas the percentage share of HMT sales during 2010–2014 stands at 35% only.

The invasion of many national and international brands into Indian market has changed the very face of the market. It is imperative for the company to evolve new market strategies to overcome the decline in its sales which has been most conspicuous over the past three years.

This chapter gives an overview of the sales statistics during the years 2010–2014 under separate classes classified on the basis of the price range of the watches.

2.1 Elite Class
Price: ₹50,000 and above

HMT watches belonging to this class fall in three different models—Festive, Sports, and Tinytots. The sales of the renowned *festive* models remained relatively constant over the years 2010, 2011, and 2012. However, there had been a steep decline of 8.9% during 2012–2013 although it stabilized during 2013–2014 registering a fall of 2.7% in its annual sales.

The *Sports* watches were introduced in 2012 which also recorded a drop of 11% in the year 2013-14; after this, sales figures have remained constant. The *Tinytots* model is a recent introduction in this segment.

2.2 Executive Class
Price: ₹15,000–50,000

These festive brands in this category have displayed a steady rise over the years 2010–2014. Though the increase has only been about 1.15%, it is a healthy trend.

The notable 'general' models of this category have undergone minor fluctuations. There has been an alternate rise and fall in sales of these watches over the past five years.

The 'other' category comprises the 'sports' and 'tinytots' models. The 'sports' designs were introduced in 2012; they have shown constant sales since then. The 'tinytots' brands are also well received in the market which is reflected in the sales figures for the year 2014.

2.3 Business Class
Price: ₹8,000–15,000

This class also has a wide range of festive, *general*, *sports*, and *tinytots* models. However, its sales have declined alarmingly over the past three years. An analysis of the statistics is as follows.

In festive category of this segment, it can be seen that over the years 2010–2012, the sales dropped by 16%. There was a further drop of 20.5% during 2012–2013 and 29% during 2013–2014 as depicted by the graph.

The *general* watches, on the similar lines, have shown a sharp decline in sales. These models which were 80,023 in 2010 have fallen to 28,127 in 2014, a quarter of its original figure. During 2010–2012, the sales went down by 27.5%. In the next year the percentage drop was 23.9. This decline finally grew into a shocking figure of 36.2% during the last year.

'Other' category of **'sport'** varieties managed to record a marginal increase in its annual sales of about 4.6%; the sales of **'tinytots'** brand has been exceptionally encouraging which registered an increase of 57% in the very first year of its introduction.

2.4 Economy Class
Price: ₹1500–8,000

The economy class is also a cause of worry with its sales statistics showing a steady decline over the past five years.

The **festive** watches of this class showed an overall decrease of 9% from 2010–2014, an average drop of 2.25% per year. The sales of **sports** brands have been fluctuating since their introduction in 2012. On the other hand, sales of 'Tinytots' have increased by 33.6% like their counterparts in the preceding classes.

The **general** watches however have been a cause of concern. As depicted in the graph, there has been a fall of 5.4% during 2010–2011. This grew to 9.9% in the next year. However the last year sales registered a fall of 7.5% as against the fall of 13.7% during 2012–2013.

(Contd)

(Contd)

2.5–'Bachat Class'
Price: ₹500–1500

This class comprises economic models in 'festive' and 'general' categories.

The **festive** varieties of this class have on the whole recorded a drop of 47% over a period of last five years. There has been a decline of 16% during 2010–2012.The subsequent years of 2012–2013 and 2013–2014 have shown a decrease of 10.3% and 21.9% respectively in the annual sales.

The **'general'** watches of 'Bachat class' remained relatively constant during the year 2010–2011.The subsequent period between 2011 and 2014 witnessed a steep fall of 64%. Years 2011–2012 registered a drop of 12%. The following years the figures showed 21% and 31% decline in sales for 2012–2013 and 2013–2014 respectively. The decline in sales of this class has deepened the crisis in the HMT.

Number of Watches Sold during 2010–2014 (in the North Zone)

Sales of Different Models	Year	Name of the Class				
		Elite	Executive	Business	Economy	Bachat
Festive	2010	552	14265	13256	18011	20417
	2011	576	14432	12144	17901	19200
	2012	562	14512	11070	16892	17298
	2013	512	14680	8800	16592	15518
	2014	498	14770	6225	16380	12119
Sports	2010	–	–	–	–	–
	2011	–	–	–	–	–
	2012	370	9327	8016	5556	–
	2013	327	9492	9142	6301	–
	2014	332	9484	9565	5366	–
Tinytots	2010	–	–	–	–	–
	2011	–	–	–	–	–
	2012	–	–	–	–	–
	2013	–	4500	3066	2576	–
	2014	87	5123	4831	3443	–
General	2010	–	82137	80023	108920	120255
	2011	–	83088	72068	103014	115261
	2012	–	81087	58003	92803	101320
	2013	–	83145	44136	80071	80143
	2014	–	81325	28127	74015	55296

(Contd)

Chapter Three

Reasons for Decline in Sales

This chapter deals with the analysis of the sales figures of HMT watches.

3.0 Sales Policy and After-sale Services

The Sales Policy of a company influences the customers and yields profits to the company. HMT, in its initial years, had marketing strategies that fitted perfectly into the market and consequently enjoyed monopoly for a long time. The Indian market has witnessed a lot of changes in recent times and with more national and international watch companies competing in the Indian market, the entire face of the market has changed. HMT has remained slow in changing its sales policies along with the market requirements. The whole strategy needs to be modified to suit the modern market. About 5.35% of the customers are not satisfied with the after-sale services of the company. It is mainly due to the absence of easily accessible sale-service centres.

3.1 Prospective Buyers

It is seen from the company's sales figures that the decline in sales has ranged from 40 to 50% in the Bachat Class and from 53.04 to 64.85% in the Business Class over the years 2010–2014. These are shocking figures as it can be seen that the number of people belonging to these classes have increased significantly over the past five years. But the sales fail to increase in consonance with the increase of potential buyers belonging to the respective classes. This can be so because the company could not stand up to their expectations.

3.2 Sponsorship

Nowadays, the watch companies sponsor youth festivals, games, sports meets, community gatherings, music nights, etc., for publicity purpose and betterment of their sales. HMTs Budget for the sponsorship in 2010 was ₹250000 and in 2014 it has increased only by a mere ₹50000 whereas, the budget of Bitan for sponsorship exceeds ₹1500000.

Even the budget for Advertising in HMT is found to be 40–45% less when compared to the other watch companies like Bitan, Jantha, etc. In fact, after studying carefully the policies and procedures of the company, it can be noticed that HMT has not paid attention to sponsoring the youth festivals, sports meets and community events which could have given bulk buyers for the company watches.

3.3 Dealers and Distributors

The HMT dealers as of now get a profit share of only 2% against Bitan, Jantha, and Citigen which give 5, 3, and 4% profit shares to their respective dealers. The number of dealers has therefore decreased from 1180 in 2010 to 886 in 2014, whereas the other watch companies have recorded a hike in this number. As per the survey, 30% of the distributors are unhappy with the quality of dials of our watches. Twenty per cent of them complain about the fact that our watches are not water proof. They also find that 5% of the watches do not stay in perfect working conditions even in the guarantee period.

3.4 Revised Policy Matters

Before 2012, HMT had a sales policy in which watches were sold in bulk to the employees of different firms on rebate. There was a loan scheme and salary deduction scheme as well. This bulk sale contributed to about 12.5% of the sales in the watches of executive and economy class and hence contributed to the large recorded share prior to 2012. But in the subsequent years, these schemes weren't revived, which in turn led to the decline of sales in economy class and failed to notch up the expected rise of sales in the executive class.

3.5 Design

Appearance: The appearance of HMT watches is quite traditional. Companies like Bitan and Jantha caught the pulse of the westernizing Indian society and updated their watch designs, adding a western colour to them.

(Contd)

(Contd)

HMT has not been successful in updating the old designs and has failed to bring out changes in the appearance of the watches.

Introduction of new designs: The number of designs of HMT has increased by a mere 12% against an increase of 30–40% in other brands. The other brands have also adopted to have the models which mimic the costlier watches in Bachat and Economy class as well. No proper steps have been taken to concentrate on the provision of the wide range of watches for the 'Bachat' and 'Economy' classes which constitute the major customers.

3.6 Publicity

It is a matter of pride that the backbone of HMT's sales is its age-old reputation. However nowadays, it is essential for a model to stand on its own in the market. One of the most important reasons for the success of a product is how the company projects it out in the market. Here arises the need for advertisement and publicity in new and modern forms.

It is not very encouraging to note that there has only been a marginal increase in the budget allocation for publicity. For the financial year 2012–2013, the allocation for publicity was 53 lakhs, as against 49 lakhs in 2011–2012, which is an increase of about 8%. Four per cent more was given to television, which is not an encouraging figure. The situation for 2013–2014 was no different with only about 10% increase in funds for publicity. Recently a website has been opened by HMT. This will help a lot and is a good step forward. The discouraging fact is that the budget allocated for publicity through newspapers has not increased in the past three years and there has been just 5% increase in budget allocated for publicity through television and radio.

3.7 Internal Environment

HMT is the only watch company in India in the corporate sector. Its staff is well experienced and the management is well aware of the pulse of the market. Despite this, the company is witnessing a decline in its sales. The internal environment can be partially responsible for this. It should be noted here that some leading watch companies in India have introduced 'result-oriented perks'. This boosts the staff's morale and their attitude changes. HMT has remained immune to such factors.

The enthusiasm shown to 'develop' the technical knowledge among the staff has also been quite low. There had been only 3 seminars and 2 workshops for the staff, in the year 2013–2014. Technical knowledge in its latest form is very essential for the staff, especially in this competitive world.

Chapter Four

Conclusions

Though HMT has always given high quality products to its buyers in the past, now it faces a stiff competition from many 'newcomers'. Following are the main findings of the report:

1. Five per cent of the watches do not stay in perfect working conditions even in the guarantee period. These are serious defects.
2. A complaint from 30% of the distributors is that the dials get rust. Twenty per cent of them have the complaint that the water-proof watches are not up to the expectations.
3. The after-sale services are also not satisfactory. About 5.35% of the customers are not satisfied by the after-sale services, as per this survey. HMT has one after-sale service centre in a radius of 200 km, but other leading companies have three such centres in the same radius.
4. The enthusiasm shown to 'develop' the technical knowledge among the staff has also been very low. There had been only 3 seminars and 2 workshops for the staff, in the year 2013 – 2014.
5. HMT has remained slow in changing its sales policies along with the market.

(Contd)

(Contd)

6. The discouraging fact is that the budget allocated for publicity through newspapers has not increased in the past three years.
7. The budget allocated for publicity is inadequate looking at the level of competition in the market.

Chapter Five

Recommendations

1. Due to stiff competition in the market, it becomes absolutely essential to revitalize the quality control department, and revise the quality standards by doing technical research on the watches.
2. More seminars and workshops should be conducted to develop the technical knowledge and skills of the staff.
3. 'Result oriented perks' should be introduced to change the attitude of the staff and motivate them further.
4. More polytechnic and fresh ME mechanical students should be taken for internship and they should be provided with requisite trainings by the company itself. This would help in doing research for bringing out new models and also in reducing the financial burden of the organization to some extent.
5. There should be no hesitation in cancelling the dealership of those dealers who do not provide prompt and efficient services.
6. More dealers can be included and the percentage of profit given to them as incentive be increased.
7. Every care should be taken to provide original spares manufactured by the company.
8. More *after-sale service centres* can be set up. The distance between one service centre and another should be reduced.

 ## RECAPITULATION

✓ A formal report is specifically designed to enable the reader to access easily and quickly, verify and/or explore key data. For that reason, a formal report is organized, written, and well formatted.

✓ Reports enable the authorities to take timely decisions. The data may also be given to the readers to let them analyse and interpret on their own and decide on the course of action.

✓ Reports are written for various purposes ranging from a simple presentation of facts to the intricate pattern of the analysis and its interpretation. Whatever the purpose maybe, it is always need-based.

✓ Not all reports are distributed within an organization, nor do they always provide possible solutions to problems.

✓ Each report should be tailored keeping in mind the audience. Therefore, while determining the elements to be included in a lengthy formal report, you need to keep the purpose and the audience in mind.

✓ A meticulous preparation will save your time and make the writing of the report easier. It will also help you keep your objective in view and realize it effectively.

✓ While writing a report, you should carefully provide proper links between and within its sentences, paragraphs, and sections which make it is easy for the reader to follow the message it conveys. For this purpose, you should use plain, simple, and familiar words rather than complex and unfamiliar ones. Also try to avoid redundant expressions and circumlocution to maintain objectivity in your writing.

WISEWELL QUIPS

EXERCISES

1. Identify which methods/sources of data collection you would prefer for preparing a report on the following:
   (i) Checking the conduct of an employee
   (ii) Market survey of a product
   (iii) Survey on availability and utility of resources
   (iv) A review of historical moments or incidents
   (v) A survey on the rehabilitation programme conducted for the flood-affected areas in Orissa
   (vi) Findings of an experiment done at the CSIR Lab, Delhi
   (vii) Finding the reasons of dissatisfaction among customers about the recent Food Processor model by Whitelines
   (viii) Survey on the reaction of parents and students regarding the changes done in the syllabus and the evaluation pattern of 12th Board exams

2. Conduct an interview with a person employed in a mobile company in your city. Try to find out, as far as possible, about the exact nature of the jobs performed, specific working conditions and opportunities for advancement.
   (i) Prepare a set of at least twenty questions that you want to be answered.
   (ii) After your interview, sit down with your own interview sheet and see if you could collect the desired data from all your questions.

3. Think of some problem related to your campus. Perhaps the students want some classrooms to be opened earlier or kept open after college hours for preparation of extracurricular activities; maybe the library doesn't have an appealing list of casual books for reading. Identify a person on campus who should be able to give you some valid information regarding the problem, and make an appointment to see that individual. Before you interview, prepare a list of 15 questions that you want answered. After your interview is over, tabulate the data to be used for your report.

4. Select the information you gather in either question number 2 or 3 above, and organize the material into a carefully unified 500 word theme.

5. Sometime during the semester, your instructor asks you to attend a nearby meeting. Submit to the instructor the following information:
   (i) When was the meeting held? Where? What dates?
   (ii) Who was the sponsor?
   (iii) How many persons were in attendance?
   (iv) Was the program largely composed of general sessions with important speakers, or was it a workshop program with smaller study groups?
   (v) Exactly what sessions of the meeting did you attend?

(vi) Who was the keynote speaker?

(vii) Write a brief summary of the speaker's important statements.

(viii) What are his or her qualifications?

(ix) Write a short reaction to the statements: Did you agree with the speaker? Did you find him or her convincing? Was he or she interesting?

(x) If you attended any small workshops, indicate the subject discussed.

(xi) Who was in charge of the workshop group? His/her qualifications?

(xii) Was the subject treated in panel discussion, with a debate or forum, or in a general discussion by everyone?

(xiii) Write a brief summary and evaluation of the discussion. Did the subject receive fair treatment? Did any opposition present have an adequate voice? Was the subject handled in such a way that it gave you any new ideas or changed any that you already had?

6. What are the sources and methods that you can use for collecting data for your report? Discuss them in detail.

7. Discuss the advantages and disadvantages of the telephonic interview and questionnaire as methods of data collection for reports.

8. What do you mean by a formal report? Discuss its various features at length.

9. How important are the business reports in the business world? Provide examples to substantiate your answer.

10. How are formal reports different from informal reports?

11. Enumerate the various differences between a letter report and a memo report.

12. How do recommendations differ from conclusions? What things you will keep in mind to bring out the distinction between the two?

13. Assume that in the capacity of the office manager of an electronic company, you have to submit an inventory report to your controlling officer. Now prepare the pro forma and fill it up to be sent to the officer.

14. Imagine that you are working in a construction company, and have been assigned to look after the construction of a five-storied apartment which is a three year project. Since the second year is over, prepare the progress report for this purpose to be submitted to the managing director of the company.

15. You have been asked to study the reasons for increasing smoking and drinking habits among the university students. Now prepare a questionnaire to be sent to the students of various universities to obtain their views on this issue. Your questionnaire should consist of twenty questions apart from the personal details.

16. With a view to rescheduling its programmes, the Director of Zevit TV Channel has asked its Publicity Manager to submit a report. Assuming yourself to be the publicity manager, plan to conduct a survey among two thousand families in the four cities in north India namely Delhi, Chandigarh, Lucknow, and Jaipur. Prepare a mail questionnaire to be sent to the viewers for this purpose. Your questionnaire should have a cover letter, personal details and at least 20 questions to elicit the relevant information.

17. Two years ago, Reliance Retails, Mumbai, started a chain of petrol pumps across the country with a plan for future expansion. After two years, the Chief Executive Officer of the company feels concerned about the slow growth of the company. It has been found that customers still prefer Hindustan Petroleum related petrol pumps. You, being the Chief Marketing Manager of the company, have been asked to study the reasons of slow growth. Assuming that you have studied the problem, prepare only the *Introduction* of your full length analytical report in about 250 words which you would submit to the CEO of your company. Invent necessary details.

18. The Government of India is concerned about the steady increase in the outbreak of diseases among working class. In order to work on the possible solutions, the Health Minister has asked the Chief Medical Officer, IMSI, Delhi, to find out the possible causes of these diseases, the problems faced and the other related factors. On the basis of the following data, as CMO, IMSI, Delhi, prepare a report to be submitted to the Health Minister.

Table 1 Data Showing Diseases and the Percentage of Suffering People

Serial No.	Name of the Disease	Percentage of People Suffering from It
1.	Diabetes	25%
2.	Blood Pressure	20%
3.	Stress	35%
4.	Migraine	10%
5.	Asthma	5%
6.	Slip disc	5%

Now prepare a full-length report including the elements such as the Title page, Introduction, Discussion, Conclusions and Recommendations.

19. The Centre for Social Research, Chennai, is conducting an independent research on 'Reasons of Increasing Use of Violent Means among Youths' to press for their demands and the findings of this research are likely to be published in national and international news magazines. As part of this research, views of people of different age groups from ten metropolitan cities—Delhi, Chennai, Hyderabad, Jaipur, Pune, Bengaluru, Thiruvananthapuram, Chandigarh, Mumbai, and Kolkata—have been ascertained through mail questionnaire. The tabulated data are presented below:

Age Group	10–20	20–30	30–40	40–50	50–60
Corruption	20	21	18	27	20
Unemployment	5	35	35	11	07
Faulty Legal System	10	20	16	24	33
Easy Access of Arms	10	10	11	09	08
Cinema	22	04	17	12	10
Unitary Family	33	10	18	17	22

(The given data are in percentage.)

Assuming yourself to be the chief investigator, prepare an analytical report to be submitted to the Director of the Centre, using the given data and inventing other necessary details. Your report should contain the Title Page, Introduction, Discussion, Conclusion, and Recommendations only.

20. The Chairman of Hindus Computers Ltd., Dispur has felt the need for an intensive HR training programme for its entry-level and middle-level managers. Hence he has asked you, being the Director of the company, to analyse the various areas in which training is required for them. You have collected data for this purpose which is tabulated below:

Analyse and interpret the data and draft a letter report to be submitted to the Chairman so as to enable him to organize an effective training programme.

Table showing training needs in percentage

Level of Managers	Soft Skills	Mangerial Skills	Software Skills
Middle Level	39.8	26.2	34
Top Level	34.4	23.3	43.3

21. The recent flood that hit in the Vijayawada and Guntur areas of Andhra Pradesh has caused a large-scale damage to the government and the public properties. The state and the central governments want to assess the damage before they start the reconstruction work. Assuming yourself to be the Executive Engineer of the Public Works Department, Vijayawada, prepare a *letter report* inventing necessary details on the devastation caused by the recent floods and submit the report to the Chief Secretary, Andhra Pradesh.

22. India Airways has been constantly incurring losses due to a sharp decrease in the number of passengers traveling by this airway. The Civil Aviation Department, Govt. of India, is serious about this problem and wants to know the real causes for the passengers' lack of interest in this airways. You, being the Chief Commercial Manager, have been asked to look into this problem and suggest suitable measures to attract more passengers. Now write the *discussion, conclusion, and recommendation* (only) of your full-length report. Invent necessary details.

23. After the successful launch of the cyber café, the Chief was shocked to notice a steep decline in the number of customers. To retain the original number of customers, the owner has asked the Marketing Manager to conduct the survey and find out the

reasons so that corrective measures can be taken. The results of the survey are given below.

Options	Satisfactory	Poor
Seats Availability	48%	52%
Internet Connection	23%	77%
Supporting Staff	34%	66%
Proper Ambience	25%	75%
Charges	45%	55%

Now, write a report to be submitted to the Chief of the café. Your report should have the following components:

(i) Introduction   (ii) Discussion

(iii) Conclusions   (iv) Recommendations

24. Siddhartha Softdrinks Ltd wants to introduce a new soft drink. The Director of the company has assigned the Production Manager to make proper investigation about the taste, advertising, transport, customers' needs and cost of the product to make it an instant hit. Now assuming yourself to be the Production Manager, write a report to be submitted to the Director. Your report should have the following elements:

(i) Introduction   (ii) Discussion

(iii) Conclusions   (iv) Recommendations

25. Assume that you are the Chief Librarian of the Radha Krishnan Community Centre, Chandigarh. Nirman Organization, a nonprofit group, raises funds and provides volunteers to support your centre. Every February, you send a report of the previous year's activities and accomplishments to this group, as it provides an annual grant of fifty lac rupees. Now write a letter report to be submitted to the Director of the Nirman Organization giving the details of the previous year's activities and informing the new activities you are planning to introduce in the coming year.

26. The spread of H1N1 virus causing Swine Flu has created panic in general populace of the country. Patients with even minor cold and fever are doubted for Swine Flu by their neighbours. The central and the state governments have realized the gravity of the problem and therefore have asked NGOs working on health issues to launch a mass awareness campaign on Swine Flu. You being the Chief Medical

Officer, Narayana Hospital Group, Pune, have been asked to submit a report on the recent efforts being made in this direction in the district. Prepare a letter report to this effect. Invent the necessary details.

27. The Senate of Comp. Edu. Institute has decided to conduct a survey on the functioning of its Library. The survey will include issues like management of the library and the problems faced by the users. As the student member of the Senate, you have been asked to write an interpretive report on the same. Now write an outline and Introduction for this report. Your outline should have around 8–10 topics and sub-topics up to third level.

28. Assume that the Kinetic Udyog Limited is paying the cost of your education. The agreement is that you will serve them for five years after doing your MBA. They want a report on the progress of your work and the quality of training you have received so far. Write this report, which will be circulated to the members of the board of directors. Your report should contain the following elements only:

(i) Introduction   (ii) Discussion

29. Rewrite the following sentences to make them simple, clear and precise:

(i) Our experience suggests that who listen to others with dwindling attention, fail to ever speak properly. It is so because listening is the mother of all speaking.

(ii) Though we all pretend to listen to others while sitting in a meeting or attending some oral presentation, we usually are subtly occupied with the impulsive idea of speaking at the earliest opportunity.

(iii) It is commonly observed as charging on their emotional intensity, youngsters are prone to hasty and often miscalculated guess usually much to the chagrin of their officers.

(iv) Taking feedback from employees helps them in the removal of the unnecessary cobwebs arising out of the ills of hierarchy and achieve a commonality of purpose within an organization.

(v) The task of laying off employees is a daunting one and needs subtleties of expressions to be employed deftly and strategically in order to keep the right-minded-employees hooked to the goals of the organization.

(vi) In professional environment, listening acquires a monumental importance as without listening to the others' views, no corporate communication would attain the precision and authenticity desired in such circles.

(vii) Inspiring the employees to come out with a genuine feedback, processing the public perceptions received and emitting a positive signal continuously requires monumental communicative talent.

(viii) Most of us have had experiences as harrowing, traumatic and distressing as this one, to some extent at least.

(ix) He seemed at first to conscientiously and carefully try to carry out the captain's instructions.

(x) Michael successfully underwent on the third of last month a surgery at the Ray Hospital.

30. Rewrite the following passages to make them simple, clear, and precise.

(a) There have been failures and shortcomings in various sectors of the economy. It is also true that avoidable mistakes have been made, which have adversely affected the country's economic development. Apart from the consequences of human failures, other factors beyond our control have also had their effect in slowing down the pace of economic progress. Perhaps, to an extent, all this was inevitable and unavoidable for a country trying to rise above the limitations imposed by time and history.

(b) Nevertheless, it will be churlish to deny that the country has made tremendous progress in many directions, notably in agriculture and industry, in education and technology, in health and housing. But we have yet to solve the twin problems of unemployment and illiteracy. There is no blinking the fact that economic independence is still a chimera for millions of people. The planners and government are fully aware of this. They are equally determined that the battle of economic independence should go on till every citizen is able to lead a full life and face the future with hope and confidence.

(c) No doubt some temperaments take much more kindly to a regular routine than others. There are many who shy away from the self-regimentation of a weekly time-table, and dislike being tied down to a definite fixed programme of work. Many able students claim that they work in cycles. When they become interested in a topic they work on it intensively, passionately and aggressively for three or four days at a time. On other days they avoid work completely. It has to be confessed that we do not fully understand the complexities of the motivation to work. Most people over about 25 years of age have become conditioned to a work routine, and the majority of really productive workers set aside regular hours for the more important aspects of their work. The 'tough-minded' school of workers is usually very contemptuous of the idea that good work can only be done spontaneously, under the influence of inspiration.

(d) It is almost impossible to escape from advertisements. Hoardings stare down at us from the sides of the roads; crude neon signs wink above shops; jingles and slogans assault our ears; in magazines, pictures of washing machines and custard-powders take up more room than the letter press. All these are twentieth century developments which have grown side by side with the spread of education and technical advances in radio and T.V. Advertising assaults not only our eyes and ears but also our pockets. Its critics point out that in this country 1.6% of national income is spent on advertising and that this advertising actually raises the cost of products. When a housewife buys a pound of flour, 5% of what she pays goes to some advertiser or other, even if she has not bothered to ask the shopkeeper for a particular brand.

(e) In good old days during ragging seniors used to tease the timid and nervous juniors and in the process all became good friends, even life-long friends. All that is bygone and over now. Ragging has become a nightmare, dreadful and horrible experience for the juniors and their parents. Consequently, some even dropout. According to a press release a boy in Vijayawada committed suicide as he was teased for wearing cheap clothes and his-complexion was dark. Other day there was a news in the Hindu that a girl in

Hyderabad took her life since she was unable to bear the ragging that was done to her in her college. All these years I was under the impression that girl students were spared of the torture. But women's demand for equality is squarely met in ragging which has become synonymous with the fitness, strength and robust tests conducted by the seniors, a vulgar display of exploitation and abuse.

ANNEXURE 23.1

Checklist for a Business Report

Having prepared a business report, answer the following questions before submitting it to the primary recipient of your report:

Q.1. Does the report have a cover and a title page?	Yes ( ) No ( )
Q.2. Does the report have acknowledgements?	Yes ( ) No ( )
Q.3. Does the report have an abstract?	Yes ( ) No ( )
Q.4. Is the abstract of the report written properly?	Yes ( ) No ( )
Q.5. Does the report have a proper introduction?	Yes ( ) No ( )
Q.6. Does the introduction have a reference to the methods of data collection?	Yes ( ) No ( )
Q.7. Does the introduction refer to the objective and scope of the report?	Yes ( ) No ( )
Q.8. Does the introduction highlight the significance and limitation of the report?	Yes ( ) No ( )
Q.9. Does the introduction orient the reader to the report properly?	Yes ( ) No ( )
Q.10. Is the discussion section of the report divided in different parts?	Yes ( ) No ( )
Q.11. Is each of these sections given a proper heading and number?	Yes ( ) No ( )
Q.12. Does the report include all the data relevant to the interpretation?	Yes ( ) No ( )
Q.13. Does the discussion sound convincing and worthwhile?	Yes ( ) No ( )
Q.14. Does the discussion probe the situation or analyse the problem efficiently?	Yes ( ) No ( )
Q.15. Is the analysis in the discussion suitably supported by figures, tables, charts, graphs and other illustrative material?	Yes ( ) No ( )
Q.16. Are the conclusions given in the report logical and convincing?	Yes ( ) No ( )
Q.17. Do the conclusions fall in consonance with other elements in the report such as the introduction and the various parts in the discussion section?	Yes ( ) No ( )
Q.18. Are the recommendations appropriate and relate to the rest of the discussion in the report?	Yes ( ) No ( )
Q.19. Is the back matter in the report proper?	Yes ( ) No ( )
Q.20. Are the footnotes and references properly cited?	Yes ( ) No ( )
Q.21. Are the references in the report properly cited and documented and conform consistently to a standard format?	Yes ( ) No ( )
Q.22. Is the bibliography in the report alphabetically arranged?	Yes ( ) No ( )
Q.23. Are the bibliographical details arranged in proper sequence in all the references and entries in the bibliography?	Yes ( ) No ( )
Q.25. Is the appendix neatly divided into various parts and numbered?	Yes ( ) No ( )
Q.26. Are the various elements in the report arranged in a conventional or desired form?	Yes ( ) No ( )
Q.27. Is the report written in an objective, concise, specific, and direct style?	Yes ( ) No ( )
Q.28. Are all the pages in the report numbered and sequenced correctly?	Yes ( ) No ( )
Q.29. Are all the tables, figures, charts, and graphs in the report numbered and titled?	Yes ( ) No ( )
Q.30. Do the items in the table of contents and index match the contents in the report?	Yes ( ) No ( )
Q.31. Does the report have an overall consistency in its style, order, layout, and presentation?	Yes ( ) No ( )

Technical Proposals

Learning Objectives   After reading this chapter, you will be able to

- understand what a proposal is and how it differs from a business plan or a report
- understand the purpose and importance of technical/business proposals
- learn in detail about the structure of a formal full length proposal
- equip yourself with the various elements required to prepare a winning proposal
- learn how to write different proposals including a letter proposal and a short proposal

24.1 INTRODUCTION

As a professional, you may face countless situations in which it is essential for you to prepare technical documents. These could range from reports, notices, letters, and memos to proposals. In the preceding chapters, we have dealt with all other elements except proposals. So now let us discuss what a proposal is, its features, structure, and style in this chapter.

When a company thinks of a merger with another company, it may ask its personnel to submit a proposal reviewing the possible industrial and commercial developments that might make the merger desirable. If some multimedia laboratory wishes to approach various colleges and universities for setting up multimedia lab, they are required to understand the needs of the specific organization and submit a proposal based on their requirement. Thus, a *technical proposal* is a written offer from a person who has a selling idea to a prospective buyer. It intends to elicit business from a prospective buyer. Proposals have important place in business growth and professional relationships. The cut-throat competition in global business has made it essential for companies to look for new ideas to sustain their market and growth. Often, it is confused with *business plan*. However, a business plan is different from a *business proposal*. A business plan is usually drafted to get capital for the start-up venture. These plans cover your business structure, your products and services, your market research and marketing strategy, and your complete budget and financial projections. A technical or business proposal however is written by some professional or expert who proposes to sort out a technical, business, or managerial proposal of a prospective buyer. A proposal also identifies the prospective buyer's need to implement new ideas for enhancing efficiency, increasing productivity, and improving performance in various professional operations.

Before we talk about the features, structure, and style of a proposal, let us understand the differences between a report (read in Chapter 23), a proposal, and a business plan. Though they are similar to some extent, there exist various differences between them which are worth noting (Table 24.1).

Table 24.1 Differences between a Report, a Proposal, and a Business Plan

Proposal	Report	Business Plan
It is written to someone who needs to sort out a technical, management, or business problem or needs to implement new ideas to enhance efficiency and productivity in various professional operations.	Written to someone with authority to know the causes of the problem and possibly take a decision.	It is written to someone who needs to make a decision for profit making or strengthening its operational aspects.
It identifies a particular need, explains it thoroughly, and recommends how this need can best be met.	Written to identify a specific problem, explain it, and recommend action that will lead to a solution.	Like proposals, it identifies the workable idea, explains all aspects related to financial needs, target market, demographics and other information.
Formal but persuasive style.	Formal in style.	Persuasive.
It involves cost for the execution of proposed idea.	It only highlights the reasons based on the interpretation of data and make recommendations.	It is written to get capital for a start-up venture.
It is written to someone whose decision will directly benefit the writer in some way.	It is written to people who can take action or affect outcome.	Both the applicant and the approving authority aim at making profit.
It also deals with future professional possibilities. Since it is often put forth by experts and experienced professionals, there are less chances of failure.	It deals with some event or situation that occurred in the past.	Like proposals, it deals with future action but there are chances of failure as taken up by budding entrepreneurs.

24.2 PURPOSE

As defined in the preceding section, whenever a new idea or workable solution for a problem or advice is proposed, it is called a proposal. However, in your professional life you will come across various nomenclatures which describe the purpose for which that proposal is written. Some of these are given below for your clear understanding:

1. When the aim of the proposal is to modify or create something that requires a good understanding of technical knowledge and skills, it is called a *technical proposal*.

2. A *business proposal* is a document that you submit to your company or another enterprise proposing a business arrangement dealing with any aspect of business, commerce,

or industry. These proposals help to appraise and improve the existing products and services to meet the ever changing demands of market. Sometimes these are also termed as *sales proposals* when they focus only on increasing the sales of a product or service of a company.

3. When we intend to undertake a systematic research, we are required to submit a proposal outlining broadly the basis of the proposed research, its purpose, scope, significance, limitations, etc. Such proposals also include a reference to the procedures, methods, and theories to be followed in conducting the research. Academic in nature and written with scientific objectivity, such proposals are known as *research proposals*.

24.3 IMPORTANCE

In this section, we will learn about the importance of technical proposals.

1. Proposals serve as an indicator of the growth or progress of a company or organization.
2. These proposals help to invite other companies or industries for strategic alliances, joint ventures, acquisitions, and mergers.
3. If your proposal presents your business ideas effectively, it may help the company improve its products and services to have the competitive edge.
4. By sorting out a technical problem, proposals enhance productivity and improve performance.
5. Proposals help in securing technology partnership, fundraising, donation, event sponsorship, tenders, or inviting others for participating in an event.
6. Successful proposals usually ensure financial gains too for companies.
7. Research proposals help in creating new methods and procedures, and opening new dimensions of concepts which in turn expands the horizon of knowledge.

24.4 TYPES

There are two main categories of proposals:

- Solicited (invited)
- Unsolicited (uninvited)

Whenever a proposal is drafted in response to an advertisement or demand from an authority in a company or organization or outside the organization or agency, it is termed as a *solicited proposal*. These are invited proposals. For example, when the government and large corporations wish to purchase services or products for constructing dams, bridges, providing parking facility, etc., a large number of proposals bidding for the deal are received. In such a scenario, we are left competing with all other bidders that noticed the opportunity and responded. In this case, a proposal that finally gets picked up from the massive lot not only has to provide the most effective solution to the problem or make the most lucrative deal, but should also have stronger arguments in staking claims for the bid than other competitors.

Usually solicited proposals are to be written in the format defined by the agencies that invite such offers.

In comparison, unsolicited proposals are more demanding and require greater imagination on the part of the bidder. In an *unsolicited proposal*—a non-invited proposal, we might have an idea for a product or service that would be of benefit to a particular organization. We submit a proposal to that particular organization suggesting how we can provide some service or develop a product in exchange for funding or some other consideration. In this case, we do not know if the company is open to our proposal or not. There is every possibility that the company may not like our proposed idea. In this situation, a person may submit a proposal on his/her own initiative. Here your proposal has to convince the client that not only is the service/product potentially valuable to them, but you and your company are reliable and stable.

Whether invited or uninvited, our proposal must be well researched, well written, and must contain a properly worked out realistic budget.

24.5 STRUCTURE

Like reports, proposals too are written in a conventional or usable form. In case of solicited proposals, the structure is determined by the person or the company asking for it. In unsolicited proposals you can use the conventional sequence of element and if any other element or information is required, you have the freedom to put it forth at the right place. Since proposals are written for different purposes and for different organizations, there is no single format that suits all. According to the requirement and the nature of the proposal, you have to choose the elements to be included in the unsolicited proposals. The three main parts of a formal full length proposal are as follows (Fig. 24.1):

- Prefatory
- Main body
- Supplementary parts

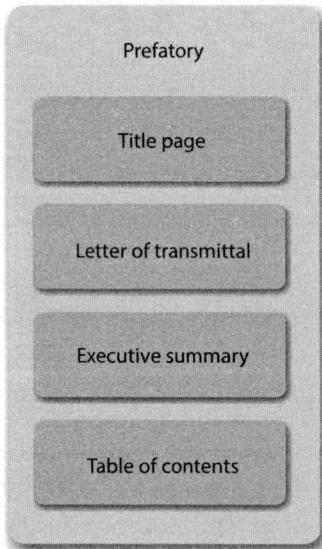
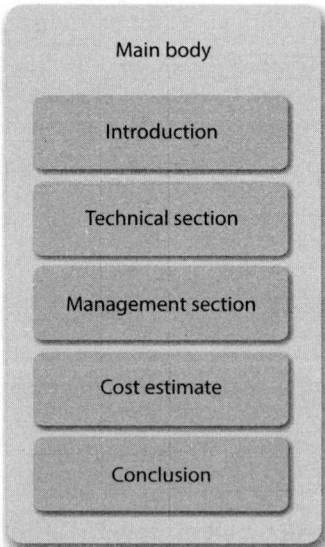
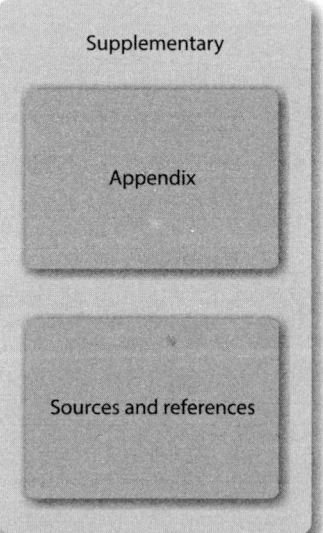

Prefatory
- Title page
- Letter of transmittal
- Executive summary
- Table of contents

Main body
- Introduction
- Technical section
- Management section
- Cost estimate
- Conclusion

Supplementary
- Appendix
- Sources and references

Fig. 24.1   Sections of a Formal Full Length Proposal

Title page   The *title page* of a proposal serves as a cover and if laid out well with proper spacing in between the various elements of information, it creates a good impact on the reader. The title page consists of information such as the title of the proposal, the name of the organization for which it has been prepared, the proposer's name and designation, the name of the organization to which he/she belongs, and the month and year when the proposal is being submitted. Look at the following sample:

Sample Title Page

A Proposal
on
Setting up a Recreation Community Centre at Yumiko Telecom Ltd
Submitted to
General Manager
Yumiko Telecom Ltd, New Delhi
by
Project Manager
Salina Builders, Mumbai
Salina Builders, Mumbai
March 2015

Letter of transmittal   As the title of this element suggests, a *letter of transmittal* transmits the proposal to the primary recipient. It reveals the topic, the purpose, its highlights and makes an appeal towards a favourable response. Like in case of reports, the letter of transmittal can be either sent along with the proposal or it can be bound immediately after the title page.

Table of contents   In lengthy proposals (more than 10 pages) for quick location of major elements, the table of contents is given immediately after the draft contract. It is prepared the way the table of contents is prepared for a report, as discussed in Chapter 23. Short proposals do not include this element. Similarly, if there are too many illustrations, we need to give a separate list of illustrations; otherwise it is not required.

Executive summary   The recipient of the proposal goes through the executive summary very carefully. He/she tries to assess whether the proposal is worth considering. The obvious reason is that it not only provides a brief background, purpose, scope, and methodology, but also gives a budget summary besides re-emphasizing the significance of the project. A proposal may not have a table of contents but it will surely have an executive summary to give a quick view of the entire proposal. This includes the following elements:

- Background
- Technical details
- Purpose
- Significance
- Scope
- Re-emphasis
- Infrastructure facilities

Let us look at a sample executive summary for a proposal:

Sample Executive Summary

Today's world, rightly known as 'The Computer Age', has taken a huge leap in the field of computers and information technology (IT). Earlier the task which used to take hours can now be done in a few minutes with the help of computers. It won't be wrong to say that today, software runs the world. Since computer technology has revolutionized our world, there is hardly any sphere of professional endeavour, including the hotel industry, that can hope to achieve a smooth and consistent growth.

Background

(Contd)

(Contd)

We are aware of the fact that your Food Pride Internationals (FPI) is one of the most prestigious low-cost chain of hotels in India. Over the years, FPI has carved a niche for itself for the quality of food offered. However, the great customer rush has created a demand for more efficient and prompt delivery and service methods. Moreover, with the newer technological advancements, it makes a sound business strategy to utilize these advancements for efficient functioning of the organization, for a positive and emphatic impression on the clientele and for maintaining in future too the high standards of quality and operations set in the past. With the help of the software tools offered in the proposal, we propose to make the functioning of your popular chain hotels smooth and more customer friendly.

Infrastructural facilities

This proposal aims at providing high quality software tools and very efficient maintenance personnel to your chain of hotels. The services offered include installing the software tools, ensuring that the software is functioning perfectly well, and deputing competent personnel for any future troubleshooting purposes. However, the proposal does not include providing the platforms which will be required for proper functioning of our software tools. These platforms (authorized versions of Microsoft Windows XP with specific memory requirements) will have to be bought by you.

Technical overview

The computers with the above-mentioned specifications will be used by our company to install the software tools. Our team will contact you to get the specific minute requirements of the software you want to have from us. This team will stay for some time in your hotel for this purpose.

The technical standards and strength of the project emanate from the world-class hotel management software tools that we plan to use for this project. Moreover, as per your requirements, all the requisite changes will be made to make the software amenable to your system. With the installation of the software, any customer can login to the hotel website and book any room as per the availability. The website created, will be certainly user-friendly so that a customer who knows only a little about computers can also use the system. The software will also be user-friendly from the perspective of your hotel staff. Further, proper training will be provided to your staff; the training would orient your operating staff to the functioning of the software without any additional charges. Even after the entire software tool has been deployed and training is imparted to your staff, you will have 24 x 7 assistance from our side. Our technical experts will always be there to help you use the software in the best possible manner.

Significance

By adopting the proposed software, your chain of hotels will be equipped with the world-class, customer-friendly, and round-the-clock services. All this will be made available to you at a lower cost compared to other in-line companies.

Introduction   Since an emphatic statement of purpose and a clear understanding of the problem are the most important aspects of your proposal, an introduction to the proposal should specify what your project intends to do. Also provide a brief review of the background information that gives rise to the present need. This helps the reader understand what needs to be done. If you fail to do so in the first paragraph of the introduction, all the efforts made in writing a proposal will go in vain. Let us look at the following examples and understand how a catchy beginning can ensure the attention of a prospective buyer:

1. As you read through this, you are sure to understand how to improve the efficiency of the bio-metric system installed in your organization.
2. Have you ever wondered how to ensure the full safety of the lab equipment in your institute? If yes, the following pages suggest an efficient way to achieve this objective.
3. This proposal is submitted with a view to introduce to you a proven scheme for reduction of production cost by almost 10 per cent in your plant.
4. The objective of this proposal is to help you make the cars manufactured by you more efficient in terms of fuel consumption and environment protection.

In fact, the beginning of a proposal is likely to be similar to that of a sales letter as in both these types of communication, we need to capture the attention of the reader. It is particularly so in case of unsolicited proposals which catch professionals unawares and are to be read by the prospective buyers without any apparent desire or willingness to do so. In order to keep the readers interested in your proposal, try starting on an innovative note by using strategies used for drafting sales letter. Though these have been discussed in the chapter on *business letters* in detail, here is a quick reminder of the tips that can render a beginning catchy, tempting, and effective:

1. Start on a striking statement.
2. Write a short, crisp sentence.
3. Address the need of the reader rather than the desire of the author.
4. Remind the prospective buyer of the difficulties he/she faces.
5. Bring into picture the utility of the proposed idea in the reader's world.
6. Instantly build a rapport with the prospective client with a 'you' type of approach.

Having made a captivating beginning, we can make further inroads into the reader's defence by outlining the purpose and scope of the proposal. A write-up that starts intriguingly well but fails to give a clear view of the objective, purpose, scope, utility, and benefits of the proposed idea, offends the reader. Therefore, after a good start, a proposal should be able to sustain the interest of the reader.

One sure way of keeping the prospective client hooked to the proposal is to help him/her understand and appreciate the utility of the offer. Though the details of the features of the proposed idea, research, method, plan, or scheme need to be highlighted separately, it is in the introduction that the need should be established. At times, it is also required to help the reader appreciate and comprehend the problem he/she faces before recommending the solution. Therefore, begin with a clear and empathetic understanding of the prospective buyer's need; bring in the utility of the proposed solution; sustain his/her interest by highlighting its special features such as low cost, high returns, improved services, substantial experience, and established expertise and goodwill.

Written in a persuasive manner, the introductory section of a proposal should be able to capture the interest of the reader; establish the need for the proposal; bring into view the positive highlights of the proposed solution; give a clear view of the technical, monetary, and human assistance required for the successful implementation of the idea within a specified amount of time.

Technical section In writing the technical section of a proposal, provide a solution which requires a technical plan. In this section, we need to analytically explain the technical terms, theoretical principles, and methods to be followed. This section also talks about the machines, equipment, and materials that we wish to use in carrying out the proposed task. While specifying these, clearly mention the advantages of these in bringing out the desired outcome of the whole project. Moreover, describe in detail the procedure that you plan to use. Give arguments for each step and explain how important they are for the success of the project. Use diagrams and other visual aids to emphatically and professionally present this requirement. If you require infrastructural facilities, such as an office or laboratory space, specify what those needs will be. Thus, in technical section, you should necessarily provide the technical description, methodology, and facilities required and the technical plan of action.

Management section   Once the need and efficiency of the proposed solution is established, convince the reader of your ability to deliver the goods. In the management section of the proposal, highlight the following points:

Credentials of the task force involved   Most of the proposals fail to click, simply because the readers do not feel convinced about the capability of the people who propose to solve their problems. The management section of the proposal should, therefore, highlight the credentials of the people who are involved or are likely to be involved in implementing the proposed idea. For this, give reference of the academic and technical knowledge of the people involved, mention their expertise and experience in the relevant field, and highlight their achievements in the past.

Company profile   A good company normally boasts of good personnel and services at its disposal. In fact, the people involved and the company they represent keep affecting each other's reputation. Talk about the efficiency of the system, policies, and implementation scheme, so as to convince the reader further about the company's ability to produce good results. A company's established profile, goodwill, and reputation undoubtedly create a very positive impact on the reader. However, even in the absence of an established reputation in the stated field, one can hope to create a positive impact on the reader. This can be done by focusing clearly on the efficiency of the people involved and the efficacy of the proposed solution that is likely to come through even in the absence of a staggering company profile.

Execution plan   A good proposal should always pointedly provide a clear outline of an execution plan for a proposed idea. An efficient execution plan includes a precise work schedule—the amount of time the proposed idea will take in its implementation; date/month of the commencement and completion of the task; date-wise/month-wise work plan; details regarding the reporting, maintenance, delivery, and payment system, etc. Moreover, a clear statement establishing an adherence to the work schedule in achieving the stated objective in the specified time helps the reader feel convinced about the workability of the proposed solution. At times, proposals already implemented and running in other organizations are also appended in the appendix section of the technical proposal. The purpose is to help the reader see that you have not only an intelligence to envisage the need for the proposed solution but also an acumen to execute and implement ideas within a limited time frame.

Don't Quote Unrealistic Cost Estimate in Your Proposals.

Cost estimate   A good proposal invariably takes into account the monetary variables that finally decide whether the proposed idea actually takes off. In fact, without a clear, precise, and detailed cost estimate, all our proposals are likely to remain confined to folded files. It is so because unless the prospective buyer is able to figure out whether he/she can actually support the proposed plan financially, he/she fails to make a firm assessment about the viability of the proposal received. Therefore, it is important for us to give a proper cost estimate so as to help the reader evaluate and assess the viability of a proposal. The figures quoted in cost estimate at the end of the proposal should not differ from those suggested in the draft contract that usually comes in the earlier part of the proposal. At times, the cost is calculated in terms of the number of hours spent on a particular task.

The cost estimate of a proposal includes fixed expenses such as purchase of land and equipment, and recurrent expense such as raw material and transportation.

Conclusion   A proposal usually ends with a realistic and detailed cost estimate, particularly when it is forwarded through a covering letter which is written in a persuasive and effective style. However, in order to end on an emphatic note, many bidders end their proposals with a conclusion. The purpose of a conclusion in a technical or business proposal is to remind the reader of the unique features of the proposed solution. At times, rather than writing a detailed conclusion, authors choose to end their proposals with a concluding line which, written in a persuasive manner, intends to induce action.

Appendix   Appendix is not an integral part of every proposal. However, in order to help the reader feel convinced about the author's claims, an appendix can be appended to a proposal. Usually, an appendix entails the following material:

- The description and results of a similar technical or business proposal already successfully implemented elsewhere
- The curriculum vitae of the personnel involved in the task force outlining their credentials, expertise, and achievements in similar tasks
- Approved contracts signed with other companies
- The detailed historical background of the proposed plan
- Extensive details regarding the operating system to be used and the procedure to be followed as the proposed plan gets executed
- The requisite layouts and maps, if any, and any other material that is likely to make the proposal look more comprehensive, realistic, and tempting

Draft contract   When some proposal is accepted, it is resubmitted with a *draft contract*. It gives a bird's-eye view of the most important information in the proposal. A draft contract includes the following elements:

- Title of the proposal
- Name of the proposer and designation (in case of research proposal, the name of the principal investigator and co-investigators' names and their designations)
- Name of the organization in which the project is to be carried out
- Duration of the project
- Terms and conditions
- Time required for the start of the project after approval
- Total cost to be followed by year-wise or phase-wise break-up of the total cost

Sample Proposal I

High Tech Lab India System
St. No 17, Near Birla Mandir, Ludhiana, Punjab

The Principal
Bhagwati College of Education
Moga, Punjab

Dear Madam,

SUB: Proposal for installing High Tech Digital Language Lab (HTDLL)

We wish to introduce ourselves as a well established and leading manufacturer of communication equipments for schools, colleges and other educational institutes. High Tech Digital Language Lab or HTDLL, in a nutshell, is a professionally engineered, comprehensive and user friendly language learning software brought to you by us, a dynamic IT solutions provider focused on delivering cutting-edge solutions primarily for educational institutions. Our Digital Language Laboratory System helps in enhancing the communication skills of students by two-way communication and inbuilt audio and video recording facility.

We are forwarding our most competitive quote tailored to your requirements along with HTDLL product details and optimal system requirements for your immediate perusal.

Some of our more recent clients are as follows:

- I I T, Guwahati
- I M T, Ghaziabad
- Indian Army, Dehradun
- Shrinathji Institute of Technology & Engineering, Udaipur
- Biyani Girls College, Jaipur
- UV College, Ganpat University, Gujarat
- Amity International Schools, Delhi
- Institute of Computer and Communication Technology, Anand
- Shankersinh Vaghela Bapu Institute of Technology, Gandhinagar
- Truba College of Science & Technology—2 LABS, Bhopal
- Sanjay Ghodawat Institute of Technology, Kolhapur
- Chinar Public School, Alwar (Rajasthan)

Kindly visit us at www.htdll.in for a detailed review of our range of products, services and clientele/testimonials.

Should you require any further details/clarifications, it would only be our pleasure to oblige at once.

Thank you once again and we look forward to a mutually rewarding and long-term association.

Yours truly

Mohit Bhagat
Manager, Marketing
High Techno Systems (India) Pvt. Ltd Cochin, Bangalore & New Delhi Customer Care: 0484–4141 000 to 4141 099 (100 Lines)
Encl: Brochure and Technical Literature

Sample Proposal II

Oxbridge Institute of Technology and Science, Chennai
Department of Communication and Media Studies

27 January 2015

A Proposal
For
A Three-day Workshop

(Contd)

(Contd)

on

Communication Skills for Effective Teaching

Objective

The workshop aims at providing a platform for inculcating effective communication skills among teachers through an array of innovative strategies and a wide range of interesting activities. This, in turn would enhance the effectiveness of their teaching and improve the student–teacher relationship.

Day 1	
9.00–10.00	Registration and inauguration
10.0–11.00	Innovative teaching—Sharing experiences
11.00–11.15	Tea Break
11.15–1.00	Understanding and using body language—Nonverbal agenda
1.00–2.00	Lunch
2.00–3.00	Preparing PowerPoint presentation
3.00–4.00	Mock presentation followed by discussion
4.00–4.15	Tea break
4.15–5.00	Communication activities

Day 2	
9.00–10.00	Strategies for group discussion and mock GD
10.0–11.00	GD practice followed by performance appraisal
11.00–11.15	Tea break
11.15–1.00	Oral presentations by participants
1.00–2.00	Lunch
2.00–3.00	Emotional intelligence—Skills involved
3.00–4.00	Mock presentation followed by discussion
4.00–4.15	Tea break
4.15–5.00	Emotional Intelligence—Problems and solutions

Day 3	
9.00–10.00	Viewing the recorded presentation and self assessment
10.0–11.00	Vocabulary enrichment
11.00–11.15	Tea break
11.15–1.00	Feedback and valedictory

(Contd)

(Contd)

Cost Estimate

Participants	:	25 teachers from various schools in East Chennai
Tentative date	:	29–31 March 2015
Resource Persons	:	5
Registration Fee	:	₹600/- per head
BUDGET		

Folders (40)	=	₹2000.00
Learning material	=	₹3000.00
Tea and snacks	=	₹10000.00
Certificate production	=	₹2500.00
Miscellaneous	=	₹1000.00
	Total	₹25000.00

(Dr Abha Mittal)
Department of Communication and Media Studies

Sample Proposal III

A Proposal
To Seek a Loan
For
**Establishing an Agricultural Tool
Manufacturing Unit**
Submitted to
Mr M. L. Kaushik, Director
**The State Industrial Development
Corporation (SIDC), Jaipur**
By
Mr D. RajaShekhar
**Shram Agricultural Tools
Mount Abu, (Raj.), 333032**

Proposal Summary

Project title	:	To establish an SSI Agricultural Tool Manufacturing Unit in Rajasthan
Broad subject	:	Approval for loan under the liberalized loan facility offered by the SIDC
Type of the loan	:	Liberalized loan policy under the self-employment scheme
Total cost	:	₹2,500,000
Loaner	:	Dr Raja Shekhar
Designation	:	Chairman, Shram Agricultural Tools

(Contd)

(Contd)

Company	:	Shram Agricultural Tools
Address	:	Branch Office, 23 Janak Puri, Mount Abu, Rajasthan
Telephone	:	Off.: (91) (02974) 22365 (Extn.) 142, Res.: 22366
Guaranteers	:	1. Mr XYZ, Kalyannagar, Jaipur (Rajasthan) (Father)
		2. Mr ABC, 14/56, C-block, Vasant Vihar, New Delhi (Uncle)

Number of installments for paying off loan with interest : 12
Number of years in which loan will be paid off : 6

Executive Summary

Small states are particularly vulnerable to development and confront a range of structural challenges to sustainable development. The plan is to take up a concerted action so as to bring about an enhancement in the state development by setting up an agricultural tool-manufacturing unit. Agriculture is something which can never be out of date. India, primarily being an agricultural country, has always promoted such steps which aim at raising the developmental level of agriculture in any state. This plan is a dual advantage scheme which when materialized, will be profitable, both to the common man and the state.

The establishment of this unit requires investment of handsome capital of 25 lakh rupees apart from mettlesome efforts on our part. By the setting up of this unit, we intend to boost up agricultural as well as industrial growth of the state, which in future will pave the way for fresh investments. Moreover, it is also expected to combat the existing unemployment scenario in the state. To enable the industry to modernize and acquire newer technologies, a substantial amount is needed to start this small scale venture. Capital has to be invested in land, plant, machinery, initial expenses, viz. installation expenses and recurring expenses.

Table of Contents

Appendix

1. Introduction

1.1 *Background*

The district has a good agricultural base (see Appendix A) and a reasonable good infrastructure of various facilities such as roads, electricity, and water but the industrial base is weak, resulting in a low economic return to the district. The gap between the dominating agriculture and the weak industrial base can be bridged by a coordinated approach to enrich the economy of the district. Lack of enterprise and technical knowledge in the local people are two major handicaps in the way of the industrial growth of the district.

(Contd)

(Contd)

1.2 *Origin of the proposal*

The growing foreign exchange levels and high inflation rates have given rise to a need for establishing an agricultural tool manufacturing unit which can withstand internal and external shocks. Moreover, the proposed site is well connected to the highway and can facilitate easy procurements of raw material and labour. Further, selling the manufactured tools in an area dominated by agro-dependent population, should not be a problem. Keeping in view all these factors, establishing a manufacturing unit for agricultural tools is being proposed.

1.3 *Objective*

This establishment of the agricultural tool unit aims at meeting the present challenges posed in front of the agricultural status of the state and pursuing reforms in the existing scenario by manufacturing tools of greater utility. It takes into consideration the basic needs and requirements that make tasks easier for a common agriculturist. We intend to fix the price of our products in such a way so that they cater to the demands of an ordinary farmer. Also quality products will be manufactured with cost-effectiveness.

Our plan primarily intends

- to set up a unit that initiates a revolutionary growth in the state economy keeping in view the agricultural aspects
- to manufacture tools that are not only handy in use but also satisfy the basic requirements of the farmer at nominal investments
- to solve the unemployment problem in the state to an extent
- to uniquely position our quality products in the market

The company's main thrust will be on adhering itself to quality standards such as product yield improvement, cost management, and focused attention on working capital management.

1.4 *Advantages of setting up of manufacturing unit in the state*

- As of now agricultural tools are manufactured in others states. So farmers of our state have to pay extra amount because of long distance transportation and excise duties, etc. This establishment is expected to reduce the cost by a substantial amount.
- This will bring into use the undeveloped land of the state.
- The proposed plan will also provide good employment opportunities to the locals.

2. Supporting Factors

2.1 *Infrastructural benefits*

- Industrial land provided by the state on lease
- *Power subsidy*—A subsidy on the consumption of power is granted to registered small-scale units up to 20 h.p. at the rate of ₹0.09 per unit (maximum).
- *Raw material*—The state has sufficient and easy availability of raw material for various projects in sectors such as Agro, Engineering, and Small Industry
- Telecommunication facilities are available
- Transportation facilities are available
- Skilled manpower is available

2.2 *Other Benefits*

Good amount of employment opportunities. Being one of the backward areas of the State, there is scope for it to receive more assistance and concessions from the State Government.

3. Industrial Potential

As Rajasthan is an agrarian state, modern methods of cultivation are required for the improvement in production that will improve the economic condition of the cultivators of the state. The prospects of establishing this manufacturing plant here are very promising.

(Contd)

(Contd)

4. Technical Aspects

4.1 *Set up plan*

We are planning to complete the setting up of the manufacturing unit in 2 phases:

Phase 1:

Estimated period: 6 months

- Aim: Proper levelling of the purchased land.
- Getting ready planned buildings for manufacturing tools, security, finished product storage, waste management, vehicle parking shed, etc.
- Setting up machinery

Phase 2:

- Production of the agricultural tools
- Estimated productivity: 5 units/week

4.1.1 Land plan

Required land area to set up the manufacturing unit 12500 sqm. We have chosen *Industrial Growth Center*, District Sirohi, 200 km from Ahmedabad, 490 km from Jaipur, and 27 km from Mount Abu for construction purposes. This land will be divided into following parts:

- Workshop unit area, where machines are required to be set up
- Assembly unit area
- Office area

Basic facilities required for the manufacturing plant

- Power supply for the machines as well as for the other units
- Water supply
- Telecommunication facilities
- Transport facilities

For further details refer to Appendix B.

4.1.2 Machinery details

Following machines are required for making the tools:

- Lathe machine for making the blocks of required specifications
- Electronic cutting machine are required for cutting the sheets of different gauge with proper accuracy
- Other machine tools required for assembling

4.1.3 Raw materials

- Cast iron
- Steel sheets of different gauge
- Low carbon steel and mild carbon steel
- Coal

4.2 *Cost Estimate*

Estimating the cost of the manufacturing plant is the pivotal aspect of setting up a unit. Project cost includes cost of land where the unit is to be set up, cost of building, cost of machinery, social infrastructure, etc., as we want to set up a manufacturing unit to manufacture agricultural tools such as threshers, cutting chaffs and ploughs.

4.2.1 Total budget estimate (in rupees)

Land : 5 lakhs
Building construction : 7 lakhs
Electricity set up : 1 lakh
Communication facilities : 50,000/-
Machinery : 4.5 lakhs
Raw materials : 5.5 lakhs
Advertising on products : 50,000/-

(Contd)

(Contd)

Unseen expenses : 1 lakh
Total : 25 lakhs

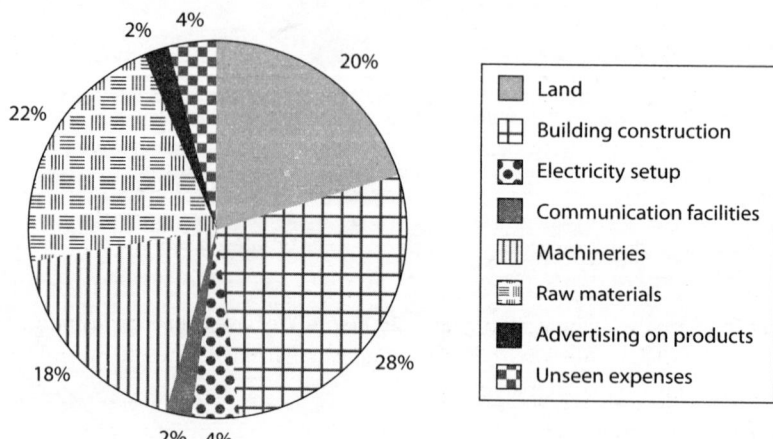

| Land |
| Building construction |
| Electricity setup |
| Communication facilities |
| Machineries |
| Raw materials |
| Advertising on products |
| Unseen expenses |

4.2.2 Monthly salary plan (in Rupees)

Skilled labour : 5,000/-   Supporting staff : 6,000/-

Unskilled labour : 3,500/-   Manager : 7,500/-

5. Repayment schedule

We are aware of the fact that loan-issuing agencies are always interested in knowing the mode of repayment of loans. We have done a considerable planning in this field too. According to the liberalized loan scheme, we will pay the loan along with interest in time to avail the rebate of 2%. The loans will be paid off in 12 equal installments that are recoverable in 6 years.

Figures

Total loan : 25 lakhs   Payable amount : ₹3967185.80736

Rate of interest compounded per annum : 8%   Rate of rebate : 2%

Number of installments : 12   Net payable amount : ₹3546297.78064

Number of years (in which loan will be paid off) : 6 yrs   Payable amount for each installment : ₹295524.81505

6. Conclusion

As our state moves into the twenty-first century, it is important that we begin to look at industrial development in the context of a new paradigm. It is understood that rapid industrialization will not be possible without requisite infrastructural development. The setting up of the unit will provide the social infrastructure, which caters to the human need for good living. Ours is a small beginning, but this is a potential market for a long term, which is expected to attract foreign investments in the years to come.

Appendix A

Rajasthan's Share in Agriculture: 2010–2014 (Lakh tonnes)

Crops	All-India	Rajasthan	Share(per cent)	Rank
Bajra	79.05	23.15	29.29	I
Rapeseed and mustard	69.42	26.22	37.77	I

(Contd)

Crops	All-India	Rajasthan	Share(per cent)	Rank
Coriander	2.24	1.55	69.20	I
Cumin	–	0.66	52.00	I
Guar	8.18	7.40	90.46	I
Moth	3.55	3.10	87.23	I
Barley	14.36	3.78	27.66	II
Gram	57.54	10.71	30.65	II
Maize	93.34	10.29	11.02	III
Soyabean	52.02	4.54	12.89	III
Oilseeds	–	–	14.92	III
Rabi pulses	–	–	22.32	III
Groundnut	72.73	2.73	3.75	VI
Cotton (Lakh bales)	142.52	13.63	9.56	VI

Appendix B

Land Plan

Industrial Growth Center, District Sirohi, 200 km from Ahmedabad, 490 km from Jaipur, and 27 km from Mount Abu.

District Sirohi

- **Location**: 200 km from Ahmedabad, 490 km from Jaipur, and 27 km from Mount Abu
- **Land**: 720 acres in Growth Center, 308 acres in Ambaji Industrial Area
- ***Rate of allotment**: Growth Center & Ambaji Industrial Area—₹200 per sqm
- **Water availability**: Underground water 0.4 MGD being stepped upto 1.0 MGD. Underground yield 2,000 to 2,500 gallons per hour. Water available at a depth of 10 m
- **Wind velocity**: 1.5 km per hour
- **Power**: Power supply from 132 KV GSS at Abu Road. 220 KV GSS is proposed
- **Tele-communications**: Telephone, telex fax, and STD available
- **Manpower**: ITI at Abu Road and polytechnic at Sirohi
- **Internal transport**: Bus services available
- **Facilities in the area**: Housing colony, bank, community centre, recreation club, restaurant and post office
- **Social amenities**: School, medical, housing, recreation, etc., available at Abu Road
- **Incentive to industries**: Investment subsidy and other benefits as per the State Policy
- **Prominent industries**: Rajasthan Polymers & Resins Ltd, Modern Insulators Ltd, Tirupati Fibres & Industries Ltd, etc. Area is suitable for marble and granite, electrical and high-technology based industries

@ Please refer to the Online Resource Centre for more samples of technical proposals.

RECAPITULATION

✓ Proposals, reports, and business plans are distinct from one another.

✓ A proposal mirrors the identified needs of the prospective customer. Proposals can be broadly classified as the solicited and the unsolicited ones.

✓ A proposal can be written to solve a technical problem, offer a managerial solution, suggest a modification of methods, procedures or services, etc.

✓ Elements such as the executive summary, introduction, technical section, management section, cost estimate, etc. form an integral part of a proposal.

✓ A proposal is always written in a persuasive tone.

✓ An effectively drawn proposal always links the plan offered to the subsequent action to be taken, is specific in details, uses effective and persuasive language to clinch the deal, and presents a realistic cost estimate for the services and expertise offered.

✓ In variegated professional situations one may write different types of proposals, such as technical proposal, business proposal, sales proposal, letter proposal, research proposal, etc., for meeting different objectives.

✓ Proposals can be written by using different structures such as short proposal, full length proposal, and letter proposal.

WISEWELL QUIPS

EXERCISES

1. India and Sri Lanka plan to commence ferry service between Tamil Nadu and the island nation, a step towards promoting tourism between the two countries. Being the Senior Secretary in the Ministry of Tourism prepare a proposal to be discussed while signing the Memorandum of Understanding with Sri Lanka.

2. You have decided to establish an Agro Product Unit in Bhiwadi, your hometown. For this purpose, you wish to avail yourself of the liberalized loan facility under the self-employment scheme. Therefore, you wish to submit an unsolicited technical proposal for seeking loan from the State Industrial Development Corporation,

Jaipur, Rajasthan. Now draft this proposal to be sent to the Director, SIDC, Jaipur, Rajasthan.

3. You are the Vice President of operations for Mirch Masala food chain, Mumbai. You plan to open a new restaurant in Jaipur on Jaipur–Delhi highway. You prepare a proposal for this purpose to be submitted to the President of your organization. Now draft this proposal. Your proposal should consist of the following elements:

(a) Title Page
(b) Draft Contract
(c) Introduction
(d) Technical Section
(e) Management Section
(f) Cost Estimate

4. Tirupati Industries, which deals in fabrics, electronics, and computers, has recently decided to venture into the production of laptops. Therefore, the company is going to set up a new unit for producing laptops on a large scale. As Finance Manager of the company, you have been asked to draft a proposal containing all the required details such as space, air conditioning and dust-proofing, hardware, peripherals, consumables, networking, power supply, technical and other staff, ancillary support system, etc. Now draft this proposal to be submitted to the Secretary, Department of Company Affairs, for approval.

5. The need for recreational activities cannot be ignored in today's stress-borne world. The Taata Group of Companies, Kolkata has decided to set up a Recreational Activity Centre (RAC) in its educational premises to provide recreational facilities to its students, faculty members, and supporting staff. As the Senior Administrator of the Educational Staff Welfare Unit, you have planned to draft a proposal to be submitted to your Chairman for setting up this centre. Draft this proposal containing all the required details.

6. A generous alumnus has established a fund in your institute for promoting research in the application oriented research for the rural people. Now try to think of an idea that you would like to explore and prepare a full length research proposal covering the background, objective, scope, methodology, national and international relevance of the project, research plan, proposed budget, conclusion, and references, to be submitted to the Director of the institute for this fund.

7. You think about establishing an agro-knowledge-cum-farma-centre where you plan to educate the farmers by showing videos, discussing their farming related problems and also selling them the modern agro-equipment and fertilizers in your village. Now prepare a proposal for setting up this centre to be submitted to the state government for funding.

8. Assume that M/s Petra Corporations, New Delhi, is planning to set up a refinery unit at Ranchi. The company has recently solicited proposals to establish a world class security system at the refinery premises. Assuming yourself to be a consultant in security services, draft a proposal containing all the relevant details such as an Executive Summary, Introduction, Technical Section, Management Section, Cost Estimate, and Conclusion.

9. New World University is the name of a university that has recently been established in your city. The university has recently sought proposals from professionals who can help it set up a computer lab on its campus. Assuming yourself to be an expert in the field, draft a proposal to be sent to the Registrar of the University.

10. The Commissioner, Traffic Control of your city is looking to set up new traffic lights at the main crossings in the city, besides making the entire traffic operation more efficient and reliable. The Commissioner's Office has recently invited proposals to carry out the task in an efficient, cost-effective, and time bound manner. Draft a proposal to this effect.

11. Given below is a poorly written Executive Summary of a Proposal. Rewrite it to make it more effective and relevant.

These days one of the most important skills to be acquired is effective spoken English. As we all know English is the world language now. In India too English has become the most important language. For all jobs and professions English is required. We have the requisite exposure in setting up Language Labs in academic institutes all over the country. In the past two decades, ever since the concept of language labs picked up momentum in India, we have set up language labs in various parts of the country. Today we have labs in Delhi, Mumbai, Lucknow, Indore, Pune, and Agra. Since you are an upcoming Engineering Institute in Jodhpur, it would really help you establish a language lab in your premises. In this lab, you would require 20 computers with a master console and an earphone set for each seat. It would cost your institute only 5 lac rupees. In this meager amount, we will provide you the entire software back up; the required training and the expert inputs from time to time. The rest of the details can be seen in the following sections of the proposal.

12. 'A technical proposal is different from a report and business letter.' Do you subscribe to this? Discuss and substantiate with proper examples.

13. 'A proposal is a combination of marketing and technical skills.' Comment on the statement and illustrate your views with the help of appropriate examples.

14. Given below are some of the extracts taken from technical/business proposals. Rewrite them so as to make them effective, precise, persuasive, and appropriate.

   (a) The proposal enclosed establishes beyond doubt our credentials as the best production unit advisors in the country.

   (b) It would cost you peanuts but would turn in blinding returns.

   (c) The entire lab can be refurbished by investing an amount which we think is not much for a corporate house such as yours.

   (d) All that we would require would be a handful of good cameras with impressive mega pixel and zoom facility; a room of average width and height; and a couple of other machines to set up a media centre at your premises.

   (e) The proposal would be executed as soon as possible. We have never kept our clients waiting and it certainly will not happen in your case too.

ANNEXURE 24.1

Checklist for a Technical Proposal

Having prepared a technical proposal, answer the following questions before its onward submission:

Q.1.	Does your technical proposal have all the required elements?	Yes ( ) No ( )
Q.2.	Does your technical proposal have a good Summary and Introduction?	Yes ( ) No ( )
Q.3.	Does the Technical Section of your proposal give all the specifications required?	Yes ( ) No ( )
Q.4.	Does the Management Section of your proposal answer the questions such as who will do, what, where, why, etc.	Yes ( ) No ( )
Q.5.	Is the Cost Estimate given in your technical proposal realistic and detailed enough?	Yes ( ) No ( )
Q.6.	Is the technical proposal written by you has the required persuasive tone?	Yes ( ) No ( )
Q.7.	Does the technical proposal prepared by you have the right tone suited to the requirement of the recipient of the proposal?	Yes ( ) No ( )
Q.8.	Does the technical proposal highlight the problem referred to with focused attention and spotlight?	Yes ( ) No ( )
Q.9.	Is your technical proposal accompanied by a covering letter highlighting its features?	Yes ( ) No ( )
Q.10.	Does the technical proposal prepared by you suggest the best solution in the best possible way?	Yes ( ) No ( )

CHAPTER 25

Email and Blog Writing

Learning Objectives After reading this chapter, you will be able to

- understand the function and purpose of email writing
- identify the advantages and disadvantages of using email communication
- understand the pitfalls in using email communication
- figure out the nuances, content, and style considered standard in emails
- learn ways to improve your efficiency in drafting effective emails
- develop techniques to write an effective blog

25.1 INTRODUCTION

A few days back we found a message in our email box. It read something like this:

> Gd 2 c u at d con'f. ur papr was nice. me too intrstd n d same area. shall v come up wth a combo paper?
>
> b,rgds
>
> A. Prashar

For a while we were simply befuddled and could not figure out what the message was all about. As it became apparent after some initial struggle that the writer had appreciated our research paper that we had presented in a conference, we smiled. It was, however, a painful smile as we were hurt to see a language being distorted to the extent of being rendered unintelligible. There was no question in thinking of writing a 'combo' paper with such a writer who treated the language, which he used for communicating his ideas so disdainfully.

A writing like this certainly gnaws both at the heart of the language and the reader alike. Having said that, we are aware of the fact that such email messages have become a norm rather than an aberration. Many amongst us, particularly the youths of our times, are quite content in composing a message like this. Have you thought of the reasons that give rise to a practice like this in email communication?

And do you find it worth practising yourself? The coming section attempts to help you understand why writing such non-standard English is not advisable and why you should not develop a habit of using such language in your personal or professional interactions with others.

25.2 EMAIL WRITINGS—REASONS FOR POPULARITY

Electronic mail, popularly known as email (also e-mail), came into existence in the late twentieth century. Quickly, this progeny of computer technology took over from all other methods of exchanging information and now it is almost impossible to conceive of a world where people do not share their views, ideas, information, and data through emails. Following are some of the factors that have made email writings such a phenomenon in our times:

Emails are fast, cheap, easy to operate One reason why emails are today preferred to the traditional means of communication is that they are easy, prompt, and fast to use. Once you compose a mail, you can send it to as many people as required, instantaneously. Practically, to an email message of only a few words, you can attach large files of data, complicated spreadsheets, voluminous reports, lengthy procedures, elaborate schedules, detailed proposals, long instructions, and anything else that you wish your readers to read anywhere in the world. Because of its sweeping reach, universal accessibility, and blinding promptness, email is certainly the easiest, most prompt, and the cheapest mode of getting across to a large number of people all across the globe instantly.

Emails capture the spirit of the age Advancements in science and technology have added frantic pace to our lives. Today, all of us are so busy in our professional and personal lives that we tend to feel irritated and impatient with anything that consumes even a little more of our time than required. We certainly hate waiting in long queues for our turn to come. That is why, we have agents and operating-round-the-clock counters for getting us tickets for our rail, road, or air transport. There are kiosks in big cities where the agents take care of our water, electricity, and telephone bills; there are some other agents who take care of our life and health insurance premiums and income tax returns; there are buses, cars, bikes, other motor vehicles, and metros running all across the metro cities taking their mobile population from one destination to another. In this fast-paced life, anything that saves any time for us is certainly most welcome and becomes immediately desirable. Emails fall in consonance with such ethos of contemporary age.

Emails offer flexibility in tone and style Just as most of us are impatient in getting our turn to arrive in a long queue, many of us get impatient not only in getting our message across once written, but also in composing it. Once we take up a writing assignment, we want to see it off as soon as possible. So rather than choosing an elaborate sentence structure, we resort to choppy prose. Owing to the crunch of time available partly and mainly because of an attitude that regards contemptuously everything that requires some effort, we apply short cuts in all that we write and speak today. Since no traditional means of communication offers this flexibility, electronic mail is lapped up most readily by us. Aided by the speed offered by the medium of expression, we intend to carve messages which are easy to compose and quick to send. Moreover, since emails are used not just professionally but also personally, a sense of informality starts creeping into our style the moment we set ourselves into composing an email.

It is here that we need to have a cautious approach while composing an email. A small amount of informality may add warmth and vigour to your writing, but an exaggerated sense of informality can actually get you across to your reader as an amateurish individual showing you in a very poor light. Hence, you need to carefully observe the email etiquette both in terms of content and style. A detailed discussion to this effect follows later on in this chapter.

Emails have become legal and valid   Though starting informally, emails are now regarded as a valid and tenable proof to someone's claims legally. The logic behind this credence can be appreciated in the sense that after all, it is a written message sent over to the reader electronically. The change of medium thus does not take away the essential presence of the receiver and the sender in the process of communication. Moreover, since the delivery of an email message is immediately confirmed, the recipients can never pose as though they never received the mail in the first place. Having added another dimension to its utility thus, emails are likely to grow in popularity and usage in the times to come. However, before we fall headlong in love with this computer progeny, let us be aware of the pitfalls of using this means of communication so that we not only utilize its immense potentialities but also overcome the challenges posed by the medium.

25.3 EMAIL WRITING—SOME COMMON PITFALLS

Under this section, we will discuss some common pitfalls of email writing.

Privacy is lost   Since emails are sent through an Internet facility, a system administrator can always reach our mails. If required, a company's system operator can intercept our mails, disclose our mail to others, and can read whatever we regard as secret and personal. A more serious fallout of this vulnerability of the system can present itself in a misfortune, as some hacker may get into our computer system and can send some undesirable mails from our account. A terrorist breaking into innocent people's account and shooting terror mails through it is not something we are unaware of now. At times, hackers and other swindlers approach us through an apparently innocuous mail in which they require us to fill-in some form and reveal to us the information related to our bank accounts, credit cards, ATMs, voter ID cards, PAN card, etc. Making any such detail available to others may turn out be extremely perilous as it may land us in some unforeseen crisis.

Casualness creeps in   Owing to an implied sense of informality that we tend to experience while communicating through an online means, a sort of casualness creeps in our attitude as soon as we plan to send an email message to someone. As already suggested, though this casualness may at times be purposeful and intended, not every message we compose online can afford to sound casual or informal in tone. However, because of the impact of the medium chosen, we see messages swarmed with emoticons, abbreviations, non-standard spellings, and choppy and chatty sentence structures. The real tragedy occurs when the authors tend to forget the distinction between a formal and an informal mail, and end up writing in a colloquial language what needs to be corresponded in a proper professional manner.

Ambiguity impairs communication   With the enormous popularity of the electronic medium of expression, the number of people writing and reading emails in their day-to-day professional interactions has grown manifold in the recent past. However, since there are no universally or traditionally accepted conventions in email writings, people tend to write the way they want. Therefore, most of us end up writing casually, informally, colloquially, and hence ineffectively when faced with the task of composing a mail. This habit of casualness is further bolstered as sending messages through mobile has also become a way of life. Because of this, *texting* a message seems like an ability required to be developed rather than a bad norm which needs to be avoided. Writing in such an anarchic *free style* however can seriously impair the efficacy of our message as reading through our colloquial, chatty, and abbreviated text, the reader feels

These unsolicited emails have clogged my mailbox.

Unsolicited Emails can Clog Your Mailbox

baffled, intrigued, and confused. A message, if not understood well by the reader, is in any case of little significance because only a partial misunderstanding can result from a message that is not clear enough to the reader.

Virtue is sacrificed to convenience Language is an expression of a culture in communication. Since the online culture promotes a racy, pushy, and jerky way of self-expression, we come across emails that seem to move without a compass. Since many of the email writers are quite used to short, quick, and abbreviated versions, they find it difficult to write in an elaborate sentence structure and we come across messages which hardly seem to make sense to us. All such writers compose their mails governed more by the fad rather than the serious intent behind writing the message.

Moreover, since a proper convention has not yet emerged in email writings, people write in acronyms; abbreviate their subject lines beyond a point of comprehension; forget to add courteous salutations in the beginning; or add a complimentary close at the end and make the entire piece of communication seem like a jigsaw puzzle. Because of this, the reader has to not just read what is written but also assume what is deliberately or inadvertently left out.

Junkyard is always full An undesirable offshoot of email communication, that is, receiving unsolicited and unrelated mails, is a very common nuisance that people encounter. Since sending an email does not take time, circulating a copy also becomes an easy option. Since everything is just 'a click away', many writers unmindfully send a copy of their messages even to those people who are not concerned with the information in any way. So, a production unit manager may get a mail asking him to 'furnish the sales details at the earliest', the quality control head to 'provide a list of those who fail report in time more than thrice a month in the sales division', and so on and so forth. All such unrelated mails infuriate business professionals and can even lead to loss of temper or open confrontation between individuals.

25.4 EMAIL WRITING—GUIDING PRINCIPLES FOR COMPOSITION

As suggested earlier, emails are pieces of communication with quite a few chinks in the armour. Moreover, due to the lack of a settled convention, it becomes a challenge to compose a mail that is proper in all respects. Following are some of the principles following which we can cast our emails in an effective manner:

Avoid being abrupt Read the beginning of the following email:

> Hi,
>
> Meeting likely to postpone. Call before you start.
>
> V.K.Sarkar

A message like this might bring relief to someone who does not want to attend the meeting! However, reading this gives us a feeling as though the author wrote it sitting on the edge of his chair before leaving for something else. It does not leave a proper impact on the reader and also leaves him/her wondering about the information which such writings generally do not convey. Consider an alternative way of communicating the same message:

> Dear Members,
>
> Due to some other pressing matter, the meeting regarding revision of draft proposal is likely to be postponed. Our office will call you up to let you know about the revised schedule for the meeting somewhere around 5.00 p.m.
>
> Sorry for the inconvenience.
>
> V.K. Sarkar

Use subject lines effectively While composing an email, we normally do not give any importance to the subject line at all. Hardly anyone thinks about the subject of their message before shooting it off from their mail box. Many a time, it is the automatically generated reminder 'would you like your message to be sent without a subject?' which draws the attention of the author to the fact that he/she should also attach a subject to the content he/she plans to send. This itself suggests what little worth we assign to choosing an appropriate subject line to our message. Observed seriously, however, we realize that the subject line of the message is as important as the message itself. A subject line lets your reader form a proper idea about the message he/she is likely to see in the mail. Since the reader is likely to see the subject line as soon as he/she opens the mailbox, a proper subject line is likely to stimulate the reader while a message introduced through a poorly worded subject line is most likely to be ignored by a busy professional. Also keep in mind that a proper subject line helps the recipient of the mail in the filing process. Here are some of the subject lines extracted from professional emails. See how the revised subject lines add to the worth of the message:

Original	Revised
Meeting	Schedule for the Meeting Tomorrow
Fancy Dress Competition	Inviting Entries for Fancy Dress Competition
Retail Outlet	Announcement Regarding Retail Outlet
Just Like That	Exchanging Pleasantries
My Mobile Number has Changed!	Change in Mobile Number
Re: Delay	Re: Delay in Consignment Delivery
Absenteeism	Overcoming Absenteeism

Start courteously   Regarding emails, there is a serious misconception that they are informal pieces of interaction and hence courtesy is not much desired. It is a misconception of this sort that leads to an email message like this:

> Madam
>
> Have you received any communication from marketing guys? If yes, please kick the mail back. You know I am dying to listen to them.
>
> MKN

A message like this is only going to be scorned by the reader, particularly if he/she is senior to the person who has sent him/her the mail and expects him/her to 'kick the mail back', as he/she is 'dying to listen to them'. The whole message is shorn of courtesy and decorum and evokes confusion as well as contempt on the part of the reader as it hurts a person's professional dignity, besides bemusing him/her.

See the revised version of this mail and observe how the same message seems upright and dignified:

> Dear Madam,
>
> Kindly let me know if you are aware of the schedule for the talk by the Marketing personnel. Being from a similar field, I am very keen to hear what they have to say.
>
> Thanks and regards,
>
> Mani Kumar Nayak

While writing an email, keep in mind that though a degree of informality is essentially part and parcel of such communications, disregarding courtesy and politeness is certainly not desirable.

Add a warm-up sentence   Consider the beginning of an email:

> Dear Chaitnya,
>
> No issues with the proposal sent. We'll clear it in the meeting this evening.
>
> Aninda

A message such as this hardly appears to be professional. While composing an email communication, we need to understand that though it is sent over an electronic medium, it is after all a professional piece of writing if used in professional communication situations. Therefore, the style of an email should not essentially differ much from that of a business letter. Though email writings make room for more informality and seem quite intimate and personal at times, it is not advisable to start in a manner that seems abrupt, unofficial, and jerky. In fact, it is more appropriate to start, particularly a professional email, with a warm-up sentence to fix the context and set the tone of the message. Take a look, for instance, at the revised beginning of the email we have just read and realize how by providing a proper beginning assigns to the message, the tone and tenor that it needs to display:

Dear Chaitnya,

Thank you very much for sending the proposal. It seems fine and will most likely be approved in the meeting scheduled this evening. I'll let you know about the final decision of the committee after the meeting.

With best wishes,

Aninda

Go through another such instance where an introductory sentence is almost unavoidable and the whole message struggles in the absence of it:

Dear Mr Chairperson

Keeping quite busy; can't attend the meeting scheduled next week.

I.C. Sharma

Board Member

A message like this certainly conveys the message but rings like a blunt, crude, and bad news to the reader. In professional situations, starting without fixing the context leads to crudity of expression and a message devoid of it seems far too sudden to be appreciated. See the revised version of the same email:

Dear Mr Chairperson,

Thank you very much for informing me about the Board Meeting scheduled in the next week. However, due to some other pressing urgent matters, it would not be possible for me to attend the same. Nevertheless, I would be available online and can be consulted for any crucial decision that requires my view.

Wish you all the best for the meeting and look forward to receiving the Minutes for the same.

Yours truly,

I.C. Sharma

Board Member

While writing an email, we need to understand that it is not in any way a less human interaction. In fact, simply because it combines the features both of a written and oral communication, it needs to sound all the more personal, warm, and interactive. Of course, we need not talk like a machine—something that we should not do even while writing a business letter, a formal report, a research article, or a technical proposal—in composing an email. However, rather than starting with an intrusive statement, we need to build a rapport with the reader through an introductory, warm-up sentence or fix the context in some other way before communicating the intended idea. Given below are some such introducers which, when used, are likely to make a message sound more professional, decorous, and appropriate:

1. It was as usual a pleasure talking to you. Your suggestion for introducing...
2. Thank you very much for showing an interest in our agency. However, because of...

3. We have gone through the brochure sent to us. We believe that we can go for...
4. It was nice to have heard from you after such a long time. It seems you have shaped up really well as a...
5. Thanks indeed for informing me about the concert. It would be an added delight if...
6. As you are aware of the fact that we are in the process of changing our nomenclature, it may take some time for us to look into...
7. Thanks for calling last evening. We are going to sit on this in the afternoon...
8. I hope this finds you in the best of health and spirits...
9. Thanks for your enquiry. However, as per company policy...
10. The meeting for rescheduling the conference is being planned. It would be appreciated if you can return the enclosed form with details...

Avoid all capital letters At times the mails received read like this:

> ALL CONCERNED ARE REQUIRED TO TURN OFF THE LIGHTS BEFORE LEAVING THE CONFERENCE ROOM AFTER THE MEETING. TIME AND AGAIN IT IS OBSERVED THAT THE FANS AND TUBELIGHTS ARE ON WELL AFTER THE MEETING IS OVER. LOSS OF ELECTRICITY IS LOSS OF NATURAL RESOURCES AND INCURS COST TO BOTH COMPANY AND NATION.
> CHIEF (MAINTENANCE)

Reading such a mail gives you a feeling as though the writer is actually screaming while writing this mail. Though the highlighted matter is really worthy of a concern and the author of the mail deserves our appreciation for bringing it to the notice of others, the tone in which it is written is certainly not appropriate. Moreover, the use of capital words all through the text makes the message seem rather intimidating. Again, lack of courtesy in the language and absence of salutation in the beginning instills in the message a sense of bluntness.

Use proper spellings A large number of emails are written in the spirit of a message on a mobile. Just as on a mobile a message can accommodate a certain number of words, writers composing emails perform under a compulsive urge to be concise and brief. Of course, writing in a brief and concise manner is certainly an achievement. After all, brevity is the soul of wit. However, overdoing this can lead to confusion; particularly so, when people start chopping letters from certain words and trimming their spellings beyond a degree of acceptability. Remember, many of the emails we write are part of our professional communication and carry the image of the organization we represent. Writing in a fashionable 'email shorthand' can get us in some serious professional embarrassment. Imagine, for instance, the impression that an email like this is going to create on the reader:

> Hi all,
> V r intrstd in buying the latest heatconverter launched by ur company. r u ready with supplies? Pl specify schdl of delivery 4 v 8t waiting wons the deal is thru.
> VKR
> for Toto Syringe Pvt Ltd, Hyderabad

Frankly speaking, except the title of the company at the bottom, hardly anything else seems worthy of a professional communication. Writing in such a fashion is most likely to create only a comic effect on the reader. Such crippled spellings can never create a robust impact on the recipient of the mail. Writing in such a manner can lose one's credibility as well as show in poor light the organization one represents. See the revised version of this email to steer clear of such obnoxious trappings:

Dear Sir/Madam,

We are interested in purchasing the latest Heat Convertor (Model CCS 1232), launched recently by your company. We would like to understand whether the product is ready to be supplied. Kindly let us know how much it would take for you to send the consignment once we place an order.

Thanks and soliciting an early reply,

Yours truly,

Vikram Reddy

Purchase Coordinator

for Toto Syringe Pvt Ltd., Hyderabad

Avoid acronyms   Acronyms are shortened replacements for certain words, terms, persons, and objects. Youths take fancy to such abbreviations as for them it is really exciting to refer to Instructor-in-Charge as IC; Comprehensive Examination as COMPREE; All Night Canteen as ANK; Students Activity Centre as SAK; Sarvjanik Hospital as SAARVI and so on and so forth. Truncating a name gives our youth a sense of omnipotence and usually such terms become popular in all segments of society. After all, who is not aware of the meaning of ISBT, and ITO in Delhi? Using a well known acronym is not a taboo in emails. However, most of our young communicators do not wait for an acronym to become well recognized and easily understood and almost everyday push new coinages replacing a large number of expressions. Since the meaning and significance of many such acronyms are known only to a few 'chosen ones', it is not possible for all those who receive an email which is full of such acronyms to make sense of something that leaves them completely baffled at reading something like this:

Hi Sid,

Wud u mind passing me FAQ? I'll wait at BC 4 u EOL. BTW, REPREE Tuts were gr8.

TVM & BFN

KK

A message like this is not less than a crossword puzzle for a person who is not aware of the unnaturally and forcibly formed acronyms. See the revised version and observe what it was all about:

Hello Siddharth (Sid)

Would you mind passing on to me the list of frequently asked questions (FAQ) distributed in the class yesterday? If possible, please come to the Bank Canteen (BC) after the lecture. I'll wait for you there at around 2.00 PM. By the way, thanks for notes on Report Writing; they were quite useful.

Thanks very much and bye for now.

Karthikeyan

Use emoticons and smileys sparingly Besides crippled spellings, jargons, and acronyms, what abound emails these days are the various types of emoticons and smileys. As we know, emoticons are short symbols carved with or without the help of some keyboard letters. Used in different shapes and shades, emoticons are largely perceived to assign the features of oral communication to the written communication that an email essentially is.

Given below are some of the commonly employed emoticons:

☺   A happy face suggesting joy and humour
☹   A sad face suggesting unhappiness or sadness
@   An expression suggestive of shock and surprise
O   An expression suggesting realization of an error

There is no harm in using such emoticons and smileys while writing mails to our friends and to those with whom we share an intimate relation. Such fads however need to be restricted in professional communication as they are too personal to find a place in a formal piece of communication. Remember to keep the following before using any emoticon:

- Relationship with the reader
- The gravity of the subject matter
- The occasion

Take care of punctuation marks Taking a leaf from mobile messages where many find it difficult to locate and apply punctuation rules, many writers of email communication write without giving any thought to punctuation marks. Therefore, it is not just that letters are omitted, acronyms are chosen, and abbreviated forms are written, but also that many letters are written in lower case even after a full stop. An example of this sort would put the matter in proper perspective:

dear mohit,
I heard dr chawla speak the other day in a workshop. the guy talks sense. looks like we can get him to talk to our staff and even workers. after all stress related issues r getting complex it would be nice to have a psychiatric addressing such issues. kick it back with what u see in this.
malhotra

It seems the author has thrown overboard all the rules learnt in the grammar class. A message like this only seems like a truncated mess. As the author does not choose to write the letters that follow a full stop in capital form, the whole text appears to be haphazard and amorphous besides being ungrammatical. Hope one chooses to write the same thing something like this:

Dear Mohit,
I heard Dr Chawla speak the other day in a workshop on Stress Management. It was a very informative and impressive talk as Dr Chawla brought to the fore the complexities of work-related environment that lead to stress in professional situations. I would suggest arranging a talk by Dr Chawla in our

(Contd)

organization as well. We all realize the mounting stress that all our employees and workers experience from time to time. Though we keep counselling our staff, a talk by Dr Chawla, a practicing psychiatrist, would help all of us negotiate the complex issue of stress management in a more effective manner. Please let me know your views on this.

Thanks,

Smriti Malhotra
Sr Manager—HR

Use salutations and complimentary close   A growing sense of casualness generally accounts for the absence of proper salutations in the beginning and omission of complimentary close at the end of an email. However, without all such professionally mandatory etiquette, an email sounds abrupt, curt, and shorn of hospitality. Here are some of the suggested salutations and complimentary close which generally make our diction appear decent and graceful:

Some common salutations

1. Dear Dr...
2. Dear Mr/Mrs...
3. Dear Ms...
4. Dear Sir/Madam
5. Dear Colleagues
6. Dear Friends
7. Dear Customer
8. Dear Reader
9. My Dear...

Some common complimentary close

1. Thanks and regards
2. Best wishes
3. With best wishes
4. With deep personal regards
5. With warm regards
6. Best regards
7. Bye for now
8. Good bye
9. With love
10. Love
11. Look forward to receiving a mail soon
12. Look forward to seeing you at the meeting
13. Soliciting an early reply
14. Please call me, or send a mail, if I can be of some help to you
15. Please let me know how I can be of some help to you
16. Your enquiries are always welcome
17. Feel free to call anytime
18. Call me up; should there be a need

Identify yourself   Keeping with an undesirable habit prevalent in email writings, at times some writers do not identify themselves properly. We come across short, abbreviated names or sometimes some initials and nicknames as can be seen at the end of the email message cited below:

Dear Sush,

Pl return the files marked with corrections and suggestions. This needs to be incorptd. in the bulletin today.

PK

Despite an empathetic understanding on the reader's part, such choppy and amateurish prose can hardly be seen as a piece of professional writing. Moreover, addressing the recipient of the mail with a nickname and rounding off the mail with initials suggests a scorn for convention and a callously perfunctory approach towards the entire process of reading and writing.

Moreover, as already suggested, professional emails are now used for official and legal purposes. So, not writing a proper name at the end of a message that you compose is like sending a cheque without signing it. Therefore, let us not be in a hurry to dash off emails without bothering about whether the reader understands what we write and, of course, who we are in the first place. Take a look at the revised version to understand how a serious message had been reduced to a juvenile farce in the original version:

Dear Sushmita,

As discussed telephonically, please return the files marked with corrections and suggestions as soon as you complete the task. As all these suggestions and corrections are to be incorporated in the Bulletin today itself, please give top priority to it treating it as most urgent.

Thanks and regards,

Prabhat Kashyap

Editor-in-Chief

25.5 EMAIL WRITING—MAINTAINING COMMON ETIQUETTE

Besides following the guiding principles while composing an email, we also need to observe some common etiquette while maintaining and operating an email account.

Reply immediately Generally emails are used to suggest that the matter needs urgent attention and action on the part of the recipient. Therefore, those who send an email expect getting a reply within a day or two, if not immediately. Hence, it is advisable that the mails we receive as a professional are answered immediately or as soon as possible. If the matter needs a detailed perusal, consideration, or some other time consuming action, we need to send an interim reply acknowledging the receipt of the message and assuring detailed response later on.

Avoid circulating emails to everyone Since it does not take much to shoot the copy of a mail already composed, it is seen that at times we end up circulating our emails even to those who are remotely or hardly concerned with the information. A tendency such as this is certainly uncalled for as receiving an unnecessary mail is a needless nuisance and it generally irritates a busy professional. Therefore, avoid sending a copy of the mail to all and sundry. Send the copy of a mail only to those who have something to do with the mail. Also remember to click 'bcc' instead of 'cc' in case you intend to forward the copy of your mail to someone without letting other recipients know about it.

Avoid attaching unnecessary files At times, email writers do not really judge what the reader actually requires to read and attach unnecessary files to the mail. One reason for this practice is the fact that attaching a file is again just 'a click away' on the Internet. Moreover, by sending all the material, some writers believe that they can dispose of the matter without exerting much. However, just as receiving an unnecessary mail is a nuisance, so is an unrelated file that requires the reader to read through the material that has little relevance to him/her. Sending an irrelevant file not only suggests a sloppy attitude on the part of the writer, but also offends the reader as it consumes time on his/her part unnecessarily. In the jet age of ours, robbing someone of their precious time is almost like committing a crime and the writers who attach redundant material to their mails are seriously despised.

Answer all queries   At times, we get mails seeking clarification on certain issues. While answering such mails, we should try and answer as exhaustively as possible. It would be further advisable that besides answering the questions asked, we also anticipate other possible queries that the reader may raise in future, and answer all such questions. Receiving an answer for questions that are answered even before they are asked makes a reader appreciate the concern and emotional intelligence shown by the writer. This certainly helps us display our caring attitude for others and generates a sense of gratitude in them. All this strengthens our professional image and adds to the goodwill of the organization we work for.

Avoid sexist language   Gone are the days when one would happily harp on a saying like *man is mortal* and hope to suggest a universal human frailty. In modern times, saying 'man is mortal' can also imply that 'woman is not so', for today what applies to man may not necessarily apply to woman. It is so because in an age of feminist quest for equality, a woman aspires to pursue and maintain a distinct identity of her own. Therefore, writing in a sexist language could turn a female reader off. Consider, for instance, an email such as this:

Dear Customer

Remember to use your ID while contacting us. Enquiries from someone who does not use his Unique ID No. (UIN), would not be entertained.

Manager

Customer Care

In earlier times, it was a general practice to refer to everyone with a masculine pronoun 'he'. Now of course, we write in a tone and style that is suggestive of gender neutrality on our part. Therefore, avoid such sexist overtones in email writings too. Choose either 'he/she' or use the plural forms of pronouns such as they, them, their, etc.

Be aware of email jargon   Jargon refers to the technical language used in a particular field of study. Like every other domain, email writings have also developed a jargon of their own. However, using a jargon is not something that makes us understood well. Thus, we need to restrain ourselves and keep its use to the minimum. Nevertheless, we need to be acquainted with the common jargon used by email writers so that we are aware of the meaning of the terminology employed. Given below is the list of the jargon commonly used in email writings:

Bot: A part of software that acts on behalf of a human.
Mailbot: A piece of software that automatically replies to an email.
Listbot: A piece of software that manages the distribution list.
Bounce: A message that returns to the sender because of some configuration problem or because of some error in typing the address of the receiver's address.
Ping: A test to see whether the other person is available online.
Lurk: To read messages anonymously
Spam: Unsolicited emails sent to many people simultaneously; used mainly as commercial advertisements.
Flame: Emails that contain hostile message.

Keep your mail box uncluttered   Just as a writing desk needs to be kept spic and span, we need to pay careful attention to the fact that we delete the junk mails regularly from our mail account. A cluttered mail box is difficult to navigate and becomes unwieldy at times for looking through an email box running into thousands of messages consumes time unnecessarily and also leads to confusion at times. Ideally we should delete all the redundant messages. However, the messages that we need to refer to in future and are likely to be used for record purposes should not be deleted.

Read and edit your mails   One reason why we get to read sloppy, choppy, and ungrammatical structures in emails is lack of revision and editing on the writer's part. Since most of us believe email writing to be a part of informal piece of writing, it is customary that a writer shoots the mail as soon as he/she finishes composing it. This attitude leads to perfunctory and casual writing, which helps neither the writer nor the reader. We must bear in mind that email writings too are a part of our professional communication and just as every other professional writing needs to be revised and edited, so do the emails that we write. With revision and editing we can really compose emails which are polished and effective.

25.6 BLOG WRITING

Blog, also known as *weblog*, is a very popular platform for expressing some ideas, views and opinions. Blogs, like other social networking sites such as Facebook or Twitter, provide a common place where everyone can put his/her views on any topic. It allows individuals to write quickly; express freely and publish their views on any subject via the Internet; and, in the process, connect with thousands of people from all over the world.

For blog writers, there are a number of free websites available, such as *Wordpress*, *Tumblr*, *TypePad, Myspace* and *Blogger*. All of them offer free design themes so that you can customize your own blog. They have now been converted into a hugely effective marketing and communication tools for academia and businesses as well. In fact, teachers use blogs for teaching effective writing, critical thinking and analytical skills to students. Recruiters also look at your blogs and try to assess your personality and skills.

After discussing the relevance of blogs, let us pick up some tips for successful blogging.

- Choose a blogging platform and customize your own blog. And if you are not sure how to use them, a plenty of online video tutorials can be found on the web which can help you start your own blog.
- When you start a free blog, make sure you provide the weblinks connecting to your artistic portfolio such as photostoart, *valuemystuff, valsparpaint* etc.
- While choosing a topic to write about, try and find a niche for yourself. Make your blog about something very specific and you will please both the search engines and your readers.
- Find a topic you are comfortable with and write about that.
- If you want your blog to be read by many people, offer value by talking about your views on current issues, your perspective on life or your experiences.
- Get people talking on your blog by asking them to add their comments.
- Make commenting easily possible and write about the stuff that gets people talking positively. If people can interact on your blog, they'll keep coming back for more.
- Occasionally you may get odd or negative comment on your blog; in that case, you should not feel put off or stop blogging.
- While writing a creative blog, remember to use relevant images to make your web page look attractive.

- Try to keep your articles as quickly readable as possible.
- Assign such titles to your articles that are appealing and catchy; also when people search for them through search engines, your titles get easy and quick attention.
- Try to keep the layout of your blog clean, fresh and uncluttered. Avoid adding lots of unnecessary features.
- Do not neglect social media if you are a blogger. Update your followers and fans with your latest post on Twitter and Facebook as well.
- Since people want to follow a person, and not a brand, include an "about" section if possible, and a photograph of yourself. By offering a personal slant, you will be encouraging people to like your blog posts.
- Promote your best blog posts in your email newsletter.
- Try and keep your blog updated with atleast two articles in about a week.
- Finally, before you hit the "publish" button, it is crucial to check the spelling and proofread your article again and again as bad spellings and grammar will not only ruin your credibility but will also turn many a reader away from your blogs.

 ## RECAPITULATION

✓ Email writings have become quite popular in our everyday interactions. Even in the professional world, emails are now widely recognized as an important tool of official communication.

✓ As an email is regarded as an informal piece of writing, many a time the emails composed and received by us are ungrammatical, abrupt, blunt, and unprofessional in many other ways.

✓ Since emails are also as important as other forms of professional communication, we need to observe the principles of effectiveness such as clarity, courtesy, specificity, grammatical accuracy, and proper beginnings and endings in our emails.

✓ In order to draft effective and professional emails, we need to overcome common pitfalls such as abruptness, lack of punctuations, poor spellings, abundance of abbreviations, acronyms and jargon, use of sexist language, etc.

✓ Blogs like Twitter and Facebook provide a platform to express your views on any topic. Effective blogging adds a valuable dimension to your personality.

WISEWELL QUIPS

EXERCISES

1. State whether the following statements are true or false.

   (a) Email writing is informal and no grammatical rules are required to be observed while composing it.

   (b) Email writing is quite frequently used in professional communication these days.

   (c) The expression 'Yours truly' is an example of complimentary close in an email writing.

   (d) The expression 'Dear Sir/Madam' is an example of a salutation in email writing

   (e) Since an urgent email requires immediate attention, we must choose to write the entire text in capital letters.

   (f) In order not to sound abrupt, one may consider starting an email with an introductory, warm up sentence.

   (g) In email jargon 'flame' refers to the mail that doesn't reach the recipient well in time.

   (h) While composing an email, we must choose different cases of masculine noun such as *he*, *him* and *his* for those persons whose gender we are unaware of.

   (i) Since email writings are informal in nature, we need not worry about grammar and punctuations while composing them.

2. Assume that as the Cultural Secretary, you are organizing a flute recitation programme in your Institute/College/University. Draft an email informing all the teachers, students and staff members of your college about the event and inviting them to attend the programme. Invent the necessary details.

3. Assuming yourself to be a student who aspires to join the Communication Skills Advancement Programme offered by Oxbridge Institute of English and Communication Skills, New Delhi. Draft an email seeking the relevant information such as the duration of the programme, course fee, batch timings, quality of the material used and the type of audio-visual aids to be employed by the Institute in the programme. Invent the other necessary details.

4. Assume that you are associated with an NGO that engages itself in the animal protection activities. As a Secretary of the NGO, draft an email to be sent across. Urging people to come forward and join hands in the noble cause, your email should have the required elements of emotional appeal and persuasion.

5. Assume that recently you have purchased a dishwasher from Modern Electrical Equipments, Bhopal. The appliance has not performed upto your expectations. Draft an email informing the company about the cause of your dissatisfaction and seek an appropriate replacement/claim in this regard.

6. Assume that the bank that you work for has decided to freeze all the savings accounts which have remained inoperative for the past two years. Assuming yourself to the Manager (Customer Care), draft an email to be sent to all concerned.

7. Given below is an email. Inventing the other necessary details and keeping in mind the principles of effective email writing, rewrite this mail so as to make it look professional, specific, and effective:

Hi rakki

Hope you rember 2 days be4 v organized a PAARTY to honor the retiring Taneja Sir. It was fun no dout. But Acounts Guys want an auditing on the expenses. Hope u have kept all vouchers, bills intact. Shoot back a reply detailing all the expenses incurred. Would appreciate if the detailing includes ref. to exact figures, rcpt. number, date and budget head for all expenses.

b.rgds

salvi

8. Discuss the features of an effective email. Highlight writing techniques required to draft effective professional emails. Provide appropriate examples to substantiate your answer.

9. What are the common problems encountered by email users? Provide examples to support your points besides suggesting ways to overcome these.

10. Given below are some of the extracts taken from professional emails. Keeping in view the principles of effective email writing, rewrite them so as to make them precise and effective:

   (i) Wud you just let me have your marketing proposal by the dawn tomorrow?

   (ii) Good news! Company is planning a trip to Goa. Those interested may contact the undersigned.

(iii) Queried some respondents. Seems they had problems filling out the detailed query sheets. Let's keep the queries to the minimum and resend.

(iv) Find enclosed in the attached files the new customer friendly on-line travel guide. Kick back the same to us with your ticked options and we'll arrange the rest.

(v) Meeting sched at three. Drop in with sales recds.

(vi) SORRY TO ANNOUNCE THE SAD DEMISE OF MR BLOCKHEAD JEJUNE. CONDOLENCE MEETING TO BE HELD AT 11 PM IN THE PORTICO IN FRONT OF PLANT C.

(vii) Proposal returned herewith. Redo cost estimate. Call back if wanted.

(viii) don't send employee review package plan. mgt following promotion scheme of empallo.

(ix) We introduce ourselves as one of the leading sports company in the company. Have a look at the attached catalog and mail us about ur dream sport paraphernalia today!

(x) read ur mail and appointment letter. thanks.

11. Given below is the text of an email. Revise it so as to make it effective in terms of grammar, usage, capitalization and punctuation marks:

love is sth that makes mee happeee ... u know it makes u think its ok if nothing else is going rite for u but u have got someone to live for...that's the grandee feeling we need...isn't it? ...but sometimes love make you feel it is one of the magic of god to forgot the real meaning of life...anyways...who bothers... iam happy in love and feels great to hv found my soulmate...i don't care abt meaning of life and all that serious crap people use their life in and keep a serious face all the time.

12. Prepare a draft in about 150 words on any of the following topics to post on your blog:

(a) Recent floods in Kashmir

(b) World Cup 2015

(c) Your first meeting with your beloved

(d) Cultural Diversity in India

(e) Thousand Invisible Strings in life

(f) Life beyond consciousness

(g) Your views on the recent book you read or movie you watched

(h) Recent earthquake in Nepal

ANSWER KEY

1. **(a)** False   **(b)** True   **(c)** True   **(d)** True   **(e)** False   **(f)** True   **(g)** False   **(h)** False   **(i)** False

26

Other Business Writings

26.1 INTRODUCTION

In the preceding chapters, we have discussed how to write formal business reports, memo reports, letter reports, technical proposals, business letters, essays, and emails. Besides these, there are other professional writings too, such as itineraries, notices, agenda, minutes, memos, circular, etc. that professionals are quite often required to write. Let us learn how to write these pieces of professional communication one by one.

26.2 ITINERARY WRITING

People engaged in the professional world are supposed to travel quite often. Most of these visits are planned. A plan that lists a professional's engagements for a proposed trip is known as an *itinerary* of that particular trip. In an itinerary, the professional's engagements on a trip are listed on an hourly or daily basis. A well written itinerary helps a professional save time and remain focused during a trip. It also minimizes the risk of missing out on an important engagement or running into uncalled for aberrations during a professional trip.

For preparing an itinerary, keep in mind the following points:

1. Begin by creating a word processed document. It is always good to prepare a neatly printed itinerary rather than scribbling the engagements in a casual way. A neatly drawn layout helps the professional maintain interest in the engagements planned. Though an itinerary needs to be followed conscientiously, keeping a soft copy also helps us make alterations and changes in case of a change in the programme.

2. Use a layout that suits your taste. Some people like seeing their programme listed in a vertical fashion by aligning items on the left hand margin and providing details leaving equal space through tabs, while others find it more convenient to see it in a spreadsheet.

3. Assign a particular title to the trip and write the header at the top. Giving a title to the trip is quite essential as it captures the essence of a visit. A title thus written should be appropriate and well thought out. For example, writing 'Mumbai Visit for Site Survey', 'Negotiation Trip Down South', 'Singapore Visit for Signing up the Contract', 'Trip to Agra in December'10' is certainly more specific than by not defining the visit or by writing something perfunctorily. A defining title at the top also serves as a useful indicator for filing purposes.

4. Divide the trip into days and hours. Enter the start-up time for the day's activities. Mention all the important engagements by splitting them on an hourly basis. Highlight the time in terms of hours (1600 hrs, 1800 hrs, etc.) a particular engagement is likely to consume by mentioning both the beginning and the ending times (1800 hrs to 2100 hrs, etc.). While listing the engagements, we need not arrange them as per their significance. The purpose of an itinerary is to remind a professional about his/her engagements on a particular day and unless it is chronologically arranged, it would not serve the purpose.

5. Avoid mentioning trivial affairs of daily routine, for instance 'taking an afternoon nap from 2.30 to 3.00 p.m.' need not be mentioned. Similarly, an itinerary is an account of your professional engagements. Therefore, keeping some reference to a personal task such as 'calling home at 9.30 in the night' makes little sense in an itinerary. As already suggested, an itinerary may be used for future reference as you may be asked to produce your itinerary for an official trip in the office or to your superior besides a report highlighting the main business of your trip. Don't leave your defences wide open by listing your personal predilections, such as 'watching soccer' or 'catching up with latest share market trends' at the opening of your itinerary!

6. Though we should carefully plan the items that need to be listed in the itinerary, writing only vague abbreviations and ambiguous terminology—something that we may not recall while referring to our itinerary—hardly serves any purpose. Therefore, use clear phraseology while listing your items. For instance, avoid writing 'vale. speech at 5' for 'delivering valedictory speech at 5.00 p.m.' A clearly worded itinerary not only helps you avoid last minute confusions but also reflects a picture of clarity in your mind.

Given below is an example of a senior professional's itinerary:

Sample I

Trip to Plant C at Jamnagar				
Nov. 21'10	Catching Flight from Delhi	1300 hrs	Reaching Ahmedabad	1630 hrs
Nov. 21'10	Starting for Jamnagar	1700 hrs	Reaching Jamnagar	2100 hrs
Nov. 22'10	Leaving for Plant Visit	1000 hrs	Reaching the Site	1100 hrs
Nov. 22'10	Trip to Production Unit			1100 hrs to 1230 hrs

(Contd)

(Contd)

Nov. 22'10	Visiting Quality Control Dept			1230 hrs to 1300 hrs
Nov. 22'10	Consultation with Works Manager			1315 hrs to 1400 hrs
Nov. 22'10	Lunch			1430 hrs to 1500 hrs
Nov. 22'10	Attending Sales Briefings			1515 hrs to 1600 hrs
Nov. 22'10	Meeting with Unit Heads			1630 hrs to 1800 hrs
Nov. 22'10	Reviewing Security Concerns with VP Security			1830 hrs to 2030 hrs
Nov. 22'10	Reaching Auditorium and Addressing Officials			2030 hrs to 2130 hrs
Nov. 22'10	Attending Official Dinner			2130 to 2300 hrs
Nov. 22'10	Leaving for Guest House			2330 hrs
Nov. 23'10	Leaving for Ahmedabad			0800 hrs
Nov. 23'10	Catching Flight from Ahmedabad			1300 hrs
Nov. 23'10	Arriving Delhi			1630 hrs

Sometimes in an itinerary, some vital details or clues for conducting the task efficiently are also listed besides mentioning the type of engagement. Given below is a sample itinerary:

Sample II

Itinerary of February Visit to Jaipur Office					
Feb. 12, 2011	Leaving for Jaipur	1100 hrs	Reaching Jaipur and checking in at Sheraton Rajputana	1700 hrs
Feb. 12, 2011	Leaving for Seminar at Press Club	1800 hrs	Attending the Seminar; Recording Proceedings; and Returning to the Hotel	1800 hrs to 2200 hrs
Feb. 13, 2011	Leaving for Study Centre Conference Room	1100 hrs	Attending Presentation on Marketing Strategy; Observing Trends, Future Projections; Seeking Clarifications; Making Suggestions	1130 hrs to 1400 hrs
Feb. 13, 2011	Lunch at Rajasthani Restaurant with Fellow Officials	1430 hrs to 1530 hrs		

(Contd)

(Contd)

Feb. 13, 2011	Attending a Meeting with Sales, Marketing, and Media Heads; Discussing Grey Zones; Proposing Alternative Plans	1600 hrs to 1830 hrs		
Feb. 13, 2011	Attending Cultural Evening at Jawahar Kala Kendra	1900 hrs		
Feb. 13, 2011	Dinner at Officers' Mess and Leaving for Hotel	2130 hrs to 2300 hrs		
Feb. 14, 2011	Leaving for Conference Hall at Blitz	1000 hrs		
Feb. 14, 2011	Making a Presentation on Reinventing Effective Marketing Strategies; Highlighting Past Trends; Bringing into View Socio-cultural-economic Factors; Suggesting Iinnovations and Improvisations	1100 hrs to 1230 hrs		
Feb. 14, 2011	Discussing Future Projections with Sales, Marketing, and Media Heads at Blitz Meeting Room	1300 hrs to 1430 hrs		
Feb. 14, 2011	Sharing Inputs with Young Executives during Lunch at Chokhi Dhani	1500 hrs to 1700 hrs		
Feb. 14, 2011	Visiting Proposed Site for establishing Production Unit at Sitapura Industiral Area	1700 hrs to 1830 hrs	Visiting Birla Temple Followed by Dinner at Rajasthali and Leaving for the Hotel	1900 hrs to 2200 hrs
Feb. 15, 2011	Leaving for Delhi	0900 hrs	Reaching Delhi	1400 hrs

26.3 INTER-OFFICE MEMORANDUM (MEMO)

An inter-office memorandum (memo) is a document written to pass information between people and departments within an organization. Memos are extremely important for smooth running of an organization because they provide a written record and history of a company's decisions. Memos also serve as a record for all the background, variables, and alternatives which are considered, viewed, and weighed before arriving at such decisions. Memos are critical in the sense that they keep a record of responsibilities assigned to people within an organization. Conveying information about various operations and influencing decisions, memos handle the flow of information up, across, or down in an organization.

An Itinerary should not Look like a Laundry List

Though normally the word 'memo' connotes a reprimand of some sort, a memo can be written to carry out different functions in an organization. Moreover, a memo is not written essentially *by a superior to warn or scold a subordinate*. In professional situations, one may be expected to write a memo to one's superior to make a routine recommendation; to a fellow colleague or an associate to confirm an agreement; or to a subordinate to announce, explain, or remind.

Since memos are written to deal with many an official matter, one may be expected to write a memo to do any of the following in a professional organization:

To confirm A memo can be written to confirm the details of a meeting, a conversation, or a telephonic call. The purpose behind writing one such memo is to have a written record of decisions that were made and the points of terms agreed on.

To suggest We can write a memo to recommend solutions for various business problems. Memos are also written to offer alternatives and improved services besides making suggestions for using new procedures and methods for approaching an official task.

To request A memo is often written to make a written request for taking an action; looking into a matter; taking up a complaint; or passing a piece of information. By writing a memo for such routine matters, one can ensure a focused attention and speedy action.

To explain A memo can be written to explain or define clearly what had not been understood initially. A memo thus written explains a procedure or method often considered complicated or newly introduced in an organization.

To announce Memos are commonly written to make announcements about changes in the company's policy; timings of an office; functions of a department; transfer of equipment from one branch to another; change of address; transfer of responsibility, etc.

To report Memos are also written to give an account of a journey; to highlight the trends in sales and production; to analyse a situation; to present an evaluation of a visit to some site; to define or establish a fact, phenomenon, situation, etc.

To caution or warn   Memos are also written to remind people of their jobs and responsibilities. A memo can also caution and warn people in case they do not keep time; ignore their work; delay the completion of the tasks assigned, perform below expectations, etc.

For these and several other professional functions, a memo travels within an organization. In fact, no other type of written communication reaches so many people at so many levels as does a memo in an organization. The larger a particular organization, the greater is the number of memos written in it.

26.3.1 Structure of a Memo

The structure of a memo is the same as that of a memo report. Please refer to Chapter 23 on *business reports* to understand the different components that comprise the structure of a memo report. A memo too has the same components such as the letter head (header) of the organization; a centralized tag; the designation of the sender and the recipient of the memo; the date; reference number; subject line; the body of the memo; the name and designation of the sender, etc.

26.3.2 Style of a Memo

Since memo is a piece of inter-office communication, its style is generally informal and even conversational at times. In fact, as regards the style of memos, the relationship between the sender and the recipient of the memo, the ethos of the company, and the environment of the organization decide its features. In some organizations, a formal style is expected; in others, an informal style is what is desirable. Some professionals draft their memos in a tone and style which is detached, objective, and official, while some others write in an informal and conversational style. While writing a memo therefore, we need to choose the style that suits our purpose and defines our relationship with the recipient the way we want. Therefore, we can easily come across a memo that is written in an intimate, friendly, and warm style, whereas we may also discover a memo that maintains a detached and matter-of-fact tone. In any case, since one of the desirable traits of a memo is its informality, we need not draft a memo displaying the same degree of formality as we maintain while writing a formal report, a research paper, an essay, or a technical proposal. Generally, a memo is written in an informal manner and the tone adopted is more or less conversational and shorn of formalities. While drafting a memo, bear in mind the tips given below:

1. Don't ramble. Since a memo is generally a short piece of communication, writing a memo that runs into several pages is hardly appreciated.
2. Announce your purpose immediately. Nothing annoys memo readers more than having to read through lines and paragraphs and then coming to grips with the real intent of the memo.
3. Be sure that you have a point to make and state it with clarity. If a reader of your memo has to ring you up to understand the meaning of your message, the memo that was written and sent has failed in its purpose.
4. Stick to making one point in a short memo. If you have to talk about more than one subject, draft separate memos for each of them.
5. Call for action. Unless a memo is written to share information, it needs to end by calling for an action on the reader's part. In such a case, the memo should clearly spell out what needs to be done, when, how, and where.
6. Write your memos using the standard format generally used for the purpose.

Following are some sample memos:

Sample I

Simplex Technologies Pvt. Ltd, Lucknow
Inter-office Memorandum

To: All Departmental Heads
From: Office Manager

Reference: RD/2014/23
Date: 25 November 2014

Subject: Travel Arrangement

Due to an increased travelling activity, at times it becomes difficult for our office to provide company vehicle from the campus area to the railway station/airport to our staff. It becomes increasingly hard for us to arrange a company vehicle particularly for the travel plans taking shape a couple of hours before the take-off or landing. To facilitate the travel within and outside campus therefore, we are planning to sign up a contract with a private travel agency to deal with all such exigencies.

Please let me know your views on this. All your suggestions would be seriously considered before signing up the deal. As the contract is likely to take shape in a month or so, please send your feedback and suggestions within a fortnight.

Akhil Bajpayee
Office Manager

Sample II

Power Cycles Limited, Sonepat
Inter-office Memorandum

To: All Employees
From: Manager (HR)

Reference: HR/T/12
Date: 18 August 2014

Subject: Office Hours

Let me remind you that the official timings of the company are from 10.00 a.m. to 6.00 p.m. Exceptions resulting in late arrival or early departure must be reported to the Time Office. Employees not reporting such aberrations and constantly found irregular in keeping with the office hours will be asked for a written explanation.

Parul Gupta
Manager (HR)

Sample III

Knitfare Woolens and Garments Pvt. Ltd, Ludhiana
Inter-office Memorandum

To: General Manager
From: Sr Sales Officer

Reference: S/Pun/2014/13
Date: 16 January 2015

Subject: Sales Figures for the month ending December 2014

As desired, given below are the sales figures for the month of December in the last two years as reported by the district supervisors in Amritsar, Jalandhar, Ferozepur, Gurdaspur:

District	December sales (2014) in Rs	December sales a year ago
Amritsar	5,75,383	6,70,000
Jalandhar	4,18,998	8,35,345

(Contd)

(Contd)

District	December sales (2014) in Rs	December sales a year ago
Ferozepur	2,09,320	2,08,987
Gurdaspur	1,34,540	1,30,560

As can be seen, sales dropped by almost 50% in the Jalandhar district and by about 20% in Amritsar. The other two districts held their own. This may be partly attributed, in my view, to the fact that the senior sales supervisors in both these districts were new and they took charge only by the first week of December. Moreover, both these supervisors had to deal with the workforce that consisted of fresh graduates. This, however, does not mean that they are to blame for the decrease. I believe they should be given a fair chance to prove their worth. Let me see what I can do to help them.

Ashok Arora

Sr Sales Officer

26.4 CIRCULARS

Circular is a brief piece of professional communication that goes to everyone concerned in an organization. Circulated both within and outside an organization, a circular is written to promote a new product, to inform policy holders of movements in the insurance field, to inform shareholders of market trends, or to make matters of general interest known to several persons.

Depending upon their function, circulars can be divided into various categories, which are briefly discussed below.

26.4.1 Informative Circulars

These circulars are written in the style of business letters and consist of an introductory paragraph, other information paragraphs, and a closing paragraph. In such circulars, the content is factual and the information contained is relevant.

26.4.2 Public Circulars

Public circulars are written and circulated mainly by public bodies, associations, and institutions. These circulars contain the matters of general public interest, awareness, and welfare. Such circulars are sent not only to a close, select group of persons but also to other groups of individuals and societies that would like to know more about the activities of such a society. For example, when the Woman's Commission of a country sends a letter to all its state bodies and also to the International Commission of Woman and to other social bodies on human rights interest, it becomes a public circular.

26.4.3 Circulars of Partnerships and Companies

The circulars of partnerships and companies have the same purpose as that of their business circulars. The difference is that these letters sent out by partnerships or companies contain information and particulars which are of specific interest to shareholders and business partners.

26.4.4 Official Circulars

Out of all the various types of circulars, we are most likely to write official circulars more frequently. Official circulars contain information sent out by the head or senior members of a department or the members of other relevant departments.

When written as an interdepartmental or inter-office piece of communication, an official circular appears to be similar to an inter-office memorandum that we have discussed earlier. However, both memos and circulars can be properly discriminated on the basis of the need felt by the sender and the number of people involved in it. In a situation where the number of interacting people consists of one sender and one recipient, the question of writing a circular does not arise. For example, if the production manager has to apprise the general manager of his/her company about the production trends in a particular plant, he/she has to draft a memo and not a circular. However, when the production manager has to announce a general policy regarding the incentive scheme which has recently been revised, he/she may have to draft a circular to be circulated among the related staff.

Generally, when a wider population is to be addressed, we choose to write a circular and when the number of receivers is small, we rely on memos. Therefore, a memo is more personal and subjective, whereas a circular is objective and detached. Owing to its formal nature, a circular is not written with a degree of informality and warmth that a memo usually displays. Similarly, the individual names and designations of the recipients do not appear. A circular is hence meant for a wider, formal circulation of a message that is relevant to all those who receive it.

Sample

New Age Information Technologies, Bengaluru

Reference: Ptg/C/24

Date: 23 October 2014

Office Circular

As the Review Board is planning to meet shortly to assess and evaluate the candidature of staff members for further increments and promotions, all are requested to collect the self-appraisal form from the HR Department and return the same to the undersigned by 31 October 2014.

While filling in the form, remember to mention all your professional attainments in the past one year. For this purpose, the special projects undertaken, new designs developed, latest technologies adopted, and the models patented may be mentioned besides the specific goals achieved during this period. In case of a query or clarification, please contact the office between 2.30 and 4.00 p.m. in the next three days.

Please note that only the self-appraisal forms duly filled-in and countersigned by the Controlling Officer will be considered by the Review Board.

S. Krishnan
Jr Manager (HR)

26.5 NOTICE, AGENDA, AND MINUTES

Meetings are a form of formal interaction and are held in all organizations, small or big, public or private, government or semi-government. According to a survey, the top-level executives of a company spend about 23 hours per week in meetings, whereas middle level managers spend about 11 hours in meetings. Meetings are considered to be a routine phenomenon for the simple reason that everyone calls for meetings yet everyone is critical of them. This form of communication requires a lot of planning and preparation. There are three major components, namely *notice*, *agenda*, and *minutes*. Notices are sent to the prospective participants along with agenda well before a meeting, whereas minutes are taken down during the meeting. Now let us understand each of these in detail.

26.5.1 Notices

Notices are written information about the day, date, time, and venue of a meeting. Generally, these are sent a few days before the meeting. Notices are not sent long before the meeting because the participant might forget and they are also not sent at too short a notice as the prospective members may have some other prior engagement due to which they might fail to attend. Notices are sent to all those who are entitled to attend it. In case of a general body meeting, a notice is circulated to all the employees whereas in case of board of directors' meeting, the intimation of the meeting is sent to all the directors. Do the following after you have decided to call a meeting:

1. Prepare a notice which includes the date, time, agenda, and venue of the meeting. Send the notice to all the participant members five to seven days before the meeting.
2. Attach the minutes of the previous meeting (if there has been one). This gives the members a chance to bring up anything they do not understand or agree with.
3. Send the agenda with the notice.

26.5.2 Agenda

As the cornerstone to any successful meeting in an organization, an effective meeting agenda provides structure and focus and clearly indicates the purpose of the meeting. The agenda serves as the road map for the meeting.

A well-constructed and thought out agenda is an indispensably valuable tool for achieving the desired meeting results in a reasonable amount of time. A good meeting agenda always serves as a guide to the participants, thus making the meeting more efficient and productive.

An agenda is a list of the topics you will address to get to that objective, with a time limit to keep you on track. For example, if you are writing agenda for the fourth meeting, write that as suggested below:

4.1 Confirmation of the minutes of the last meeting
4.2 Review the status of last quarter's goals
4.3 Appointment of a new sales manager and three project engineers
4.4 Reporting and reviewing the ongoing construction of new factory site. etc.

Significance of an agenda   Following are the uses and significance of preparing an agenda:

1. It forces the convener of the meeting to think about what needs to be accomplished.
2. If it is sent ahead of time, the agenda lets participants know what to expect and allows them to prepare as required.
3. It provides a blueprint for the meeting to follow.
4. It helps the concerned members to think of what is left uncovered and this can help in adding those issues with the permission of the chair.

Tips for preparing an agenda   Given below are a few important tips to make your effort result oriented:

1. Send a preliminary meeting agenda and solicit any further agenda topics (be sure to include a strict deadline for additional topic suggestions).
2. Include only those additional topics which assist in achieving the meeting objective.
3. Your agenda should include all the topics and allow the participants to begin preparing for the meeting.

4. If you have special guests attending the meeting, find out whether they have any issues that can be combined because they are related, similar, or even the same in terms of means or ends. If so, arrange them under one agenda item. Organize the order of events according to their time and importance.

5. Generally, the first item on every agenda is *confirmation of the minutes of the previous meeting* and the last *any other matter with the permission of the chair*.

6. In case there are only a few points to be discussed, the agenda can be written on the notice itself.

7. Check the agenda for errors.

8. Send the agenda along with the notice or email it to all attendees. You should do this as near to the actual meeting as possible.

How to prepare an agenda?   As discussed earlier, you need to be careful while preparing an agenda for a meeting. Mainly there are two parts:

1. Header
2. Body—list of items

Header   The header is particularly useful if the participants belong to various groups/organizations, or if the agenda will be made public record. Your header should include the following:

- Name of the organization
- Group meeting agenda
- Location
- Date
- Starting and ending time

Body   The body of the agenda lists the actual business to be transacted during the meeting. When possible, use action words such as approve, discuss, adopt, develop, assign, conceptualize, brainstorm, review, and announce so as to let the participants know what is expected of them. Against each item is a suggested time, but in reality the time allotted will depend on the nature of issue/agenda item being discussed. Allocate a reasonable and realistic amount of time to each agenda task. This keeps the meeting focused, helps it to proceed on time, and ensures the smooth conduct of the meeting. Place important tasks at the beginning of the agenda. It is so because in the beginning of the meeting, energy levels are higher and participants are focused. Adopt a strict policy of not discussing any topic not listed on the agenda. Designate a presenter for each agenda task.

Take Down the Minutes Correctly to Avoid Ambiguity

26.5.3 Minutes

Minutes are the written proceedings of the business transacted during a meeting. Since the minutes will serve as an official record of what took place during the meeting, you must be very accurate in writing them. The minutes are generally recorded sequentially by the secretary of the concerned group or organization. However, at times, any other member attending the meeting may also be required to draft the minutes. In any case, the minutes of a meeting include the main points of the discussion held and the decisions taken. At times, minutes are written in the prescribed format of the organization and are regarded as an important record in the organization. *Minutes, thus, are a written record of committee meeting times, attendance, topics covered, discussion on topics approved.* Besides, it includes all the important decisions taken, and methods and motions adopted.

The minutes of a meeting form the basis of future actions and decisions related to matters discussed, such as promotion of staff, determining the incentive, procedural changes, increase in the membership fee, etc.

Here are some examples of notice, agenda, and minutes for a meeting in an organization.

Pink Square Mall
Vaishali Nagar, Jaipur 302009

20 December 2014

Notice

The Eighth Meeting of the Executive Committee will be held as per the following schedule:

Date: 24 December 2014
Day: Friday
Time: 6 p.m.
Venue: Seminar Hall, Ashoka Hotel, Jaipur

The agenda for the meeting is attached.

Anubhav Nagpal
Secretary

To: The Members of the Executive Committee

Pink Square Mall
Vaishali Nagar, Jaipur 302009

Agenda for the Eighth Meeting of Executive Committee to be held at 6 p.m. on Friday, 24 December 2014 at Seminar Hall, Ashoka Hotel, Jaipur.

8.01 Confirmation of the minutes of the last meeting

8.02 Appointment of the Manager at Mansarovar Branch

8.03 Opening a retail outlet comprising major brands of shoes and sports goods

8.04 Announcement of the festive discount

8.05 Decision to be made regarding decoration of Pink Square branches across the city

8.06 Date of next meeting

8.07 Any other matter with the permission of the Chair

Anubhav Nagpal
Secretary

Pink Square Mall

Minutes of the Eighth Meeting of the Executive Committee held at 6 p.m. on Friday, 24 December 2014 at Seminar Hall, Ashoka Hotel, Jaipur.

Present

Shri Narain Das Baweja   Chairperson

Shri Satish Girotra
Shri Ajit Agrawal
Shri Ashok Saxena   Executive Committee Members
Shri Ravi Arora
Shri Raghav Dixit

In Attendance

Shri Anubhav Nagpal, Secretary
Shri Utkarsh Sinha, People's Officer

No. of Minutes	Subject of Minutes	Details of Minutes
8.01	Confirmation of the minutes of the last meeting	The minutes of the previous were distributed and approved by the members with consensus.
8.02	Appointment of the Manager at Mansarovar Branch	Mr Arun Lohiya presented the details of the interviews held for the selection of manager for the Mansarovar Branch and read the recommendations of the interview panel to appoint Mr Akash Jain to this post.
8.03	Opening a retail outlet comprising major brands of shoes and sports goods	Mr Satish Girotra, one of the executive members, came up with the proposal of opening a retail outlet comprising major brands of shoes and sports goods of companies like Adidas, Nike on the second floor; the proposal was accepted as a positive step for the growth of the mall.
8.04	Announcement of the festive discount	The committee decided to declare the new year festive offer as proposed by Raghav Dixit, Sales Manager of Vishali Nagar Branch. Details given in the attached sheet.

(Contd)

(Contd)

No. of Minutes	Subject of Minutes	Details of Minutes
8.05	Decision to be made regarding decoration of Pink Square branches across the city	As discussed and approved by the committee, decoration of all the branches was given to Glitters and Sparkles Decorators, Ajmer Road, Jaipur.
8.06	Date of next meeting	The next meeting was scheduled for 17 March 2011.

26.6 WRITING INSTRUCTIONS

Being able to write clear instructions is a valuable skill that professionals are quite often required to display at their workplace. Instructions play a vital role in various areas—the technical manuals which students frequently use in labs; a reference while using a new device, equipment, or instrument; or in the preparation of a recipe in the kitchen. Given below are a few tips for preparing and drafting clear, concise and effective instructions:

- Before attempting to write an instruction, know exactly how to do the task yourself. Instructions are written about clear, tangible steps to be followed by the reader. So, for writing precise and clear instructions, one must be familiar with each step involved in the process.
- Inform the reader about how to begin the process. This may include helping him/her understand the material, place, or equipment that is required for the procedure.
- Explain the steps in a logical order. Steps are needed to be explained in a chronology of occurrence. Normally, the steps that are easy to follow are listed first followed by those that are a bit complicated.
- Begin instructions with a verb. It is so because verbs denote action, and, thus, specify the step that is required to be taken up by the reader.
- Write each step as a small piece, so that it is easier for the reader to grasp it at once. It should not contain multiple things for the reader to do at the same time.
- Express steps in the affirmatives. Sometimes we get to see instructions such as "don't forget to press the button after the woollens are firmly lodged in the washing machine." Instead of this line, we can say, "put the woollens in the machine and press the button." However, if we intend to share some warnings, they should be listed with a negative "don't", for example, "don't try to stop the spindle when the rotation is on."
- Avoid offering choices to the reader. It may confuse them and can lead to some accident or failure. Minor choices, if any, can be stated after adding an "or" in the instructions. Remember that instructions are not about personal choices, therefore, "just try fixing the nail in the wall and then putting the device up for better frequency," is not precise enough to be appropriate.
- For technical, scientific, engineering, or any other mechanical process, try to provide an image for each step. It can be a photo, drawing, or sketch. Make sure it is large and clear enough for the reader to understand the exact process.
- Help the reader see what his/her effort may result into. For example, in the instructions for a recipe, one can add "the cake must turn brownish before you turn the power off."
- Review all your instructions carefully before finalizing them and passing them on to the reader.

Courtesy: McDonald's India. Used with permission.

26.7 ADVERTISING

Advertising is a form of communication intended to persuade an audience (viewers, readers, or listeners) to purchase or take some action upon products, ideas, or services. It includes the name of a product or service and suggests as to how that product or service could benefit the consumer. Effective advertising is an extremely important aspect of generating business for a company.

26.7.1 Purpose

There are several reasons for advertising; some of which are as follows:

- Increasing the sales of the product/service
- Creating and maintaining a brand image
- Communicating a change in the existing product line
- Introduction of a new product or service
- Increasing the market value of the company

26.7.2 Types

Print advertising describes advertising in a printed medium, such as a newspaper, magazine, trade journal, pamphlet, billboard, or hoarding. Billboards are large structures located in public places, which display advertisements to passing pedestrians and motorists. Most often, they are located on main roads with a large amount of passing motor and pedestrian traffic; however, they can be placed in any location with a large number of viewers, such as on mass transit vehicles and in stations, in shopping malls or office buildings, and in stadiums. These are termed as *physical/outdoor advertising.*

Surrogate advertising is prominently seen in cases where advertising a particular product is banned by law. Advertisement for products like cigarettes or alcohol which are injurious to health are prohibited by law in several countries and hence these companies have to come up with several other products that might have the same brand name and indirectly remind people of the cigarettes or beer bottles of the same brand. Common examples include liquor brands like Kingfisher, Hayward, which are often seen to promote their brand terming it as soda.

Coffee cup advertising is any advertisement placed upon a coffee cup that is used in serving beverages in an office, café, or coffee shops.

Digital advertising such as *online* and *TV commercial advertisements* are today considered the most effective mass-market advertising format, as is reflected by the high prices TV networks charge for commercial airtime during popular TV events. This is known as *digital advertising.*

Guerrilla advertising is another form of advertising that is also becoming increasingly popular with a lot of companies. This type of advertising is unpredictable and innovative and persuades consumers to buy a product or an idea. It involves unusual approaches such as staged encounters in public places, and giveaway of products such as cars that are covered with brand messages.

26.7.3 Tips

Given below are some simple writing tips to create attractive business advertisements:

1. A powerful headline, expression, or caption is one of the most crucial aspects of an advertisement. It is also possible to use a headline that centres on a problem that this particular product of the company can solve. This will attract the attention of people experiencing that specific issue. One can easily recall the brands popularized through catchy lines such as the following:

    (i) *Zindagi Ke Saath Bhi— Zindangi Ke Baad Bhi* (LIC)

    (ii) *Happy to Help* (Vodafone)

    (iii) *How do I fulfill my daughter's precious dreams*? (Bharti AXA Life Aajeevan Anand)

2. Be sure to include all information, but in a concise manner. The advertisement should have complete information about products and services so that the potential customers understand what it is exactly about.

3. Always know how much space is available for the advertisement before beginning to write it. Cut the advertisement down accordingly.

4. Write in a persuasive style.

5. A song or an apt jingle helps listeners relate to the product.

 RECAPITULATION

✓ An itinerary is a written document that lists all the major engagements of a professional's official visit on a daily or an hourly basis.

✓ An itinerary includes only the important official tasks and leaves out sundry, trivial, and non-professional engagements.

✓ A well prepared itinerary helps a professional stay focused during a trip and utilize the time at his/her disposal in a planned manner.

✓ An inter-office memorandum (memo) is written to carry out various functions such as making an announcement, making a short request, informing people about a new policy, procedure, and scheme, reporting briefly about an accident or event, cautioning or warning people on lack of performance, negligence, absenteeism, etc.

✓ Essentially a short communication, the memo is written in a style that is somewhat informal and interactive and frequently uses expressions such as 'I feel...', 'In my view...', 'Let me see...', 'I'll keep you informed...', etc.

✓ A circular is a short piece of communication that is circulated among a company's staff members.

✓ Concisely worded, a circular generally gives the intended information in a single paragraph that contains an information, suggestion, or announcement of a general interest in a particular organization.

✓ The notice of meeting is sent beforehand and includes details regarding the date, time, place, and venue of the meeting. Sometimes, in case of being arranged sequentially, the notice also refers to the number of the meeting being proposed.

✓ The agenda of the meeting refers to the vital points of business to be transacted during the meeting. Generally forwarded with the notice itself, an agenda lists all the items that need concentration, consideration, and deliberations during the meeting.

✓ The minutes of a meeting are the official proceedings of the important business transacted during the meeting.

✓ Starting generally with a reference to the approval of the minutes of the last meeting and ending on a vote of thanks to the Chair, all the important items transacted during the meeting are numbered, titled, and described.

✓ Leaving out minor and trivial details, the minutes of a meeting record the important details, discussions, and decisions taken on the points of vital significance during the meeting.

✓ For writing clear, concise and effective instructions, one must explain the steps in a logical order and also provide clear Dos and Don'ts list.

✓ Since advertising is meant for persuading an audience, the copy writer should create a catchy tag line and use an appropriate jingle for visual advertisement.

EXERCISES

1. Keeping in view the environmental hazard that plastic causes, the management of your company has decided to ban the use of plastic carry bags in the organization's campus area. Assuming yourself to be Public Relations Officer of your company, draft a circular to be sent across the organization to this effect.

2. Assuming yourself to be the Purchase Officer of the Budding Brains Incorporations, New Delhi, prepare a memo to be written to the Section Heads of your organization informing them about the new procedure they should follow for sending their departmental requisitions. Invent other necessary details.

3. Visions Unlimited India, Pune, is planning to start an NGO that will look after the welfare of blind people in the city. Assuming yourself to be the Coordinator for the project, draft a memo to be sent to the Vice President. Your memo should bring into focus the highlights of your project.

4. Your company has recently established a medical centre on its premises. Draft a circular to be sent to all employees of the organization informing them about the medical facilities to be offered and the timings of the centre.

5. As the Student Coordinator for TechnoBlitz—the cultural festival of your institute—you have been asked by the Professor-in-Charge (Cultural Programmes) to do a survey of the preparations done by various student clubs and departments such as theatre, photography, paintings, dance, and music. Prepare an itinerary for your proposed visit.

6. Spinballs Sports Pvt. Ltd, Jalandhar, is planning to set up a manufacturing unit at Bhatinda where the company already has a branch office and a sales outlet. As the Head, Production Unit, you are required to visit the proposed site travelling by the company's car; listen to the presentations by the company's personnel at Bhatinda office; conduct meetings with them while exploring the viability of setting up the proposed venture. Listing all the important professional engagement and dividing your trip in hours, draft an itinerary for the proposed visit.

7. Given below is an itinerary. Rewrite it so as to make it seem professional and methodical:

21 Dec. '10	Flying by Air	3.00 p.m.
21 Dec. '10	Reaching Hyderabad	7.00 p.m.
21 Dec. '10	Checking Out of Airport	8.00 p.m.
21 Dec. '10	Checking in the Hotel Arranged by the Company	9.00 p.m.
21 Dec. '10	Calling Home	9.30 p.m.
21 Dec. '10	Taking Dinner	10.00 p.m.
21 Dec. '10	Going to Bed	10.30 p.m.
22 Dec. '10	Getting Up and Getting Ready	7.30 a.m.
22 Dec. '10	Taking Breakfast	9.00 a.m.
22 Dec. '10	Leaving for Conference	9.30 a.m.
22 Dec. '10	Discussing Issues of Vital Importance in the Conference	10.00 a.m. onwards
22 Dec. '10	Coming Out of Conference Hall	12.00 noon
22 Dec. '10	Calling Office and Giving a Feedback on the Conference	12.05 p.m.
22 Dec. '10	Going for Lunch	12.45 p.m.
22 Dec. '10	Enjoying Lunch and Relaxing at Pakiza Restaurant	1.00–2.30 p.m.
22 Dec. '10	Leaving for Meeting at Hotel Teej International	2.45 p.m.
22 Dec. '10	Attending Meeting with Advertising Head	3.00–5.00 p.m.
22 Dec. '10	Reviewing and Taking Down Important Points Discussed	5.15–6.00 p.m.
22 Dec. '10	Calling Boss and Giving Him the Feedback of the Meeting	6.00–7.00 p.m.
22 Dec. '10	Relaxing and Hanging Around	7.00–8.00 p.m.
22 Dec. '10	Taking Dinner at Food Café	9.00 p.m.
22 Dec. '10	Going Back to Hotel	10.00 p.m.
23 Dec. '10	Getting Up and Getting Ready	5.00 a.m.
23 Dec. '10	Leaving Hotel for the Airport	6.00 a.m.
23 Dec. '10	Boarding Plane	8.00 a.m.
23 Dec. '10	Reaching Delhi	11.30 a.m.

8. At the tenth meeting of the Board of Directors of Nixon Electronics Co. Ltd, Jodhpur, the following business was transacted:

   (i) Minutes of the previous meeting

   (ii) D.A. to the employees, purchase of furniture for the common room

   (iii) Creation of five posts of travelling salesperson

   (iv) Opening a cultural centre in the company's premises

   Assuming yourself to be the secretary of the company, write the minutes of the meeting. Insert the necessary details.

9. As the Secretary of the Staff Association of your organization, write the minutes of the seventh meeting of the General Body of the association held at 4 p.m., on Tuesday, 30 December 2010, at Gadia House, Industrial Area, Sitapura, Jaipur.

   The agenda was as follows:

(i) Confirmation of the minutes of the previous meeting

(ii) Celebrating important festivals

(iii) Organizing slogan writing competition on Environment Day

(iv) Membership drive

(v) Extending indoor games facilities

(vi) Increase in subscription fee

(vii) Any other matter

10. (a) Prepare the agenda for a meeting of the Jaipur Chamber of Trade, to be held at 7 p.m. on 28 September 2011, arranging the following items in proper order:

   (i) Organization of shopping week

   (ii) Any other business

   (iii) Increasing the efficiency of the exchange of Report of the Finance Committee

   (iv) Nomination of two delegates to the seminar on marketing management

   (v) Minutes of the last meeting

   (vi) Estimate for decoration of the building

   (vii) Deciding the chief guest of the seminar

   (b) Assuming that you were the secretary in attendance, write the minutes of the meeting.

11. The University Grants Commission (UGC) has appointed a twelve-member committee to study the quality of research and technical education in the country and its relevance to the social needs and national requirements. In its seventh meeting held at 4 p.m. on 25 April 2000, at Vigyan Bhawan, Coppernicus Marg, New Delhi–110006, this committee transacted the following business.

(i) Confirmation of the minutes of the previous meeting

(ii) Identification of the points on which information is to be sought from the institution

(iii) Constitution of four sub-committees for personal interaction with central research institutes, IITs, regional engineering colleges, government technical institution, and private engineering colleges

(iv) Appointment of four research assistance for collection and organization of data

(v) Appointment of two office assistants and one accountant to handle the increased volume of work

(vi) Any other matter with the permission of the Chairperson.

Assuming yourself to be the Secretary of the Review Committee, write minutes of this meeting.

12. Prepare the copy of advertisement for any of the following products:

   (a) Tea (b) Plastic furniture (c) Face Cream (d) Ready made garments

Invert the necessary details.

13. Imagine you are working in an electronics company and you have been assigned the task of preparing manuals for the following products: Television, Fridge, Mixer-Juicer, Blender, Washing machine.

Now draft instructions to be published in the manual. Invert the necessary details.

14. Imagine that you are working in a mobile company. Your company is going to launch a new mobile set this Diwali.

Now draft instructions to be included in the manual prepared for this purpose. Invert the necessary details.

27

Movie and Book Review

Learning Objectives   After reading this chapter, you will be able to

- understand what book and movie reviews are
- know the various steps following which you can write an effective movie review
- get familiar with the features of a good book review
- learn how to write an effective book review
- discover how writing book and movie reviews can help improve our communication skills

27.1 INTRODUCTION

Though a piece of art is not created or produced by an artist for others' critical observation, people often appreciate or criticize it. This is called reviewing that piece of art, whether it is a movie, painting, or book. The key to writing a good review or critique of any media—books, films, music, etc.—is to be able to express what you think and why you think it. Though all manifestations of art are open for review, in order to have a useful opinion, we must be able to explain our opinion clearly.

27.2 MOVIE REVIEW

Most people enjoy watching movies but not many write a movie review. The main reason is that writing a movie review involves more than just storytelling. It also requires understanding the movie thematically and analyzing its content and form, both from literary and cinematic perspectives. Though the purpose of most movie reviews is to help the reader determine whether the movie is worth watching or not, writing movie reviews can also help students develop their written expressions by composing effective passages in this craft. The steps for writing an effective movie review are given here.

Watching the Movie at Least Twice

The first step in writing a review is to watch the movie with focus and intent. For this purpose, watching a movie just once may not suffice, as it is likely that you will miss some key points. Watching the movie a second time will help you notice more details about it. Taking notes while watching the movie is also important for writing an effective review.

Discussing the Theme, Script, Dialogues, and Characterization

Analyzing a movie requires us to look at the movie thematically as well as stylistically. While looking at a movie, one needs to look at its theme and relevance in a given context. It becomes even more essential if we are analyzing an artistic or socially-oriented movie that often has more purposive content than those that are mainly designed to entertain people and cater to popular taste. Most of the socially-committed movies take up seminal issues related to our social, political, emotional, and psychological realities. Subjects such as woman emancipation, discrimination between the rich and the poor, child labour, problem of untouchability, evil of dowry system, political corruption, etc. have often figured in movies. To be able to appraise the work effectively, we need to take in view the ideological stance of the film-maker.

Besides looking at such thematic concerns, we should also pay attention to the strength or weakness of the script that plays a very important role in films. Dialogues too play a significant role in assigning a conversational ability to the movie. Since characters take the action further in a film, analyzing their individual role and performance becomes crucial. In movie reviews, it is also important for us to talk about whether the character is fully developed or not, whether it is flat or round, and if any particular character emerges as the spokesperson for the director's point of view on a particular issue.

Appraising the Mise-en-Scène Elements of the Movie

Since films are a visual phenomenon, looking at the visual elements such as the setting, props, lighting, costumes, colours, acting, background music, and camera movements, angles, focus, distance, and height require our close attention. Let us look at each of them briefly.

Setting and location   While shooting a film, it is mandatory for the director to shoot it in a location where its plausibility is automatically ensured. For example, we often see a movie regarding the underworld shot in crowded, narrow streets. Similarly, movies related to social discrimination are mostly shot in rural settings; romantic movies are many a time shot in natural settings; movies on Indian diaspora are pictured in the agrarian beauty of Punjab and the corresponding affluence of the West; those related to dacoits create a make-believe world in the rural Rajasthani deserts; while horror movies require sequences shot in the night. Movies such as *Ankur*, *Company*, *Lagaan*, *Namastay London*, *Raat*, *Sholay*, *Mumbai Meri Jaan*, *Fashion*, etc. come to our mind when we talk about the plausibility of location and setting.

Props and colours   Just like setting and location, props too play a very significant role in establishing the credibility of a storyline and its celluloid presentation. For different movies, different types of props are chosen. For instance, it becomes hard for us to imagine a detective without a hat on his head simply because of its association with the legendary character Sherlock Holmes. Even small and seemingly inconsequential items such as a smoking cigarette, a violin, a notebook, or a suitcase can confirm the socio-economic status of a character. Therefore, while interpreting a movie, we need to pay attention to such small details that eventually characterize the work.

Costume, makeup, and hairstyle   To create the appropriate mood, moviemakers choose costumes that suit their subject. One can easily recall the costume worn by the male characters in *Mughle-e-Azam* where the armoury is an essential part of their dress. We often see actors dressing up as farmers or soldiers, all of which is done to add to the plausibility of the drama that unfolds on the silver screen. Just like costume, hairstyle and the rest of the appearance too helps us view a character appropriately. In the 1981 classic *French Lieutenant's Woman*, Meryl Streep appears before us in two contrastive costumes and hairstyles. As Sarah, she is seen in a black mourning garb while as Anna she adorns herself in professional attire with a stylish haircut.

Lighting and shades   Just as colours and costumes, lighting and shades help a director build an environment suited to the mood of the film. Mainly we come across high-key and low-key lighting. In high-key lighting, we see light being diffused evenly across the shot resulting in a lower contrast between the brighter and darker areas, whereas in low-key lighting we see a higher contrast between the brighter and darker areas of focus. High-key lighting is often used to convey a mood of optimism whereas the low-key lighting adds to the mood of anxiety, uncertainty, menace, and foreboding. Similarly, directors use back-lighting, side-lighting, under-lighting, and top-lighting to emphasize an appropriate mood and at times they acquire decisive connotations. Guru Dutt's *Pyaasa* is one such movie where different patterns of lighting and shades add to the message of the movie.

Background sound and music   In films, background sound and music acquires significant worth. Sometimes, it is the background score that characterizes a movie. One cannot, for instance, imagine a horror movie being able to create hair-raising fear in the audience without appropriate background sounds. Even in movies of other genres, background score plays an important role. It is important, therefore, to pay attention to the background sound and music while watching a movie. Recall the sounds emanating from the gargantuan engine of the Titanic and gushing waters that the movie *Titanic* so powerfully conveys.

Camera work   Since we get to see the entire movie through the eye of the camera, it holds a central point in the process of moviemaking. In fact, watching a movie critically requires observing its various nuances such as its distance, height, focus, angle, and movement. In movies we often see interplay of long shots, medium long shots, medium shots, medium close-up, and close-up shots. Such movements of the camera are chosen to establish at times the relative size of an object. For example, in some of the shots of *Titanic*, the relative smallness of an otherwise huge ship as compared to the expansive Atlantic Ocean has been pictorially depicted.

Just like distance, the artistic use of height, angle, and level of the camera results in emphasizing a point of view. Therefore, we come across high angle, straight-on angle, and low angle shots in movies varying in their content, moods, and shades.

Acting   Amidst the mechanical manoeuvring, acting holds a crucial, and at times, decisive position. Many movies have become immortal because of memorable acting by legendary actors whereas some others have been spoiled owing to the lack of it. Acting in cinema includes an actor's on-screen facial expressions, eye movements, gestures, hand movements, positioning, and dialogue delivery. Most of the good actors are considered so because of their convincing portrayal of the characters they are playing.

Being Objective

While writing a film review, one needs to remain objective and impartial. Reviews should also give impartial details so as to allow the reader to make their own mind over an issue. Giving a review that is the result of a bias or prejudice is not worthwhile. A film review that seems like a personal attack on an actor or director hardly serves as a review. It is also observed that a reader of a review, who shares a similar taste in films, confidently follows that reviewer's recommendations. Therefore, being objective is suggestive of being morally upright and critically penetrating.

Documenting the Review Properly

Since writing reviews is a professional exercise, one needs to do it like an expert. Therefore, a reviewer needs be careful about the mechanics of writing. While writing a review, remember to italicize the title of the movie. Dialogues of characters, if any, need to be cited within inverted commas. Begin with a small outline of the entire work, mentioning the title, names of the director, producer, music composer, and lead actors and gradually begin to focus on its thematic and stylistic features. Again, like a true professional, avoid being a spoilsport by divulging vital details if the movie essentially thrives on building up a suspenseful ending.

Edit and Proofread the Review Appropriately

Read and check your review thoroughly. It can be embarrassing to find errors in your work after it has been published. Good write-ups are rarely composed flawlessly in the first draft. Therefore, develop the habit of revising your reviews so that they are well-received by the audience.

Sample Movie Review

Film	*The Namesake*
Director	Mira Nair
Based on	*The Namesake* by Jhumpa Lahiri
Language	English
Year	2007
Main cast	Tabu, Irrfan Khan, Kal Penn, Jacinda Barrett, Zuleika Robinson, Sahira Nair

After remarkable movies such as *Salaam Bombay!* (1988) and *Mississippi Masala* (1991), Mira Nair comes up with her poetically surcharged *The Namesake*. The movie focuses on the story of a young Bengali couple Ashoke (Irrfan Khan) and Ashima (Tabu) who have an arranged marriage in Calcutta and move to New York, where they attempt to keep their cultural roots alive. Gradually however, the story shifts to focus on their son Gogol (Kal Penn). Based on Jhumpa Lahiri's much acclaimed novel by the same title, *The Namesake* subtly unfolds the plight of two generations of immigrants in America—the parents who have come to a foreign land to realise their dreams and their offspring who grow up almost as strangers, often torn and conflicted between the two diverse cultures.

Mira Nair deftly handles the cultural conflict waged at Gogol's soul. The movie continues to focus on his angst without ever degenerating into sentimentality and melodrama. Alongside, she also develops Ashima and Ashoke's own rootlessness amidst an alien culture. Gogol's own cultural and existential vacuum is also resonantly evoked as he falls in and out of a relationship with Maxine (Jacinda Barrett), an American girl with whom he fails to acquire a significant cultural space owing to his own conflicted sense of duty and characteristic in-betweenness. His painful attempts at shrugging off the cultural appendage by changing his name from Gogol to Nikhil have been sympathetically drawn. During the course of the movie, Ashoke's death places an onerous responsibility on his tonsured head as he is expected to harmonize both the Eastern and Western values for the rest of his life. Falling out with Maxime during the mourning period, he eventually marries a Bengali immigrant Moushumi (Zuleikha Robinson) but since both of them carry a fractured legacy, their marriage turns out to be a fiasco, further accentuating the protagonist's cultural discomfiture. In order to heighten the intensity of his struggle, Mira Nair delineates his sister Sonia's (Sahira Nair) conflict in lighter shades.

Besides the profound thematic concerns, the film also scores owing to its technical richness. The entire spectacle of the movie has been presented with poetic elegance. Scene, sequences, and characters and their names acquire symbolic connotations. The tentative and tender lovemaking between Ashima and Ashoke, followed by the contrastively pitched tempestuous love scenes between Gogol and his girlfriend, the tentative slipping into alien shoes by Ashima, the sterility of the frozen snow and the evocative silences, to name just a few, add to the poetically surcharged canvas of the movie.

The movie spans over a period of 25–30 years. What holds this long stretched edifice together is the subtle but powerful performances by Tabu, Irrfan, and Kal Penn. All of them let their identities merge with those of their characters. Irrfan's profound silences and Ashima's vulnerability besides Kal Penn's conflicted psyche leave on us a profound and lasting impact. Though a movie adaptation, *The Namesake* by Mira Nair offers enough artistic cinematic delight to the viewer independently of its fictional moorings.

27.3 BOOK REVIEW

Once we have understood how to write a movie review, writing a book review may not be a challenge. Writing an effective book review, however, requires dealing with some aspects of the book that are found primarily in a book and may not essentially figure in the discussion when it comes to writing a movie review. For writing an effective book review, consider the following points.

Provide the Basic Information

Book reviews often begin with the details about the book, namely title of the book, name of the author and publishers, price of the book, and its page extent.

Discuss the Theme

Thematic concerns of a book often characterize its other aspects. Therefore, telling the reader what the book is all about will establish its essential nature. In case of non-fiction books, we can focus on the issue raised by the book and its relevance in the contemporary context. If the book is a work of fiction, establishing its thematic concerns helps the reader establish his/her interest in the book. For instance, an initial thematic introduction will not let a reader confuse R.K. Narayan's *Talkative Man* with Amartya Sen's *The Argumentative Indian*.

Elaborate the Plot Line

Plots and subplots often hold the structure of a fictional work. Therefore, while reviewing one, we need to let the reader get a feel of the main plot of the work, besides its subplots. Good structure of a novel is the result of the seamless fusion of its plot and subplots. An able reviewer therefore usually talks about the structure of the work and comments whether the book is able to sustain the reader's interest or not.

Talk about the Art of Characterization

At times, a work of fiction becomes monumental owing to the great skill with which the writers sculpt their characters. A well developed and rounded character can instil in readers a sense of wonder and delight. While judging a character, we need to evaluate whether the characters created by the writer seem convincing and authentic. Since characters derive their conviction from the milieu they are created in, reviewers often discuss the appropriateness of a character in a given context.

Critically Appraise the Language and Stylistic Features

In a work of fiction, it is often the language and style of a book that finally decides its worth. Therefore, it becomes imperative for a reviewer to comment on the stylistic devices employed by the writer. Often writers use analogies, metaphors, symbols, images, epigrammatic expressions, etc. to help them unfold their narratives with poetic and literary elegance. A good style often redeems even a theme on a mundane issue while a book on a significant theme may fail to click with the reader owing to its poor style. Style, therefore, holds the key to the reader's interest in the book regardless of it being a work of fiction or a non-fiction book.

Refer to the Relevance of the Book

As suggested earlier, books become classics owing to their continued relevance over the ages. On the other hand, books that are essentially temporal and lose their relevance over the years, are eventually forgotten. A reviewer should judge a book on the crucible of its relevance in the present and the coming ages. For example, a book that focuses specifically on social or political issues of a particular time will fizzle out of its relevance once that issue is resolved. A book concerning human predicament on the other hand, is likely to sustain the interest of the reader over many years. Novels at times are suffused with autobiographical elements or the author's point of view on some issue; a reviewer can help the reader relate the work to the author by dropping a hint about that.

Cite Peer Testimony by Other Reviewers

Often we come across various comments on the blurb of a book. These are mentioned to induce interest and curiosity in the reader's mind. Such comments may also heighten the impact of a book review as they may add to the significance of our own comments about the work under discussion.

Sample Book Review

Book	*The White Tiger*
Author	Aravind Adiga
Country	India
Language	English
Genre	Fiction
Published	Atlantic Books (UK) (2008), HarperCollins (India) (2008)
Media type	Print (hardback)
Pages	318
ISBN	1-4165-6259-1

The White Tiger is the debut novel by Arvind Adiga. Published in 2008, the novel created ripples in the literary world as it saw Adiga bagging the coveted Man Booker Prize. The novel explores the journey of Balram, a 'tea boy' from a small, inconsequential village ignored in the tapestry of the country. Indigence draws Balram to Delhi where he finds employment as a chauffeur. With passage of time, he wins the faith of his master Ashok and his wife Pinky Madam. To everyone's shock however, he kills his kind master, steals his money, and flees to Bangalore.

Composed in an epistolary form, the novel presents an honest and frank onslaught on the complexity of Indian society that is firmly rooted in an age-old class-based feudal hierarchy. Through a unique narrative structure, Balram Halwai, the protagonist, writes to the premier of China and in the process, lays bare the underbelly of the divided, discriminating, and lopsided Indian social set-up. During its course, the novel throws open the seminal issues related to poverty, darkness, corruption, caste politics, feudalism, perennial social exploitation, etc. Negotiating such sociopolitical filth, Balram transcends his humble origins and shrugging aside his 'tea boy' tag, he metamorphoses himself into a successful entrepreneur, setting up his own taxi service.

In this ingeniously layered texture, Adiga makes Balram narrate his life-story, recounting how he got to where he now is—a successful entrepreneur in Bangalore. And from early on, we learn that he is a wanted man, as he writes about a poster describing him and alluding to his misdeeds. The novelist flawlessly captures his bohemian spirit as coming from a tiny rural Indian town Laxmangarh and having suffered the pangs of discrimination, Balram eventually devises an escape from its tyrannical clutches, break the crippling rooster coop, and rewrite his own destiny. Essentially a prototype of the underprivileged lot from the society, Balram seems to herald a 'tomorrow', when the subdued voices from 'the Darkness' will eventually be heard, albeit in a tragic manner.

Besides such thematic intensity and concerns, the novel offers a confluence of rich literary feast as Adiga composes his multi-layered text in poetic elegance. The novel abounds in symbols, imageries, metaphors, and analogies between the human and the animal world. Employed dexterously, all these stylistic devices run interwoven in the text. Most of the characters find an equivalent of themselves in some animal with Balram emerging as the White Tiger, surviving in starkly dark realities on his own terms.

Besides all the prizes, the movie has generated considerable critical speculation both in India and abroad. Some such remarks have been added for you to warm you up before you run into Adiga's literary world of intensity, guile, mendacity, and reckless hilarity.

Some other observations on *The White Tiger* are as follows:

'As Balram's education expands, he grows more corrupt. Yet the reader's sympathy for the former tea boy never flags. In creating a character who is both witty and psychopathic, Mr Adiga has produced a hero almost as memorable as Pip, proving himself the Charles Dickens of the call-centre generation.'
– *The Economist*

'Aravind Adiga's first novel is couched as a cocksure confession from a deceitful, murderous philosopher runt who has the brass neck to question his lowly place in the order of things. His disrespect for his elders and betters is shocking -- even Mahatma Gandhi gets the lash of his scornful tongue. (…) Balram has the voice of what may, or may not, be a new India: quick-witted, half-baked, self-mocking, and awesomely quick to seize an advantage. (…) There is much to commend in this novel, a witty parable of India's changing society, yet there is also much to ponder. (…) My hunch is that this is fundamentally an outsider's view and a superficial one. There are so many other alternative Indias out there, uncontacted and unheard. Aravind Adiga is an interesting talent and I hope he will immerse himself deeper into that astonishing country, then go on to greater things.'
– Kevin Rushby, *The Guardian*

'Adiga's message isn't subtle or novel, but Balram's appealingly sardonic voice and acute observations of the social order are both winning and unsettling.'
– *The New Yorker*

'(A)t once a fascinating glimpse beneath the surface of an Indian economic "miracle," a heart-stopping psychological tale of a premeditated murder and its aftermath, and a meticulously

conceived allegory of the creative destruction that's driving globalization. (...) That may sound like a lot to take in, but The White Tiger *is unpretentious and compulsively readable to boot.*
— Scott Medintz, *The New York Sun*

'His voice is engaging -- caustic and funny, describing the many injustices of modern Indian society with well-balanced humour and fury. But there's little new here -- the blurbs claim it's redressing the misguided and romantic Western view of India -- but I suspect there are few to whom India's corruption will come as a surprise. As social commentary, it's disappointing, although as a novel it's good fun.'

— Francesca Segal, *The Observer*

'Adiga's training as a journalist lends the immediacy of breaking news to his writing, but it is his richly detailed storytelling that will captivate his audience. (...) The White Tiger *contains passages of startling beauty (...). Adiga never lets the precision of his language overshadow the realities at hand: No matter how potent his language one never loses sight of the men and women fighting impossible odds to survive. (...)* The White Tiger *succeeds as a book that carefully balances fable and pure observation.'*
— Lee Thomas, *San Francisco Chronicle*

'(E)xtraordinary and brilliant (....) Talk of "lessons" should not be taken to suggest that The White Tiger *is a didactic exercise in "issues", like a newspaper column. For Adiga is a real writer -- that is to say, someone who forges an original voice and vision.'*
— Adam Lively, *Sunday Times*

'What Adiga lifts the lid on is also inexorably true: not a single detail in this novel rings false or feels confected. The White Tiger *is an excoriating piece of work, stripping away the veneer of "India Rising". That it also manages to be suffused with mordant wit, modulating to clear-eyed pathos, means Adiga is going places as a writer.'*
— Neel Mukherjee, *The Telegraph*

'Balram's cynical, gleeful voice captures modern India: no nostalgic lyricism here, only exuberant reality.'
— Kate Saunders, *The Times*

'Aravind Adiga's The White Tiger *is one of the most powerful books I've read in decades. No hyperbole. This debut novel from an Indian journalist living in Mumbai hit me like a kick to the head (.....) This is an amazing and angry novel about injustice and power.'* — Deirdre Donahue, *USA Today*

RECAPITULATION

✓ The purpose of a movie review is to help the readers determine whether they wish to watch the movie. Besides this, writing effective reviews can help students improve their overall expression in language.

✓ It is advisable to start a movie review with a summary of how the story moves and what it essentially deals with. This will help a reader gauge his/her interest in the review further.

✓ A movie review also briefly touches upon the mise-en-scène elements of a film as it helps the reader appraise the technical features of a movie.

✓ A book review should be brief and specific. It needs to highlight all the salient features of a book without unnecessary repetitions and elaborations. One can begin with a brief idea about the story followed by discussion on some other essential elements such as theme, plot, structure, language, point of view, etc.

✓ Through book reviews, some readers also wish to know the opinions of other writers who too have read and evaluated the information contained in the book.

✓ Revise and edit your review uncompromisingly so as to send a neatly drawn, correct, and appropriate write-up.

WISEWELL QUIPS

Because winter is drawing up, I am going to purchase a jacket.

Say 'As winters are *drawing* in, I am going to *buy* a jacket. But you are correct; the days are drawing in.'

What *days drawing in*? What do you mean?

I mean that it is becoming dark earlier in the evening so the days seem shorter and nights longer.

So is there a phrase like 'drawing out' also.

Yes, there is and it means the opposite of drawing in. You could say summer is drawing out and winter is setting/ drawing in.

EXERCISES

Critical Review Questions

1. How does a movie review help the reader? Is it useful for students to learn how to write good movie reviews?

2. Write a note on the significance of a book review. Give appropriate examples to support your ideas.

3. What do you understand by the term *mise-en-scène*? How do *mise-en-scène* elements of a movie help the reviewer compose better reviews?

4. Write book reviews on the following books:

	Title of the Book	Author
(a)	The Kite Runner	Khaled Hosseini
(b)	The Alchemist	Paulo Coelho
(c)	The Da Vinci Code	Dan Brown
(d)	What Young India Wants	Chetan Bhagat
(e)	Cathedral (Stories)	Raymond Carver
(d)	Ignited Minds	A.P.J. Abdul Kalam
(f)	The God of Small Things	Arundhati Roy
(g)	The Guide	R.K. Narayan
(h)	The Namesake	Jhumpa Lahiri
(i)	The Road	Cormac McCarthy
(j)	The Gathering	Anne Enright

	Title of the Book	Author
(k)	Life of Pi	Yann Martel
(l)	The Sea, the Sea	Iris Murdoch
(m)	Princess	Jean Sasson

5. Write movie reviews on the following films:

	Title of Movie	Director
(a)	Changeling	Clint Eastwood
(b)	Donnie Darko	Richard Kelly
(c)	Jurassic Park	Steven Spielberg
(d)	Titanic	James Cameron
(e)	Spider Man	Sam Raimi
(f)	King Kong	Peter Jackson
(g)	A Beautiful Mind	Ron Howard
(h)	Schindler's List	Steven Spielberg
(i)	Udaan	Vikramaditya Motwane
(j)	Chakde! India	Shimit Amin
(k)	Swades	Ashutosh Gowariker
(l)	A Wednesday	Neeraj Pandey
(m)	Rang De Basanti	Rakeysh Omprakash Mehra

Index

About the Authors

Sanjay Kumar, a PhD from BITS, Pilani, is currently a Freelance Author, Public Speaker, and Consultant in English and Soft Skills Development Training. In an academic career spanning over 23 years, he has been a Lecturer, BITS, Pilani, Reader and Chairperson, Department of English, Chaudhary Devi Lal University (CDLU), Sirsa and Associate Professor of English and Head, Department of Humanities, JK Lakshmipat University (JKLU), Jaipur. He has delivered invited lectures and conducted a number of workshops for teachers, students, and professionals. Dr Kumar also has to his credit several research articles, poems, short stories, and eight other books.

Pushp Lata is currently Associate Professor of English at Department of Humanities and Social Sciences, BITS, Pilani. With a PGDTE from CIEFL, Hyderabad, and PhD from the University of Rajasthan, Dr Pushp Lata has to her credit a number of research articles in national and international journals and eight books from reputed publishers. She has delivered talks, lectures and conducted several workshops on effective communication, public speaking, group discussions, and interviews.

Related Titles

TECHNICAL COMMUNICATION, 3E [9780199457496]

Meenakshi Raman, *Professor and Head, Department of Humanities and Social Sciences, BITS Pilani – Goa Campus*
Sangeeta Sharma, *Associate Professor, Department of Humanities and Social Sciences, BITS Pilani*

The third edition of *Technical Communication: Principles and Practice* is a comprehensive textbook specially designed to meet the needs of undergraduate students of engineering. The text material has been restructured to provide a more balanced and exhaustive coverage of the subject.

Key Features
- User-friendly approach with simple and easy-to-understand language
- Practical tips on pronunciation, accent, intonation, grammar, and vocabulary basic to effective technical communication
- Discussion on the topics essential for enhancing the skills required to communicate through the web

EFFECTIVE TECHNICAL COMMUNICATION: A GUIDE FOR SCIENTISTS AND ENGINEERS [9780195682915]

Barun Mitra, *Formerly Professor of English, Indian Institute of Technology, Kharagpur*
Effective Technical Communication is designed to serve as a practical guide and useful resource for scientists, engineers, and researchers. It addresses the need of practitioners engaged in the exchange of technical information to effectively share their ideas with, and make an impact on, their peers.

Key Features
- Acquaints readers with key communication techniques
- Includes all forms of communication including technical papers, reports, and proposals
- Provides illustrative examples of concepts and forms of communication
- Explores emerging trends in communication, including email and voice mail

PERSONALITY DEVELOPMENT AND SOFT SKILLS [9780198066217]

Barun Mitra, *Formerly Professor of English, Indian Institute of Technology, Kharagpur*
The book aims to provide crucial insights into various facets of developing one's personality, as well as to improve written, verbal, and non-verbal communication skills. Special attention has been paid to the specific needs of a job aspirant, such as writing of effective CVs, participation in group discussions, tackling job interviews, and to hone one's public speaking and speed-reading skills.

Key Features
- Provides inputs on avoiding common mistakes in speaking English
- Provides several case studies, examples, and illustrations to elucidate the concepts discussed
- Contains several classroom-based activities for students to develop their personalities and enhance their soft skills

FUNDAMENTALS OF TECHNICAL COMMUNICATION [9780199457472]

Meenakshi Raman, *Professor and Head, Department of Humanities and Social Sciences, BITS Pilani – Goa Campus*
Sangeeta Sharma, *Associate Professor, Department of Humanities and Social Sciences, BITS Pilani*

Fundamentals of Technical Communication is designed as a comprehensive textbook specifically aimed at the undergraduate students of engineering.

Adopting a functional and practical approach, the book discusses the basics of technical communication—listening, speaking, reading, and writing (LSRW).

Key Features
- User-friendly approach with simple and easy-to-understand language
- Model test papers at the end of the book